SECONDARY BATTERIES

SECONDARY BATTERIES

Recent Advances

Robert W. Graham

NOYES DATA CORPORATION
Park Ridge, New Jersey, U.S.A.
1978

Library of Congress Catalog Card Number: 77-94228
ISBN: 0-8155-0696-1
Printed in the United States

Published in the United States of America by
Noyes Data Corporation
Noyes Building, Park Ridge, New Jersey 07656

FOREWORD

The detailed, descriptive information in this book is based on U.S. patents, issued since July 1975, that deal with secondary batteries and their commercial technology.

This book serves a double purpose in that it supplies detailed technical information and can be used as a guide to the U.S. patent literature in this field. By indicating all the information that is significant, and eliminating legal jargon and juristic phraseology, this book presents an advanced, technically oriented review of secondary batteries, their manufacture and use.

The U.S. patent literature is the largest and most comprehensive collection of technical information in the world. There is more practical, commercial, timely process information assembled here than is available from any other source. The technical information obtained from a patent is extremely reliable and comprehensive; sufficient information must be included to avoid rejection for "insufficient disclosure." These patents include practically all of those issued on the subject in the United States during the period under review; there has been no bias in the selection of patents for inclusion.

The patent literature covers a substantial amount of information not available in the journal literature. The patent literature is a prime source of basic commercially useful information. This information is overlooked by those who rely primarily on the periodical journal literature. It is realized that there is a lag between a patent application on a new process development and the granting of a patent, but it is felt that this may roughly parallel or even anticipate the lag in putting that development into commercial practice.

Many of these patents are being utilized commercially. Whether used or not, they offer opportunities for technological transfer. Also, a major purpose of this book is to describe the number of technical possibilities available, which may open up profitable areas of research and development. The information contained in this book will allow you to establish a sound background before launching into research in this field.

Advanced composition and production methods developed by Noyes Data are employed to bring these durably bound books to you in a minimum of time. Special techniques are used to close the gap between "manuscript" and "completed book." Industrial technology is progressing so rapidly that time-honored, conventional typesetting, binding and shipping methods are no longer suitable. We have by-passed the delays in the conventional book publishing cycle and provide the user with an effective and convenient means of reviewing up-to-date information in depth.

The Table of Contents is organized in such a way as to serve as a subject index. Other indexes by company, inventor and patent number help in providing easy access to the information contained in this book.

Some of the illustrations in this book may be less clear than could be desired; however, they are reproduced from the best material available to us.

15 Reasons Why the U.S. Patent Office Literature Is Important to You –

1. The U.S. patent literature is the largest and most comprehensive collection of technical information in the world. There is more practical commercial process information assembled here than is available from any other source.

2. The technical information obtained from the patent literature is extremely comprehensive; sufficient information must be included to avoid rejection for "insufficient disclosure."

3. The patent literature is a prime source of basic commercially utilizable information. This information is overlooked by those who rely primarily on the periodical journal literature.

4. An important feature of the patent literature is that it can serve to avoid duplication of research and development.

5. Patents, unlike periodical literature, are bound by definition to contain new information, data and ideas.

6. It can serve as a source of new ideas in a different but related field, and may be outside the patent protection offered the original invention.

7. Since claims are narrowly defined, much valuable information is included that may be outside the legal protection afforded by the claims.

8. Patents discuss the difficulties associated with previous research, development or production techniques, and offer a specific method of overcoming problems. This gives clues to current process information that has not been published in periodicals or books.

9. Can aid in process design by providing a selection of alternate techniques. A powerful research and engineering tool.

10. Obtain licenses – many U.S. chemical patents have not been developed commercially.

11. Patents provide an excellent starting point for the next investigator.

12. Frequently, innovations derived from research are first disclosed in the patent literature, prior to coverage in the periodical literature.

13. Patents offer a most valuable method of keeping abreast of latest technologies, serving an individual's own "current awareness" program.

14. Copies of U.S. patents are easily obtained from the U.S. Patent Office at 50¢ a copy.

15. It is a creative source of ideas for those with imagination.

CONTENTS AND SUBJECT INDEX

INTRODUCTION

Reversible electrochemical cells, i.e., secondary cells, storage batteries, and accumulators, are characterized in that the chemical relationships of the electrodes and the electrolyte may be restored to a condition of substantially full charge by causing current to flow into the cell. Exemplary of such cells are lead-acid cells, nickel-iron-caustic soda cells, and nickel-cadmium cells.

Rechargeable or secondary batteries such as the ordinary automobile battery, and other batteries useful in radios, tape recorders and other electronic equipment, have found a wide acceptance as a useful source of energy. However, each of these presently known secondary batteries have their own limitations such as, for example, the great weight of the lead-acid batteries used in automobiles.

As the world increases its desire for electrical energy and as increasingly sophisticated electronic equipment having its own self-contained energy source is developed, the limitations of the conventional and well-known secondary batteries become more important. Many new applications for electronic devices could be developed if a power source were available which would be significantly lighter in weight, have a higher energy density, and operate over a wider temperature range.

Thus, while lead-acid storage batteries, known for over 100 years, continue to be the subject of research effort around the world, many other rechargeable battery systems are being developed to meet the ever-increasing needs of industry and commerce.

The nickel-cadmium and silver-zinc batteries are most commonly known and are produced in a variety of shapes and sizes ranging from small button-type cells to very large rectangular cells with large ampere-hour capacities.

One source of electrical energy which has recently become quite important in the electrical power generation industry is the family of batteries known as lithium batteries. Lithium batteries, which are broadly defined as any nonaqueous battery employing lithium as an anode, have become important sources of electrical energy in the form of primary batteries. Primary batteries, of course, are those batteries which are capable of producing electrical current by an electrochemical reaction. These cells are capable of operating only in the discharge mode. Development of lithium secondary batteries, which are cells capable of operating both in the discharge mode and in the charge mode so as to permit multiple reuse of the cell, is rapidly progressing and a large number of processes are described in the literature.

Another rechargeable energy source which is being developed is the solid electrolyte battery employing a sodium anode and sulfur cathode.

The extensive research effort for secondary batteries, on a worldwide scale, is most evident in the patent literature of the United States. This book describes over 260 processes relating to the design, fabrication and performance characteristics of all phases of battery technology as presented in the recent patent literature. These processes, representing the recent technological advances of Germany, Japan, England, Sweden, France and many more countries in addition to the United States are presented with over 100 detailed process illustrations as well as specific formulations for all battery components.

In principle, virtually all electrochemical reactions can be reversed and many cell compositions can be used either in the primary or secondary mode. This book is largely dedicated to batteries where the rechargeable nature of the system is demonstrated. Since many cells now considered primary may well be developed in the future as secondary cells, the reader is referred to another very timely publication of the Noyes Data Corp.:

Primary Batteries–Recent Advances, 1978

LEAD-ACID BATTERIES

GRID PLATES AND ELECTRODES

Rod-Type Units

According to a process described by *J. Brinkmann, G. Trippe and W. Heissman; U.S. Patent 3,959,015; May 25, 1976; assigned to Varta Batterie AG, Germany* the vertical rods of the battery plate grid are formed partly of whole and partly of half rods. The steps at which transitions from whole to half rods take place are positioned to optimize current flow patterns.

Referring to Figure 1.1a, there is shown a rectangular grid frame formed of vertical members **4,6** and horizontal members **5,7** terminating in plate lug **9** leading to pole-bridge **10**. Plate rests **8** are provided at the bottom of the frame. The frame encloses vertical grid rods, different longitudinal portions of which are designated by reference numerals **1** and **2**, respectively. These vertical rods are intersected by horizontal rods, different longitudinal portions of which are designated **11** and **12**, respectively. In addition, there are a few diagonal grid rods **3**.

By considering the elevation shown in Figure 1.1a, together with the cross-sectional views of Figures 1.1b through 1.1e, it becomes apparent that all of these grid rods exhibit variations in thickness along their length. There are two kinds of variations. One is a gradual thinning of the rod cross section. The other is an abrupt transition on one side of which there is a whole rod, while on the other side there is only a half rod. As shown in Figures 1.1a through 1.1e, the transitions between the upper whole and lower half rod portions **2** and **1**, respectively follow a family of curves **20, 21** and **22** (Figure 1.1a) of generally parabolic shape, open towards the bottom, and having apices generally located beneath pole bridge **10**.

Adjacent half-rod portions **1** starting at any given parabolic curve are preferably laterally displaced from each other, alternating between half-rod portions on one side of the plane of symmetry **23** (Figure 1.1e) of the grid frame and half-rod portions on the other side of that plane of symmetry. This serves a variety of purposes. For example, the undulating paste mass configuration which results yields a larger interface between grid and mass, leading to better mass adhesion. Also more uniform mechanical forces are developed during pasting and more trouble-free filling of the grid is achieved.

Adjacent transitions between whole rods **2** and half rods **1** along any particular parabola are preferably uniformly spaced from each other in the horizontal direction. Moreover, the end points of whole-rod portions **2** terminating at consecutively higher parabolic curves are respectively shifted by one compartment in the direction of the apices of these curves.

In a square grid plate, such as shown in Figure 1.1f to which reference may now be had, the apices of the individual parabolic curves are somewhat horizontally displaced relative to the pole bridge and those of the different vertically displaced parabolic curves (shown by the different dot-dash lines) are vertically aligned with one another. For reasons connected with the producibility of the grid, and also to give it greater strength below the plate lug **9**, vertical rod **2a** in Figure 1.1f is carried as a whole rod to the point shown in that figure.

FIGURE 1.1: PLATE GRID

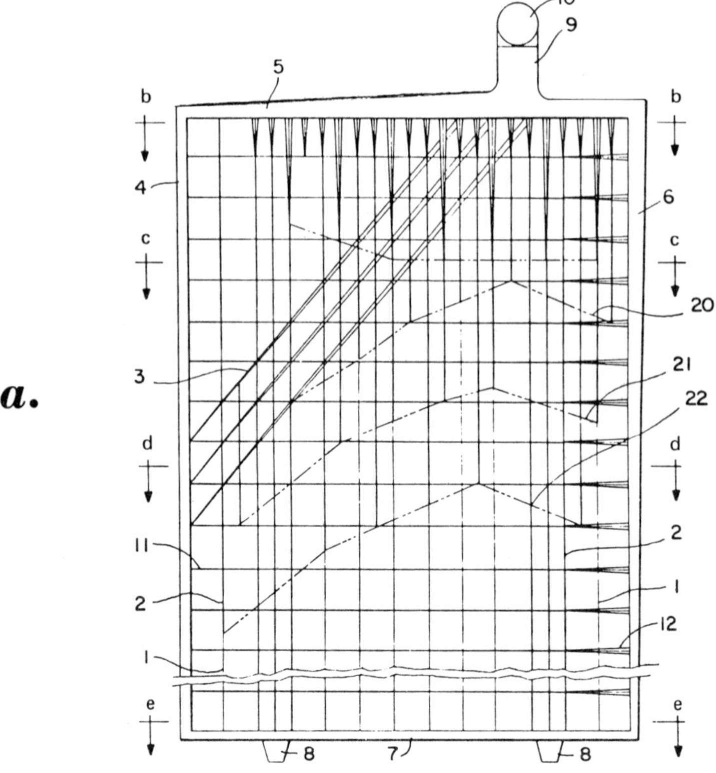

(continued)

FIGURE 1.1: (continued)

b.

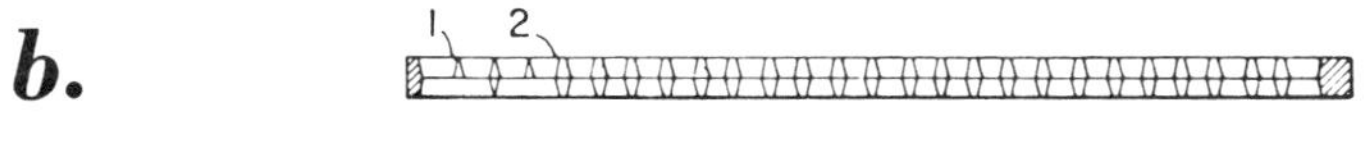

c.

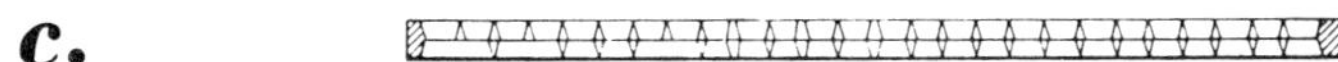

d.

e.

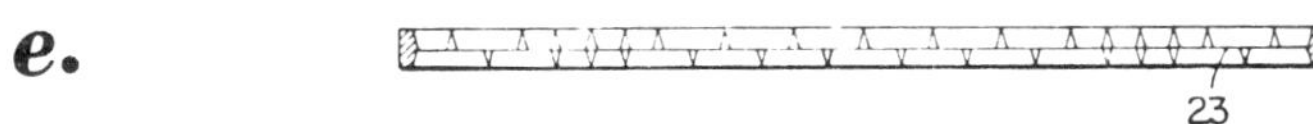

f.

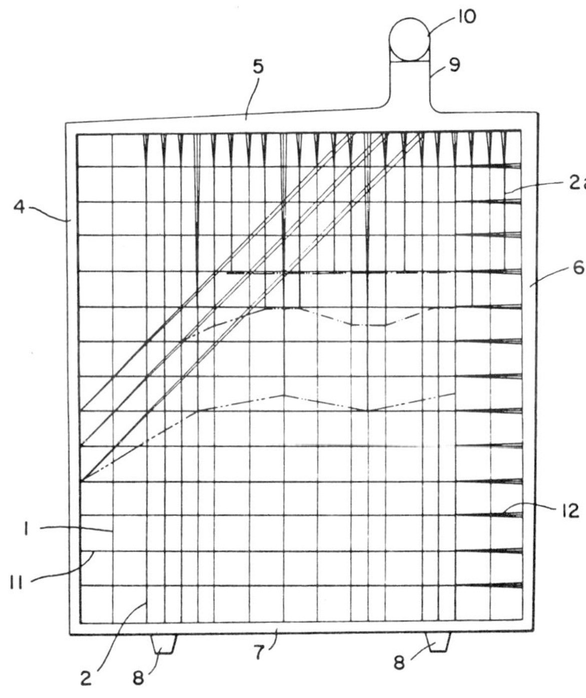

(continued)

FIGURE 1.1: (continued)

g.

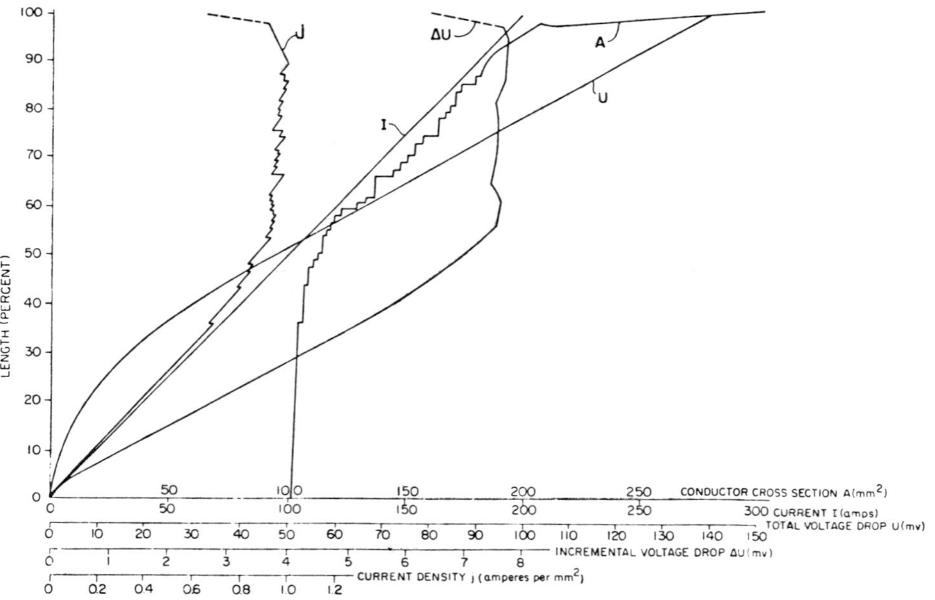

(a) Diagrammatic elevation of a rectangular grid plate
(b)-(e) Cross sections taken through the grid of Figure 1.1a along lines **b-b** through **e-e**
(f) Diagrammatic elevation of a square grid plate
(g) Principal electric characteristics of unit

Source: U.S. Patent 3,959,015

This gives the two parabolic curves the appearance of having downward inflections, but conceptually they retain their downwardly open parabolic form, similarly to curves **20** to **22** of Figure 1.1a. In accordance with the process, the thickness of whole rods **2** diminishes, starting from the upper edge **5** of the plate over at least one grid compartment. Also in accordance with the process, there extend from the upper plate edge **5** a plurality of diagonal rods **3** which join individual intersections of the grid compartments and continue all the way to the lateral plate edge **4**. Preferably the thickness of diagonal rods **3** also diminishes starting from the upper plate edge. In a preferred case, as shown in Figures 1.1a and 1.1f, those rods which are directly above the plate rests **8** are entirely in the form of whole rods **2**.

If plate lug **9** is positioned in the middle of the upper edge **5** of the grid plate, then it is desirable that the cross sections of side members **4,6** of the plate frame increase in the directions of the plate lug. On the other hand, if plate lug **9** is close to one of frame side members **4,6**, then in accordance with the process the cross section of that side member **4,6** diminishes from top to bottom. In either case, the horizontal rods extending from the thinning side members **4,6** of the plate frame can include portions **12** of diminishing thickness and portions **11** of uniform thickness.

The grid has the following distinctive electrical characteristics. The lead cross section is generally greater in the direction of current flow than perpendicularly to the current flow direction. As is particularly discernible in the lower portion of the grid, this results from the fact that grid compartments are bigger in the direction of grid height for a given conductor cross section than in the direction of grid width. (A grid compartment is a rectangle with its long dimension parallel to the grid height.) This contrasts with prior grids which generally have square compartments.

The conductor cross section in the direction of current flow increases progressively in the upper portions of the grid. This results from the fact that, at a predetermined height, half rods become whole rods (thereby doubling the rod cross section), and at a predetermined height, the rods become conical. In those portions the rod cross section increases linearly.

The stepwise increase in cross section of the individual rods, through transition from half to whole rods, is provided in accordance with the process because it is conducive to improved production of the grid casting form. From the electrical standpoint, a rod of extremely small cross section would be desirable at the foot of the grid. On the other hand, from the standpoint of casting technology at least a certain minimum cross section must be maintained. For this reason the grid rods are made with uniform cross section up to a predetermined height.

Optimization calculations show that, from a predetermined height on up, a linearly increasing cross section is desirable. However, in practice this cannot be achieved within the production capabilities of casting forms. The same result, namely a grid cross section which increases linearly from a predetermined height on up, is achievable for the grid as a whole by positioning the rod steps (transitions from half to whole rods) at various heights.

The magnitude and position of this step (transition from half to whole rods) is so chosen that the current density remains substantially constant from a predetermined grid height on up. As is well known, a conductor exhibits minimum voltage drop when the current density is constant over its entire length. These relationships are further explained by means of Figure 1.1g, which corresponds to a structure such as shown in Figure 1.1a. In Figure 1.1g, various characteristics are shown as a function of grid height (length).

(1) The curve designated **I** shows the variation in current flowing in the grid from bottom to top and indicates that this current rises linearly from the bottom grid rest to the plate lug.
(2) The curve designated **A** shows the variation in conductor cross section. As shown in that curve, up to about 50% of full grid height this cross section is substantially constant due to production technology considerations. In contrast, from 50% of full grid height on up, the grid cross section increases essentially linearly in small steps.
(3) The curve designated **ΔU** shows the incremental voltage drop. This incremental voltage drop increases substantially linearly in the lower portion of the grid (up to about 50% of its full height). On the other hand, from 50% of full grid height on up, the voltage drop is essentially constant.
(4) The curve designated **J** represents the current density. Essentially the same comments apply to it as to the voltage drop shown in curve ΔU and discussed in (3) above.
(5) The total voltage drop is shown in curve **$\bar{U}$**. It follows a quadratic function in the lower portion of the plate, while being essentially linear above 50% of full plate height.

In accordance with the process, the ideal variation of current density, incremental voltage drop, and total voltage drop can be closely approximated. Only the step-like character of the grid, and the resulting steps in current density and voltage drops, represent concessions to the producibility of the grid casting mold. Likewise, the fixed cross section in the lower portion of the grid represents a concession to ready castability of the grid. However, neither of these concessions causes the grid characteristics to depart significantly from the ideal.

The various specific features of the grids discussed above provide two significant improvements: (1) For a given weight of lead, the grid has minimum voltage drop across the grid rods and (2) for a given voltage drop the weight of the grid is at a minimum. Thus, the voltage drop of a grid according to the process was able to be reduced to 140 mV, compared to a conventional grid of the same lead weight. This represents a 39% improvement. This improvement in voltage drop leads not only to improvement in the output voltage of the battery, but also to equalization of the current density on the plate surface. This is known to produce a significant increase in the load capacity of the battery in high current operation.

Depressed Lattice Design

C.H. Smith and K.G. Dunning; U.S. Patents 4,016,633; April 12, 1977; and 4,004,945; January 25, 1977 describe a battery plate grid which comprises a perimeter frame, a lattice network within the frame, the perimeter frame having a greater thickness than the lattice network, the principal plane of the lattice network being between the spaced apart planes of the opposed face surfaces of the perimeter frame, protuberances on one side of the lattice network of the grid, the protuberances having a height of about half the difference in thickness between the lattice network and the perimeter frame.

The battery plate grids are manufactured according to the process by casting a series of grids between the continuous cylindrical patterned end surface of a rotating drum and a smooth surface cooperating shoe, the rotating drum being patterned to form the grids with one smooth side, with a perimeter frame and with a lattice network within the perimeter frame and with the perimeter frame of a greater thickness than the lattice network.

Referring to Figure 1.2, **10** generally refers to a battery storage plate grid of lead or lead alloy. This grid plate is "pasted" with a lead oxide paste to form a battery plate of a lead and acid-type storage battery of the type commonly used for the ignition of automobiles, boats, etc. and Figures 1.2c and 1.2d are sectional views showing the grid as cast and as used respectively. It will be noted that the grid **10** comprises a perimeter frame **12**. The bars of the lattice network **14** preferably extend diagonally across the frame and they are of a lesser thickness than the perimeter frame **12** as illustrated in Figure 1.2d. The principal plane through the center of the lattice network and parallel to its two flat surfaces is disposed between the spaced apart planes of the opposed surfaces of the perimeter frame as illustrated in Figure 1.2d (Figure 1.2c is a view illustrating a stage in the manufacture).

FIGURE 1.2: BATTERY PLATE GRID

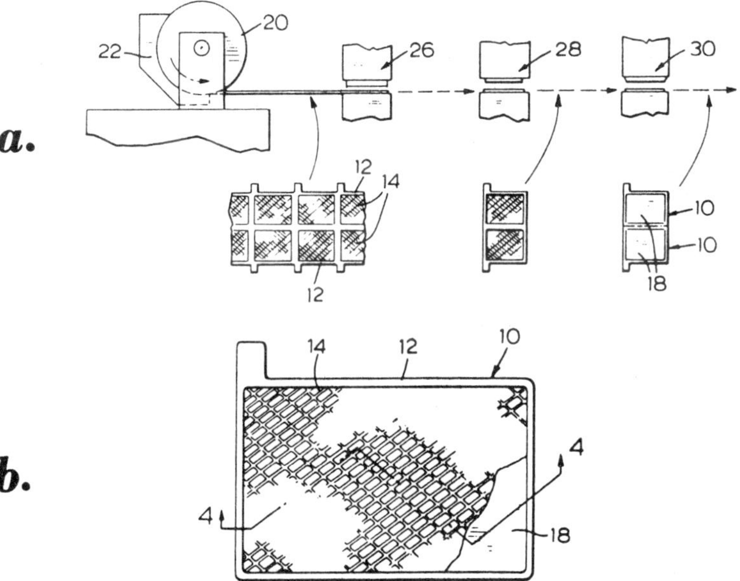

(continued)

FIGURE 1.2: (continued)

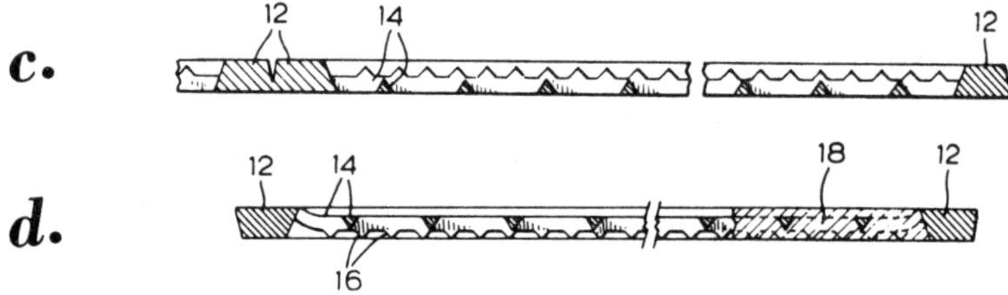

(a) Schematic illustration of steps in the making of a battery plate
(b) Battery plate with portions of the lattice network omitted for drafting convenience, as viewed from the smooth side of the grid
(c) Cross section of a battery grid showing its condition as cast and before the lattice network is raised to locate the plane of the lattice network between the planes of the faces of the perimeter frame
(d) View along line **4-4** of Fig. 1.2b showing disposition of paste

Source: U.S. Patent 4,016,633

Protuberances **16** extend from one side of the lattice network **14** at the intersections of each of the bars of the lattice network. The outer extremities of these protuberances lie substantially in the plane of one side of the perimeter frame **12**. The protuberances serve as locating devices in the manufacture of the grid and they serve to rigidify the grid as the lead oxide paste **18** is applied as will be explained later. When the grid is pasted the only parts of the lattice network of the grid that are visible are the tips of the protuberances.

The bars of the lattice network preferably have a thickness of about $^{25}/_{1000}$ of an inch and the perimeter frame portion of the grid preferably has a thickness of about $^{40}/_{1000}$ of an inch. It has been found that a grid of these dimensions will hold an equivalent amount of lead oxide paste to a conventional grid to about $^{58}/_{1000}$ of an inch thickness and expose substantially less lead at the surface of the oxide paste. The overall dimensions of the grid are not critical. $5\frac{1}{4}$ inches high and $5\frac{7}{8}$ inches long is a common size. They can be varied as required but the important thing about the grid is that it is capable of holding an equivalent amount of paste as a conventional grid with less lead in its construction and with less lead exposed at the surface of the paste.

Figure 1.2a schematically illustrates the method of making battery grid plates and for pasting them. The grid plates are cast by a continuous casting method between a continuous cylindrical patterned surface of a rotating drum **20** and a smooth-surfaced cooperating shoe assembly **22**. The casting machine for achieving this operation is one where molten lead is continuously fed into the molding space defined by the patterned surface of the rotating drum and the smooth shoe. A machine capable of achieving this operation is described in Canadian Patent 934,522 dated October 2, 1973.

The output of the molding machine is a continuous series of grid formations two abreast and interconnected at their perimeter frame portions by a weakened line.

The molding machine is operated on a continuous basis. The grid structures continuously progress from the output thereof to a cropping station **26** where they are severed transversely of their direction of flow to separate pairs of battery plates by means of an intermittently operating press.

The grid structures then proceed to station **28** where they are turned over and indexed under a press that has a boss thereon that is of a size to substantially cover the lattice network portion. It is caused to move downwardly against the lattice network portion of the grid structures to depress the grid structure to the position shown in Figure 1.2d. The downward movement of the press is limited by the engagement of the protuberances **16** with the underlying support surface for the grid structures.

When the grid structures leave the casing machine the cross section of the grid structure is as illustrated in Figure 1.2c; it will be noted that the bottom surface of the lattice network is coplanar with the bottom surface of the perimeter frame. This is the surface that is cast against the smooth shoe. This surface is turned to be face up as it passes under the press **28** and is depressed to achieve the disposition illustrated in Figure 1.2d wherein the principal plane of the lattice network lies substantially between spaced-apart planes of the opposed faces of the perimeter portion of the grid.

As noted, the protuberances **16** are designed with a height of about one-half the difference in thickness between the lattice network portion and the perimeter frame portion of the grid so that when depressed, the principal plane of the lattice network will lie about midway between the face portions of the perimeter frame. With a casting machine of the type indicated in Canadian Patent 934,522, the grid structures are preferably turned over before they are pressed. This turning-over operation would not be necessary with a machine of the type where the molding machine drum rotates in the opposite direction and the grid structures come out from the top of the drum with the smooth side down.

After the principal portion of the lattice network has been depressed, the grid structures are conveyed to station **30** wherein paste is applied to the grid within the perimeter frame. The disposition of the paste is indicated by **18** of Figure 1.2d. The protuberances **16** also serve to locate the grid within the oxide paste. By applying a force on a pasted plate towards the upper face of the lattice work network of the grid as shown in Figure 1.2d, and supporting the lower face on a flat surface the protuberances **16** remain aligned as shown in Figure 1.2d. This ensures that the lattice work remains aligned after the pasting operation.

The cropping operation at station **26**, the pressing operation at station **28** and the pasting operation at station **30** are not described in detail. The important thing about the process is the sequence of operations as a whole which consists of the forming of the grid structure in a molding machine of a type indicated wherein the output is flat on one side and irregular on the other side, depressing the lattice work grid portion to locate its principal plane within the frame and subsequently filling the space between the frame with paste.

The grid can be manufactured with less lead, exposes less lead at the surface of the paste, gives better support to the lead oxide because the lattice work support structure is better embedded in the oxide paste and is cheaper to manufacture. The protuberances are an aid to accurate location of the grid structure between the perimeter frame but the location might be achieved by cooperating platens of a press.

Extruded Spines and Yoke

A process described by *H. Krug, H. Niklas and R. Thörnblad; U.S. Patent 3,973,992; August 10, 1976; assigned to Varta Batterie AG, Germany* relates to a tubular electrode for electrode storage batteries, comprising grid rods interconnected by means of a grid yoke. In making such grids by a casting-on process, the extruded spines, which have been cut to predetermined lengths, are cast within a grid yoke in a casting mold. A casting arrangement suitable for the production of such a tubular electrode is described below.

Referring to Figures 1.3a and 1.3b, the grid spines **2**, made by extrusion and cut to a predetermined length, are clamped in a rack **1**. The rack containing the spines is inserted into one side of the open form **3**. The form is then closed and liquid lead is introduced into the form. After the melt has solidified, the form is reopened and the tubular electrode can be lifted out by means of the rack. After opening of the clamps of the rack, the finished grid can be taken out, at which time spines which may have become bent can be straightened.

Form **3** contains a suitable recess **6** for the current takeoff, as well as for ceramic inserts **9** which provide heat insulation for the extruded spines **2**. In addition, cooling ducts **8** can be provided, as well as a heater body **5** for heating the two halves **3a** and **3b** of the form (mold). By quenching the grids in water, the hardness of the yoke can be further enhanced. The arrangement described above is especially suitable for automatic manufacturing techniques. For example, four mold pairs can be arranged in a carousel, and spines can be inserted automatically into these molds.

A special advantage derives from the fact that virtually any desired lead alloy can be used for casting the yoke, provided only its strength is sufficient for the subsequent processing. For example, antimony-poor alloys with antimony contents of only about 2.5 to 3.5% are suitable. Yokes of soft lead can easily be damaged during subsequent filling with active mass. The temperature of the melt used for casting depends on the antimony content of the extruded grid spines. At an antimony content of 3% this temperature should be between 450° and 500°C. For an antimony content of 3.5% it should lie between 400° and 450°C.

To improve the characteristics of the casting and the quality of the cast piece, the form can be provided with the usual mold release coatings. For example, it is advantageous to apply graphite-containing coatings to a coating of low heat transmissivity. Be means of heater body **5** the form can be maintained at optimum temperature, which lies between 180° and 230°C. Care should be taken that the ends of the spines to be surrounded by the casting extend only a small distance into the form cavity, and come into contact with the melt only in a narrow region. For example, it may be desirable to allow the ends of the spines to extend only 2 mm into the cavity.

In principle, the extruded grid spines can also be made to advantage of any desired alloy which is suitable for extrusion molding. If spines of soft lead are used, there is the danger that the junction zone may not fuse completely. These difficulties decrease with increasing antimony content, leading to trouble-free fusing of the spines with the yoke. To obtain still further improvement of the junction it may be advantageous to treat the cut ends of the spines with a flux.

FIGURE 1.3: TUBULAR ELECTRODE

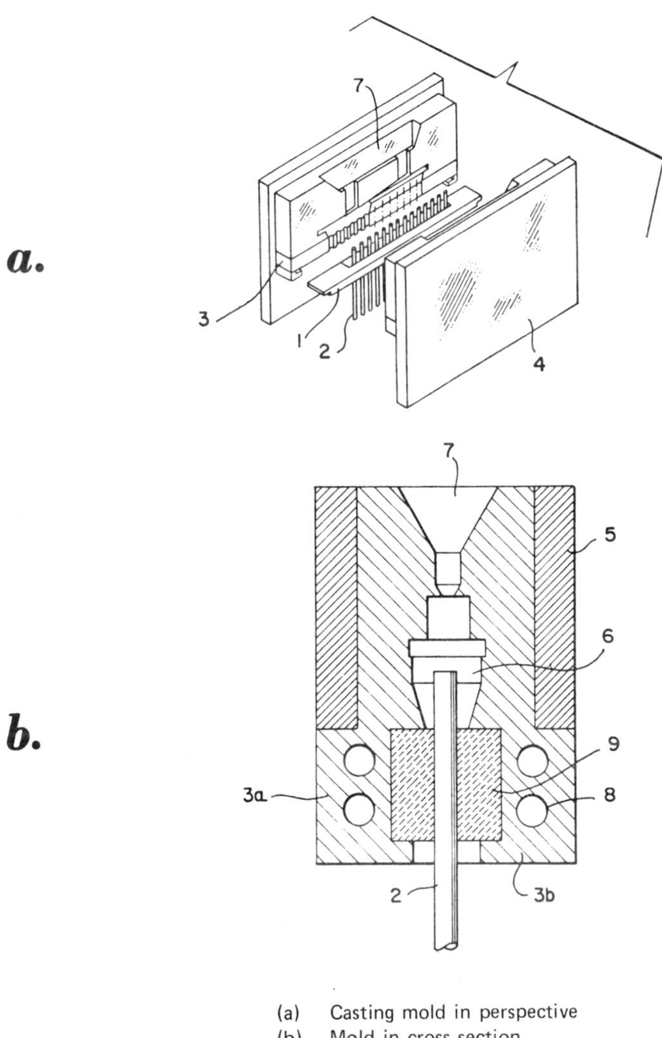

(a) Casting mold in perspective
(b) Mold in cross section

Source: U.S. Patent 3,973,992

Use of extruded lead spines further makes it possible to use spines which contain wire-like inserts of a material providing better conductivity than lead. Any highly conductive metal can be used as such conductor material as, for example, aluminum, silver, copper, titanium, nickel, or alloys of several metals, brass or bronze. The choice may be made so that the lead forms an alloy with the conductor material at their interface. In most cases, however, an intimate mechanical contact suffices. Before cladding the conductor with lead, adhesion-enhancing coatings may be applied galvanically or in a thin layer from the melt which join the base material with the lead cladding in an alloy. To produce such spines with a core of more highly conductive material, processes can be used which are conventional in the cable industry. To this end, the conductive wires are introduced through suitably shaped injection nozzles into a cavity containing molten lead and then drawn out again through a second, wider aperture having a cooled nipple. In this way, the conductive wire becomes surrounded by a thick seamless coating of lead.

The particular advantages of a tubular electrode grid according to the process reside in the fact that the junction region between the extruded wire and the cast yoke is defect-free even when different lead alloys are used for the two parts. In addition, grids of any desired length can be produced, and any desired lead alloys may be used. In particular, those combinations of antimony-free and antimony-poor portions suitable for the application may be used. The tubular electrode is also well-suited for mass production, since the manufacturing process described above can be automated.

Wire Elements

A process described by *S. Ikari, T. Hayakawa, Y. Kamikawa and F. Imai; U.S. Patent 3,944,431; March 16, 1976; assigned to Keishin Matsumoto, Japan* provides a plate grid for use in a lead storage battery wherein the weight of the plate grid is substantially reduced to improve the effectiveness of the lead storage battery. The plate grid comprises a plurality of spaced longitudinal wire members, the longitudinal wire members each including a wire element of acid-resisting material and a wire element of electrically conductive material selected from a group consisting of lead and lead alloy.

The electrically conductive wire element is wound around the acid-resisting wire element. A collector of electrically conductive material selected from a group consisting of lead and lead alloy is attached to one of the ends of the longitudinal wire members whereby the collector is electrically connected to the electrically conductive wire elements of the longitudinal wire members.

Layered Band-Shaped Portions

K. Yonezu, M. Tsubota and K. Asai; U.S. Patent 3,981,742; September 21, 1976; assigned to Japan Storage Battery Co., Ltd., Japan describe a lead-acid battery suitable for high-power application, having elements comprising a plurality of positive and negative plates which are composed of grids formed with band-shaped portions at one end facing to the side wall of container, and layered one upon another with separators put between in such a way that the respective band-shaped portions for the positive plates and negative plates respectively face to the opposing side walls of the container. The positive plates and the negative plates are respectively connected at least at a part of the length of the band-shaped portions to form straps.

A preferred form of an assembled element having grids that are manufactured suitably based on the process is shown in Figure 1.4a, wherein **1** denotes a positive plate formed with a band-shaped portion **1'** along the entire length of one side thereof and connected to a positive strap **2** for electrically connecting a plurality of such positive plates together. Each of a negative plate **3**, band-shaped portion **3'** for the negative plate and a negative strap **4** has the same construction as corresponding members in the positive side, and these members **3'** and **4** are situated opposite to members **1'** and **2** respectively.

FIGURE 1.4: LEAD-ACID BATTERY

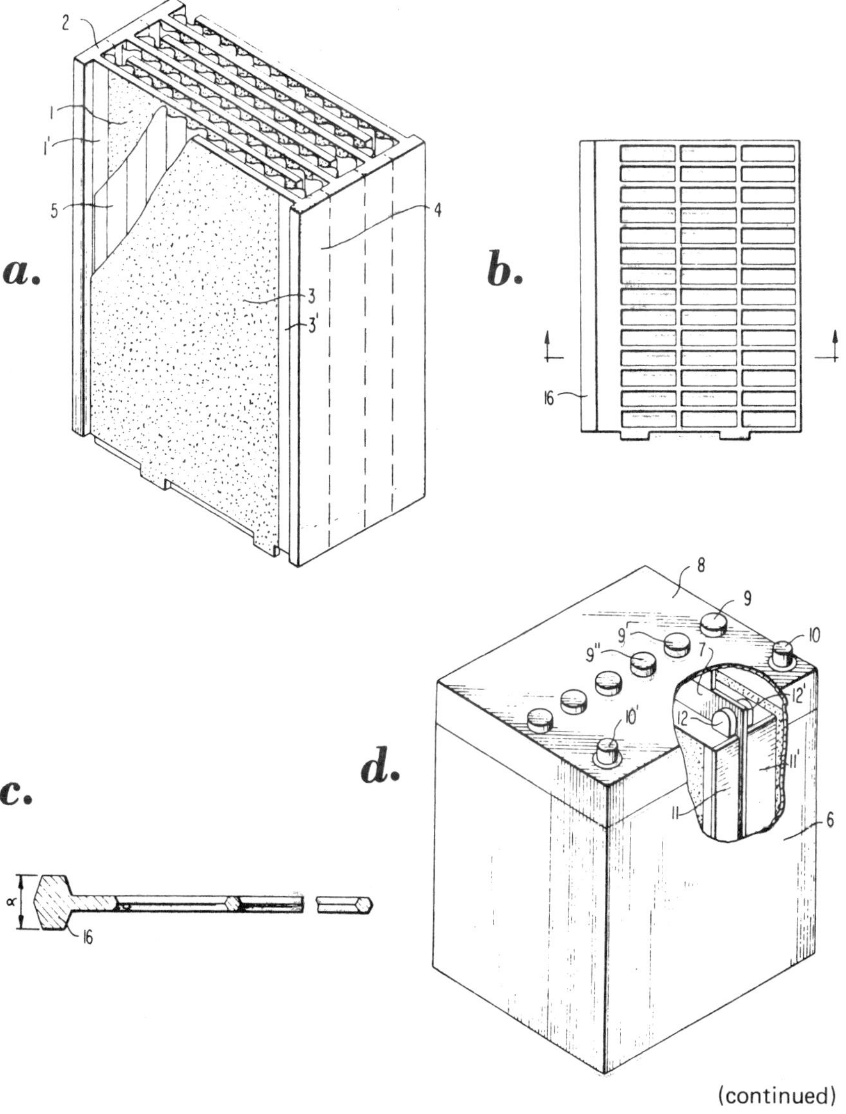

(continued)

FIGURE 1.4: (continued)

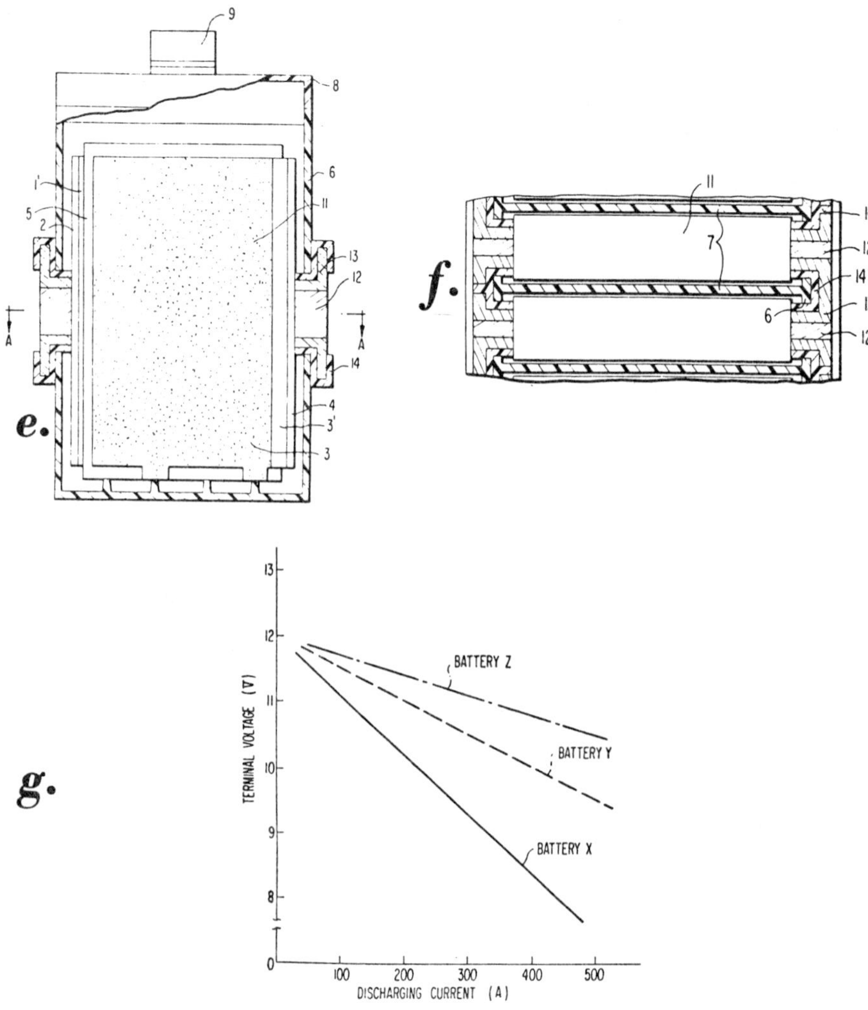

(a) Partially cut-away perspective view showing an assembled element of a lead-acid battery
(b) Plan view of one form of a grid for use with a lead-acid battery
(c) Partially sectional view
(d) Partially cut-away perspective view showing lead battery having assembled elements shown in Figure 1.4a
(e)-(f) Various forms of process elements
(g) Relation between discharging current and terminal voltage of a lead battery of this process

Source: U.S. Patent 3,981,742

Positive strap **2** or negative strap **4** may be formed by laminating the positive plate **1**, separator **5** and negative plate **3** in order and then adding additional lead alloy by way of melting and solidification according to the conventional method. Alternatively, it may be formed by previously making an enlarged portion at the transverse end **16** of the grid as shown in Figures 1.4b and 1.4c for the strap, layering plates and separator and then partially or entirely melting and solidifying the enlarged portion. In the latter method, the thickness **d** of the portion **16** of the grid is desirably greater than that portion filled with active substance but less than the sum of the thickness of a sheet of positive plate, a sheet of negative plate and two sheets of separator since the separators can then press the surface of plates and, therefore, effectively prevent the active substance from shedding.

An example of a battery structure is shown in Figure 1.4d, in which assembled elements shown in Figure 1.4a are used and are connected to each other by the connector at the current takeout ports provided at the upper parts thereof. In the figure, there are shown container **6**, partition wall **7** between cells, cover **8**, vent plugs **9**, **9'**, and **9''**, terminals **10** and **10'**, assembled elements **11** and **11'**, and poles **12** and **12'** that are integrated with a connector passing through the partition walls **7** in an airtight manner. In these elements, straps may only be formed at least in those areas containing the upper parts of the grids.

In Figures 1.4e and 1.4f, another structure of the battery is shown in which assembly elements shown in Figure 1.4a are used and are connected with each other at portions projected from the side walls of the container by way of poles provided at the middle of the elements' sides. In the figure, there are shown band-shaped portion **1'** for a positive plate, positive strap **2**, negative plate **3**, band-shaped portion **3'** for the negative plate, negative strap **4**, separator **5**, container **6**, partition wall **7**, cover **8**, vent plug **9**, assembled element **11**, negative and positive poles **12** and **12'** of the adjacent cell, connector **13** for the connection of assembled elements, and sealing parts **14** for closing the ports of the container.

The positive pole **12** and the negative pole **12'** are connected to the positive and negative straps **2** and **4** respectively, e.g., by way of welding, and project out of the side wall of the container, and these poles are electrically connected to each other by the connector **13**. The connector **13** may be lead alloy but it is preferred to constitute its portion which is not directly contacted with the electrolyte, i.e., the portion which projects out of the side wall of the container **6**, with such metals as copper, aluminum, silver, nickel, tin, zinc or the like as well as alloys mainly composed of these metals, for the reduction of weight and the electrical resistance.

The above-described projected portion may be plated or coated with lead, if required. Sealing parts **14** for the ports of the container serve as receivers for the connector **13** as well as sealant for the openings in the side wall of the container. The parts **14** consist of plastic molded products, sealing adhesives, etc. The battery shown in Figures 1.4e and 1.4f has electrode poles provided at the middle of the elements' sides and, consequently, has projected portions on the side wall of the container. This means that a greater space is required for disposing batteries when a plurality of them are arranged side by side.

A battery X of a conventional type and batteries Y and Z having the structure according to this process and shown in Figures 1.4d, 1.4e, and 1.4f respectively are assembled for a test. All of the batteries were designed so as to have a weight of 12.5 kg as filled with electrolyte, and discharge capacity of 30 Ah at 5 hour rate. They were discharged at various levels of discharging current under the ambient temperature and electrolyte temperature of 25°C, and the terminal voltage was measured five seconds after the initiation of the discharge. The results are shown in Figure 1.4g. The internal resistance was determined based on the slants of terminal voltages versus discharging currents shown in Figure 1.4g and they were 9 mΩ/6 cells for the **Battery X** of the conventional type, 5 mΩ/6 cells and 3 mΩ/6 cells respectively for **Batteries Y** and **Z** according to this process. It can be seen from the above results that the batteries according to this process have less internal resistance and, hence, excellent advantages.

Tapered Grid Members

R.S. Margulies and R.E. Biddick; U.S. Patent 3,923,545; December 2, 1975; assigned to the U.S. Environmental Protection Agency describe a grid structure which improves the power-to-weight ratio of the lead-acid battery. It has been found that the use of tapered structural elements with the thickest part of the element in the region of highest current density will substantially improve the efficiency of a battery plate.

The grid design which has been found to have the optimum power output comprises a plurality of parallel current-carrying members with a fewer number of parallel current-carrying members perpendicularly intersecting one another in the same plane. One edge of the grid structure is tapered from the corners to the widest portion in the middle. A current-collector tab from which power is taken off the plate is situated in the widest portion of this edge.

Intersecting the tapered edge and perpendicular thereto in the same plane, are at least two tapered current-carrying support members which are equispaced. The taper of the support members runs from the narrowest end at the opposite side of the grid to the broadest end intersecting with the tapered edge of the grid. By placing these tapered elements in the interior of the grid, they provide maximized mechanical support as well as greater current-carrying capacity.

Casting Process

T. Okura, T. Tatsumi, H. Watanabe, H. Murayama and Y. Kobayashi; U.S. Patent 4,034,793; July 12, 1977; assigned to Shin-Kobe Electric Machinery Co., Ltd., Japan describe a process for casting a plate grid for a lead-acid storage battery wherein a thin grid can be produced having neither fins nor discontinuation of the longitudinal and latitudinal members of the grid.

The process for casting a plate grid for a lead-acid storage battery comprises the steps of preparing a casting die assembly including two interengageable die halves between which is formed a casting cavity to cast a plate grid and pouring a molten metal of lead or lead alloy into the casting cavity to form the plate grid. The process is further characterized as comprising the steps of venting air in the casting cavity through gaps between the die halves and then through a narrow air passage in at least one venting member in either of the die halves so that the venting member is flush with the casting surface of the corresponding die half, the venting member at the top surface having cavity-forming portions operatively associated with the die halves.

Lead-Base Alloys

According to a process described by *J.C. Duddy and E.R. Hein; U.S. Patent 4,035,556; July 12, 1977; assigned to ESB Inc.* an alloy comprising lead with small additions of zinc and tin is used to make grids for lead-acid cells. The alloy as produced is softer than is desirable and is hardened by mechanical working. When properly worked, it has a strength and stiffness approaching that of other low-alloy lead materials currently in use as storage battery grids. In the process, lead is alloyed in the molten state with tin and zinc to form an alloy having 0.5 to 2% zinc and 0.5 to 2% tin.

First, individual ingots of the material are cast in a size suitable for rolling. The ingots are then rolled out into sheets. The sheet material is further manipulated to form finished storage battery grids by any of several known means. Among the methods for producing battery plates from sheet material that have been practiced are: (a) slitting and expanding to form an open grid; (b) punching out an open grid; (c) forging an interlocked type of grid; and (d) combinations of (a) or (b) with (c).

It is to be noted that methods (a) and (b) above do not add additional working to the metal; methods (c) and (d) do. It has been determined that the hardness of the alloy is dependent upon the degree of working that the alloy receives. Therefore, in the cases of (a) or (b), the initial rolling process will include greater reductions than (c) or (d) in order to get the same total degree of working.

Example: A lead alloy containing 1.5% zinc and 1.0% tin was used to cast automotive battery-type grids. These were processed into positive and negative plates by conventional processing steps and then assembled into cells. The cells were tested for capacity and cycle life as follows:

Initial capacity at 20 hr rate	72 Ah
Charge voltage at 30 amperes	2.78 volts per cell
Capacity at 20 hr rate after 2 weeks stand at 95°F	69.7 Ah
Capacity at 150 amperes 0°F	8.3 min
Cycle test, 4 hours charge at 6 amps	
2 hours discharge at 10 amps	
at 61 cycles	
capacity at 150 amperes 0°F	8.5 min
at 210 cycles	
capacity at 150 amperes 0°F	6.5 min
at 325 cycles	
capacity at 150 amperes 0°F	3.1 min
Charge voltage at 325 cycles	2.75 volts

From this testing, it can be stated that the alloy gives the desirable high charge voltage during cycle life found with lead-calcium alloys and considerable better than antimony alloys (one would expect a charge voltage of 2.50 volts at 325 cycles). The life on cycle test is comparable to calcium alloy life but shorter than expected cycle life for antimony alloys (500 cycles).

According to a process described by *K. Peters; U.S. Patent 3,912,537; Oct. 14, 1975; assigned to Electric Power Storage Ltd., England* an electric storage battery grid made from an antimonial lead alloy containing 0.002 to 0.5% of selenium with

0.25 to 0.5% arsenic and up to 4.0% antimony is disclosed. The alloy has improved casting properties for making the grid and the grids have a reduced tendency to release antimony in use.

Cadmium-Tin-Lead Alloy

P. Rao and G.W. Mao; U.S. Patent 4,007,056; February 8, 1977; assigned to Gould Inc. describe a cadmium-tin-lead alloy composition containing at least 0.4 wt % of both cadmium and tin which is used for forming the straps employed which weld together the lugs of the grids in lead-acid storage batteries. The relatively small amount of cadmium and tin provide a large volume fraction of a cadmium-tin eutectic phase that is low melting and serves to facilitate the welding. This alloy composition may also be used to form intercell connectors and terminal posts for such batteries.

Titanium Support Plate

F. Beck; U.S. Patent 4,001,037; January 4, 1977; assigned to Badische Anilin- & Soda-Fabrik AG, Germany describes electrodes for lead batteries which contain lead salt solutions and/or their corresponding acids, exhibit improved conductivity, do not contaminate the electrolyte, and even after repeated charging and discharging undergo no changes in shape. This is achieved when the positive base plates are made of graphite or of graphite powder embedded in a binder which is stable to the electrolyte.

To obtain particularly adherent layers of lead dioxide on the titanium plates the surface of the titanium may be mechanically roughened or degreased by conventional methods. It is also possible by an oxidative treatment in the presence of compounds of metals of subgroups I, VI, VII and VIII of the Periodic Table and of aluminum, vanadium and bismuth to produce a layer of titanium dioxide provided with the oxides of the metals onto which can be deposited the layers of lead dioxide which adhere firmly during charging.

Graphite or graphite-filled plastics based on polyolefins, fluorinated or chlorinated polyolefins, polystyrene, polyamides, polyesters or crosslinked polycondensation products of a graphite content of from 30 to 85 wt % and a particle size of from 5 to 500 microns are very good as plates for the deposition of lead on the negative side. Obviously a dispersion of finely divided graphite in a suitable binder, e.g., based on an acrylate or epoxide, can be applied to a suitable substrate, e.g., of chromium-nickel steel, nickel or copper and the dispersion dried and hardened after application to the substrate. Completely uniform layers of lead are obtained which are dendrite-free, have a matt surface and adhere well to the substrate.

Example 1: A cell is composed of a positive electrode consisting of a 1-mm titanium sheet which is seeded with magnetite and provided with a 20 micron layer of lead dioxide on one side from a lead nitrate solution, and a negative electrode consisting of a plate of synthetic graphite having a thickness of 3 mm. Between the electrodes there is a frame of polished polymethyl methacrylate having a thickness of 5 mm which leaves open a rectangular useful electrode surface of 14.5 cm^2. Two openings for filling and venting are provided at the upper small side of the frame. The backs of the electrodes are in contact with brass plates by which the entire arrangement is held together.

At the commencement of the experiment 15.3 g of a mixture of 65 wt % of lead perchlorate, 32 wt % of water and 3 wt % of perchloric acid is filled into the cell. The useful surface area of the electrodes is completely covered by the electrolyte. The cell is charged with 145 mA, equivalent to a current density of 10 mA/cm^2. The potential is 1.96 volts at the beginning, 1.99 volts after one hour, 2.02 volts after two hours, 2.05 volts after three hours and at the end (after 3.8 hours) 2.10 volts.

At this point the electrolyte has the approximate composition: 10 wt % of lead perchlorate, 45 wt % of $HClO_4$ and 45 wt % of H_2O. The change in volume is quite small because the higher density of the active material is compensated by the lower density of the electrolyte in the final condition. No gas bubbles are formed at the negative plate and only a few gas bubbles at the positive plate during charging. The layer of lead is smooth, devoid of dendrite and adheres well.

After standing for 24 hours at room temperature (the cell potential changes from 1.88 to 1.84 volts) discharge is begun at 145 mA, i.e., 10 mA/cm^2. The initial voltage is 1.68, after 1 hour, 1.62; after 2 hours, 1.56; and after 3 hours, 1.51 volts. After 3.2 hours the discharge is stopped because the potential has fallen below 1 volt. An energy weight ratio of 33 watt hours/kg is calculated with reference to the weight of the electrolyte and the utilizable basic electrodes. After passing through a cycle of 20 chargings and dischargings no significant change could be found in the current voltage data found at first.

Example 2: A cell having bipolar electrodes consists of: (1) two end plates of graphite seeded with magnetite and coated with 20 microns of PbO_2 as in Example 1 and (2) nine bipolar electrodes of titanium sheeting having a thickness of 0.05 mm onto which a polypropylene sheeting filled with 70 wt % of graphite and having a thickness of 0.15 mm has been stuck by means of a graphite-filled epoxide adhesive.

The graphite side is roughened, and the titanium side is seeded with graphite and provided with a 20 micron lead dioxide layer from a lead nitrate solution. The electrodes are stuck in grooves in a polyethylene case at a distance of 5 mm from one another so that a free area of exactly 1 dm^2 is formed. A solution of 70 wt % of lead perchlorate and 3 wt % of perchloric acid is then introduced so that the electrodes are just completely submerged. Charging is at the rate of 1 amp. Charging is stopped after 4.5 hours.

After one day, the cell is discharged at 1 amp. The potential falls from 17.0 volts to 15.2 volts during 3.6 hours and then rapidly to lower than 10 volts. A total of 57.5 watt-hours is taken from the battery. The weight of the cell is made up as follows:

	Grams
Electrolyte	1000 g
Two end plates	45
Nine bipolar plates	50
Cells	240
Terminals, leads to end plates	50
Total	1385

There is therefore an energy storage density of 41.5 watt-hours/kg.

According to a process described by *H.P. Fritz, D. Wabner and R. Huss; U.S. Patent 4,019,970; April 26, 1977; assigned to Rheinisch-Westfalisches-Elektrizitatswerk AG, Germany* in making a lead electrode for a lead-acid accumulator, a supporting body of titanium is pickled in a boiling solution of oxalic acid and then adsorptively coated with a titanium(IV) layer in a treatment bath. This treatment is followed by the anodic coating of the titanium body with PbO_2 in an electrolyte containing the ions of lead(II), in the presence of a lead(II) nitrate or a lead(II) salt of an amido, imido, nitrido or fluoro sulfate or phosphate.

The treatment bath may be an aqueous solution of an organic titanium salt, the pickling solution, or the electrolyte. If the pickling solution is used for this purpose, the titanium body is anodically connected or the solution is oxygenated. If the electrolyte is utilized, Ti(IV) ions are introduced into that liquid in which the titanium body is immersed for an extended period before closure of the anodizing circuit.

Thus, it has been found that oxide formation can be effectively inhibited by the adsorptive coating of the titanium substrate with a titanium(IV) layer prior to the deposition of PbO_2 thereon from a suitable electrolyte. The presence of an adsorptive Ti(IV) coating in a titanium substrate establishes an electrochemical surface potential substantially different from that of a body covered with Ti(III). As a result of this difference in potential, the flow of anodizing current in the electrolyte does not cause an initial surface oxidation of the substrate but brings about an immediate deposition of lead oxide thereon, possibly accompanied by a lead titanate.

Titanium and Zirconium Supports

W.R. Jacob; U.S. Patent 4,037,031; July 19, 1977; assigned to Imperial Metal Industries (Kynoch) Limited, England describes a lead-acid battery including at least two cells containing sulfuric acid electrolyte. A bipolar electrode comprises an electrically conducting impervious sheet of metal having on one side a first surface of titanium or a titanium alloy, and on the other side a second surface of zirconium or a zirconium alloy, the sheet forming a barrier wall between two adjacent cells. The first surface forms a support structure for the lead dioxide active mass of the positive electrode of one cell, and the second surface forming the support structure for the lead active mass of the negative electrode of the adjacent cell.

The first surface may be directly connected to the second surface, e.g., by a metallurgical bond which is preferably formed by hot rolling or explosive cladding. Alternatively, there may be a further metal between the two surfaces. Preferably, the further metal has a higher electrical conductivity than the titanium or titanium alloy or the zirconium or zirconium alloy. The further metal may be chosen from the group copper, aluminum and iron.

There may be a layer of lead between the first surface and the lead dioxide active mass. There may also be a layer of lead between the second surface and the lead active mass. There may be a foraminate structure on the first and second surface to retain the lead or the lead dioxide paste. The foraminate structure may be in the form of wire having an external surface of the same metal as the surface to which they are secured and may be secured to that surface, e.g., by spot welding.

The separators are microporous layers which suppress the formation of trees or dendrites which tend to grow outwardly from the negative active mass layers, to short out the cells. Separators also prevent shedding from the positive lead dioxide layers. The pastes which would be used would be proprietary and conventional paste formed basically from a mixture of PbO, Pb_3O_4, H_2SO_4, Pb for the positive paste and with the addition of suitable expanders such as lamp black, etc., for the negative pastes.

When it is stated that the fibrous material may be formed of melded fibers, it is meant that the fibers may have a higher melting or softening point core, and a lower melting or softening point sheath. These fibers are matted together by conventional means, and are then heated to a temperature at which the sheath becomes tacky to bond touching fibers together, but at which temperature the core is unmelted. After cooling, a bonded fibrous material is formed. Such structures are commercially available.

The accumulator (shown in Figure 1.5) is in the form of a container **101** of plastic or rubber or other suitable material which has a dividing wall **102** to form two compartments **103** and **104**. The dividing wall is a bipolar electrode having one face **105** of titanium and a second face **106** of zirconium. The compartment **103** forms a first cell and immersed in sulfuric acid located in the cell is a second electrode **107** which is a zirconium-base electrode. The second compartment **104** also forms a cell, also contains sulfuric acid and an electrode **108** having a titanium base. The bipolar electrode **102** is secured to the walls of the container **101** in a watertight manner.

FIGURE 1.5: TWO-CELL LEAD BATTERY

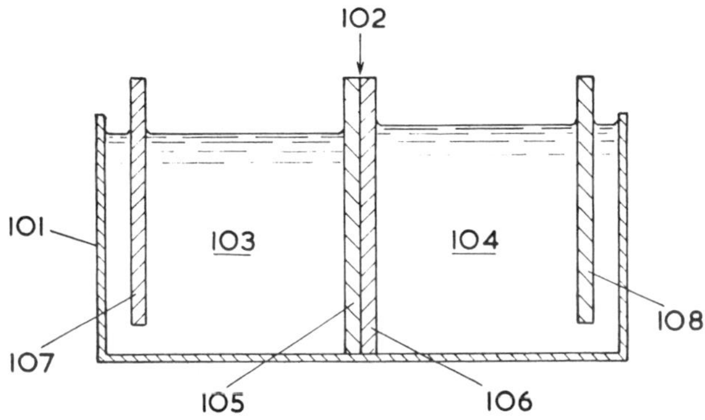

Source: U.S. Patent 4,037,031

The bipolar electrode **102** is manufactured by roll-bonding under an inert atmosphere a sheet of titanium to a sheet of zirconium. This produces a metallurgical bond between the two components which has a very low electrical resistance.

The bipolar electrode is coated with lead on both sides. Lead is electroplated onto the zirconium and the titanium surfaces after the surfaces have been etched. Subsequently, the electrode working pastes are applied to the surfaces. The titanium surface is pasted with a conventional $PbO + Pb_3O_4 + H_2SO_4 + Pb$ paste and the zirconium electrode is pasted with a basically similar paste which contains, in addition, suitable expanders such as lampblack, etc., which are well known in the art. The pastes are then formed by the passage of a suitable electric current to produce a substantially porous lead dioxide on the titanium positive surface and porous lead on the zirconium negative surface.

In use, the electrodes **107** and **108** are connected to an external load and the surfaces **105** and **106** act as independent electrode working surfaces with a current transfer taking place through the thickness of the electrode. If required, there may be a metal between the components **105** and **106** to provide strength and to reduce the need for too great a thickness of the expensive metals titanium and zirconium. Such a metal may be aluminum or copper or iron. On the surfaces of the bipolar electrode there may be a foraminate structure such as a mesh of wires secured to the surface to enhance the surface area of the electrode.

The importance of this is that it increases the keying of the lead or lead dioxide on the surface to enhance the electrical efficiency and electrical capacity of the accumulator. The wires may have an external surface of zirconium in the case of the zirconium surface and titanium in the case of the titanium surface. Alternatively, the foraminate structure may be an expanded metal mesh of titanium or zirconium. The zirconium or titanium surfaces may be formed of suitable alloys having comparable electrochemical properties to the base metals.

Drying Technique Using Inductive Heating

J. Brinkmann and W. Saul; U.S. Patent 3,965,321; June 22, 1976; assigned to Varta Batterie AG, Germany describe a method and apparatus for the drying of storage battery plates through inductive heating utilizing a magnetic field, which makes possible the rapid drying of plates of various dimensions. The process involves placing the plates side by side on a carrier, and then continuously moving the carrier through a drying tunnel traversed by a magnetic flux perpendicular to the plate surface.

An alternating electromagnetic field induces in the plate grids electrical voltage which causes heavy current to flow, due to the short circuit formed by the grids. Because of the low heat capacity of the lead, this current causes rapid heating of the grids. The heat is transferred from the grid to the surrounding active mass, whose temperature is thereby raised. The moisture in the active mass is thereby caused to evaporate.

This inductive drying of storage battery plates yields several advantages. The heat is produced within the plate grid and therefore heats the active mass from the inside. This causes first evaporation of the moisture near the grid. In this way, the moisture can be rapidly driven out from the interior of the active mass. At the same time the moisture in the outer regions of the active mass is forced to the plate surface. Thus, the drying process can be performed relatively rapidly and with economy of time. Since the inductive drying has high efficiency, it uses less energy than conventional methods.

Another considerable advantage is that the above-described shrinking process of the active mass, caused by drying out, begins right at the grid. The active mass is thereby firmly shrunk onto the grid, resulting in good mechanical and electrical contact between the active mass and grid surface, with low internal plate resistance and favorable battery terminal voltage.

Still another advantage is the following. Since the shrinkage starts at the grid, moist and malleable mass remains available in the middle and on the surface of the mass. This mass yields during the entire shrinking process, so that no appreciable mechanical stresses arise, and the formation of cracks in the plate mass is averted, again leading to good electrical conductivity of the mass bodies.

This shrink-on process and the resulting absence of cracks also considerably increases the mechanical stability of the active mass and its mechanical connection to the plate grids. This appreciably reduces the crumbling away of active mass during manufacture and during battery operation. In addition, inductive drying of storage battery plates, immediately following pasting, leads to a speed-up of the curing process. With prior methods, the curing process for lowering the Pb_{met} content of the active mass to a few percent has required a period of 4 to 6 days. This made necessary a lot of interim storage and also had other advantages.

Plates which are inductively dried, immediately after pasting, show a remarkably rapid decrease of the Pb_{met} content to the same low values in the subsequent curing process even though everything else is the same, so that the process is completed in a substantially shorter period. It thereby becomes possible to eliminate the usual interim plate storage in the curing chamber after pasting and to transport the plates, by means of chain conveyors or the like, in the space of a few hours through the curing chamber on a continuous basis. This saves much expense.

Rod Shaped Electrode

C.-E. Granqvist; U.S. Patent 3,922,175; November 25, 1975; assigned to Tudor AB, Sweden describes a plural cell battery composed of a plurality of cells wherein the electrodes in adjacent cells are rod-shaped comprising a core of a plastic insulating material surrounded by active material in which several electrical leads of conductive material are imbedded. Electrodes of this type may be interconnected by conducting lines which are arranged to pass straight through the common wall between adjacent cells, or may be connected at one end to a pole bridge and thereby form a tubular plate electrode.

Figure 1.6a shows a battery composed of three cells **10**, **11**, and **12**. Each cell has its own bottom and side walls, and the battery has a common cover member. To simplify the discussion, only two rod electrodes have been shown in Figure 1.6a in each of three cells **10**, **11**, and **12**, namely, in cell **10** the positive electrode **13** and the negative electrode **14**, in cell **11** the positive electrode **15** and the negative electrode **16**, and in cell **12** the positive electrode **17** and the negative electrode **18**.

The cells **10**,**11**, and **12** contain, in the usual manner, electrolytes **19**, **20**, and **21** which are held in insulating material containers **22**, **23**, and **24**. It is desirable that at each end of the battery, outside walls **25** and **26** respectively, form separate pockets for carrying the vertical conductors which provide the terminal connections **27** and **28**. As will become apparent from consideration of Figures 1.6f and 1.6g, each cell contains a large number of positive and negative electrodes.

Each of the electrodes **13** through **18** is supported individually by a holder **29**, **31** made in one cell wall and by a feedthrough bushing **30**,**32** in the other cell wall. For example, electrode **14** is supported on its left side, as shown, by a feedthrough bushing **30** in the wall of the cell container **22** and on its right side by a holder **29**. The feedthrough bushing connects electrode **14** with the conductor to the negative terminal **27** in the battery. In a similar manner, electrode **13** is supported on its left side by a holder **31** in the container wall and is fastened with the feedthrough **32** in one wall in each of two receptacles **22** and **23** for cells **10** and **11**.

FIGURE 1.6: PLURAL CELL BATTERY

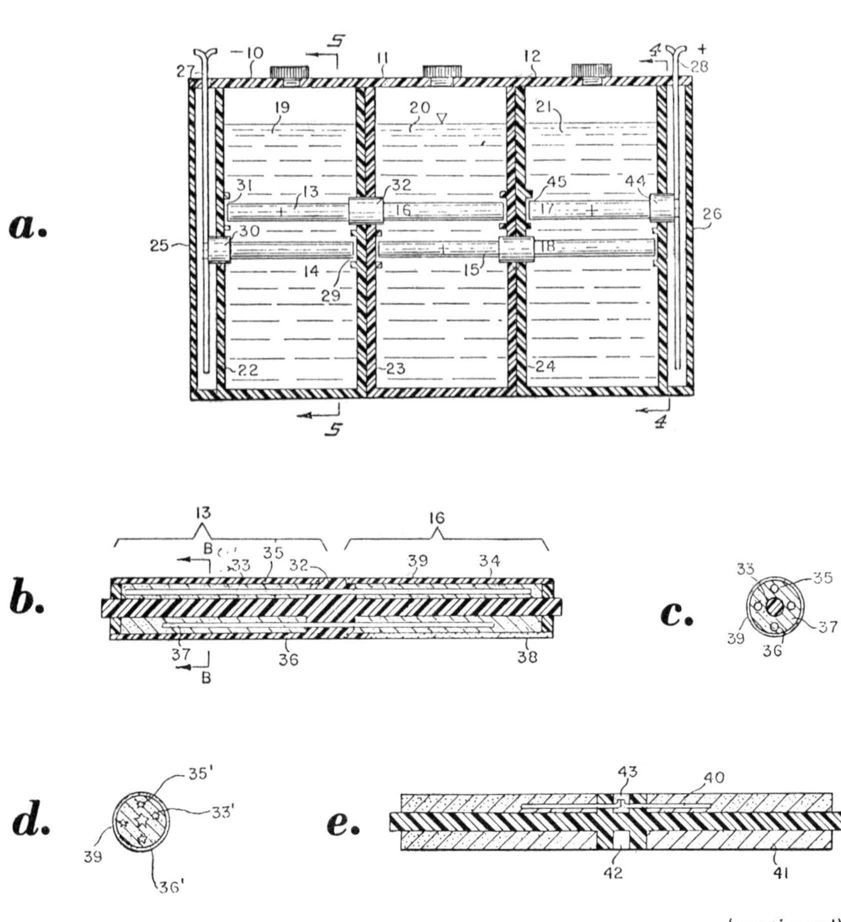

(continued)

FIGURE 1.6: (continued)

f.

44

45

g.

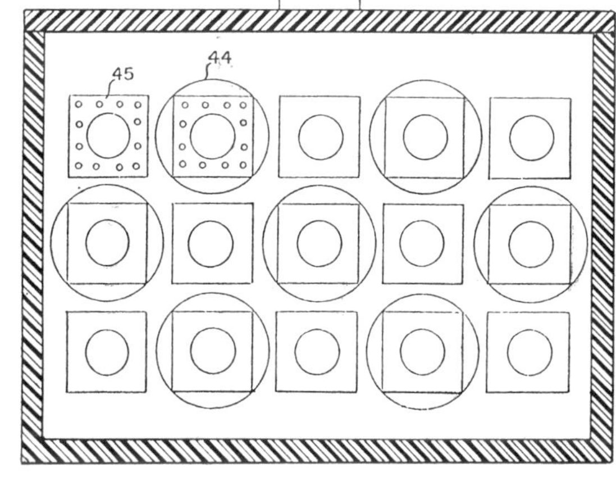

(a) Front elevation in section through a battery, consisting of three series coupled storage cells
(b) Front elevation in section of the electrode in the cells of Figure 1.6a
(c) Section taken along line **B–B** in Figure 1.6b
(d) Section similar to Figure 1.6c but of a rod core having a star-shaped cross section
(e) View similar to Figure 1.6b but showing a rod electrode which is especially suitable for use with cells of the iron-nickel type
(f) Side elevation taken along line **4–4** of Figure 1.6a
(g) Side elevation taken along line **5–5** of Figure 1.6a

Source: U.S. Patent 3,922,175

With closing of an external electrical circuit, the current from the positive terminal **28** flows to negative terminal **27** and in succession through electrodes

14-13-16-15-18-17 to positive terminal **28**. Current passes through the electrolytes in the space between electrodes **14** and **13**, **16** and **15**, and between electrodes **18** and **17** in the three cells.

Each rod electrode **13,14** is therefore mounted separately in a holder of the type shown, e.g., at **29** and **31**. Preferably, the electrodes are mounted in these holders with an electrolyte-resistant glue that can withstand the corroding influence of the electrolyte. The rod electrodes also pass through bushings **30** and **32**. Bushings **30** and **32** must be tightly sealed and, because of this, it is usually suitable to make them in part of the same material as the casing **22** through **24**. Also, it is desirable to use the same kind of glue that is used to fasten the electrodes in holders **29** and **31**, respectively. Through this construction, the rod electrodes **13** through **18** will serve as supporting elements inside each cell. As a result, the cells can be made with considerably thinner walls while maintaining sufficient stability, whereby the total battery weight is decreased.

In the arrangement shown in Figure 1.6a, both rod electrodes **13** and **16** and the rod electrodes **15** and **18** are made in one integral part. One such rod electrode is shown in longitudinal section in Figure 1.6b; and in cross section in Figure 1.6c. The bushing **32** that lies embedded in the container walls is shown in detail in Figures 1.6b and 1.6c. Bushing **32** preferably is made of plastic or other suitable insulating material and is provided with a pair of rod cores **33,34**, which extend outwardly in opposite directions so that the rod core **33** will carry positive electrode **13** (Figure 1.6a), while rod core **34** will carry negative electrode **16**. The rod cores **33,34** may have any suitable cross section so long as sufficient strength is provided for supporting the electrode. The rod cores may be porous or have channels, and may have any suitable diameter.

One or more pairs of collecting conductors **35** and **36** pass through bushing **32**. Two pairs of collecting conductors are shown in Figures 1.6c and 1.6d. Conductors **35,36** serve to collect the current from the active material **37** and **38**, respectively, which is contained around rod cores **33** and **34**, respectively. An electrolyte permeable sleeve **39** holds the active material, which is different to the electrodes of different polarity, physically in place.

The best balance between the space that is needed for the collecting conductors **35,36** and the space that is needed for the active material **37,38** is theoretically obtained for constant current density through the collecting conductors **35,36**. Since the current is fed to the collecting conductors **35,36**, a lesser total conductor area is required as the distance increases from the collecting conductor feed-through point in bushing **32**, and therefore these conductors advantageously can be made of a reduced length, as indicated by conductor **36** which is somewhat shorter than conductor **35**.

The current is collected from the active material **37,38** at the outer surface of the collecting conductors **35,36** and, if the cell is made for extremely high impulse currents, this surface area should be as large as possible, which suitably can take place if the collecting conductors are made of extruded profile material with, e.g., star-shaped cross section profile wires **35′** and **36′**, as shown in Figure 1.6d.

Each of the two rod electrodes **14** and **17** may be made with one as one half of the rod shown in Figure 1.6b and the other as the second half of the same rod.

In these half-rod electrodes **14** and **17**, the conductors extend beyond the end surface of the bushings **30** and **44**, and can either be elongated in the shape of the upwardly extending conductors rising to the terminals **27** and **28** shown in Figure 1.6a, or can be connected to these rising conductors in any suitable manner, such as by soldering or welding. In the case of a lead storage cell, conductors **35** and **36** must be made out of lead, and where it is important to decrease the internal resistance, the rising conductors located outside the elements should be made of material with better conductivity, e.g., copper or aluminum.

In Figures 1.6b and 1.6c, core rods **33** and **34** have a circular cross section which is desirable especially from a manufacturing point of view. If, however, the quantity of active material is increased further in order to improve the relationship between the charge and the weight of the storage cell, then a significant improvement can be obtained where the rods **33** and **34** are produced to include the star profile, as shown in Figure 1.6d. In the case of a lead cell, the active material consists of Pb and PbO_2, respectively, and in the case of an alkaline cell, the active material consists of nickel hydroxide and potassium or iron oxide, respectively. In both cases, this pertains to secondary cells.

Figure 1.6e shows a rod electrode for a battery of cells which work as secondary cells according to the iron-nickel procedure. The conductors **40** have been compressed with the powder-like active material **41** before being sintered, after forcing it down into a groove **42** in the seal bushing, which groove thereafter was filled with plastic.

The collecting wires or conductors **40** at the feedthrough of the seal bushing and eventually along its entire length, suitably should be made of rather small dimensions. Hereby, the risk of leaks is reduced, in that the axial length of the holes through the seal bushing is long compared with the diameter of the hole. Preferably, the conductors **40** are covered with active material all the way through the container wall, but if this should prove unsuitable in some cases, the conductors can be covered with a suitable semiconducting sleeve.

If, in spite of these precautions, a leak should develop, then a known method used in other technical fields to cover the conductors with one or possibly several successive layers of a material can be used, whose expansion coefficient lies between that of the conductor and that of the wall, or, where several such layers are used, successively varying the expansion coefficient step-by-step between the two. In this way, the resulting corrosion of the conductors can be decreased and the useful life of the battery increased. Among the materials with which the conductors can be covered for this purpose are such materials as lead-rich glazes and plastics enriched with quartz, carbon, and/or metal powder.

Another method to accomplish a good seal for the seal bushings as they pass through the container wall is to enclose them with an elastic material, e.g., thiocarbon polymerization products (i.e., synthetic rubber), which eventually can be made to cure and/or expand when it comes into contact with the electrolyte.
A preferred placement of the different rod electrodes is made in a diagonal pattern, as shown in Figures 1.6f and 1.6g. Figure 1.6f shows the outside of a cell container of a storage battery, e.g., container **24**, Figure 1.6a, where the broken lines outline the inside of the wall located on the other side of the same container.

The seal bushings of the type shown in Figure 1.6a are at **44**, again alternating with smooth surface areas, but corresponding to these smooth surface areas are holders **45** made for supporting the other ends of the rod electrodes. Through this arrangement, it will be possible to use only one common casting for the same enclosure. Housings **22** and **24**, Figure 1.6a, have been turned the same way, while container **23** in between is turned the opposite way. Thus, the correct fit between cutout **45** and the feedthrough hole **44** is obtained automatically. With this design constituting the basis for the presentation in Figures 1.6f and 1.6g, the rod electrodes have been shown with square cross sections that differ from those in Figures 1.6b, 1.6c, 1.6d and 1.6e. Each such rod electrode contains, therefore, twelve collection conductors.

To obtain the symmetrical placement of the rod electrodes, which are shown in Figures 1.6f and 1.6g, it is necessary that the rod electrodes be arranged in regular horizontal rows, whereby each horizontal row includes an odd number of rod electrodes; for arrangement according to Figures 1.6f and 1.6g, therefore, five rod electrodes are shown such that horizontally as well as vertically alternate rod electrodes belong to one or the other of the terminal systems.

Tube Plate Type Electrode

E.G. Sundberg; U.S. Patent 4,025,701; May 24, 1977; assigned to Tudor AB, Sweden describes a tube plate type electrode for a storage battery where the metal rod of the tube is held centrally in a surrounding sheath by spacing fins. The surface of the spacing fins is coated with a lacquer to reduce the peak effect caused by high current density to prolong the life of the tube sheath.

In Figure 1.7b, **1** designates the tube sheath which surrounds the active material **2** in the center of which there is a conductor **3**. On the conductor **3** there are suitably seated spacers or spacing fins **4** and **5** in pairs which center conductor **3** within the sheath **1**. The conductors **3** are joined at the top with a pole bridge **9** and at the bottom with a support **10**, usually made of a nonconductive material such as plastic, which together form the positive electrode. On either side of the tube sheath **1**, there is a negative electrode **7a** and **7b** (see Figure 1.7d), and between the electrodes there are separators **6a** and **6b**. The remaining free space is filled with electrolyte, as is conventional.

The spacing fins **4** and **5** on an electrode rod according to the process are given a coating **8** of an electrical insulating material. The insulating material coating **8** is desirably thin, about a millimeter or less, in order not to take up unnecessary space and reduce the amount of active material. The insulating coating **8** must not react with the other materials in a way that interferes with the normal operation of the battery. While the insulating materials should have a high resistance to resist the attack of the medium, this is not a necessary requirement, because even a layer that gradually becomes porous and possibly disappears in part, still has a considerably improving effect on the life expectancy of the electrode and therefore of the battery.

The insulating material is preferably a thin coating of an electrolyte-resistant lacquer or suitable polymer such as polyester. Polyester coatings are known as and may be made, e.g., from neopentyl glycol (1.1 mol), dicyclopentadiene (1.1 mol), maleic anhydride acid (1 mol), hexahydrophthalic acid (1 mol), and monomer styrene (0.6 mol) and divinylbenzene (0.6 mol). The coating of insulating material is applied only to the spacers **4** and **5** by painting, spraying, dipping, or any other suitable method to give a thickness which will give effective protection and yet not reduce the amount of active material by a noticeable amount.

FIGURE 1.7: TUBE PLATE TYPE ELECTRODE

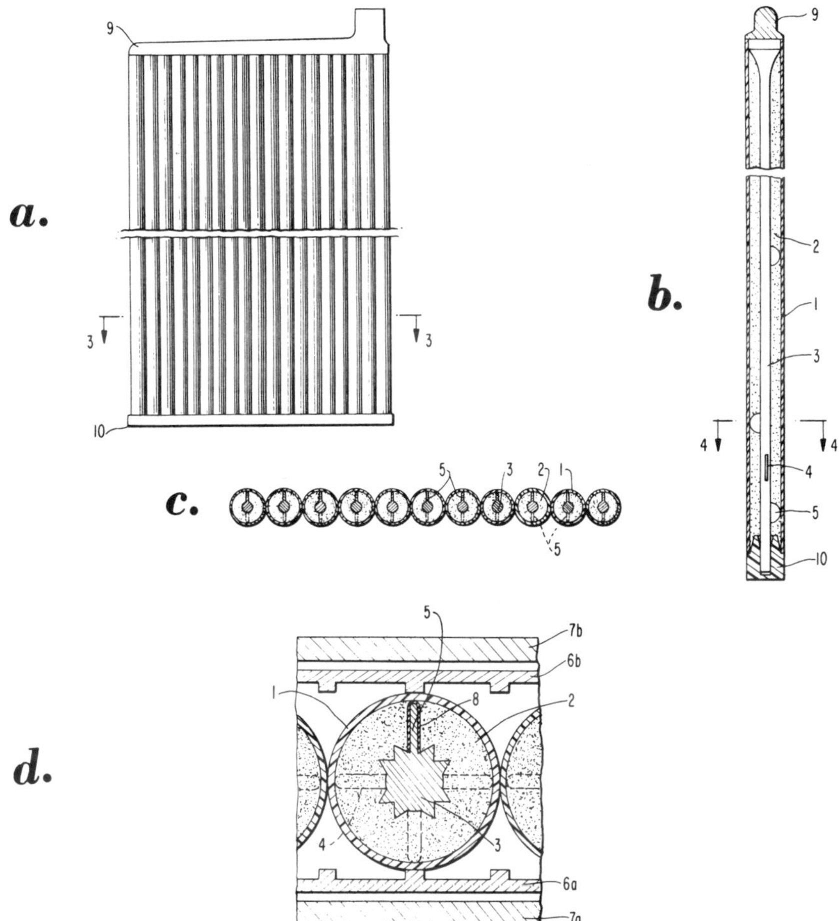

(a) Front elevation of a tube plate of conventional construction employing the process electrode

(b) End elevation in section, to an enlarged scale, of the tube plate of Figure 1.7a

(c) Top plan view in section, to an enlarged scale, taken along line **3-3** of Figure 1.7a

(d) Plan view in section to an even greater enlarged scale of a single electrode tube taken along line **4-4** of Figure 1.7b, together with negative electrodes and separators on opposite sides

Source: U.S. Patent 4,025,701

One disadvantage that may arise with electrodes is that the subsequent conversion of the active material into lead dioxide is made somewhat more difficult. Customarily, the forming process begins at the spacers, because of the peak current effect previously mentioned, and thereafter spreads out to the active material all along the conductor **3**.

This disadvantage can be obviated to some extent by giving the conductors **3** a configuration such that there is still some kind of peak effect. For example, the conductor can be made with a cross section other than circular, e.g., a many-sided star, as shown in Figure 1.7d. Other geometric cross sections which have sharp edges or provide increased contact surface with respect to the active material can also be effective. These cross sections may be obtained either directly in the casting of the conductor or in a subsequent forming treatment.

When the insulating coating is applied to the spacers, care must be taken that other parts of the conductor **3** are not insulated. This would lead to a poorer utilization of the active material and a corresponding reduction of the capacitance of the storage battery. On the other hand, it is not necessary or desirable to cover the spacers entirely with an insulating coating **8**. An uncoated region on the spacers **4** and **5** nearest the conductor would have no marked importance. The insulating coating can therefore suitably be applied in such a way that the conductors, joined in suitable numbers, are fixed in a jig such that only the surface of spacers **4** and **5** to be coated would protrude. Thereafter, the rods can be inserted in their respective sheaths **1** and the sheaths filled with active material. Finally, the tubes are mounted in the upper and lower frames **9** and **10**.

Laminated Electrode Using Synthetic Fibers

J.P. Cestaro and U. Sokolov; U.S. Patent 3,973,991; August 10, 1976; assigned to NL Industries, Inc. describe a lightweight lead-acid battery which is prepared using a laminated electrode structure. The laminated electrode contains at least three plies, one of the plies being a conductive sheet and the other of the plies being porous, compacted and sintered sheets comprising a homogeneous composite of synthetic fibers and active powder material.

This lead-acid battery is light in weight since the battery plates are composed of the three-ply laminated electrode structures described above. These plates are approximately 30 to 50% lighter in weight in comparison to the prior art pasted battery grids containing the same amount of active material. In addition, these composite electrodes being much lighter in weight and more flexible than the normal battery grids, may be formed into various irregular shapes such as bent, curved, spiral, cylindrical, and the like, which is not possible with the known pasted battery grids.

The center ply of the three-ply laminated structure is a thin sheet of conductive material, such as a thin metal sheet or paper sheet containing the conductive material, preferably a thin perforated lead sheet, sandwiched between two lightweight compressed and sintered composite sheets, composed of synthetic fibers and active powder material, such as PbO, Pb, tribasic lead sulfate or tetrabasic lead sulfate.

Centering Fins

E.G. Sundberg; U.S. Patent 4,011,369; March 8, 1977; assigned to Tudor AB, Sweden describes a tube plate for a storage battery formed from a grid of conductive rods that are surrounded by active material held in place by a sheath of insulating material. The conductive rods are centered in their respective sheaths by centering fins solely lying in the plane of the electrode. The conductive rods are centered in the perpendicular direction by placing the electrode in a jig which has cylindrical channels for receiving the sheaths. In each channel, two rows of needle-like elements are provided which penetrate the sheath and engage the rod surface during the period when the active material in powder form is loaded into the electrode.

The tube plate shown in Figure 1.8a traditionally consists of an upper strip of conductive material **10** having a joining lug **12** and being connected to a group of tubes **14** that are also supported on the lower end by a bar **16**, which may be of either conductive or insulating material. The number of tubes **14** in a plate generally ranges between about 10 and 25 per electrode.

Figure 1.8b shows in section the interior of a tube which includes a sheath **18**, active material **20**, and the center conductive rod **22**. On the conductive rod are guide fins **24** and **26** which lie in the plane of the electrode. The dashed lines **28** indicate the position of a centering fin which has been used in the earlier electrode construction and disposed in a plane perpendicular to the plane of the electrode, but which is omitted in accordance with the process. It is a characteristic of the process that there are guide fins **24** and **26** on the conductive rod **22** only in the plane of the electrodes.

In the manufacture of the electrode, the conductive rods **22** and bar **10** with lug **12** are cast as an integral unit. Since the centering fins **24** and **26** are traditionally a homogeneous part, the molding operation is greatly simplified by the deletion of the fins customarily used, which lie in a plane perpendicular to the plane of the plate and indicated in dotted lines **28**. After removal of the grid from the mold, the next step is to place sheath **18** over each of the rods **22** and active material to be charged in the space between the rods **22** and the sheath **18**.

FIGURE 1.8: ELECTRODE GRID PROCESS

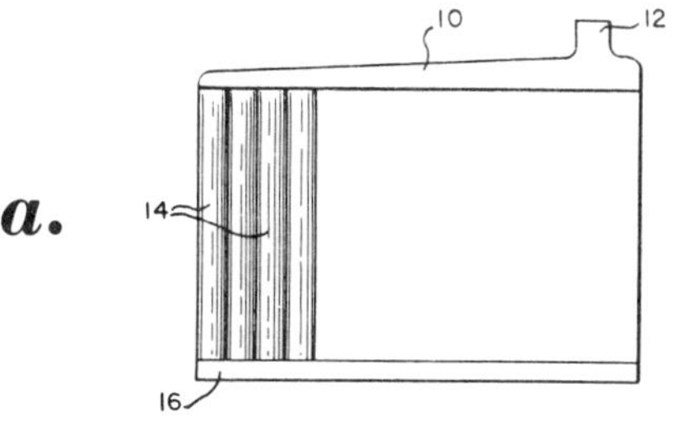

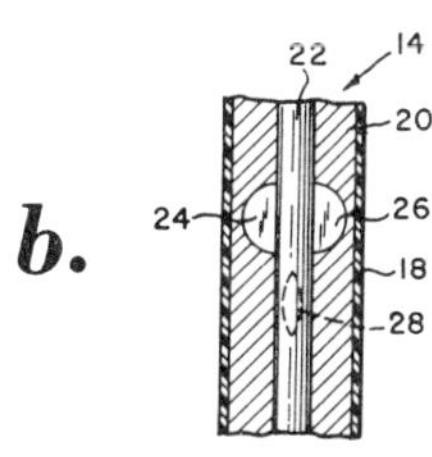

(continued)

FIGURE 1.8: (continued)

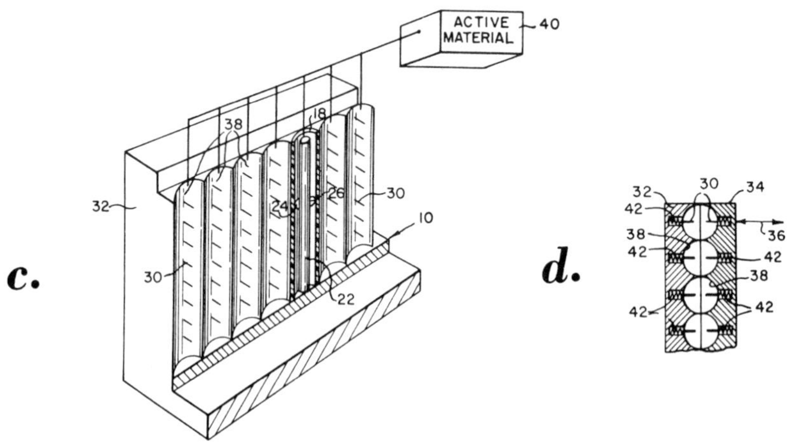

(a) View in elevation of an electrode
(b) Elevation in partial section which shows the interior construction of a tube as used in the electrode of Figure 1.8a
(c) View of part of the jig used in the manufacture of the electrode of Figure 1.8a showing rows of sharp needles which penetrate the electrode sheath and engage the surface of the rod for centering the rod within the sheath during the filling of the sheath with active material
(d) Partial top plan view in section of the jig of Figure 1.8c

Source: U.S. Patent 4,011,369

If the rods **22** are not located centrally in the active material, uneven electrical current loads in different parts of the electrode result and there is a consequent poor utilization of the active material. In the manufacture of electrodes with guide fins **24** and **26** lying only in the plane of the electrode, it is therefore necessary to center each rod **22** in its sheath **18** in the plane perpendicular to the electrode. The centering of rod **22** is accomplished by use of thin, sharp needles **30** which are thrust through sheath **18** to engage the surface of conductor **22**, including fins **24** and **26**.

Referring to Figures 1.8c and 1.8d, the electrode with each rod **22** surrounded by sheath **18** is located in the apparatus which consists of a jig having two parts **32** and **34**, with one part being movable relative to the other, as illustrated in Figure 1.8d by the arrow **36**. Figure 1.8c shows diagrammatically only half of the jig which is essentially symmetrical. In each of plates **32** and **34** there are a number of semicylindrical surfaces or grooves **38** located to correspond to the position of an electrode tube **14** when installed in the plate of Figure 1.8a. The electrode may advantageously be placed in an inverted position with the bar **10** at the lower end and the open end of the tubes **14** exposed to active material from a supply **40** which is adapted for use in filling each tube **14** with the powder material.

The electrode with rods **22** and sheaths **18** in surrounding position is placed in the apparatus of Figure 1.8c and the two plates **32** and **34** are brought together so that the thin, sharp needles **30** are thrust through sheath **18** to engage opposite sides of rod **22** without damage to or alteration of the form of the sheath. This assures centering of the conductive rods **22** in the sheath **18** during charging of the active material.

It has been found that, on occasions, the needles strike portions of the sheath which are too hard to be easily penetrated. To avoid this disadvantage, it is therefore desirable to have the needles **30** movably mounted in the plates **32** and **34** so that at a specific force, they can move longitudinally in a direction away from the intended position of the electrode. This is also desirable to avoid a situation where the force from a fixed needle might be so great as to cause deformation or deflection of the sheath **18** against conductive rod **22**. Therefore, in Figure 1.8d, each of the needles **30** is shown mounted with its one end supported by a suitable resilient member such as a small spring **42**.

The length of the needles **30** is chosen so that they are somewhat less than the radius of the groove **38** since the sharp end goes only to the surface of rod **22**. The axis of each needle **30** is perpendicular to the plane of the electrode plate **10** so that the two rows of needles **30** lie in a plane that passes through the axis of rod **22** and is perpendicular to the plane of the electrode plate.

To achieve complete filling of the tube cavities, it is customary to cause the electrodes to vibrate during charging with active material, a substance or powder with different properties. The term powder here means any material with such particle size that it can pass down into the space between the rod **22** and the sheath **18**. Since the rheological properties of this material are often relatively poor, it is customary to shake the electrode during the charging. The electrodes as a rule consist of material with good vibration-damping capabilities, however.

It has therefore turned out that in the manufacturing process, there is an improvement in the uniformity of the charging of the electrodes. This is due to the fact that the vibrations are transmitted to the electrode rods **22** which are in contact with the powder while the rods **22** are directly connected by needles **30** to the jig and vibrating base. The vibration is transmitted to the electrodes also at the constant surfaces between electrode and jig. Since the needle-like rod centering elements are fixed in the jig, the transfer of vibrations occurs at a large number of joints.

Whether or not the electrode is vibrated, it has nonetheless been shown to be desirable to have the needle-like elements fixed in the jigs, because there is greater precision in electrode manufacture this way. It has likewise been shown to be desirable to have the needles movably fastened in the jigs so that at a specific force, the needles can move longitudinally in a direction away from the intended position of the electrode.

Spaced Spirally Wound Electrodes

W.N. Frazier; U.S. Patent 3,904,433; September 9, 1975 describes a lead-acid storage battery having spirally wound electrode cells, in which the positive electrode plate appears only in one cell and the negative electrode plate appears in the second cell which is axially spaced from the first cell.

The electrode strips are made from a conventional lead grid having holes and covered by lead oxide paste. However, the electrode plates in both cells may be structurally identical and capable of receiving either a positive or a negative charge. Thus, the charge on any electrode plate depends upon what charges are impressed upon it, the electrode in the other cell having the opposite charge.

It is also within the scope of this process to provide an elongated insulator strip to be spirally wound with each electrode plate in which one or both surfaces of the insulator strip have circumferentially spaced axially disposed ribs. Grooves or valleys are defined between the ribs to facilitate the axial passage of the liquid electrolyte between the respective surfaces of the insulator strip and their adjoining electrode plates.

The spirally wound cells are preferably mounted coaxially within a solid casing or housing and immersed in sulfuric acid. The cells are separated from each other and from the bottom wall by spacers. The housing or casing is also provided with a detachable lid so that either or both cells may be easily removed, inserted, replaced or repaired. The disposition of the spirally wound electrode cells are such that there is substantial economy in space, construction, time and labor, not only in the original construction of the battery and the cells, but also in the replacement of the cells. There is also a more efficient transfer of ions between the cells in the battery made in accordance with this process and this battery can be charged and discharged faster than conventional batteries. Since the electrode plates are identical in each cell, any cell can be replaced by another cell, regardless of whether the electrode in that particular cell functions as a positive or negative electrode after it is charged.

Slurry Filling Technique Using Porous Envelope

S.C. Foulkes; U.S. Patent 4,037,630; July 26, 1977; assigned to Chloride Group Ltd., England describes a method for preparing enveloped plates for batteries which involves introducing an active material composition containing water into the porous envelope of the plate, when the envelope is assembled on the current conducting element of the plate. The technique is characterized by using an active material composition with a water content such that active material is filtered out by the porous envelope, and a bed of active material is built up in the envelope, the bed building up from the end remote from that at which the composition is introduced back to the end at which the composition is introduced, liquid issuing through the walls of the envelope throughout the period that the bed is being built up. The active material is preferably a lead oxide active material and the ratio of solids to water in the composition is preferably from 2.5:1 down to 0.4:1 by weight.

The term envelope covers arrays of separate tubes as well as arrays of tubes joined together or formed from sheets of material in addition to covering any envelope effective to form a bag or pocket around the current collecting element or elements of the plate and effective to filter out active material as a bed around the current conducting element or elements.

Slurries useful in accordance with the process are readily pourable and the solids settle out rapidly from the liquid phase, namely in less than 15 minutes on standing. Thus, the compositions are characterized by having a rotating vane viscometer torque value of less than 0.006 preferably not more than 0.004 lb/ft at 20°C.

The suspension half-life of the slurries is preferably not more than 15 minutes preferably in the range 1 to 10 minutes. The slurries have viscosities substantially independent of shear rate, i.e., they are not thixotropic gels and while the viscosities fall with increase in shear rate this fall is not pronounced and a gel does not form when the shear force is removed.

The slurry can contain conventional fillers and additives for the active material such as hydrophobic or hydrophilic silica, e.g., 0.1 to 0.5 wt % based on the oxide. The introduction of the slurry into the tubes is preferably carried out under gravity, i.e., at zero pressure or at a pressure of less than 5 psi until the tubes are filled with the composition, the pressure then being allowed to rise to a value not in excess of 100 psi and the pressure thereafter being released.

In one arrangement, the tubes are allowed to fill substantially under gravity and then the pressure is allowed to build up to apply pressure to the active material in the filled tube for only a fraction of the time taken to fill the tube. Thus, the pressure may be in the range 5 to 50, e.g., 10 to 30 psi, applied, e.g., for one-tenth to one-half the time taken to fill the tube or for a time equal to that taken to fill the tube. Thus, the tube may take 5 to 15 seconds to fill and the pressure may be applied for 1 to 5 seconds.

In another arrangement, the pressure is applied for longer periods of time. In this arrangement, the tubes are allowed to fill substantially under gravity by the slurry being pumped into the tubes under zero back pressure; once the tubes have filled the pumping is continued and the back pressure allowed to build up to a value not in excess of 70 psi. Thus, the pressure may be in the range 5 to 50, e.g., 10 to 30 psi. The weight of oxide in the tubes can be adjusted by adjusting the pressure buildup as it is indicated in the examples. Usually the pressure is merely allowed to build up to a set value at which point the pressure is released.

In contrast to the prior proposals where the whole filling operation is carried out under high pressure, which results in the active material being stratified, the paste being more dense nearer the inlet, this arrangement enables the density of the active material in the tube to be increased evenly throughout the tube without stratification occurring. The material of the tube as indicated above is selected to have a filtering action on the active material used. However, this does not mean that all the active material is removed from the liquids issuing from the tubes, merely that a proportion is retained within the tubes.

A balance must be struck between the need for the material to have a high water permeability to provide good conductivity in use in the battery and the need for the material to have a good filtering action so as to enable filling to be carried out rapidly and the active material to be retained in the tubes over long periods of use and conditions of shock and vibration. One suitable material is made from a nonwoven batt of polyester fibers which is 0.5 to 0.7 mm thick and weighs 120 to 160 g/cm^2. This is not perforated, its porosity being derived from the various gaps between the fibers from which it is made. It has a nitrogen permeability of 8.0 liters/cm^2/min, and water permeability of 1.5 liters/cm^2/min.

Referring to lead-acid systems the lead oxide preferably has substantially all of its particles having particle sizes less than 100 microns, e.g., less than 1 wt % are above 200 microns in diameter. In addition, less than 1% is below 0.0001 micron in diameter. Typically, at least 50%, e.g., 95 wt %, is less than 50 microns, 50 wt % is less than 10 microns and 5% is less than 1 micron. The oxide may comprise a blend of gray lead oxide of average particle size 20 microns and red lead oxide of average particle size 5 to 10 microns. The ratio of gray to red lead may be in the range 95:5 to 5:95, though 90:10 to 50:50, e.g., 33:67 is preferred.

The apparatus for carrying out the method preferably comprises at least one filling station comprising means for supporting the envelope of a plate assembled on its current conducting element in a substantially vertical plane and a filling manifold adapted to introduce slurry into the envelope of a plate located in the supporting means, the apparatus further comprising a slurry storage tank provided with agitating means adapted to contain a supply of active material slurry, and delivery means for delivering the slurry from the storage tank to the manifold of a selected filling station.

Molding Technique for Connectors

R. Gaide; U.S. Patent 3,944,436; March 16, 1976; assigned to Société Fulmen and Manufacture D'Accumulateurs et D'Objets Moules, France describes a method for making an electrical connection between groups of lead plates for electric accumulator elements. The process involves making such connections by an appreciably modified method, using, more particularly, either an elementary mold having a different structure, or parts molded onto the accumulator box.

The process thus provides a method of forming lead plates on an electric accumulator having several elements, and for interconnecting the plates of the elements by melting the plate lugs, in which the plate lugs arranged in an element of the accumulator are inserted in the openings made in the bottom of at least one elementary mold. The plate lugs then are connected together by rods produced by melting the lead, and the connection of the rods associated with the plates of opposite polarities of two adjacent elements of the accumulator is made. These operations are carried out preferably before the fixing of the accumulator lid on its box and is characterized in that the elementary mold comprises two channels, each of these having the plate lugs of same polarity fitted in them.

LEAD PASTES

Electrolyte-Resistant Fibers

A process described by *J. Brinkmann and J. Sucher; U.S. Patent 4,039,730; August 2, 1977; assigned to Varta Batterie, AG, Germany* relates to a method of filling the lead core-containing tubes of tubular electrodes, and to the introduction into the tubes under pressure of paste-like active mass which includes a mixture of lead oxide, sulfuric acid and water.

Thus, an intimately mixed mass of lead oxide, water and sulfuric acid is produced, which contains a quantity of electrolyte-resistant fibers. Preferably glass fibers are used as the electrolyte-resistant fibers but plastic fibers may also be used. Even in the mixed mass these fibers already tend to prevent precipitation of the lead oxide, which would lead to separation of the ingredients of the paste. Therefore, the mixed mass may be kept considerably more fluid, which makes possible its pressing into longer tubes. During pressing, a large part of the liquid escapes to the outside through the tube wall, while the mass is retained inside. During the escape of the liquid, a portion of the fibers which have been added in accordance with the process penetrate into the pores of the tube wall and are trapped there. Thereby, they produce additional occlusion of the pores of the tube wall. Finished tubes have therefore caught on the tube wall and in its pores, a high proportion of added fiber material, which proportion decreases toward the interior. The fiber portion which remains in the interior of the tube, i.e., in the active mass, materially contributes to increasing the strength of the active mass. This also makes that mass less inclined to become extruded. Extrusion is materially reduced by the fibers which have penetrated into the pores.

The tube itself can generally be any of the tubes known in the storage battery art. Preferably, however, tubes are used which are made of a double-woven material which forms the adjacent tubes. This woven material preferably is of glass fibers, or of glass fibers mixed with plastic fibers such as polyester fibers. If necessary, the tube material can be further stiffened by treatment with an impregnating solution containing a hardening plastic material.

In order to give the active mass sufficiently low viscosity, sulfuric acid and water is added. Mixtures have proven suitable which contained between 100 and 600 cm^3 H_2O per kilogram of lead oxide, and preferably between 200 and 400 cm^3 H_2O, as well as 5 to 130 cm^3 H_2SO_4 of 1.4 g/cm^3 density. Such extrudable masses are, for example, described in U.S. Patent 3,228,796.

This mass emulsion additionally contains a proportion of fibers, preferably glass fibers or plastic fibers such as polyester fibers. These have lengths from 1 to 6 mm, preferably 1.5 to 3 mm and thicknesses of 5 to 30 microns, preferably 8 to 20 microns. The proportion of fibers is between 0.1 and 2 wt %, relative to the lead oxide quantity, and preferably is 0.2 to 1.5 wt %. Especially preferred is the use of thermoplastic fibers of plastic material, particularly polyethylene fibers, which exhibit a rough, jagged and scalloped surface, whereby the active surface of the fibers is increased relative to their geometric surface.

It has been found that precipitation of the lead oxide in the emulsified mass is appreciably slowed down by this type of fiber. Due to its rough and jagged surface this fiber holds the active mass back more strongly than do fibers with smooth surfaces when it is positioned in or in front of the meshes of the pocket or tube fabric. The fibers with rough and jagged surface become anchored in the active mass better than fibers with smooth surfaces.

The active mass is therefore better held together and contained. As a result, the extrusion rate of the active mass is also appreciably improved. A polyethylene fiber of this type has a diameter of from 0.5 to 5 microns, and preferably 1 to 3 microns. Its length lies between 0.5 and 6 mm, and preferably between 1 and 3 mm. The quantity added of such fibers should be from 0.1 to 2 wt % relative to the lead oxide, and preferably 0.2 to 1 wt %. To fill the tubes which contain the lead cores, apparatus can be utilized which is known for the filling of tubular voids. For example, injection molding devices may be used which are suitably adapted to the materials and operating parameters involved.

It has been found that tubular electrodes can be produced by means of this method which have a particularly low extrusion rate and therefore a particularly long life. This is attributable to the fact that a portion of the fibers contained within the fiber-containing paste deposits preferentially in the vicinity of the wall, and penetrates into the pores of the tube wall, whereby occlusion of the tube wall pores is achieved. In particular it follows that the meshes of a fabric which are distended during filling with active mass are then, during the filling itself, closed again by the partially occluding fibers. As a result, even the enlargement which is unavoidably produced by the filling does not give rise to any future disadvantages.

Lithium Sulfate

D.F. Taylor and E.G. Siwek; U.S. Patent 3,926,670; December 16, 1975; assigned to General Electric Co. describe a cohesive lead oxide paste which consists of at least one battery lead oxide, lithium sulfate and water. It has been found that the use of lithium sulfate rather than sulfuric acid in the paste for the preparation of both positive and negative electrode plates for lead-acid batteries results in increased ease of handling, ease of paste mixing, and plate integrity. Additionally, the positive plate provides improved electrochemical performance on deep cycle duty.

Lithium sulfate can be mixed dry with both leady and nonleady battery oxides and then mixed with water as opposed to mixing such lead oxides with water and sulfuric acid which is often employed. An acceptable quantity of lithium sulfate in the paste is equivalent in sulfate/oxide ratio to the ratio used in conventional paste formulation.

The paste does not set up or harden as rapidly as a conventional paste, and spreads more smoothly. Lead acid battery plates made with the paste are initially lower in active material utilization, but gradually build in capacity and outlast conventionally made plates pasted on nonantimonial lead-alloy grid on deep discharge cycle service by a factor of two or more. The paste made with nonleady oxide appears to be slightly superior to those in which a leady oxide, 20 to 30% metallic lead and 70 to 80% lead oxide, is used.

Example 1: A battery lead oxide mixture was prepared in which one lb of battery lead oxide and 36.6 g of lithium sulfate monohydrate, $Li_2SO_4 \cdot H_2O$, were mixed together by dry blending for a period of 60 minutes. The resulting battery lead oxide mixture is adapted to form a cohesive paste for use in the preparation of lead-acid storage batteries.

Example 2: A cohesive battery lead oxide paste was prepared which employed the battery lead oxide mixture of Example 1. 40 g of the mixture from Example 1 was mixed with 4.5 cm^3 of water and stirred for 5 minutes to provide a cohesive paste for use in the preparation of lead-acid storage batteries.

Examples 3 and 4: Two positive lead-acid battery plates were made from the lead oxide paste of Example 2. The finished plates are identified as plates 1 and 2. Each of the plate grids was a calcium-tin-lead alloy. After the paste was applied to the grids, the resulting positive plates 1 and 2 were cured and formed in a conventional manner as set forth in the above-referenced volume.

Examples 5 and 6: Two positive lead acid battery plates were formed from a conventional paste consisting of battery lead oxide, water and sulfuric acid. The finished plates are identified as plates 3 and 4. Each of the plates 3 and 4 had a calcium-tin-lead alloy grid, and was cured and formed in a conventional manner.

Examples 7-10: Each of the above plates or electrodes 1-4, which were Examples 3-6, respectively, was cycled in identical fashion by employing identical charging time, charging current and discharge current. The table sets forth the results of this cycling which is expressed as discharge capacity in ampere-hours per gram of paste after cycle numbers 1, 9 and 60.

Positive Electrode Number	Discharge Capacity in Ah/g of Paste After ....Cycle Number............ 1	9	60
1	0.0710	0.1005	0.0610
2	0.0710	0.1005	0.0610
3	0.0940	0.1000	0.0469
4	0.0940	0.1000	0.0469

The above data show that electrodes 1 and 2, while poorer in initial performance, improve in capacity after repeated cycling. Electrodes similar to electrodes 1 and 2 have been successfully cycled in excess of 120 cycles. The deep discharges used in these examples were more severe than conventional usage would prescribe in that they were completely discharged during each cycle and recharged at constant current with considerable overcharge.

Low Dusting Paste

According to a process described by *E.Y. Weissman and B.K. Mahato; U.S. Patent 3,905,829; September 16, 1975; assigned to Globe-Union Inc.* leady dust which usually results from the handling of pasted lead-acid storage battery plates is substantially reduced by coating the surface of the pasted plate with a water-soluble polymeric compound which does not interfere with the plate's electrical capabilities. Preferably, the water-soluble polymeric compound is a polyvinyl alcohol, polyethylene oxide, alkyl cellulose, acrylic resins or water-soluble starches which are applied either as an aqueous solution or a fine powder or film of the compound directly to the wet surface of the pasted plate. Additives such as viscosity modifiers, plasticizers, extenders and antioxidants can be added with the water-soluble polymeric compound.

ELECTRODE SHEATHS

Closing Technique for Bottom Ends

According to a process described by *G. Eckerbom; U.S. Patent 4,011,370; Mar. 8, 1977; assigned to Tudor AB, Sweden* a battery electrode includes a plurality of sheaths. The base portion of a bottom strip is secured to bottom open ends of the sheaths. The bottom strip also includes movable closure sections which can be swung open to allow filling of the sheaths with active material, and which can be swung closed to close the bottom ends of the sheaths.

In a preferred case, the base is common to all the tubular sheaths of an electrode, with the two movable sections disposed on each other alongside the base. The movable sections are joined with the first part and are of such configuration that together they can cover the openings in the base. Advantageously the movable sections are identical. The extent of the movable parts in a plane perpendicular to the plane of the electrode should equal or somewhat exceed half the extent of the first part fixed in the tube ends, in the same plane. In Figure 1.9a the bottom portions of the tubular sheaths **1** are depicted in an inverted position for being filled with electrolyte. On the sheaths **1** is mounted the bottom strip which comprises a first part or base **2** and a pair of movable closure sections **3a,3b**. The base **2** is fixed to the ends of the sheaths by conventional coupling techniques.

FIGURE 1.9: CLOSURE TECHNIQUE FOR BOTTOM ENDS OF ELECTRODE

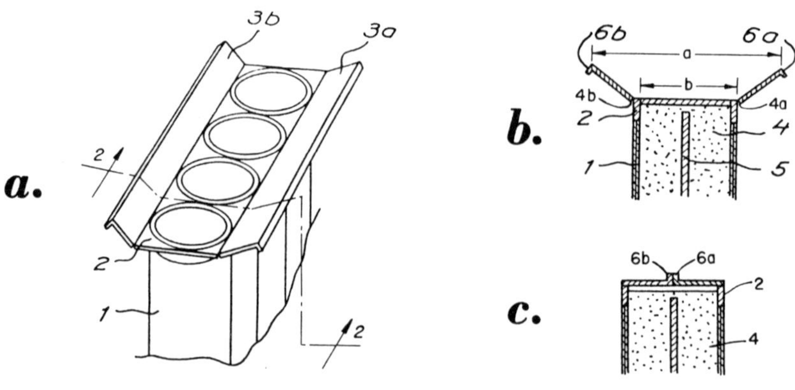

(a) Lower part of a tubular electrode with four tubes, with a bottom strip being in an open position

(b) Cross section through a tube of the same electrode along line **2-2** of Figure 1.9a with bottom strip being in the open position

(c) Cross section diametrically through a tube with the bottom strip being in a closed position

Source: U.S. Patent 4,011,370

For instance, the bottom strips can be made as pieces separate from the sheaths and fixed on the tubular sheaths in some previously known way, yet with the difference that the fixation occurs before the sheaths are filled with active material. The base can be provided with holes through which the sheaths extend (Figure 1.9a). Suitable bonding agent or mechanical fasteners can be employed to connect the base to the sheaths. The fastening of the bottom strips may be effected before or after the tubular sheaths are pulled over the electrode grid.

A second preferred method is to form them directly on the tubular sheaths (Figure 1.9b). This can suitably be done by placing the desired number of sheaths in a transfer mold whereafter the bottom strip is cast of thermoplastic material. In this way, there is a fixed joint between the bottom strip and the tubular sheaths. Also, when individual tubes are used, all tubes that belong to one electrode would be joined with such an arrangement. This is an advantage in subsequent handling of the electrode.

In the tubular sheath there is active material **4** and electrically conductive cross bars **5**. The base **2** and the movable closure sections **3a,3b** are preferably integral and include scored edges **4a,4b** which define a hinge axis for the movable sections. In length, movable sections are preferably each one-half or slightly larger than one-half the width of **b** of the base so that when they are swung about the hinge axes **4a,4b**, they will abut at their free ends **6a,6b** to close the sheaths (Figure 1.9c). The sections will thus be connected to the base, and fit tightly against one another at **6a,6b** to close all of the sheaths.

The active material that is to be charged into the tubular sheaths is generally pulverous. It is poured into the sheaths, whereafter the electrodes are vibrated to compact the active material. As a general rule, the electrodes are fixed one by one in fixtures for this purpose. For various reasons, it is desirable to have as little of the active material as possible spilled outside of the electrode. The charging devices and fixtures are therefore usually designed with this in mind. The process is very advantageous in this regard.

That is, the bottom strip can be set on the tubular sheaths before the latter are charged with active material, and the sections **3a,3b** of the bottom strip can be placed in a position where they point out obliquely upward from the electrodes (Figure 1.9b). Thus, it is possible to practically double the area to which the active material can be delivered. That is, dimension **a** is practically double that of dimension **b** (Figure 1.9b). In this way, the closing strip can serve as a funnel to substantially increase the rate-of-feed capacity of the charging device.

Also, compared with previously used methods involving fixing of the bottom strip after charging of the active material in the sheath, the bottom strip of the process provides a construction that readily allows automation of the closing of the tubular electrodes. This is an essential advantage, in aspects of both industrial hygiene and economy.

Thermoplastic-Glass Fiber Unit

E.G. Sundberg and E. Westberg; U.S. Patent 3,972,728; August 3, 1976; assigned to Tudor AB, Sweden describe tubular electrode sheaths which comprise two layers, namely, an inner layer of fibrous material and an outer layer of thermoplastic material covering the inner material. The two materials are so united that only a thin layer of the surface of the fibrous material is pressed into the surface of the thermoplastic material facing it. In this way, one obtains a mechanical union between the two materials.

For various reasons, it is not desirable to have the fibrous material penetrate more deeply into the thermoplastic material than is necessary for a satisfactory joining. The intention, thus, is not to produce a reinforcement of the thermoplastic material. Of course, the different layers may comprise a number of different materials, but experience has shown that it is best to use fibrous material consisting wholly or chiefly of glass fiber, and it is, therefore, preferred for the purpose. Glass fiber is a well-tried material for accumulator electrodes, which has proved capable of giving good qualities to the electrodes. The thermoplastic material preferably is a polyolefin. Polyethylene and polypropylene are especially preferred. In order to provide for effective penetration of electrolyte, the thermoplastic outer layer should be readily pervious to aqueous electrolyte solutions. In general, union of the fibrous and thermoplastic layers is achieved by heat-softening the thermoplastic surface adjacent the surface of the fibrous inner layer and pressing the two surfaces together. Heating can be done directly or indirectly, in any of several ways, as will be described in greater detail below.

The fibrous material can be made, in the familiar manner, of a braided glass fiber tube, while the thermoplastic material is formed of a flat strip or tape, or in the form of a net, as will be described in greater detail below. This tape can consist of perforated foil or some other open surface material. The tape is then given in the desired geometric form at the same time that it is fastened to the fibrous material. The geometric form does not need to have a circular cross section; many other forms are possible, such as, oval, rectangular, and the like. The thermoplastic material can be distributed on at least two tapes, which are joined together in a known manner simultaneously by means of the thermoplastic material. The joining together can then be carried out so that axially extending ribs are formed on the outside of the sheath.

Mandrel Technique

A process described by *E. Sundberg and E. Westberg; U.S. Patent 4,042,436; August 16, 1977; assigned to Tudor AB, Sweden* concerns the manufacture of a type of sheath for tube electrodes used with electrical accumulators or batteries; it comprises two layers. One of the two layers, the actual or inner sheath wall, comprises a layer of fibrous material, preferably glass fiber. The sheaths comprise another layer outside the fibrous material. This other layer is attached to the fibrous material and is intended to give mechanical strength to the sheath.

According to the process, production is carried out by combining two sheath materials as they are conveyed over a number of mandrels. Each tube in the completed sheath is associated with a mandrel, and these mandrels each include a connector section with a cross-sectional configuration which is elongated. Preferably the configuration is that generally of an ellipse with flattened sides so that this section of each mandrel is band-shaped or tape-shaped, with the mandrel sections being coplanar. This arrangement facilitates the joining portions of the tube together into one unitary piece as they travel along these sections of the tube.

The downstream section of each mandrel is then altered in cross section so that it corresponds to the desired final form of the product, whereby the distance between mandrels situated adjacent to each other is constant along both of the aforementioned sections of the mandrels. By bringing together the two portions and uniting them on the band-like sections of the mandrels, one greatly facilitates the obtainment of a good connection between the two sheath layers. At the same time, the method and apparatus makes possible the application of a large number of different materials and material combinations.

Referring to Figure 1.10, an electrode tube is formed around a grid with a top frame **1** and bars **2**. On the top frame there is also a so-called flag or vane **5**. Nearest the top frame the bars include a thickened section **4**, intended to provide location for the tube sheaths. The bars can also be provided with fins **3**, intended to center the bars in the tube sheaths. All these details belonging to the grid are formed of a piece of lead-base alloy adapted to this purpose. The bars are surrounded by active material **6** and tube sheaths **7**. The tubes are closed by a bottom strip **8**, which is also provided with devices **9** for location of the tube sheath and bar. The bottom strip is made of polyethylene.

The apparatus shown in Figures 1.10b and 1.10c has three mandrels **10**. The number of mandrels must conform to the number of tubes in the tube electrode and usually varies between 10 and 20. The essential construction of the apparatus, however, is not changed by any variation in the number of mandrels. The mandrels rest on three supporting rolls **11,12,13**, which are provided with guide grooves. All of the supporting rolls can, be provided with driving gears, not shown in the figure. There are also present cylinders **14,15,16,17**, intended to guide the sheath material toward the mandrels and possibly exercise pressure on the material in the direction of the mandrels. Supporting rolls **18,19** can also be used to guide sheath material onto the mandrel.

FIGURE 1.10: MANDREL TECHNIQUE FOR PRODUCING SHEATHS FOR ELECTRODES

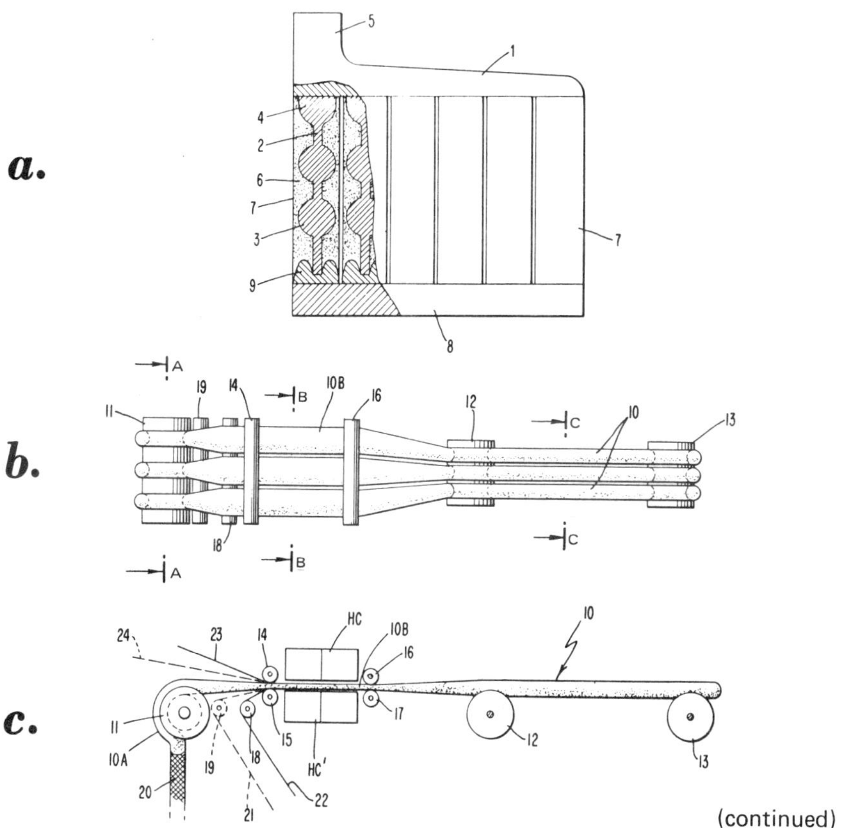

(continued)

FIGURE 1.10: (continued)

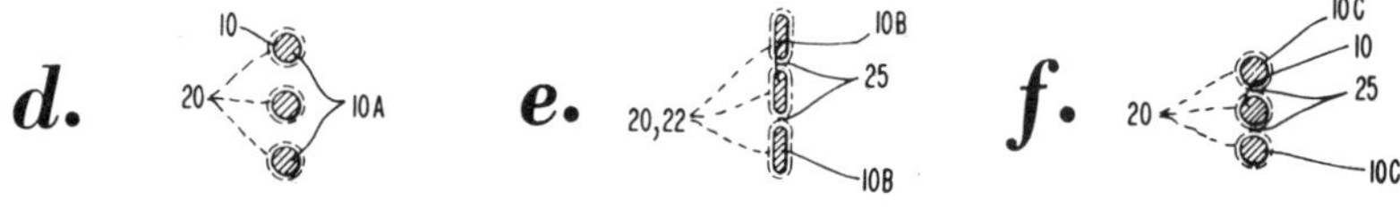

(a) Tube electrode which can be provided with a sheath
(b) Plan view of the process
(c) Side elevational view of the apparatus of Figure 1.10b
(d)-(f) Cross-sectional views through sheath-carrying mandrels employed at different places in the apparatus, in accordance with lines **A-A**, **B-B**, and **C-C**, respectively, in Figure 1.10b

Source: U.S. Patent 4,042,436

The apparatus shown can be used for the production of a sheath with different material combinations. If one starts out from braided tubes made of glass fiber which are to form the inner, fibrous layer in the sheath wall, an inlet end **10A** of the mandrel, shaped like a crutch, receives each such layer **20**. The feeding or threading of the inner layer material over the mandrels is carried out by having the cylinder **11** driven in clockwise direction by a motor. The outer sheath layer, which may consist of perforated plastic sheeting or netting made of plastic material, is applied as bands intended to cover the entire electrode surface. These bands **22,23** are held against the mandrel and on the threaded layer **20** with the help of cylinders **14,15**.

Farther on in the direction of the movement there also takes place a compression of the tube and band materials with the help of the cylinders **16,17**. With this material combination, there takes place the connecting of the two materials with each other through heating, so that the plastic material melts or fuses into the glass along the longitudinal partition lines between the tubular portion of the sheath. This heating as well as the following cooling off must occur in the section between the pair of cylinders **14,15** and **16,17**. This can be accomplished by conventional heating and cooling units **HC,HC'**.

Importantly, during this step the cross-sectional configuration of the connector sections **10B** of the mandrel is elongate, as depicted in Figure 1.10e. This configuration has been found to greatly facilitate the fusing together of the opposed surfaces and edges of the sheath material to define individual pockets within the unitary electrode tubing, which pockets will surround an associated electrode. The sheath layers have been interconnected by high-quality joints as a result of this technique. Preferably, the sections **10B** are in the general shape of an ellipse with flattened sides so as to be band-shaped.

Then there takes place a change in the cross-sectional shape of the mandrels to form a reshaping section **10C** of the mandrel so that after the passage of the supporting roller **12** they have a cross section that corresponds to the one desired for the finished product. The reshaping section **10C** of the mandrels which is present in the zone between the cylinders **12** and **13** is intended for the final formation of the product, and this can take place, e.g., through a careful heating of the sheath in this zone, so that the plastic material softens and then is recooled. After the sheaths have left the mandrels, the cutting to the desired length is carried out. Another possibility is that the outer sheath material consists of a fabric of woven polyester fiber. The two layers of material can then be combined by means of binding agent in the form of thermosetting resin or by having one or the other sheath material provided with a coating of thermoplastic material which is melted and thus combines the two sheath layers.

Another possibility is to apply an initial layer which is comprised totally of relatively wide bands or tapes rather than a tube. This fibrous material can comprise nonwoven mats or sheets of glass fiber **21,24** as depicted in phantom in Figure 1.10c. In this case, the mandrel can be attached in another way than with the crutch shown in the figure, and the propulsion of the sheath material can be carried out with the help of the cylinder **12** or the supporting rollers **14,15.**

In Figures 1.10d through 1.10f are shown the different cross sections of the mandrels which are used in the above example. In these figures material in the form of braided tubing has been indicated by broken lines. As shown in Figure 1.10d, the initial mandrel sections **10A** have a suitably circular cross section. In the production of certain sheath types, however, this section of the mandrel can be unnecessary and is replaced with another form of guide. Essential for the process, on the other hand, is section **10B**, where the mandrels have the form of bands or tapes.

These bands must have as little thickness as possible, and their total circumference must correspond to the circumference of the mandrel in that part where the forming of the material takes place. Through the shaping of the mandrels shown in Figure 1.10e, it is made possible that the sheath materials have large, even contact surfaces at the stage when they are to be connected with one another, wherefore a good union between the materials can easily be achieved. Then occurs a transition to the cross section which is desired for the finished product, an example of circular mandrels being shown in Figure 1.10f.

It is essential that the distance between the mandrels be substantially constant during mainly the whole time that the sheath materials pass along sections **10B** and **10C**. By distance is not necessarily meant the distance between center axes but the free distance between two adjacent materials which would tend to rupture the connecting joints.

Certain difficulties in uniting the materials may arise in the longitudinal dividing or joining line between two separate tubes. It has therefore been found advantageous to introduce threads **25** in each space between two adjacent mandrels, which threads become encased within the sheath layers as they are fused together along longitudinal joining lines. These threads or wires can comprise any appropriate material. If to resist a tensile stress is an essential quality of these threads, then it may be advisable to make them of glass fiber.

If their main function is to unite different material layers with one another, the threads may comprise a net or perforated film of thermoplastic material, which is melted in the production process, so that the desired connection is obtained. It can also contribute to uniting the outer layers on both sides of the tube sheath. These threads **25** are indicated in Figures 1.10e and 1.10f. The threads may comprise different materials; they can be made, e.g., of intertwined thermoplastic fibers and glass fibers. It should be noted that in the event that it becomes desirable to produce sheaths of elongate cross-sectional configuration, then the sections of the mandrels downstream of the sections **10B** will have substantially the same elongate shape as the sections **10B** (rather than being circular). In that case, a cross-sectional view taken along line **C–C** in Figure 1.10b will appear essentially as shown in Figure 1.10e.

Fabric Spacer

A process described by *H. Breidenbach; U.S. Patent 3,950,186; April 13, 1976; assigned to Varta Batterie AG, Germany* provides an envelope for tubular electrodes suitable for lead-acid storage batteries which reduces the requirement for expensive spacers and which exhibits lower resistance. This process involves making the individual tubes of a hose-like woven fabric. Two oppositely positioned spacers extending beyond the diameter of the tubes are formed by tying a plurality of warp threads together by means of weft threads.

The production of such an electrode pocket may take place as follows. First, a hose-like fabric is woven in which, depending upon the desired size of the spacers, several warp threads are so tied off by weft threads that a hose-like fabric with two ears is produced. The shaping can then be performed in the usual manner by insertion of cores followed by impregnation and drying, or by impregnation before insertion of the cores. It is also possible to impregnate the threads prior to weaving and then to accomplish shaping by heat setting. The ears, or spacers so formed are preferably simultaneously impregnated with a material resistant to the electrolyte, such as polymethacrylate, polystyrene or similar synthetic resins.

A specific example of the process involves making those warp threads which are to be tied off, and which are to serve as spacers, of a different material than that used for the remainder of the hose-like fabric. For example, thermoplastic fibers may be used for this purpose, such as polyethylene and polypropylene. In that case, the ears can then be stabilized by heat treatment followed by cooling.

To eliminate entirely an impregnation process from the manufacture of these tubes, it is also possible to additionally provide the hose-like fabric, which is then preferably made of glass fibers, with an internal supporting fabric of a thermoplastic material. In that case, the shaping can be accomplished by heat setting. The supporting web and the warp threads to be tied off may be made of the same material, such as polyethylene or polypropylene.

POLYOLEFIN SEPARATORS

Surfactant Wetting Agent

According to a process described by *N.I. Palmer and N. Sugarman; U.S. Patent 3,933,525; January 20, 1976; assigned to W.R. Grace & Co.* a battery separator is produced from a wettable fiber formed from a plastic resin having a degree of undesirable hydrophobia, that is solid at below 180°F, extrudable as a hot melt and resistant to degradation by at least either acids or alkalies. The fiber is comprised of the resin and a wetting agent. The resin and the wetting agent are preferably soluble in admixture with one another at the extrusion temperature for the resin and insoluble at room temperature. It is very important the wetting agent be developed. In one of its most preferred forms the wetting agent is comprised of two surfactants, one of which is relatively soluble in battery electrolyte and the other of which is relatively insoluble in battery electrolyte. The preferred fiber has a diameter of 0.05 to 50 microns and is contained in a mat of such fibers. The mat preferably is compacted and has a thickness of 5 to 50 mils, a porosity of more than 40%, a pore size of less than 40 microns, and enhanced resistance to delamination in the sulfuric acid environment of a lead-acid storage battery and produces only negligible foaming.

The mat when removed from a battery and washed after initial battery charging and dried at about 200°F for 30 minutes should preferably rewet sufficiently to provide an electrical resistance of no more than 25 milliohms after treatment in a sulfuric acid solution having a specific gravity of 1.20 at 160°F for 1 hour and being thereafter removed and washed in cold running water for ½ hour and then in stagnant water at 160°F for 1 hour and thereafter dried at 200°F for ½ hour. The mat should provide a battery with cold start performance such that in a Group 24 53 AH battery a cell will provide at least 1.00 volt at 280 amps at 0°F after 30 seconds and no readily observable delamination of the mat should be present after 6 months and in actual practice for more than 3 years.

By another aspect of this process a method is provided for producing battery separators using wettable fibers produced from polymeric resin having a degree of undesirable hydrophobia. The method includes mixing the polymeric resin and a wetting agent and extruding the mixture as a hot plastic through a die opening and forming a fiber. The mixing preferably includes melt-blending the resin and wetting agent to form a solution of wetting agent in the polymer. Fiber formation preferably includes attenuating the extrudate beyond the die opening in a hot gas stream to a very small diameter fiber. Afterwards, the fiber is cooled and the wetting agent thereby made incompatible with the resin and its blooming encouraged.

In the preferred method it is very important to develop the wetting agent after the fiber has been cooled. Cooling need not be all the way to ambient temperature. Developing usually includes both heating and compression. The die opening is preferably one of a series arranged in a row and the attenuation is carried out essentially in a plane away from the die openings, and the attenuated fiber is collected on a continuously moving take-up device in a web with other fibers. The fibers are preferably collected in a web thickness of between 1 and 200 mils and formed into a thickness of 20 to 200 mils. Thereafter the web is compressed and the compression is not stopped until the web's permanent thickness is reduced to between 5 and 50 mils, its permanent porosity retention is greater than 40% and its permanent maximum pore size is less than 40 microns.

Particularly preferred surfactants are C_8 and C_{18} phenol surfactants having 1 to 15 mols of ethylene oxide, more preferably 1 to 6 mols of ethylene oxide and most preferably 1 to 3 mols of ethylene oxide. These surfactants are relatively water-insoluble but oil-soluble. Their ethylene oxide values are lower than some that have substantially higher water solubility and therefore a more uncontrolled retention. The surfactants are added in the preferred amounts 0.5 to 20 wt % of the weight of the resin, and most preferably 1 to 3 wt % of the resin.

Ethylene-Acrylic Acid Copolymer Wetting Agent

W.R. Wszolek and J.A. Cogliano; U.S. Patent 3,951,691; April 20, 1976; assigned to W.R. Grace & Co., describe a process for imparting permanent wettability to a battery separator comprising a nonwoven mat of polyolefin fiber. The process involves impregnating the battery separator with an α-olefin/α,β-unsaturated acid or anhydride copolymer or terpolymer dissolved in water by reaction with a base and thereafter causing the copolymer or terpolymer to revert to a water-insoluble substantially free acid form by either heating the impregnated battery separator to drive off the base or exposing the impregnated separator to acid. In the following examples, parts and percentages are by weight.

Example 1: To a Chemco reactor equipped with high speed stirrer and heating coil was charged 400 g of distilled water, 50 g of ammonium hydroxide (28.8% NH_3) and 50 g of particulate (-100 mesh) of commercially available acrylic acid/ethylene copolymer (20% acrylic acid) having a density of 0.950 g/cc and a melt index of 0.3. The reactor was sealed and stirring was carried out for 5 minutes without heating. Heat was then applied with stirring until the temperature reached 95°C at which temperature the reaction was continued for one-half hour with stirring. The reactor was cooled and the ammoniacal solution of the acrylic acid/ethylene copolymer was removed. The solution had a pH of 10.4. This 12.4% solids solution will be referred to as copolymer solution A.

Example 2: A portion of solution A from Example 1 was mixed with 934 g of water with high speed stirring. This 4.1% solids solution will be referred to as copolymer solution B.

Example 3: A portion of solution A from Example 1 was mixed with 5,634 g of water with high speed stirring. This 1.0% solids solution will be referred to as copolymer solution C.

Example 4: Nonwoven polypropylene mats were produced by the melt-blowing process set out in U.S. Patent 3,773,590 under the following conditions:

Polypropylene resin melt flow rate	33.6
Die temp, °F	580
Air temp, °F	644
Polymer rate, g/min	7.9
Air rate, lb/min	1.27
Collector distance, inches	6.0
Revolutions per minute	1.0

The resultant polypropylene fibers in the mats had a diameter of 2 microns or less. The basis weight of mats varied from 256 to 270 grams per square meter. The thus formed mats were compacted on a heated calender to a thickness of 21 mils.

Example 5: Polypropylene mats from Example 4 (6 inch x 8 inch) were immersed in solutions A, B and C from Examples 1, 2 and 3 after 1.25 weight percent of a surfactant, i.e. an ester of a sulfonated dicarboxylic acid (Aerosol OT, American Cyanamid) had been added to each of the solutions. After 3 minutes the mats were removed, drained dry and heated in an oven for 1 hour at 75°C to drive off the ammonia and insolubilize the impregnated copolymer on the polypropylene mat. The mat immersed in copolymer solution A had a dry weight percent pick-up of 18.4%, the mat immersed in copolymer solution B had a dry weight percent pick-up of 6.7% and the mat immersed in copolymer solution C had a dry weight percent pick-up of 2.3%. The mats were then tested subjectively for wettability by immersion in 1.25 specific gravity H_2SO_4 maintained at 65°C for 2½ hours. The mat in copolymer solution A had excellent wettability, the mat in copolymer solution B had very good wettability and the mat in copolymer solution C showed good wettability. The mat in copolymer solution A had an electrical resistance of 1.6 milliohms/mil. A polypropylene mat from Example 4 which had not been subjected to impregnation as set out above had poor wettability and when tested had an almost infinite electrical resistance.

Nonwoven Thermoplastic Fiber Mat

According to a process described by *R.R. Buntin and W.A. Morgan; U.S. Patent 3,947,537; March 30, 1976; assigned to Exxon Research & Engineering Co.,* battery separators are produced from nonwoven mats of thermoplastic fibers by wetting the fibers with a surfactant-water mixture to modify the surface properties of the fibers in the nonwoven mat, vaporizing the water while depositing the surfactant on the fibers, heating the nonwoven mat prior to compressing and then compressing to increase the fiber-to-fiber bonding as well as to form the desired structure. During compressing, ribs may be formed by using appropriate embossed rolls or press plate patterns. After the compressing step, slitting, cooling and cutting steps are carried out to produce a battery separator of the desired dimensions.

The nonwoven battery separators are useful in the lead-acid batteries of the SLI-type (starting, lighting and ignition) and the industrial type. These batteries have positive electrodes and negative electrodes which are separated by the battery separators. It has been found that a nonwoven mat having a basis weight between 60 to 300 grams per square meter is preferred for producing battery separators of this type.

According to a process described by *J.S. Prentice; U.S. Patent 3,907,604; September 23, 1975; assigned to Exxon Research and Engineering Company,* the strip tensile strength of nonwoven mat battery separators of polypropylene fibers having diameters from about 1 to 10 microns is increased by fuse-bonding, as by point-bonding or by calendering, at least a portion of the fibers of the mat at a temperature within the range from about 250°F to about 325°F, while maintaining pressure on the nonwoven mats sufficient to prevent shrinkage of the fibers while they are exposed to the fuse-bonding temperatures. The nonwoven mats so treated have fusion-bonds throughout their thickness, and may have strip tensile strengths greater than 4,000 m, preferably, greater than 5,000 or even 6,000 m. The battery separators are assembled in batteries between the positive and negative plates.

Microporous Polymeric Sheet

W.M. Versteegh; U.S. Patent 4,024,323; May 17, 1977; assigned to Evans Products Company describes a battery separator comprising a microporous sheet formed substantially of a mixture of polymeric constituents. One of the polymeric constituents is an ultra-high-molecular-weight polyolefin having a standard load melt index of substantially 0 and an intrinsic viscosity greater than about 3.0. The other polymeric constituent is either a copolymer of an olefin and an ethylenically unsaturated monocarboxylic acid selected from the group consisting of acrylic acid and methacrylic acid, or a blend of a low molecular weight polyolefin and a polymer of acrylic or methacrylic acid.

Example 1: In this example a polyolefin copolymer (Dexon XEA-7) was employed. This copolymer is an ethylene-acrylic acid copolymer having a melt index (ASTM 1238-70) in the range of 20 to 25 grams per 10 minutes and contains 7% acrylic acid by weight. The ultra-high-molecular-weight, high density polyethylene homopolymer employed (Allied Chemical 1220) has a melt index of 0.

A blend was prepared containing the following components: 15% by volume polyolefin (50% weight Allied Chemical 1220 and 50% by weight Dexon XEA-7, melt index of 0.40 g per 10 minutes); 15% by volume silica (Hisil 233); 2% by volume carbon black (United 3017); 68% by volume mineral oil (Shellflex 412); and 0.5% by volume antioxidant (Ionol). The polyolefin copolymer (Dexon XEA-7) and the silica were preblended. The dry ingredients were then blended together in a Henschel high intensity mixer, and the mineral oil sprayed in while the blending was in progress.

The blend was then fed to the hopper of a twin screw extruder (Colombo) and extruded through a die to form a continuous web 42 inches wide and 20 mils thick. The web was smooth on both sides, i.e., had no ribs. The continuous web was then fed through a bath of trichloroethylene at room temperature (20°C) and for an average residence time of 15 minutes to extract the mineral oil. The extracted web was next dried in a hot air oven at a temperature of 75°C. Finally, the web was cut into battery separator sheets $5\frac{3}{8}$ x $5\frac{13}{16}$ inches in size. The resulting battery separator had an electrical resistance of 30 milliohms per square inch, a tensile strength of 955 psi and an elongation of 160%.

Example 2: The procedure of Example 1 was repeated except that 50% by weight of the polyolefin copolymer (Dexon XEA-7) was substituted with a low molecular weight polyethylene homopolymer (Super Dylan 7180) having a melt index of 18 g per 10 min. The resulting polyolefin mixture was thus comprised of 50% by weight ultra-high-molecular-weight polyethylene (Allied 1220), 25% by weight polyolefin copolymer (Dexon XEA-7) and 25% by weight polyethylene (Super Dylan 7180). The effect of this substitution was to lower the acrylic acid content to 3.5% by weight of the mixture of Dexon XEA-7 and Super Dylan 7180. The melt index of the total polyolefin mixture was 0.24 g per 10 min. The battery separator produced from the foregoing had an electrical resistance of 25 mohm/in^2, a tensile strength of 500 psi and an elongation of 80%.

Porous Body and Fused Rib

A process described by *N.I. Palmer and D.O. Grammer; U.S. Patent 4,003,758; January 18, 1977; assigned to W.R. Grace & Co.* relates to mats for use as battery separators and, more particularly, to such mats with degradation resistant spacer embossments formed therein. The rib is differentially compressed to lower its porosity relatively. The method preferably includes forming the web to have a region with an outer area extending beyond a second region. The web is treated to substantially fuse the fibers and at least substantially eliminate the interstices of the outer area of the first region while maintaining at least a substantial portion of the second region open between the fibers. In a preferred form the fibrous web joining the first and second regions is treated to fuse the fibers to a progressively diminishing degree from the outer area that is substantially fused toward the second region where at least a substantial portion of the region is open. The preferred manufacturing procedure involves utilizing a web that is formed of nonwoven fibers.

The initial web is preferably 20 to 200 mils thick and the fiber diameter is 0.05 to 50 microns. The basis weight of the preferred web is 10 to 500 grams per square meter and this initial nonwoven web is compressed until the second surface area is 5 to 50 mils thick with a porosity retention of greater than 40% and a maximum pore size of less than 40 microns. The first region is preferably formed as a plurality of spaced apart linear ribs extending across the web from one edge to an opposite edge.

In a particularly preferred form of the process, there is a region of transition between the fused portion and the intersticed portion of the mat with a steadily increasing fused state extending toward the fused portion from the intersticed portion. The fibers preferably have a diameter of 0.05 to 50 microns and the fused portion is 1.5 to 25 mils thick. The intersticed portion preferably has a porosity retention of greater than 40% and a maximum pore size of less than 40 microns.

The mat should provide an initial resistance of no more than 25 milliohms after 24 hours. The mat should provide a battery with cold start performance such that in a Group 24, AH battery a cell will provide at least 1.00 volt at 280 amps at 0°F after 30 sec and no readily observable delamination of the mat should be present after 6 months and in actual practice for more than 3 years.

To prepare the separator, an apparatus is provided having a first pair of calender rolls with evenly spaced calendering surfaces and a second pair of calender rolls with a plurality of raised ridges opposed by a plurality of grooves with opposed lands on both of the rolls separating the opposed ridges and grooves. The gap between the second pair of rolls is less at the opposed ridges and grooves than at the opposed lands. Generally all of the ridges are on one roll of a pair and all of the grooves on an opposite roll of a pair but this is not necessary. The nonwoven battery separator has the characteristics of good stiffness, good resistance to oxidation (particularly in the rib area), and good resistance to delamination.

Calendering Process

B. Hollenbeck; U.S. Patent 3,917,772; November 4, 1975; assigned to W.R. Grace & Co., describes a method for continuously producing a sheet suitable for conversion into a battery separator and having two principal opposite surfaces, at least one of which has a series of spaced apart projections. These projections may be ribs extending transversely across the sheet. In other instances the projections may be mounds projecting upwardly from the primary surface of the sheet. In any event the process involves delivering a mass of readily deformable plastic material, preferably in self-supporting sheet form having relatively plain opposite surfaces, to a calendering means where the plastic material is calender molded. Preferably, the molding takes place between two calendering rolls at least one of which has depressions formed therein that are interrupted circumferentially and spaced around the circumference of the roll. Preferably, the thus-formed sheet is then sized by being subjected to pressure, preferably by being passed between two sizing rolls that have a preset sizing gap therebetween. At least one of the sizing rolls presses the projections inwardly of the sheet to set the final sheet thickness.

In the preferred operation of the process, a very small bank of material from the extruded sheet is maintained at the nip of the calendering rolls. Preferably, this bank is maintained at a low level. It is also generally necessary to transport the calendered sheet around at least a portion of the circumference of the depression containing calendering roll and to partially harden the sheet. Then the partially hardened projections are withdrawn substantially straight out of the depressions to prevent any substantial deformation or picking of the projection structure. The process will produce a sheet having extremely uniform thickness dimensions so as to enable the use of the thinnest possible separator commensurate with the necessary separation of the battery plates and provision of adequate channels for the escape of gas formed within the battery.

Porous Layered Structure

S. Honda, H. Hata, Y. Simura and S. Une; U.S. Patent 3,967,978; July 6, 1976; assigned to Sekisui Kagaku Kogyo Kabushiki Kaisha, Japan have found that a resin sheet having a porous layered structure which is produced from a mixture comprising an olefin resin as a base and at least one substance selected from water-swellable thermoplastic resins, and water-insoluble or difficulty water-soluble hydrophilic nonionic or anionic surface active agents is very suitable as an electrode separator for electric cells. The following examples illustrate the process. In the examples, all the parts are by weight.

Example 1:

	Parts by Weight
High density polyethylene, (melting point 131°C)	100
Water-swellable resin of the polyethylene oxide type (Aquaprene)	30
Silica powder, (particle diameter less than 10 microns)	30
Diatomaceous earth powder, (particle diameter less than 10 microns)	60

The above ingredients were kneaded in a Banbury mixer, and the kneaded mixture was fed into a vent-type extruder and extruded into a sheet having a thickness of 0.5 mm through a flat die fitted to its tip. The sheet was then stretched at 90°C on a tenter-type stretcher simultaneously in the longitudinal and transverse directions at a stretch ratio of 3.5 inches in each direction. This stretching resulted in the formation of a porous layered structure in the sheet.

It was found that a greater part of the micropores of the porous layered structure had a pore diameter of 0.1 to 5 microns. The maximum pore diameter of the open cells was 0.6 micron. The resulting sheet having a porous layered structure had a thickness of 0.18 mm, and an apparent density of 0.291. The proportion of the open cells was 65%. The sheet was interposed between an anode and a cathode of an acidic electric cell, and used as an electrode separator. Each sheet of the electrode separator had an electric resistance of 0.0004 ohm per 100 square centimeters in an electrolytic solution. The electrode separator was not attacked by the electrolytic solution, and was also found to be suitable for use in alkaline electric cells. The resulting electrode separator permitted superior permeation of the electrolytic solution, and had superior mechanical strength characteristics such as flex resistance.

Example 2:

	Parts by Weight
Polypropylene (melting point 165°C)	100
Ethylene-vinyl acetate copolymer	15
Water-swellable acrylic resin, (Hydron)	30
Calcium hydroxide, (particle diameter below 10 microns)	45

The above ingredients were kneaded for 15 minutes by a kneading roll heated at 170°C, and the kneaded mixture was fed into an extruder, and extruded into a sheet having a thickness of 0.35 mm through a flat die fitted to its tip. The sheet was cooled to room temperature, and reheated to 115°C. Then, it was stretched simultaneously in the longitudinal and transverse directions at a stretch ratio of 4.0 in each direction to form a porous layered structure in the sheet. A greater part of the micropores constituting the porous layered structure had a pore diameter of 0.1 to 10 microns. The maximum pore diameter of the open cells was 0.9 micron. The resulting sheet having the porous layered structure had a thickness of 0.08 mm and an apparent density of 0.223. The proportion of the open cells was 60%.

The resulting sheet was interposed between an anode and a cathode of an alkali electric cell, and used as an electrode separator. Each sheet of the electrode separator had an electric resistance of 0.0007 ohm/100 cm^2 in an electrolytic solution. The electrode separator had superior resistance to alkali and reduction. It also permitted superior permeation of the electrolytic solution, and had superior mechanical strength characteristics such as resistance to flex.

Ultrafine Porous Polymer Articles

J.L. Weininger and F.F. Holub; U.S. Patent 3,956,020; May 11, 1976; assigned to General Electric Company describe the production of solid, ultrafine porous polymer articles. In the process, a solid body of a crystalline thermoplastic polymer exhibiting at least 70% crystallinity or blends of such crystalline thermoplastic polymers and thermoplastic polymers exhibiting up to 65% crystallinity where the latter polymer is present in an amount up to 50 weight percent of the crystalline polymer is heated at a temperature of at least in its melting temperature range and a benzoate salt is incorporated therein forming a composite body.

The composite body is shaped at a temperature in the range of the initial heating temperature of the polymer, the shaped composite body is cooled to room temperature forming a solid composite body, and the salt is dissolved from the solid composite body leaving the polymer as an ultrafine porous article.

Because of their controlled porosity and physical and chemical properties, the ultrafine porous polymers of this process have many and varied uses. Sheets and laminates make ideal battery separators for both primary and secondary batteries.

These ultrafine porous polymer articles provide thin porous separators which are desirable in a lead acid battery to more closely position the plates. Additionally, the ultrafine porous polymers lend themselves readily to various configurations in designing lead acid batteries. The separators can be provided with a wide variety of filler materials to produce the hydrophilic nature of the separator. Under certain circumstances, pores of diameter from 40 to 600 Angstroms are desirable in such separators for lead acid batteries.

OTHER SEPARATORS

Glass Fiber Mat

L.C. Davis; U.S. Patent 3,918,994; November 11, 1975; assigned to Johns-Manville Corporation describes an improved battery plate retainer mat or separator comprising a layer of sliver attached at selective points to a fibrous layer containing a resin binder. Preferably, the sliver layer is attached to the resin bonded fibrous layer with stitching which is dissolved by the electrolyte in the battery, but the layers may also be attached with an inert stitching or with either spots or strips of a soluble or inert adhesive.

As shown in Figures 1.11a and 1.11b the retainer mat **2** is made up of a continuous layer **4** of loosely matted fibers attached at selected points **6** to a resin-bonded fibrous mat **8**. The continuous layer of the loosely matted fibers can comprise any fibers that are essentially inert in the electrolyte in which they will be exposed. The fibers have diameters sufficiently small and the thickness of the layer is sufficiently thick to act as a barrier to harmful particles suspended in the electrolyte and to the particles of battery paste that may be present on the battery plate.

Among the materials suitable as fibers for the continuous layer are glass fibers, ceramic fibers, and high molecular weight synthetic fibers such as polypropylene. Acid-resistant glass fibers having a composition meeting the criteria of C glass and having diameters equal to or less than about 13 microns are preferred, e.g., 11 to 13 microns. The continuous layer of loosely matted fibers **4** can vary in thickness depending on the environment in which it must operate. Typical thicknesses will range between about one-sixteenth inch and about one-half inch.

The fibers should be sufficiently physically intertangled so that the layer can be physically handled without tearing apart or deteriorating. Such a layer would not require any binder; however, a small amount of binder compatible with the electrolyte could be used if desired. It is desirable, both in the manufacturing of the continuous layer as well as in the process of making the retainer mat, to include in the layer a small amount of an antistatic agent to prevent the fibers from clinging to surfaces coming in contact therewith.

In the preferred case, the continuous layer is a layer of sliver made up of acid-resistant glass fibers having diameters of about 11 to 13 microns and containing about 1.5% mineral oil to enhance the integrity of the layer and to provide an antistatic property. The thickness of the continuous layer of sliver is about one-half inch in its uncompressed form and weighs about 28 grams per linear foot in widths of about 5.5 to 6 inches. Sliver is well known and is produced from molten glass which is pulled from small orifices in a ceramic-noble metal crucible. To make a sliver mat the resultant fibers are gathered continuously to form a loose jack-straw arrangement and passed between two soft rollers which press the fibers into a loose mat approximately 6 inches wide and about one-half inch thick.

FIGURE 1.11: BATTERY PLATE RETAINER MAT

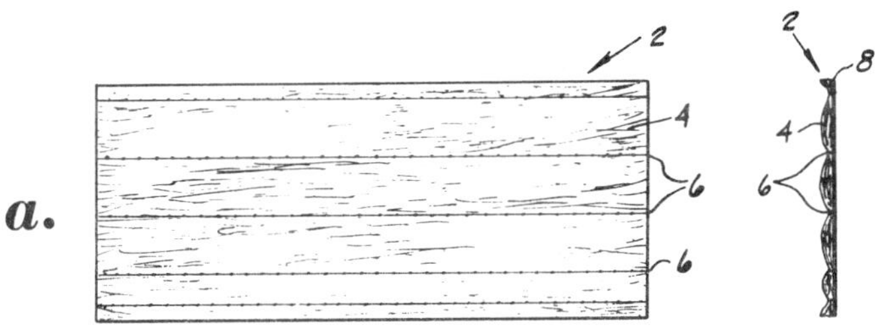

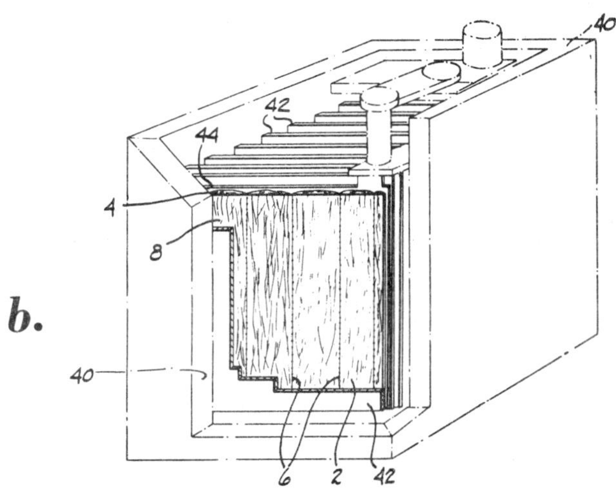

(a) Plan view of the battery separator product
(b) End view of the product illustrated in Figure 1.11a

Source: U.S. Patent 3,918,994

The resin-bonded fibrous mat **8** can be made from any fiber that is essentially inert to the electrolyte in which it is to be used and includes the materials mentioned above for use in making the fibers for the layer **4**. These fibers can be first woven and then bonded together or simply overlapped and bonded together using a resinous binder that is preferably essentially inert to the electrolyte in which it is to be used. In the preferred example, the resin-bonded fibrous mat consists of chemical grade glass fibers having diameters of about 11 to 14 microns arranged in a jack-straw configuration and bonded together with about 5 to 20% by weight of a resinous binder.

The thickness of this mat can range from about 0.01 inch to about 0.03 inch with the weight of a 0.03 inch thick mat being about 1.95 pounds per hundred square feet. Various binders could be used such as styrene, urea-formaldehyde and acrylic resins. The preferred binder is a composition containing 2,100 lb of water, 100 lb of a 45% solids acrylic emulsion available from Rohm and Haas under the designation of HA-16, and 25 ml of known antifoam agents such as the solution available from Dow Corning identified as Antifoam B. The resin-bonded fibrous mat can be produced by the process described in U.S. Patent 3,235,913.

The attachments **6** can be either permanent or temporary, i.e. the attachment means can be either essentially inert to the electrolyte or can be dissolved by the electrolyte providing that the electrolyte is not significantly affected by such dissolution. The attachments can be either continuous or discontinuous and can be either oriented or randomly placed over the interface. For example, the attachments can comprise continuous strips or small areas of adhesive, stitching, rivets or rivet-like fasteners or any other known fastening means compatible with the electrolyte so long as the fastening means do not protrude excessively above the exposed surfaces of layers **4** and **8**.

In the preferred example, the attachments are formed by stitching the component layers together using a monofilament thread, such as nylon, for soluble systems and continuous filament C glass thread for insoluble systems. The stitches shown in Figures 1.11a and 1.11b are continuous and oriented in parallel directions spaced apart at selected distances. Since the stitching is merely to hold the two component layers **4** and **8** together the type of stitch used, the spacing between stitches, and whether the stitching is continuous or discontinuous, or oriented or at random is largely a matter of choice and within the skill of the art to determine for each particular application.

As shown in Figure 1.11b, stitching of the layer **4** to layer **8** results in compression of layer **4**, particularly at the points of attachment **6**, producing a pillowing effect between the points of attachment. The advantage of using a stitching attachment that is soluble in the electrolyte is that when the soluble stitches dissolve in the electrolyte the attachment is removed allowing the compressed portions of layer **4** to expand back to at least a portion of their original thickness providing a continuous contact with the battery plate surface. The advantage of using stitching as attachment **6** compared with adhesive exists primarily in less expense and trouble in storing and applying the stitching as compared to storing and applying an adhesive. An economical process used to make the battery plate retainer mat is described in detail.

Microporous Polyvinyl Chloride Sheet

A process as described by *I. Shoichiro, T. Ohya, and S. Nagai; U.S. Patent 3,900,341; August 19, 1975; assigned to Yuasa Battery Company Limited, Japan,* relates to a lead-acid type storage battery. The battery is a lead-acid type storage battery which is made by housing a battery plate with an envelope-type separator formed by folding a microporous sheet obtained by depositing a solution consisting of a synthetic resin, a solvent dissolving the synthetic resin and a nonsolvent not dissolving it on a base, drying it and heat-sealing the sheet on the sides. The process for producing it is adaptable to mass production.

One of the important features of the process is to apply the envelope-type separators, each made of a special microporous sheet, to a storage battery. It is made, for example, by depositing a solution prepared by dissolving 7 parts of a heat-proof polyvinyl chloride resin in 63 parts of tetrahydrofuran and adding 30 parts of ethyl alcohol to the solution on a base of a polyester nonwoven fabric of a thickness of 0.08 mm or a polypropylene nonwoven fabric of a thickness of 0.15 mm and drying it. An envelope-type separator made of the thus obtained thin microporous sheet of a thickness of 0.1 to 0.2 mm is adapted to house a plate of a storage battery.

The first of its characteristics is that the electrical resistance is very low. The electric resistance of the envelope type separator of a thickness of 0.1 mm according to the process is 0.0002 to 0.0006 ohm per square centimeter per sheet and is much lower than the electric resistance of 0.0012 to 0.003 ohm per square centimeter per sheet of a conventional separator such as a rubber separator. The use of such a separator has the advantage of obtaining a high performance storage battery which has the characteristic, for example, of a good starting output (cranking ability) at a low temperature in a cold district.

The second of the characteristics is its microporosity. There are several 10,000 to several 100,000,000 micropores per square centimeter in the envelope-type separator of the process and their diameter is several 100 A to several microns and is $^1/_{10}$ to $^1/_{100}$ the diameter of the pores in a conventional separator. As is well known, the plate grid of a storage battery for automobiles is made of an antimony-lead alloy and, while the battery is being used, the antimony of the cathode plate will move to the surface of the negative plate and will be deposited so as to reduce the hydrogen overvoltage of the negative plate.

As a result, the current in the final period of the charge becomes so large as to cause an overcharge, thereby causing the electrolyte to decrease early, the self-discharge to increase and the life to become short. The envelope-type separator of the process is so much smaller in a pore diameter than a conventional separator and has a greater ability to prevent the permeation of antimony that there is an advantage of reducing the damage caused by antimony.

The envelope-type separator is high in the acid-proofness and oxidation-proofness. For example, a conventional fiber-reinforced separator of a thickness of 0.7 to 0.8 mm is usually pasted with a glass mat of a thickness of 0.4 to 0.5 mm and its oxidation-proofness is 50 to 70 hours per sheet. On the other hand, in utilizing the envelope-type separator of the process which as already described is a microporous sheet of a thickness of 0.1 to 0.2 mm pasted with a glass mat of 0.4 to 0.5 mm, its oxidation-proofness will be 100 to 120 hours per sheet. The fourth of the characteristics is that the envelope-type separator is so thin as to be 0.1 to 0.2 mm thick. As a result, there is the advantage that a battery of high performance is obtained because of being able to contain an increased number of plates in a container of a fixed capacity.

Phenol Resin Fibers

T. Ito and K. Ohtomo; U.S. Patents 3,953,236; April 27, 1976; and 3,910,799; October 7, 1975; both assigned to Kanebo KK and Ko Kondo, Japan, describe a lead-acid storage battery which comprises units of cells each having, between an alternate fibrous structure and a negative plate which are arranged in laminated fashion on each side of a positive plate which is coated on both surfaces with an anode active material, a separator comprised of a paper-like sheet having continuous microporous openings throughout the entire area and containing at least 70% by weight of phenol resin fibers each having a diameter of 5 microns or less produced by melt-spinning and curing a novolak resin.

A lead storage battery having such separators is of an improved cell capacity and output and prolonged life span. In another aspect of the process, the separators are such that the phenol resin fibers have at least 60 mol % of the phenolic hydroxyl groups of the resin capped. A battery employing these separators is of further improved duration and satisfactory cell capacity and output.

Example: A novolak resin having a molecular weight of 850 obtained by reacting phenol with an aqueous solution of formaldehyde was supplied to a spinning machine and was jetted into a nitrogen atmosphere to produce novolak resin fibers of 3.5 microns in diameter and 0.5 to 2 mm in fiber length. These fibers were cured with a hydrochloric acid or formaldehyde curing agent so that they became nonmeltable. Then the fibers were subjected to a wet-type paper-making process using a methanol-soluble resol resin having a molecular weight of 420 as the binder to produce a paper-like sheet of 0.15 mm in thickness. Table 1 shows various properties of this sheet as compared with the commercially available separators made of pulp.

TABLE 1

	Thickness (mm)	Electric Resistance* (Ω/cm^2)	PbO_2 Permeability	Porosity (%)	Tensile Strength (kg/mm^2)
Sheet of the process	0.15	0.05	Entirely non-permeable	65	0.10
Pulp sheet	1.00	0.17	Permeable	60	0.03

*Determined in sulfuric acid of a specific gravity of 1.28 at room temperature.

Then, batteries having the "positive plate–fibrous structure–paper-like sheet of this process or commercially available pulp sheet–negative plate" structure were constructed and their cell capacity was determined, and also the capacity after repeating the discharging and charging 100, 200 and 400 times was measured. The results are given in the following Tables 2 and 3.

TABLE 2

	Separator Material	Fibrous Structure Material	Thickness (mm)	Capacity (Ah)* Discharging Current (A) 1	8
Outside of this process	Paper-like sheet of phenol fibers	PAN**	0.03	3.57	1.20
This process		PAN	0.05	3.28	1.51
This process		Glass mat***	0.10	3.97	1.80
This process		Glass mat	0.50	4.00	1.85
This process		Glass mat	1.0	3.82	1.78
This process		Glass mat	2.0	3.75	1.62
Control	Pulp	Glass mat	0.50	3.30	1.41
Control	Pulp	Glass mat	1.5	3.70	1.50

*Capacity at 10 hours' rate is 4 Ah.
**Means nonwoven fabric of polyacrylonitrile (fiber diameter is 5 microns).
***Means mat of glass fibers of 25 microns in diameter.

TABLE 3

	Separator Material	Fibrous Structure Material	Thickness (mm)	Capacity (Ah)* After Charge-Discharge Repeating of Charge-Discharge (times) 100	200	400
This process	Paper-like sheet of phenol fibers	Nonwoven PAN fabric	0.5	3.84	3.81	3.77
This process		Glass mat	0.5	3.98	3.92	3.50
Control	Pulp	Glass mat	1.5	3.57	2.34	0

*Discharging current is 1 A.

As will be appreciated from Table 2 above, the fibrous structure to be used in the battery of the process appropriately should have a thickness of 0.05 to 1.0 millimeters, most preferably 0.1 to 0.7 millimeters, in view of electric capacity. In contrast to this, the battery having separators made of pulp generally is poor in cell capacity as compared with that having the separators obtained according to the process. In particular, the battery having fibrous structures made of glass mat of 0.5 mm is found to have an insufficient cell capacity.

Phenol Resin

R.T. Jones; U.S. Patents 3,926,679; December 16, 1975; and 3,893,871; July 8, 1975; both assigned to Monsanto Company describes phenol-aldehyde resins to be used in impregnation of cellulosic substrates for use as battery separators. The phenol-aldehyde resins are A-stage resoles of number average molecular weight in the range of 130 to 300 with a combined aldehyde to phenol mol ratio in the range of 1.6:1 to 2.8:1, containing less than 4 weight percent of free aldehyde and between 7 and 32 parts by weight of an antimigratory agent per 100 parts by weight of resin solids, wherein the antimigratory agent is selected from the group consisting of ethylene glycol, glycerol, erythritol, mannitol, sorbitol, and polyethylene glycols of number average molecular weight less than 200.

In the evaluation of resole resin compositions and battery separators manufactured therefrom, two tests are useful, namely, the migration test and the oxidation resistance test. In the migration test, a 30 mil cellulosic sheet of paper is saturated by dipping it in resin solution and passing it through untensioned squeeze rolls. It is then dried for 30 minutes at 170°C in a circulating air oven. At the end of this time, the sheet is removed from the oven and examined for resin migration. Examination of a cross section reveals the degree of uniformity of distribution of resin throughout the sheet. The resins are graded on a scale of 1 to 10; 1 denoting little migration and 10 denoting excessive migration.

In the oxidation resistance test, an oxidation solution containing 10 grams of potassium dichromate and 56 grams of sulfuric acid in 1 liter of water is prepared. 275 milliliters of the oxidation solution is added to a beaker for each gram of cured battery separator to be tested. The beaker is covered with a watch glass, placed on a hot plate and heated to a gentle boil. A weighed sample is carefully placed into the solution. The liquid level is marked and maintained constant by addition of boiling water at frequent intervals.

The solution is refluxed gently for three hours. It is then cooled, water is added and a preweighed 60 mesh stainless steel screen is placed over the mouth of the beaker and the solution is poured off. The beaker is carefully washed to remove all traces of residue and the washings are run through the screen. The residue on the screen is washed copiously with water to remove all traces of oxidizing solution. (Incomplete washing will give erroneous results since acid will char the residue when it is being dried.) The screen is dried in an oven for 10 minutes at 120°C, then cooled in a dessicator. The weight of the residue is determined and the weight loss is calculated from the formula:

$$\%\ \text{weight loss} = \frac{\text{initial weight} - \text{weight after oxidation}}{\text{initial weight}} \times 100$$

In addition to these tests, service life tests are run on assembled 12 volt batteries and after 350 charge-discharge cycles, the separators are examined for perforation, delamination, surface erosion, and appearance.

Example 1: A one-stage water-dilutable resole is prepared by condensing 1.5 mols formaldehyde per mol of phenol in the presence of a sodium hydroxide catalyst. The catalyst is removed and the final resin product is diluted to a resin solids content of about 40 weight percent.

Examples 2 through 8: Similarly, a series of one-stage water-dilutable resoles is prepared by condensing formaldehyde with phenol. Data for the resoles are presented in Table 1.

TABLE 1

Resole	Combined Formaldehyde to Phenol Mol Ratio	Number Average Molecular Weight	Water Dilutability	Solids, %
1	1.45	150	>50	40
2	1.60	~160	>50	40
3	1.78	167	>50	40
4	1.94	177	>20	40
5	2.35	215	>20	40
6	2.45	240	>20	40
7	1.94	~200	7	40
8	1.50	~170	9	40

Examples 9 through 16: These examples are set forth to illustrate the effect of cellulose swelling agents on the degree of migration of resole during the drying and curing of an impregnated cellulose substrate and oxidation resistance of the resulting impregnated cellulose substrate. Resin compositions of Table 1 are mixed with ethylene glycol, glycerol, sorbitol, resorcinol and urea. The mixtures are used to treat 30-mil permeable fibrous cellulosic sheets by hand-dipping the sheets in the resin compositions until the samples are thoroughly impregnated with the admixtures.

The cellulosic sheets have a Gurley porosity of about 5 seconds when measured with a 5-ounce cylinder and a ¼ in^2 orifice. The treated samples are squeezed lightly by one pass through rubber squeeze rolls and are then subjected to a temperature of about 180°C for 10 minutes. The pick-up is approximately 40% of the weight of the impregnated sheet. The samples are evaluated for oxidation resistance. The resins are evaluated for migration as previously described. The data are presented in Table 2 wherein it is observed that polyols reduce the degree of migration and improve the oxidation resistance whereas of the other cellulose swelling agents, resorcinol increases the degree of migration and urea causes a sharp decrease in oxidation resistance.

TABLE 2

Example	Resole Example Number	AMA*	AMA,* parts/100 parts Resole Solids	Oxidative Weight Loss, %	Migration Rating
9	4	None	0	15	6
10	4	Ethylene glycol	16	14	4
11	4	Glycerol	8	11	5
12	6	None	0	11	8
13	6	Ethylene glycol	16	8	7
14	6	Sorbitol	8	10	7
15	6	Resorcinol	16	11	9
16	6	Urea	16	60	8

*Antimigratory agent.

Multiple Protrusion Design

B. Hollenbeck, A.L. Stockett, J.L. Tate, and D.M. Simmons; U.S. Patent 4,000,352; December 28, 1976; assigned to W.R. Grace & Co. describe a battery separator that has a plurality of protrusions extending above the main surface. The protrusions are arranged in an entirely new pattern which permits a minimum total area of the separator to be covered with protrusions and yet assures good contact with all areas of a battery's grid. This is done by arranging the protrusions in alternating rows in a manner blocking any openings between protrusions with which an element of the battery's grid could be registered.

Referring to Figure 1.12a, a battery separator **10** is shown. The battery separator has a top edge **11,** a bottom edge **12** and a first and a second side edge **13** and **14** respectively. The side edges extend at right angles to the top and bottom edges. The edges define the outer bounds of a rectangular sheet which is shown in Figure 1.12a. The sheet is actually 5¾ inches wide and 5¼ inches high. The separator has two outwardly facing opposite faces **15** and **16** (Figure 1.12b) that have basically planar main surfaces.

FIGURE 1.12: BATTERY SEPARATOR

a.

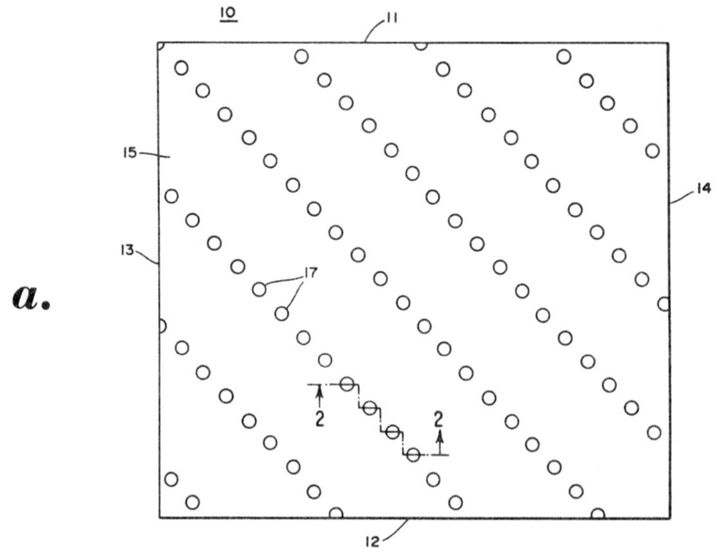

b.

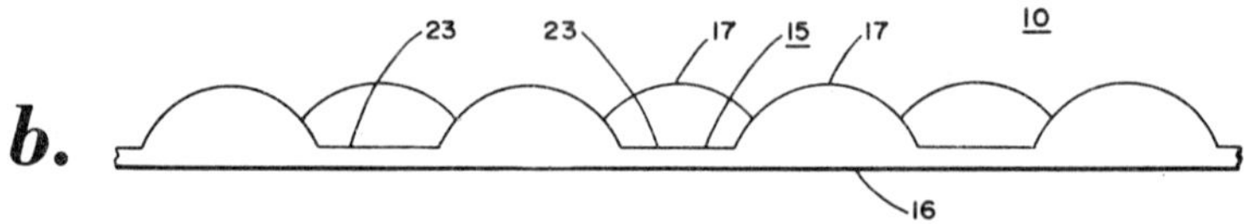

(a) Top plan view of one face of separator
(b) Enlarged edge view of one form of the separator of Figure 1.12a

Source: U.S. Patent 4,000,352

At least six rows of projections **17** are formed on and extend across the face **15** of the sheet **10** at an angle of 45°. In the example shown there are seven rows of projections. The preferred angle at which the projections extend across the sheet is 30° to 60° to each of the edges. The projections are spaced about 3/16 inch apart in the rows and are substantially circular in shape with a diameter of about 1/8 inch and a height above the main surface of about 1/16 inch. The preferred height above the surface is 15 to 62 mils with the optimal projection, according to certain aspects, having a diameter of about 1/8 inch taken from the opposite planar main surface **16**. The thickness through the main body of the separator between the main surfaces **15** and **16** is 3 to 30 mils providing a total separator thickness of 18 to 92 mils in the preferred form. The distance between the rows is preferably about 15/16 inch. It is of importance to note that the projections in one row are offset respecting the projections in an adjacent row such that no open paths exist across the face of the battery separator parallel to any edge of the separator across two adjacent rows. In other words, any straight line drawn across the separator parallel to any edge of the separator will intersect one of the projections **17** if the line extends across two adjacent rows of projections. If this line coincides with the route followed by a battery grid element, then it would clearly be engaged by one of the projections and held away from the main surface of the separator.

A feature of the process is found in the operating relationship between the battery separator and the battery grid in the sense that the two in combination form the final operating combination. As is known, battery grids are commonly pasted with an active material and these pasted members are called plates. The battery grid commonly has a plurality of elements extending along parallel axes. An additional plurality of elements extends along second parallel axes crossing the first plurality of elements and forming a cross grid. These two series of parallel elements in the most common type of automotive battery cross at right angles with a spacing of ½ inch.

The two series of elements or spaced apart battery members are usually basically linear and arranged in a common plane. However, other nonlinear elements are known and so long as they have a continuing aspect in a general direction, the rules of this process apply. This is true even when the grid is made up of a plurality of curved parts. In fact, the grid is normally cast as a unitary part and thus the linear elements are actually parts of cross elements.

The battery separator is operationally installed in normal position adjacent to the pasted battery grid with the face of the separator with the projections formed thereon positioned against the pasted battery grid. The rows of projections extend diagonally across the grid elements at an angle of 30° to 60° with the grid elements necessarily being aligned to fall on some of the projections when the separator and battery grid are shifted relative to one another so long as the diagonal relationship of the grid elements and the rows of projections remain within the angle of 30° to 60°.

In Figure 1.12b, it may be seen that the thickness and the mass of the sheet **10** is greater at the projections than at the intermediate region **23** which makes up the greater area of the sheet. The effect of this increased thickness is to block off the area. The passage of ions may be said to effectively be limited to the areas of least resistance, i.e., the thinner intermediate areas. Thus, the greater the combined area of the projections is the greater the electrical resistance and the lower the final battery electrical capacity will be.

Thus, the spacing apart of the projections is important, as well as their shape or size. It is desirable to block off the least area possible with the thickened areas or regions **17**. The sheet is cut from a long roll that is about 15 inches wide manufactured by a continuous process the preferred form of which is described in U.S. Patent 3,798,294. A 5¼ by 5¾ sheet cut from the roll stock will necessarily contain a minimum of six rows of projections **17** which gives the necessary minimum amount of separation contact or support to provide the necessary spacing between the main surface **23** of the separator sheet **10** and the battery plate. As shown in Figure 1.12a, the random cutting of the sheet may produce up to seven rows of projections. The important feature is that substantially two complete continuous coverage separator portions cross every one of the battery grid elements. In a cross grid this means two diagonal continuous coverage separator portions in each direction or across both sets or series of grid elements. It can be seen that the same projections may be considered once with respect to the continuous coverage separator portion for each series of grid elements if properly positioned. If this is done, then two hits on every element are assured which gives adequate spacing support.

Envelope Sealing Technique

A process described by *A. Sabatino; U.S. Patent 4,037,030; July 19, 1977; assigned to Gould Inc.* relates generally to storage batteries, and particularly to storage batteries which are maintainence free and/or resistant to vibration. The term "maintenance free" as applied to storage batteries refers to a battery which does not require the addition of any water or other liquid during the life of the battery, due to the use of a special alloy in the battery plates.

These maintenance-free batteries are normally constructed with the plates resting directly on the bottom of the battery container so that there is space for an extra reservoir of electrolyte above the plates, and with the plates contained in open-top envelopes to prevent active material which drops to the bottom of the container from "treeing" between positive and negative plates and causing shorting. In heavy duty storage batteries, it is normally preferred to have the bottoms of the plates bonded to the battery container to provide vibration resistance. This type of bonding is not practical in maintenance-free batteries, however, because it is the envelopes around the lower portions of the plates that become bonded to the container, and the plates are still free to vibrate within their envelopes. This process provides an improved maintenance-free and/or vibration-resistant storage battery which prevents "treeing" of active material between positive and negative plates at the bottom of the container, so that there can be no shorting between the positive and negative plates.

Referring to Figure 1.13a, there is shown a storage battery **10** having a plurality of partitioned cells **11** each of which contains four positive plates **12** and five negative plates **13.** The positive plates are all interconnected via lugs **12a** and a connector **12b** with an external positive terminal **14.** The negative plates are also interconnected by similar means as is known in the art. The illustrative battery is a maintenance-free battery, with the plates **12** and **13** all resting on short ribs **15a** on the bottom wall **15** of the battery container **16** so as to provide space for an extra reservoir of electrolyte between the top edges of the plates **12** and **13** and the top wall **17** of the container.

Furthermore, each adjacent pair of positive and negative plates is separated by envelopes **18** which surround the positive plates to prevent active material which drops to the bottom of the container **16** from "treeing" between the positive and negative plates and causing shorting. The envelopes are made of conventional battery separator material having ribs **19** on the inside surfaces and bearing against the positive plates contained therein. In accordance with one important aspect of the process, the bottoms of the plate envelopes form openings to

FIGURE 1.13: MAINTENANCE-FREE STORAGE BATTERY

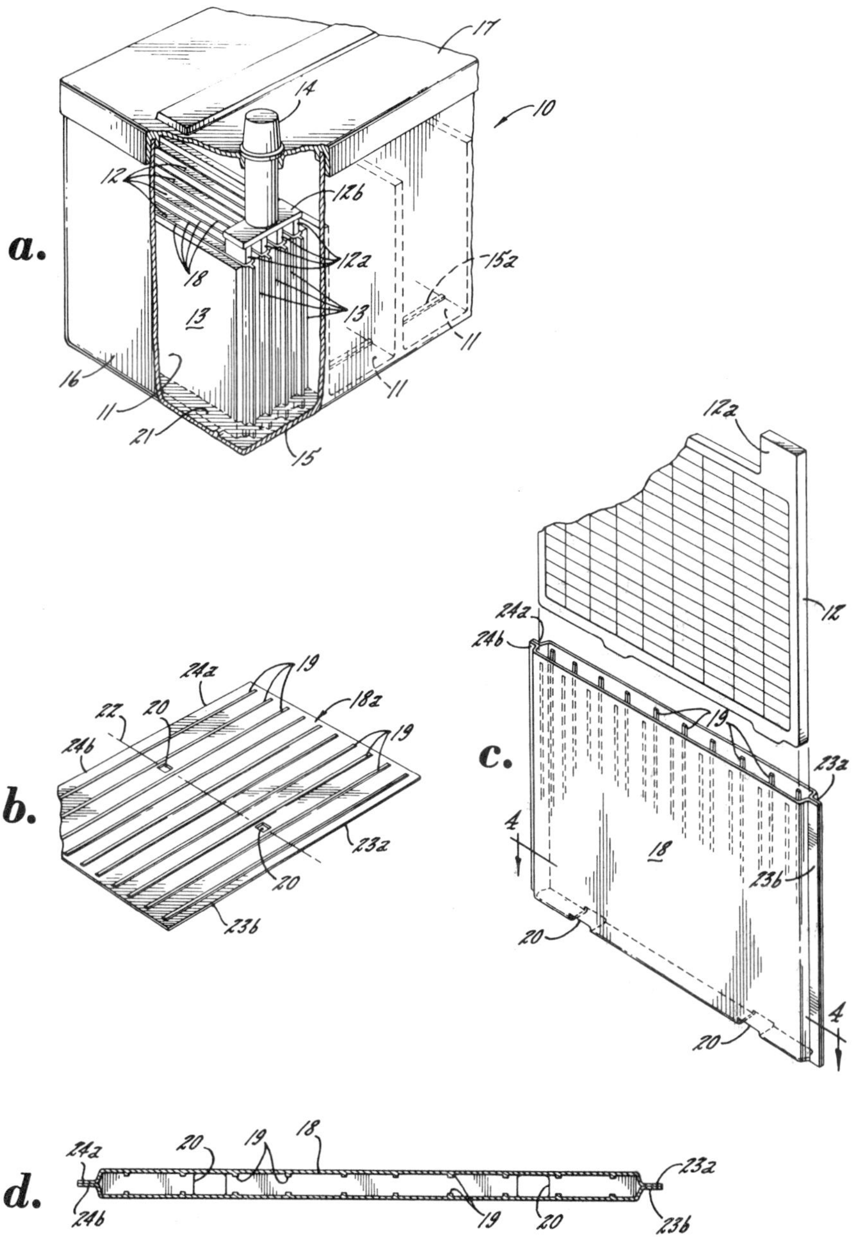

(continued)

FIGURE 1.13: (continued)

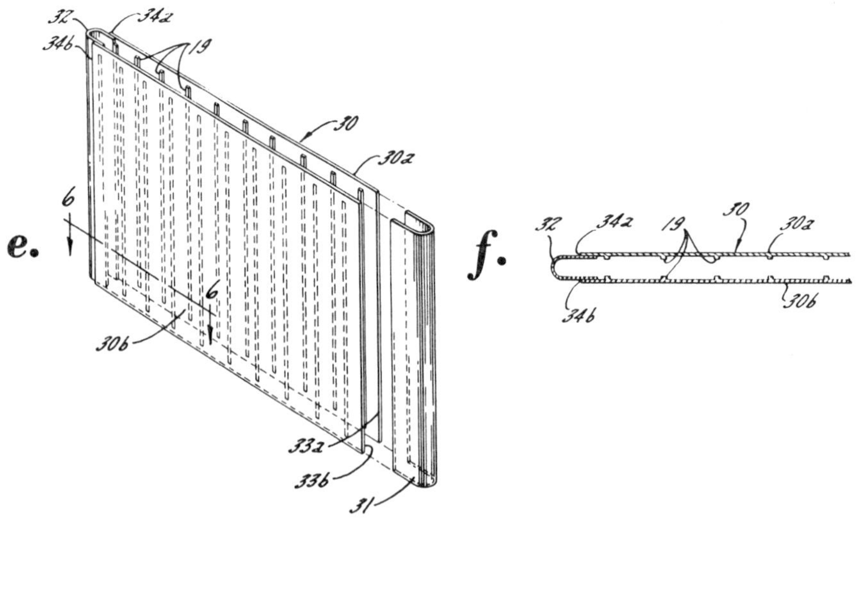

g.

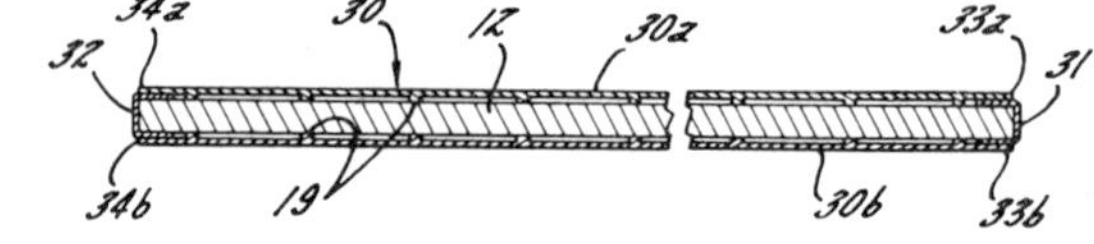

(a) Fragmentary perspective view, partially in section of a storage battery

(b) Plan view of a blank used to form one of the plate envelopes used in the battery of Figure 1.13a

(c) Enlarged exploded perspective view of one of the plates and one of the envelopes used in the battery of Figure 1.13a

(d) Section taken along line **4—4** in Figure 1.13c

(e) Enlarged perspective view, exploded at one end of a plate sleeve suitable for use in the battery of Figure 1.13a

(f) Section taken along line **6—6** in Figure 1.13e

(g) Same section shown in Figure 1.13f showing the sleeve after a plate has been inserted therein

Source: U.S. Patent 4,037,030

permit a bonding material on the bottom of the battery container to anchor the plates to the container, with the bonding material also sealing the bottom openings in the envelope to prevent the treeing of active material between positive and negative plates at the bottom of the container. Thus, in the illustrations of Figures 1.13a, 1.13b, 1.13c, and 1.13d, the envelope **18** around each positive plate **12** is sealed along the vertical edges of the plate but forms a pair of holes or slots **20** along the lower edge of the plate. When the enveloped plate **12** is placed in the battery container **16**, the bottom of the envelope **18** is submerged in a layer of liquid bonding material **21**, suitably an epoxy compound, which flows through the holes and into contact with the lower edge of the plate. The number, size, shape and location of the holes are not critical, as long as they permit the liquid bonding material to flow into contact with the enveloped plates.

After the bonding material solidifies, it adheres to both the bottom wall **15** of the battery container and the plate **12** thereby firmly anchoring the plate to the container to provide the desired vibration resistance. Furthermore, the liquid level of the bonding material is above the uppermost edges of the holes so that the bonding material also seals the holes to prevent any active material that drops down to the bottom of the envelope from escaping from the envelope and forming a conductive path between the enveloped positive plate and one of the adjacent negative plates **13**.

The particular envelope **18** illustrated in Figures 1.13a, 1.13b, 1.13c and 1.13d is suitably formed from a blank **18a** of separator material as illustrated in Figure 1.13b. The holes are illustrated as being preformed in this blank **18a** along a central fold line **22**, but if desired the holes may be formed in the finished envelope either before or after the insertion of the battery plate therein. To form the envelope, the blank **18a** is folded along line **22**, and the resulting overlapped vertical edges **23a, 23b** and **24a, 24b** are sealed to each other to form the desired envelope. The sealing of the vertical edges may be effected by heat sealing a thermoplastic separator material, by ultrasonic sealing, by the use of suitable adhesives, or by any other suitable means.

It is generally preferred to keep the width of the seal as narrow as possible while forming a reliable liquid-tight seal. The envelopes may be preformed before the plates are inserted therein, or the blanks **18a** may be folded around the plates so that the plates are enveloped during the forming of the envelopes.

In accordance with a further aspect of the process, the separator material forms open-bottomed sleeves rather than envelopes with the vertical edges of the sleeves being formed by strips of heat shrinkable plastic film so that the sleeves can be shrunk snugly around the plates therein. Thus, as illustrated in Figures 1.13e, 1.13f, and 1.13g, a sleeve **30** may be formed from two sheets **30a** and **30b** of separator material having their vertical edges joined to a pair of strips **31** and **32** of heat shrinkable plastic film. More specifically, the two heat shrinkable strips **31** and **32** are folded and sealed to the two pairs of vertical edges **33a, 33b** and **34a, 34b** of the two side sheets **30a** and **30b.**

In their original form, these two folded heat shrinkable strips **31** and **32** form lateral extensions of the sleeve, thereby facilitating the insertion of a battery plate **12** therein. However, after the plate is inserted in the sleeve, the vertical edges are heated to shrink the two strips **31** and **32** so that they contact transversely against the vertical edges of the plate, as illustrated in Figure 1.13g. The sleeve thus fits snugly over the plate to minimize the possibility of the sleeve riding up over the plate. This construction also minimizes the space occupied by the sleeve along the vertical edges of the plate, thereby permitting the width of the plate to be maximized for any given size of battery container. The open bottom end of the sleeve **30** permits access of the bonding material to the plate **12** disposed within the sleeve, with the bottom of the sleeve being sealed when the bonding material solidifies. If desired, the heat shrinkable ends can also be used in the envelope type separators illustrated in Figures 1.13a, 1.13b, 1.13c and 1.13d.

As can be seen from the foregoing detailed description, this process provides an improved maintenance-free storage battery that has the battery plates firmly secured to the battery container by means of the bonding material, so that the plates are resistant to vibration even in heavy duty applications. Because the bonding material itself seals the separator openings through which the bonding material gains access to the plate, the final battery assembly prevents treeing of active material between positive and negative plates at the bottom of the container so that there can be no shorting between the positive and negative plates. Because of the facility with which the separators can be formed, with the plates being preinserted in the separators if desired, this improved battery can be efficiently and economically manufactured at high production rates.

ELECTROLYTE ADDITIVES

Nitric Acid

W.H. Edwards; U.S. Patent 4,046,642; September 6, 1977; assigned to Chloride Group Limited, England, describes lead acid electric storage cells of the Plante type. It has been found that if a mixture of nitric acid or nitrates and sulfuric acid is placed in the container the pure lead of the positive plate can be converted to a satisfactory electrochemically active structure and the cell used without further treatment, provided the amount of nitric acid and the duration of the charging current used are such as to ensure that all the nitric acid is destroyed.

It is believed that it is largely converted to ammonium sulfate, which can be left in the cell. The electrolyte could be decanted and replaced if desired but this is not necessary. Indeed with large stationary batteries which have to be assembled on site, replacement of the electrolyte would present considerable problems and be a positive disadvantage. Thus when using nitric acid, a preferred initial composition for the electrolyte in the container is 0.8 to 1.5% by weight of 1.415 specific gravity nitric acid in 1.200 specific gravity battery purity sulfuric acid. This is equivalent to 0.2 to 4.0% weight by volume nitric acid in 1.200 specific gravity sulfuric acid.

Example 1: A cell consisting of 4 Plante positive plates and 5 negative plates each 76.2 cm wide and 119.1 cm high were assembled in a cell with microporous polyvinyl chloride separators 3.9 mm thick between each plate. The cell contained 1,420 cc of electrolyte. The electrolyte was made up from 1,411.5 cc (6.8 N) battery purity 1.200 specific gravity sulfuric acid and 8.5 cc of 1.415 specific gravity (15.8 N) nitric acid; 1.415 specific gravity nitric acid contains 1,000 grams of nitric acid per liter.

The sulfuric acid initially contained not more than 0.005% w/w ammonium ions (0.0035 M) and not more than 0.0005% w/w nitrate ions (0.0001 M). After addition of the nitric acid it contained 0.6% of nitrate (0.095 M), this being the eventual theoretical molar concentration of the ammonium ions in the electrolyte after electrolytic formation. The cell electrolyte was thus 1.200 specific gravity sulfuric acid containing 0.6% by weight by volume nitric acid. The positive plates were 7.26 mm thick overall and were plates cast with trapezoidal protuberances providing an actual surface area having a ratio to its projected surface area (762 x 119.1 sq cm) of 11 to 1.

The negative plates are flat sided and faced and are 3.9 mm thick. The negative plates are pasted grids containing 125 grams of leady oxide (40% lead, 60% lead monoxide) per plate (in the wet pasted condition). The cell has a rated 10 hour capacity of 30 ampere hours. It was charged to 180 ampere hours by being charged at 3.5 amps for 40 hours at 40°C without cooling. The electrolyte contained no free nitric acid after charging but 0.055 mol of ammonium ions i.e., about 60% of the theoretical content of 0.6 x 10/63 mol of ammonium ion i.e., 0.095 M equivalent to the original nitric acid.

Example 2: Four cells were made up as in Example 1, but using 0.25% weight by volume of 1.415 specific gravity (15.8 N) nitric acid (measured at 24°C). The cells after formation had open circuit readings of 2.06 volts at 15°C and the electrolyte specific gravity at 15°C was 1.236. The cells were then discharged at a current of 5.25 amps down to a cell voltage of 1.82 volts at 24°C. The cells had discharge durations for this, the first discharge, of 2.05 hours, 1.67 hours, 1.82 hours and 1.73 hours. The specific gravity of the cells after discharge was 1.210 at 24°C.

Metallic Sulfates

G.W. Mao and A. Sabatino; U.S. Patent 3,948,680; April 6, 1976; assigned to Gould Inc. describe a method of fashioning a lead-acid storage battery which is capable of being stored after completion of the battery processing and thereafter activated by the addition of electrolyte. The process includes adding conditioning quantities of a treating agent affording certain metallic sulfates to the formation electrolyte, a rinse electrolyte or to a separate solution to obviate the necessity for removing, as by drying, all or substantially all of the electrolytes used to process the battery. Sufficient electrolyte is removed simply by draining, and the resulting battery can be stored for extended periods of time without significantly adversely affecting the performance of the activated battery.

The following examples illustrate the process. Unless otherwise specified, all percentages are by weight. The high temperature storage carried out in the examples was used to simulate room temperature conditions that would occur over a longer period of time. The sulfates used as the treating agent were anhydrous unless otherwise specified.

Example 1: A Group 24 battery (53 amp hour capacity) was formed, and the conditioning agent was thereafter applied in varying weight percentages, either to the rinse acid or in a separate step after removal of the formation acid from the battery by inverting and dumping. Mechanical mixing was employed in all cases. After the conditioning, in all cases, the cover vent openings were sealed, and the batteries were stored at 150°F for 35 days. Following storage, the cover vent openings were unsealed, electrolyte added, a boosting charge given and the battery performance evaluated. The results are shown in Table 1.

TABLE 1

Battery Number	Treatment	5 A Boosting Following Activation with 1.300 Acid: Time into Boosting (min)	5 A Boosting Following Activation with 1.300 Acid: End of Charging Voltage	20 hr Capacities, hr (Internal Resistance, mΩ): 1st Cycle	2nd Cycle	3rd Cycle	0°F BCI Performance: 312 A Discharge 5 sec Voltage	0°F BCI Performance: Time to 7.2 V (sec)
1	2.0% Na_2SO_4 in water	221	16.72	22.05 (10.8)	21.6 (10.8)	20.4 (10.8)	7.73	61.7
2	2.0% Na_2SO_4 in rinse acid	300	16.05	21.45 (11.5)	21.5 (11.42)	20.6 (11.3)	7.63	58.1
3	10.0% Na_2SO_4 in water	221	16.68	22.60 (10.8)	21.9 (10.8)	20.77 (10.75)	7.74	60.9
4	10.0% Na_2SO_4 in rinse acid	300	15.76	22.0 (11.3)	21.6 (11.15)	20.30 (11.1)	7.63	53.3

Example 2: Example 1 was repeated, except that the boosting step was varied, and the formed batteries were stored at 110°F for 61 days. The results are shown in Table 2.

TABLE 2

Battery Number	Treatment	5 A Boosting Following Activation with 1.300 Acid: Time into Boosting (min)	5 A Boosting Following Activation with 1.300 Acid: End of Charging Voltage	20 hr Capacities, hr (Internal Resistance, mΩ): 1st Cycle	2nd Cycle	3rd Cycle	0°F BCI Performance: 312 A Discharge 5 sec Voltage	0°F BCI Performance: Time to 7.2 V (sec)
5	2.0% Na_2SO_4 in water	180	16.59	21.2 (11.9)	21.3 (11.8)	20.0 (11.7)	7.53	45.0
6	2.0% Na_2SO_4 in rinse acid	240	16.43	19.0 (12.5)	19.7 (12.5)	18.35 (12.3)	7.43	30.0
7	10.0% Na_2SO_4 in water	240	16.54	21.2 (11.7)	21.0 (12.0)	19.7 (12.0)	7.53	45.0
8	10.0% Na_2SO_4 in rinse water	270	15.80	17.7 (11.8)	17.7 (12.2)	15.8 (12.0)	7.26	20.0

Example 3: Group 24 plastic batteries (62 amp hour) were formed with a sulfuric acid aqueous formation, and the conditioning agent was added to either the formation acid or to the rinse acid. In all cases, after inverting and dumping to remove the rinse acid, the cover vent openings were sealed and the batteries were then stored at 110°F for 21 days. After opening the cover vent openings, electrolyte was added, and the batteries evaluated. The results are shown in Table 3.

TABLE 3

Battery No.	Treatment	30°F Activation 290 A Discharge: 15 sec Voltage (V)	30°F Activation 290 A Discharge: Time to 7.2 V (sec)	25 A Reserve Capacity (min)	0°F Performance 387 A Discharge: 5 sec Voltage (V)	0°F Performance 387 A Discharge: Time to 7.2 V (sec)	-20°F Performance 310 A Discharge: 5 sec Voltage (V)	-20°F Performance 310 A Discharge: Time to 7.2 V (sec)	25 A Reserve Capacity minutes: 1 Cycle	25 A Reserve Capacity minutes: 2 Cycle
1	Formed in a sulfuric acid solution containing 2.0% Na_2SO_4	8.53	83.0	95.0	7.21	6.6	7.24	7.0	93.0	92.0

(continued)

TABLE 3: (continued)

Battery No.	Treatment	30°F Activation 290 A Discharge 15 sec Voltage (V)	30°F Activation 290 A Discharge Time to 7.2 V (sec)	25 A Reserve Capacity (min)	0°F Performance 387 A Discharge 5 sec Voltage (V)	0°F Performance 387 A Discharge Time to 7.2 V (sec)	-20°F Performance 310 A Discharge 5 sec Voltage (V)	-20°F Performance 310 A Discharge Time to 7.2 V (sec)	25 A Reserve Capacity minutes 1 Cycle	25 A Reserve Capacity minutes 2 Cycle
2	Formed in a sulfuric acid solution containing 0.5% Na_2SO_4	8.71	113.0	116.0	7.40	41	7.72	50	110.0	116.0
3	Treated with a rinse acid containing 2.0% Na_2SO_4	8.61	95.0	117.0	7.44	52	7.54	34	118.0	117.0
4	Treated with a rinse acid containing 0.5% Na_2SO_4	8.66	102.0	120.0	7.47	41	7.39	35	119.0	110.0
		Specification 15 sec voltage, 7.2 V			Specification 30 sec voltage, 7.2 V		Specification 30 sec voltage, 7.2 V		Reserve capacity rating, 90 min	

Tetrabutyl Ammonium Perchlorate

H.L. Lewenstein; U.S. Patent 3,928,066; December 23, 1975 describes a storage battery which has a lead anode, a lead dioxide cathode, a sulfuric acid electrolyte and a quaternary ammonium compound wherein there has been substituted for all the hydrogen atoms an aliphatic and/or aromatic group. The quaternary ammonium compound is placed in contact with the electrolyte by dissolving the compound in the electrolyte, or by placing the quaternary ammonium compound in the battery container prior to the addition of the electrolyte or by depositing the compound on the anode, for example, via a readily evaporated solvent which leaves a residue of the compound on the anode, or for example, by thermal evaporation. The compound inhibits the evolution of gaseous hydrogen and thereby attenuates the loss of water.

One example of the process utilizing tetrabutyl ammonium perchlorate as the agent for inhibiting the loss of water from the battery is formed by immersing a lead anode and a lead dioxide cathode in a sulfuric acid electrolyte. The sulfuric acid is of a concentration customarily used in automotive batteries, and the lead anode is combined with a standard amount (approximately 5%) of antimony. Into the electrolyte there is dissolved tetrabutyl ammonium perchlorate in a concentration of one part tetrabutyl ammonium perchlorate in 10,000 parts by weight of the electrolyte, this being customarily referred to as 0.01% concentration.

This battery provides a substantial improvement in the suppression of the evolution of gaseous hydrogen and the retention of water as compared to a similar battery not having the tetrabutyl ammonium perchlorate. A convenient method of measuring water loss is to measure the quantity of gas emitted at the anode due to the breakdown of water. For example, as a control, 0.034 gram of a lead antimony anode obtained from a standard automotive battery immersed in the sulfuric acid electrolyte of the battery was found to emit 0.5 cc of gas during a four-day period at an environmental temperature of 100°F. As a test, the same amount of anodic material with 0.01% tetrabutyl ammonium perchlorate concentration in the same electrolyte emitted only 0.1 cc of the gas during a four-day period at the same temperature.

Cadmium Hydroxide

G.W. Mao; U.S. Patent 4,006,035; February 1, 1977; assigned to Gould Inc. has found that the addition of elemental cadmium (e.g. as a fine powder) or a cadmium compound to the electrolyte in certain levels in a maintenance-free battery application significantly diminishes the current draw, and thus the water consumption, so as to improve the performance of the battery. The cadmium compound may comprise cadmium sulfate or any other cadmium compound which is: (1) sufficiently soluble in aqueous sulfuric acid solutions to provide the requisite amount of the cadmium; (2) not substantially harmful to either the battery components or to the performance of the battery in use; and (3) not susceptible to producing a lead salt that would likely precipitate in sufficient amounts which would significantly reduce the porosity of the battery plates. For example, cadmium hydroxide could suitably be used.

The amount of the cadmium compound which is used should be sufficient to decrease the current draw during constant voltage overcharge. Typically, the battery at this stage will be in a fully charged condition. While the amount can vary so long as the amount of cadmium provided is sufficient to decrease the current draw to the extent required, it has been found suitable to utilize amounts in the range of from about 0.1% (or somewhat less) to about 0.3%, based upon the total weight of the electrolyte, and even up to about 0.5% by weight when the cadmium compound employed is cadmium sulfate. All that is required is for the cadmium compound to be added to the electrolyte, suitably prior to the sealing of the cover to the battery container.

Example 1: A single 70-ampere-hour capacity cell having 6 positive and 7 negative electrodes formed of a lead based alloy having 0.08% by weight calcium and about 0.25 to 0.30% tin was exposed to varying cell voltages at ambient temperatures. The current draw was compared for the cell having no additive (control cell) to the same cell containing 0.5% by weight cadmium in the electrolyte. An electrolyte comprising sulfuric acid having a specific gravity of 1.265 was used. The current draw characteristics at the various equilibrium voltages (i.e., after the cell has been at the voltage charge condition for about 1 week) are set forth in Table 1 below.

TABLE 1

	Control Cell with No Additive			Cell with 0.50% Cadmium in Electrolyte		
	Half Cell vs. Hg Ref.*			Half Cell vs. Hg Ref.*		
Cell Voltage	Positive (V)	Negative (V)	Current (mA)	Positive (V)	Negative (V)	Current (mA)
Part A, 80°F						
2.35	1.252	-1.098	24	1.215	-1.135	10
2.45	1.279	-1.171	54	1.262	-1.188	40
2.55	1.316	-1.234	176	1.305	-1.244	158
Part B, 125°F						
2.30	1.191	-1.109	65	1.172	-1.123	46
2.35	1.152	-1.198	102	1.132	-1.218	74
2.45	1.191	-1.259	296	1.173	-1.277	246
2.55	1.231	-1.318	1,020	1.211	-1.338	880

*These values were all determined experimentally.

Example 2: Two Group 24 maintenance-free batteries having capacities of 81 ampere hours were tested to show the effects of the inclusion of a cadmium compound in accordance with the process on the current draw characteristics of the batteries. Two different charge voltages were used, and the grids were formed of the lead base alloys described in Example 1. The gassing current draw was then measured after the batteries had been on charge conditions overnight. The results are shown in Table 2.

TABLE 2

Additive Concentration	. Battery No. 1 Charge Voltage. . 14.40 V	14.10 V	. . .Battery No. 2 Charge Voltage . . . 14.40 V	14.10 V
None	185 mA, 108°F	120 mA, 108°F	170-179 mA, 108°F	110 mA, 108°F
0.1% $CdSO_4$	110 mA, 108°F	73 mA, 108°F	105 mA, 108°F	65 mA, 108°F
0.3% $CdSO_4$	145 mA, 108°F	90 mA, 107.5°F	128 mA, 108°F	90 mA, 107.5°F

Immobilized Paste

J.P. Cestaro and L.J. Crosby; U.S. Patent 3,930,881; January 6, 1976; assigned to NL Industries, Inc. describe a lead acid storage battery which comprises a battery housing, positive and negative lead base electrodes, and an immobilized electrolyte paste. The immobilized electrolyte paste comprises 1.5 to 4 parts sulfuric acid having a specific gravity of from 1.050 to 1.350 and 1 part organic microparticles being inert to sulfuric acid and having a particle diameter from 0.1 to 3 μ. Organic microparticles which are inert to sulfuric acid include polystyrene, polymethyl methacrylate, polypropylene, and polyethylene. The former two organic microparticles are more readily available and, therefore, are preferred.

Example 1: In this example, a 2-volt acid battery cell was prepared in which the immobilized electrolyte of the cell contained hollow polystyrene microparticles having an average diameter of 1 μ. These microparticles were prepared by mixing polystyrene with a solvent and a nonsolvent for styrene. Both the solvent and nonsolvent were also immiscible with water. The styrene-solvent-nonsolvent mixture was then emulsified in water using high shear and a surface active agent. Next the solvent was removed from the emulsion by distillation under reduced pressure. Then the water and the nonsolvent were removed by drying, thus forming hollow microparticles of polystyrene having an average diameter of 1 μ. This method is more fully described in U.S. Patent 3,784,391.

These hollow polystyrene microparticles were used to immobilize the battery electrolyte of a 2-volt lead acid battery cell prepared in the following manner. Nine lead monoxide positive plates (1¾ x 1¾ x 0.045 inch) and ten lead monoxide negative plates (1¾ x 1¾ x 0.045 inch) were placed vertically and in alternating order in a polystyrene case (2¼ x 2½ x 2¼ inches). These positive plates and the negative plates were prevented from contacting one another by imposing between them perforated, corrugated, PVC battery separators. The positive plates were electrically connected to one another and to a positive terminal and the negative plates were also connected to one another and a negative terminal. A removable polyethylene cover for the case was provided.

To form and condition the battery cell, a sulfuric acid-polystyrene microsphere paste was prepared and used. The paste was prepared by admixing 1.7 parts of sulfuric acid having a specific gravity of 1.260 with 1 part hollow polystyrene microspheres prepared as described above. This mixture formed an immobilized electrolyte paste. The immobilized electrolyte paste was then placed in the battery cell while vibrating the cell to cause the electrolyte paste to completely cover the electrode plates. The battery cell was formed by supplying 0.455 amp to the battery cell for 48 hours. The battery cell was then conditioned by supplying 0.152 amp for 64 hours.

Next, the cycle life of the battery cell was tested. This was done by placing a 2 ohm resister across the battery cell and then allowing the cell to discharge until the voltage of the cell decreased from 2 to 1.75 volts. After discharging the battery cell was recharged. The discharging-charging cycle was continually repeated. The ampere-hours of each discharged cell was calculated. When the ampere-hours of the discharge reached 15% of the theoretical capacity of the cell, the cell was considered "dead." The theoretical capacity of the cell was 0.24 ampere-hour per gram of active material in the cell. In this example, the cell had a theoretical capacity of 14.5 ampere-hours and the cell had a cycle life of 64 discharge-charge cycles.

During the entire life of this battery cell, the hollow polystyrene microparticles in the immobilized battery electrolyte remained as a paste and did not separate from the electrolyte or degrade. In addition, the immobilized battery electrolyte remained sufficiently immobilized to prevent the spilling of the electrolyte if the battery cell were temporarily placed on its side and thus greatly decreasing the hazard of an accidental spill. The operational details are recorded in the following table:

EXAMPLE	1	2
Microspheres	Hollow Polystyrene	Hollow Polystyrene
Microsphere Diameter (μ)	1	1
Parts Microspheres in Immobilized Electrolyte Paste	1	1
Parts Sulfuric Acid in Immobilized Electrolyte Paste	1.7	2
Specific Gravity of Sulfuric Acid in Immobilized Electrolyte Paste	1.260	1.300
Forming Electrolyte	Immobilized Paste	Free Sulfuric Acid
Condition Electrolyte	Immobilized Paste	Immobilized Paste
Electrode Plate Size (in. × in. × in.)	1¾×1¾ × .045	1¾×1¾ × .045
Number of Electrode Plates	9+, 10−	9+, 10−
Discharging Voltage (Volts)	2 to 1.75	2 to 1.75
Approximate Drain Rate(Amps)	1	1
Cyclic Life	64	53

Example 2: In this example, a lead acid battery cell was prepared and tested as in Example 1 with the following exceptions. The battery cell was formed in free sulfuric acid (no microparticles present) having a specific gravity of 1.070.

After forming, the forming acid was removed and an immobilized electrolyte paste was added for conditioning and cycling. The immobilized electrolyte paste contained 2 parts sulfuric acid having a specific gravity of 1.300 and 1 part hollow polystyrene microparticles as described in Example 1. This battery cell had a discharge-charge cycle life of 54 cycles, and the hollow microparticles did not separate from the sulfuric acid or degrade, and hence, the electrolyte remained immobilized. The operational details were recorded in the table on the preceding page.

Magnetic Flux

D. Borello; U.S. Patent 4,042,754; August 16, 1977 describes a battery in which, to improve its discharging performance, the liquid electrolyte is maintained at a specific gravity that is optimum for electrical conductivity, and a magnetic flux is imposed upon the battery plates.

As shown in Figure 1.14, the device or proposed battery structure **10** has a plastic housing **12** and positive and negative terminals **14** and **16**, respectively. Within the housing are mounted porous electrodes **18** and **20**. Both electrodes are made of the usual lead battery material, but will be understood to have openings therein or a porosity which will not prevent the passage of fluids under pressure through the electrodes.

More particularly, positive electrode **18** is a porous lead peroxide, and negative electrode **20** is a sponge or porous lead. Assembled within the electrodes **18** and **20** is an insulated titanium wire **22** which is electrically connected to be energized by a chopper circuit **24** through a conductor **26** so as to alternately supply and induce current into one plate at a time, at a frequency between 60 and 300 cycles per second, during charging and discharging of the battery. It has been found that the induced magnetic flux field of the circuit **24** adds to the magnetic flux transmission and storage properties of the electrode sites during charging, and thus promotes greater saturation thereof during this phase.

This is analogous to what occurs in the operation of a generator when two magnetic flux fields cross each other, except that in the battery hereof the magnetic flux field settles upon the plates or electrode sites, to thus provide what can be called a first phase generator. The full generator cycle is then completed when current is discharged from the electrode sites. In this connection, it has been further found that the same induced magnetic flux field has the unobvious effect during the discharging phase of driving, probably by magnetic flux repulsion, the electrical components from the electrodes during the discharging phase, or when the battery is transmitting electricity.

The outer housing **12** and electrode plates **18** and **20** define three fluid chambers **28, 30** and **32**. On the right side of Figure 1.14b, chamber **28** is supplied with sulfuric acid, designated **29**, via inlet **34**, the acid being either in solution or concentrated, and preferably being supplied at a temperature of -20°F. The acid **29** is under a slight pressure and is allowed to ooze through the porous structure of electrode **18** into the center chamber or mixing compartment **32**. On the left of Figure 1.14b is water chamber **30** supplied through inlet **36**. The water **31**, more particularly, is maintained under 1 to 2 pounds pressure of pure oxygen, which is effective both to make the water oxygen-rich and also to force the water through the porous structure of the electrode **20**. The water is preferably introduced at a temperature of 150°F.

FIGURE 1.14: BATTERY DESIGN

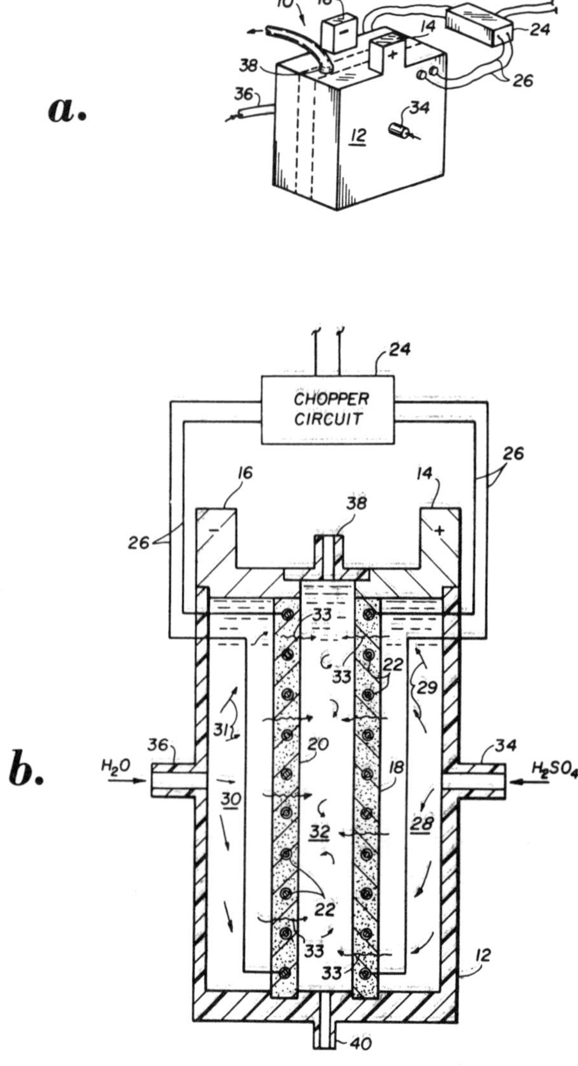

(a) Perspective view of battery structure

(b) Side elevational view, in longitudinal cross section, and on an enlarged scale, showing internal structural features of the battery structure

Source: U.S. Patent 4,042,754

From the description already provided, it should be readily appreciated that center compartment **32** acts as a mixing chamber for the continuously renewed electrolyte composed of the above noted oxygen-rich water **31** and sulfuric acid **29**. Flow of the water and acid is controlled to provide an ideal specific gravity of the mix, which is approximately 1.275, which specific gravity is known to be most favorable for electrical conductivity.

To continuously monitor or supervise the specific gravity of the electrolyte mix in the compartment, use may be made of any of the numerous models of battery hydrometers available for this purpose. Further, to maintain an optimum specific gravity of 1.275 may require appropriately either increasing or decreasing the flow of the water into the compartment, which is readily accomplished by correspondingly modifying the flow-producing oxygen pressure.

At the top of compartment **32** is overflow outlet **38**, by which any excess electrolyte is conducted to a reservoir for processing and recycling. When not in use, the electrolyte mix **29** and **31** in compartment **32** can be drained through the bottom port **40**. By virtue of maintaining the electrolyte **29** and **31** at the optimum or ideal specific gravity of 1.275, it more effectively promotes electron flow between the electrodes **18** and **20** during both the charging and discharging phases, thereby correspondingly increasing the efficiency of the battery.

VENTING TECHNIQUES AND CAP DESIGN

Vent Barrel

A process described by *D.D. Hakarine; U.S. Patent 4,002,495; January 11, 1977; assigned to Gould Inc.* relates generally to an explosion-proof vent barrel for an electrical storage battery and, more particularly, to a so-called gang vent in which two or more vent barrels are attached to a common frame for the purpose of venting multiple cells of the battery.

In Figure 1.15a, there is shown an electrical storage battery **10** having a container **11** divided into cells and closed by a cover **12** formed with tubular bosses **13** which define vent openings **14** leading into the cells. In the illustrated construction, three side-by-side openings are adapted to be closed by a corresponding number of vent plugs or barrels **15** which form part of a gang vent **16**. The gang vent includes a one-piece frame **17** (Figures 1.15b and 1.15c) having a substantially flat and rectangular top plate **19** and formed with a depending peripheral skirt **20**.

Spaced equally along the underside of the top plate and projecting inwardly from the longer sides of the skirt are three pairs of ribs **21** which serve to secure the vent barrels to the frame. The ribs of each pair are located in opposing relation to one another and each includes an inwardly projecting lower lip **23** spaced downwardly from the top plate and having a concave inner surface, the inner surface of the ribs also being concave.

The gang vent may be made from any of a variety of plastic materials. Functionally, the material should be at least relatively acid-resistant and capable of serving the intended purpose under the temperature and other conditions to which the battery will be exposed.

FIGURE 1.15: EXPLOSION-PROOF VENT BARREL

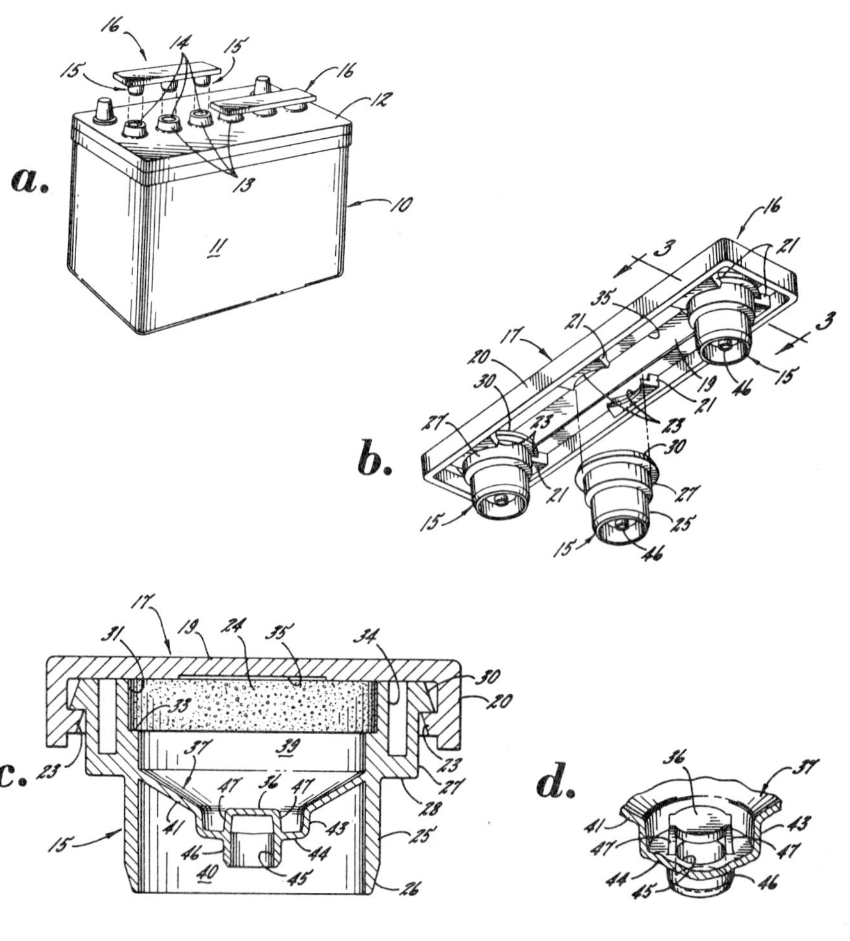

(a) Perspective view of a battery equipped with a gang vent with one of the vents shown in exploded relationship with the battery

(b) Perspective view of one of the gang vents showing one of the vent barrels in exploded relationship with the vent frame

(c) Enlarged cross section taken substantially along the line **3–3** of Figure 1.15b

(d) Fragmentary perspective view primarily showing the splash baffle of the vent barrel illustrated in Figure 1.15c

Source: U.S. Patent 4,002,495

As representative examples, polymers such as high heat polystyrenes, polypropylene and copolymers, polyvinyl chloride, ABS materials and other similar materials may suitably be employed. Each vent barrel **15** is primarily formed as an inexpensive, single-piece molding and is provided with a porous flame arrestor **24** (Figure 1.15c) which may be quickly and easily assembled to the molded part. When attached to the frame **17**, the three barrels provide individual explosion-proof venting for the corresponding cells of the battery **10** and, in addition, the top plate **19** of the frame protects the flame arrestors from grease and dirt while also insuring that each arrestor will remain in assembled relation with its associated molded part.

More specifically, each vent barrel includes a molded cylindrical tubular body (Figure 1.15c) sized to fit or telescope snugly but releasably in one of the vent openings **14**. Its lower end portion is tapered downwardly and inwardly as indicated at **26** in Figure 1.15c to facilitate insertion of the body into the opening. In addition to the primary functional requirements for the gang frame material, the material utilized for the vent barrel should allow formation of a vent barrel having a sufficient degree of flexibility to allow assembly into the frame as is described herein. The materials utilized for the frame are suitable as are various known polyethylene materials.

Formed around the upper end portion of the body **25** is an enlarged collar **27** whose lower end **28** is adapted to seat against the upper side of the vent opening boss **13**. A radially projecting flange **30** is formed around the upper end of the collar and coacts with the ribs **21** and the lips **23** to secure the barrel to the frame. As shown in Figures 1.15b and 1.15c, the flange is sandwiched between the underside of the top plate and the upper sides of one opposing pair of the lips and is embraced by the concave surfaces of the ribs while the concave surfaces of the lips embrace the outer cylindrical surface of the collar.

Assembly of the barrel to the frame is effected by shifting the barrel lengthwise of the top plate to snap the collar between the lips. The ribs and the lips permit the barrels to shift through limited distances both lengthwise and broadwise of the top plate so that the bodies **25** of the barrels may automatically center themselves with respect to the vent opening. Each flame arrestor **24** is in the form of a cylindrical disc (Figure 1.15c) and may be made from various known materials employed for similar purposes. Suitable materials include, for example, polypropylene, ultra-high-molecular-weight polyethylene and fluorocarbon polymers (e.g., Kynar, a polyvinylidene fluorine polymer).

The materials should allow formation of a porous and gas permeable flame arrestor with a relatively uniform pore size distribution and having adequate resilience and toughness to typically allow processing through conventional die-cutting and assembly equipment without the employment of special procedures. The arrestor should have a mean pore size of from about 10 microns or perhaps less to about 100 microns. It has been found that arrestors having mean pore sizes in the range of about 10 microns or perhaps less to about 20 microns or so restrict electrolyte flow which retards electrolyte spilling in the event the battery is inadvertently tipped on its side or inverted for short periods of time.

Preferably, and for normal use, pore sizes ranging from about 40 to 60 microns provide optimum venting features and explosion protection. Utilizing arrestors with pore sizes ranging from about 60 to 100 microns provides the features of venting and explosion retardation. The flame arrestor is assembled to the body simply by telescoping the arrestor into a socket **31** (Figure 1.15c) formed in the upper end portion of the body, the lower end portion of the socket having an annular shoulder **33** which defines a seat for the lower side of the arrestor. While the arrestor could be cemented within the socket, it preferably is sized to fit or telescope into the socket with a relatively tight press-fit so as to simplify assembly of the arrestor. To facilitate press-fitting of the arrestor into the socket, an upwardly opening and circumferentially extending groove **34** is formed in the collar **27**. The groove allows the circumferential wall of the socket to flex radially outwardly to accommodate the arrestor when the latter is pressed into the socket.

When the assembled barrels **15** are attached to the frame **17** in such manner, the top plate **19** closely overlies the arrestors **24** precluding any possibility of the arrestors escaping from the socket **31** and protecting the porous material of the arrestors against becoming clogged by dirt and grease. Accordingly, the arrestors will stay clean throughout the service life of the battery **10** so as to effectively diffuse the gases developed in the associated cells and retard the entrance of a flame into the cells if the battery should be exposed to an internal flame or spark.

In carrying out another aspect of the process, means are provided for maintaining a relatively small spatial separation between an underside surface of the top plate and at least a portion of the top surface of the arrestor so as to provide a positive vent for the gases which diffuse through the arrestors, while effectively retarding the entrance of the flame both into the battery cells and the space between the top plate and the arrestor. In the example illustrated in these figures a shallow longitudinally extending channel or groove **35** is formed in the underside of the top plate and extends across each of the longitudinally spaced arrestors.

The depth of the channel preferably should be sufficiently small that in the event escaping gases should become ignited a flame would not be sustained in the area of the arrestor. It has been determined that the depth of the groove should not exceed 0.015 inch, and a depth of about 0.005 inch has been found to be a desirable spatial separation both for purposes of positive venting of gases through the arrestor and flame control.

In normal use or recharging of the battery, small quantities of hydrogen and oxygen gases being liberated from the battery may accumulate under the top of the frame. If these gases should become ignited, a small but harmless explosion may occur under the vent frame and in the vicinity of the opening defined by the groove. This explosion with its resultant shock waves tends to momentarily reverse the flow of gases through the groove and arrestor, but only long enough to prevent reignition of gases. Such small explosions, therefore, additionally enhance the explosion-proof characteristics of the vent by further retarding the entrance of a flame into the groove area.

If the gassing rate of the battery should reach an unusually high level, such as might be experienced if there were a defective voltage regulator or the battery were being exposed to a faster than normal charge, it is possible that the exiting gases could sustain continuous burning. In the case of such excessive gas discharge, the relatively small orifice defined by the groove acts as a ribbon burner and effectively directs the flame away from the arrestor and barrel so as to prevent the entrance of the flame into the space between the arrestor and frame, thereby tending to protect the arrestor from early damage. The interior of the body **25** is formed with a baffle **36** (Figures 1.15a and 1.15d) for preventing liquid electrolyte from the cell from splashing upwardly against the lower side of the porous arrestor **24**. In addition, provision is preferably made of a partition or funnel **37** having a generally downwardly sloping surface which enables upwardly splashing electrolyte to readily drain back downwardly into the cell.

In this instance, the funnel divides the interior of the body into upper and lower compartments **39** and **40** (Figure 1.15c) and includes a downwardly tapered frustoconical wall **41** molded integrally with the body. Molded integrally with the lower margin of the wall is a substantially upright wall **43** whose lower margin is integral with a generally horizontal bottom wall **44**. A hole **45** is formed vertically through the bottom wall and is located at the upper end of a tubular protrusion **46** which depends from and is integral with the underside of the bottom wall. Gas from the cell is vented through the protrusion and the hole and into the upper compartment for subsequent diffusion through the porous arrestor **24**. The protrusion acts as a deflector to reduce upwardly splashing electrolyte through the hole.

The splash baffle **36** is spaced upwardly from and overlies the hole to deflect upwardly escaping electrolyte back downwardly into the cell while still venting gas from the cell and into the upper compartment. As shown in Figures 1.15c and 1.15d, the illustrative baffle is in the form of a circular disc having a diameter approximately the same as that of the hole. Two narrow webs **47** are molded integrally with the baffle and the upper side of the bottom wall **44** and are spaced diametrically from one another around the hole. The webs support the baffle above the hole but allow gas to flow through the hole and into the upper compartment.

From the foregoing, it will be apparent that the splash baffle helps protect the porous flame arrestor from becoming plugged with electrolyte. In addition, materials from within the battery cell, such as oxide particles and any other solid particles, are prevented from contacting and possibly plugging or obstructing the vent pores. Any electrolyte which does sputter upwardly past the baffle is readily drained back into the cell by virtue of the downwardly tapering wall **41** of the funnel **37**.

Since the lower compartment **40** is defined by the open ended tubular end of the body **25** it will be seen that the end of the body can be sufficiently long for securing mounting within the cell opening, and by reason of its flexibility, will reliably seal the opening so that all escaping gases will pass through the arrestor. Furthermore, the funnel partition **37** can also be located relatively high in the barrel so as to minimize the amount of electrolyte that enters the upper compartment **39**.

Vent Plug

F.A. Vengrofski; U.S. Patent 4,002,494; January 11, 1977; assigned to ESB Incorporated describes a storage battery vent plug for eliminating the formation of current conducting electrolyte paths originating at the interior of a cell and terminating at a point of different electrical potential on the surface of the cell. The vent plug includes a barrel portion having means for securing the vent plug in the annular filler neck of a battery and a sealing portion for providing a seal between the vent plug and the filler neck of a battery. The sealing portion includes an annular skirt which is constructed and arranged to depend from the vent plug and concentrically surround the filler neck of a battery with the inner surface thereof being spaced from the outer surface of the filler neck.

Referring to Figure 1.16, **10** designates a storage or secondary cell container cover which is conventional and **11** designates the normal electrolyte of the cell. The cover includes an upstanding annular vent or filler neck **12** which is adapted to accommodate a vent plug shown generally at **14.**

FIGURE 1.16: VENT PLUG

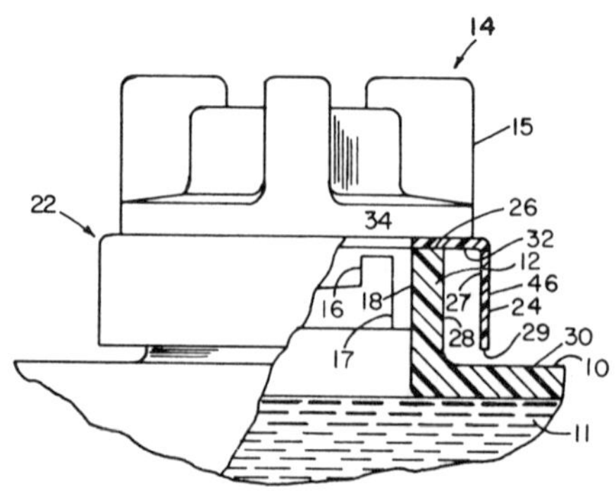

Source: U.S. Patent 4,002,494

The vent plug has a hooded portion **15** and a suitable bayonet type lug **16** integrally formed on barrel portion **17** thereof which is adapted to mate with a complementary bayonet type lug integrally formed on the inside surface **18** of the filler neck. The vent plug may, of course, be provided with a screw-threaded barrel which would screw into the filler neck which, of course, would then be provided with complementary receiving threads. The vent plug is provided with suitable baffling to prevent the escape of electrolyte while permitting the escape of gases through a vent hole. Both of these features are provided in the hooded portion, but neither has been illustrated.

In accordance with the process, an inverted cup shaped member or gasket, shown generally at **22**, is provided over the filler neck. The gasket has an annular skirt **24** which depends below the top surface **26** of the filler neck and concentrically surrounds the filler neck. The inside vertical surface **27** of skirt **22** is spaced from vertical surface **28** of the neck. The lower edge **29** of the skirt is spaced from the top surface **30** of the cell cover **10.** The gasket also includes an inwardly flanged portion **32** from which skirt **24** depends. Flange portion **32** is positioned on the top surface **26** of the filler neck and concentrically surrounds the barrel portion **16** of vent plug **14.**

In practice, the gasket **22** is fitted onto barrel portion **17** of vent plug **14** via the opening **34** provided in the gasket and the entire assembly is then secured to the vent neck **12** via the mating bayonet lugs. It should be understood that while the vent plug **14** and cup-shaped member **22** have been illustrated in Figure 1.16 as two distinct parts, they may be formed together as a unitary structure or as integral members, as for example, by suitable injection molding techniques.

Any material that is acid resistant, flexible or pliable enough to serve as seal between the vent plug and the vent neck and that does not wet readily with acid is suitable for fabricating the vent plug or cup shaped gasket **22** of the process, e.g., Teflon, polypropylene and polyethylene. Polyethylene is preferred. It has been found that use of the gasket eliminates the problems of electrolyte leakage from the interior of the battery and it has been found that electrolyte creepage does not occur on the inside vertical surface **27** or outside vertical surface **46** of annular skirt **24.** Thus, the vent plug or gasket eliminates the problems caused by the formation of current conducting electrolyte paths originating at the interior of cells and terminating at a point or points of different electrical potential on the exterior of the cell.

Multichamber Device

J. Auerbach; U.S. Patent 3,944,437; March 16, 1976 describes an explosion-proof venting device for electrical storage batteries. The device comprises a multichamber plastic receptacle which may contain catalytic material or a gas drying filter, or both of them, a disc-shaped porous flame barrier having a central aperture, a hollow plunger which serves also as a filling funnel and which may also contain an indicator device, and a cover shielding the flame barrier as well as the funnel. Acid fumes passing through this device will be made to shed their liquid contents before reaching the atmosphere by means of bringing them into contact with relatively large surfaces of the device, and also by changing the direction of gas flow several times.

According to one example, the venting device comprises a tube-like sleeve in which is fitted a hollow plunger whose bottom is completely immersed in the electrolyte at all times, just above the separators of the battery, while the inside of its top end is funnel shaped. Molded to the outside of the plunger are a number of discs and vanes to guide the escaping gases as required. The plunger may be bonded to the inside of the surrounding sleeve. Maintenance operations can be carried out through this plunger assembly in the usual manner.

Porous Ceramic Material

A process described by *J.C. Hipp; U.S. Patent 3,943,009; March 9, 1976; assigned to Globe-Union Inc.* relates to a ceramic body which is placed in the filling cap or the cover of a lead acid storage battery and serves to permit the cell gases to vent through the ceramic stone while at the same time effectively resisting any ingress of flame caused by the ignition of cell vent gases. The ceramic body is designed to effectively retard flame propagation while preventing the accumulation of large volumes of gases in the battery.

This is accomplished by the fired ceramic body which is in the form of fused homogeneous particulate matter composed of a major portion of a high melting point inorganic material having a particle size no larger than about 600 microns.

The inorganic material contains essentially aluminum and silicon oxides. A minor portion of a lower melting point glass bonding agent is employed to bind the high melting point inorganic material after it is fired. The use of the glass with the higher melting point inorganic material is critical in that the glass must wet and bond the grains of aluminum and silicon oxides without pulling them into solution and changing the particle distribution.

The fired body has an open pore porosity of about 25 to 50%. This is accomplished by providing a selected range of particle sizes for the inorganic material. In a preferred composition, the inorganic material is present in an amount of about 80 to 95% by weight and the glass bonding agent composes about 20 to 5% by weight. An optimum range of particle sizes found to achieve the desired back pressure of cell vent gases and provide a flame barrier resides in 45% by weight of the particles screened in the 600 to 400 micron range, 45% in the 400 to 300 micron range and 10% in the 300 to 200 micron range.

The preferred inorganic materials are mixtures of aluminum and silicon oxides and the lower melting point glass is a mixture of clay and feldspar. To achieve constantly controlled compaction of the green ceramic without tool wear, a wax and a lubricant are added to the high melting point inorganic material and the lower melting point glass. The green material is compacted at a compression force in the range of 2 to 3 tons per square inch and fired in a kiln at a temperature of at least 2300°F.

Another important feature of the process is the particle size of the high melting point inorganic material. None of the particles should be larger than 600 microns and a particle size distribution according to the following mesh sizes has been found to work exceedingly well:

Weight Percent	Microns
45	600-400
45	400-300
10	300-200

Blow Molded Flexible Hollow Vent

E.C. Schaumburg; U.S. Patents 3,943,008; March 9, 1976; and 4,010,044; March 1, 1977; both assigned to Standard Oil Company describes a dimensionally flexible hollow vent with vent plugs for a plurality of the filling wells of a multiple cell battery. The vent is blow molded in one piece from a thermoplastic material. Gases resulting from charging of the battery or a pressure change can be piped from the battery by means of a tube or tubes formed at one or more locations on the vent. The tube feature permits the gases to be vented at a point removed from the location of the battery and permits several gang-type vents to be connected in series.

The one-piece thermoplastic resinous hollow gang-type vent with vent plugs for a plurality of the filling wells of multiple-cell batteries comprises: a top member, a skirt peripherally and integrally attached at its upper end to the top member and extending downward from the top member, a bottom member peripherally and integrally attached to the skirt at its lower end, and a first tube extending outwardly from the vent. The bottom member comprises vent plugs of a member and location corresponding to the number and location of a plurality of the multiple-cell battery and of suitable dimensions to fit snugly into the filling wells, to form a fluid-tight fit, each vent plug having a hole in its base to permit passages of gases and electrolyte through the vent plug, and a flat region extending from the periphery of the bottom member to the tops of the vent plugs and integrally attached to both.

The vent is blow molded in one piece by the method of:

(a) either extruding a tubular parison or a pair of film sheets of heated thermoplastic resin and clamping the parison or film sheets while still pliable between mating die members having cooperating mold cavities configured to the external shape of the hollow gang-type vent to be formed, the die members being partable along a parting line of the cavities that proceeds in a single flat plane or in a step-wise fashion around the periphery of the hollow vent to be formed in the area between the top and bottom members of the hollow vent to be formed, or clamping a pair of preheated pliable preformed film sheets of thermoplastic resin between the mating die members;

(b) introducing a blow molding duct into the die members and into the parison or between the film sheets to provide communication internally of the plastic parison or film sheets at a point on the surface of the hollow gang-type vent to be formed, and introducing gas under pressure through the duct while the parison or film sheets are clamped between the closed die members in order to force the wall of the parison or of the pair of film sheets outwardly into contact with the surfaces of the mold cavities until the plastic has cooled sufficiently to rigidify; and then

(c) parting the die members and removing the molded hollow gang-type vent from the die cavities. The one-piece hollow gang-type vent thus requires no assembly work and being joint-free, provides added protection against leakage of battery acid or corrosive fumes.

Microporous, Hydrophobic Filter

J.P. Badger; U.S. Patent 3,904,441; September 9, 1975; assigned to Eltra Corporation describes a venting system for a service-free storage battery which allows escape of generated gases but not of battery liquid. The system includes a strip of microporous, hydrophobic filter media extending across vent apertures in the battery cover and a rigid guard over the filter strip which permits gas to escape through passages between the guard and cover and which protects the filter strip from external damage.

Microporous polypropylene of suitable porosity has been found to be most satisfactory as material comprising the filter strip. Because automotive batteries are commonly subjected to high rate chargers in the field at service stations, and controls to prevent overcharging are usually nonexistent, the filter strip must have sufficient surface area to be able to pass the substantial amount of gas generated under such high charging conditions. For example, a lead-acid battery at 25°C and at atmospheric pressure, will evolve gas at the rate of about 312 cubic centimeters per cell per minute, when overcharged at 30 amperes. Thus, the system must be capable of passing this amount of gas, preferably with no more than one-half psi pressure differential across the vent. It has been determined that nonwoven polypropylene filter material with an effective average pore diameter of 4.5 to 5.0 microns, a thickness of 0.050 inches or less and a total porosity of 60% or more is capable of this capacity when the total exposed filter area is 0.44 square inch per cell. At the same time, the filter material retains its ability to prevent the passage of liquid. In addition, the hydrophobic nature of this material will also prevent the material from being flooded by splashed liquid which could deleteriously block off its gas filtering capacity.

Gas Permeable Diaphragm

S. Mermelstein; U.S. Patent 3,909,302; September 30, 1975; assigned to Tyco Laboratories, Inc. describes a vent cap for a liquid acid electrolyte storage cell. The vent cap comprises a diaphragm or membrane that is permeable to oxygen and hydrogen gas and substantially impermeable to the liquid electrolyte. Means are provided for fixedly securing the diaphragm in the vent cap assembly.

In a preferred case, the membrane is made of a microporous polytetrafluoroethylene sheet having an average pore size in the range of from about 2 to 10 microns (preferably from about 5 to 10 microns), and a thickness in the range of 0.005 to 0.030 inch. Such a membrane has been found to permit relatively rapid venting of oxygen and hydrogen gas, and to be highly resistant to attack by sulfuric acid electrolyte. Additionally, and depending upon pore size, it is substantially impermeable or only slightly permeable to sulfuric acid electrolyte in the concentrations normally employed in lead-acid batteries. By way of example, the membrane may comprise a 0.005 inch thick sheet of Zitex (polytetrafluoroethylene). Such material has a pore size in the range of about 5 to 10 microns.

Rotatable Disc Valve

According to a process described by *J.E. Bell and R.P. Laczko; U.S. Patent 3,892,595; July 1, 1975; assigned to ESB Incorporated* a multicell lead acid storage battery is fitted with a one piece cover. A rotatable disc valve is mounted on the cover, the valve being of such a size that it covers a portion of all cells. A filling well and a vent opening for each cell are located under the disc valve. In a first angular position of the valve, the filling wells are exposed and the vent openings are sealed, and in a second position, the wells are closed and the vents are opened.

When the wells are exposed, the passages into the cells are clear, expose the top of the cell elements and permit the insertion of a hydrometer into the cell. In order that the top of each element is exposed, the locations of the center cell filling wells and vent openings are nearer the axis of the valve than the openings in the two end cells. Rotation of the valve may expose a legend relating to the valve position.

Flame Barrier Cap

R.R. Melone; U.S. Patent 3,907,605; September 23, 1975; assigned to Illinois Tool Works Inc. describes a flame barrier vent filter for a battery cap. The filter has an internal domed-shape configuration which makes the cross-sectional area in the center of the filter thinner than that of the edges so as to tend to confine any flame that results from ignition of vented gases to the center of the filter. The filter is made of a material having a low heat conductivity such as polyvinyl fluoride which resists melting of the cap if a flame does break out. The filter may be locked into place by an annular bead formed in the cap, or it may be spun weld to the cap by applying torque and pressure to the filter. In order to form a seal between the filter and the cap on the spin welded embodiment, a bead, preferably of a triangular cross-sectional shape, is formed on the cap so as to melt when it engages the spinning filter.

J.B. Godshalk; U.S. Patent 3,992,226; November 16, 1976; assigned to Ultra-Mold Corporation describes an inexpensive antiexplosion, antiflame cap for openings in storage battery casings. The cap is formed, as by injection molding, from polymeric material in such fashion as to provide at least one critically dimensioned and located vent orifice effective to prevent establishment of a persistent flame at the orifice when hydrogen and oxygen are evolved during charging of the battery, and further effective to prevent propagation of flame through the orifice into the interior of the caps. Caps, according to the process also include internal mist chambers so arranged that liquid electrolyte is prevented from reaching the vent orifice or orifices and does not "pump through" the cap as a result of vibration of the storage battery in use.

Visual Observation for Electrolyte Control

According to a process described by *A. Sakamoto; U.S. Patent 3,895,964; July 22, 1975; assigned to Daiden Co., Inc.* each of the battery caps comprises a cylindrical body having its lower end immersed in electrolyte and which contains a colored spherical float having a specific gravity of about 1.2 which indicates the level or specific gravity of the electrolyte due to reflection of the color of the float from the bottom of the float chamber. According to the process, the measurement of the electrolyte may be carried out in a short time by simply looking through the peephole and examining the color of the bottom end of the cylindrical body without removing the cap of the storage battery. Accordingly, the desired condition of electrolyte can be determined in dim light or in a storage battery of small size. Additionally, it is entirely practical to place the measuring device within the storage battery because the device has no effect on the storage battery.

Splash-Proof Device

A process described by *I. Sano; U.S. Patent 4,031,294; June 21, 1977; assigned to The Furukawa Battery Co., Ltd., Japan* relates to a splash-proof device for use in a storage battery provided integrally with a cell cover for purposes of preventing a liquid spray from being leaked from a cell. The splash-proof device is constructed such that a helical splash-proof baffle is provided inside a cylindrical member formed integrally with a cover of the cell. In this splash-proof device, the cylindrical member and helical splash-proof baffle are conveniently formed by integral molding. It is also possible, however, to insert and fix a separately made helical splash-proof baffle within the cylindrical member.

In the case where the splash-proof device is formed by molding, the cylindrical member and helical splash-proof baffle should be designed in thickness so as to become smaller in a direction in which a mold is withdrawn, and in addition the splash-proof plate should be so designated in thickness as to become smaller also toward the center of the cylindrical member. The splash-proof device may be made of a conventional chemical resistant and thermoplastic resin such as polyvinyl chloride, polypropylene or polyethylene. The splash-proof device is made integral with the cell cover to which a cell container or case is heat sealed or solvent sealed during manufacture of the storage battery.

TERMINAL CONNECTORS

Truncated Cone Shape

A process described by *V.Y. Norman; U.S. Patent 4,033,664; July 5, 1977* pertains to a battery terminal connector adapted to manually connect a battery cable to the rod-shaped terminal popularly found on vehicular storage batteries. The connector comprises a metallic housing to which the cable is electrically connected, a longitudinal hole passing through the housing, having a gradually diminishing diameter tapered hole at one end.

A hollow rod-like conducting resilient material having a truncated cone-like shape whose apex is adjacent the cylindrical end has two or more longitudinal notches located in the truncated conical surface, permitting the walls to be drawn inwardly radially as the conical exterior surfaces are drawn into the tapered hole of the longitudinal opening in the housing. The battery post or terminal is grasped by the opening at the base of the conical section. A spring and bolt assembly causes the cone-shaped grasping insert to be urged into the tapered hole.

Figure 1.17a shows a bolt **1** adapted with a push button-like cap **2** and external threads **21** on the free end of the cylindrical shaft thereof. The helical spring **4** has a longitudinal hole which permits the insertion of the cylindrical body of bolt **1** therein. Retaining cup **5** is shown with an external flange-like lip **7** at one end. The other end is adapted with an internally directed lip **6** surrounding an opening through which the cylindrical body of the bolt may pass. Metallic housing **9** is fitted to connect electrically to one end of a battery cable **8** and has an opening **11** through which the retaining cup may pass. Flange-like lip **7** of the retaining cup is designed to rest on the lateral surface surrounding opening **11**.

Wings **10** are designed to be grasped by the four fingers of the user permitting the thumb to exert an inward force upon the top end of the push button-like cap of the bolt. A tapered hole **12** at the other end of the housing is adapted to receive the gripping element **13**. Internal threads **14** threadingly engage threads **21** on the bolt. The widest opening **15** at the base of the conical end of the gripping element is interrupted by a number of notches **17**. An inwardly turned lip **16** is located at edge **15** corresponding to the base of the conical section whose slope matches the taper at opening **12**.

Figure 1.17b illustrates the housing **9** and a partial length of an electrical cable **8** protruding therefrom. Retaining cup **5** has been inserted into the opening as shown in Figure 1.17a. The flange-like lip **7** is shown resting on the uppermost lateral surface of the housing.

FIGURE 1.17: BATTERY TERMINAL CONNECTOR

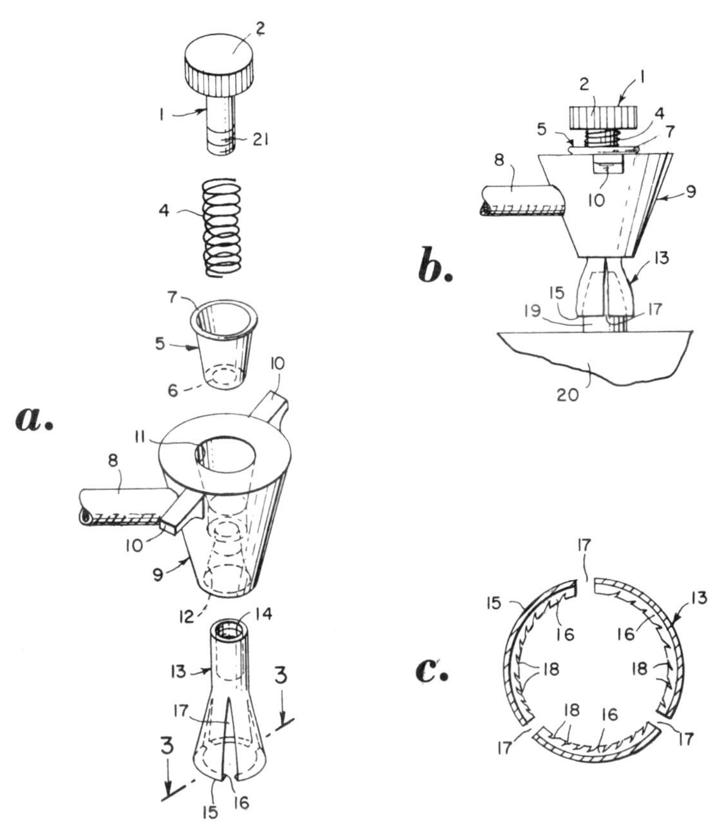

(a) Composite illustration of the perspective views of the battery terminal connector showing a retaining bolt, helical spring, retaining cup, housing attached to a partial length of a battery cable, and a battery gripping element

(b) Side elevation view of an assembled battery terminal connector electrically connected to a battery terminal shown on a portion of the uppermost lateral surface of a battery

(c) Cross-sectional view taken along line **3–3** illustrating the interior serrated lip at the base of the conical end of the battery terminal connector gripping element as shown in Figure 1.17a

Source: U.S. Patent 4,033,664

Spring **4** is illustrated surrounding the exposed shank of bolt **1** with its uppermost end resting on the shoulder of push button-like cap **2**. Gripping element

13 protrudes outwardly from the lowermost lateral surface of the housing **9**. A fragmented section of a battery **20** is shown with its post **19** inserted into the open end of gripping element **13**. Figure 1.17c illustrates the cross-sectional view taken along line **3–3** viewed in the direction of arrows **3–3** of the finger gripping element as shown in Figure 1.17a. Figure 1.17c illustrates notches **17** interrupting edge **15** upon which inwardly turned lip **16** is located. Serrations **18** are formed on the innermost surface of the lip and are utilized to facilitate good electrical connection to battery post **19** as shown in Figure 1.17b.

One of the advantages is a battery terminal connector which does not require tools to effectively electrically connect or disconnect a battery cable to a battery post. Another advantage is a battery terminal connector with increased inwardly directed contacting forces as the housing thereof is forced to move towards the battery housing. A further advantage is a battery terminal connector which enables the battery cable to be disposed above and away from the uppermost surface of the battery housing precluding thereby corrosion attacking the battery end of the battery cable.

Split Connector Cap

According to a process described by *R.R. Wening; U.S. Patent 4,012,104; March 15, 1977* a storage battery terminal post receives a split connection cap formed of compatible metal. A rear extension of the cap is secured by a clamping set screw to an exposed end of a battery cable. A screw-threaded pressure sleeve is received by internal threads of a housing and has its forward end bearing on a full circle flange portion of the connector cap. The force exerted by the pressure sleeve on such flange produces closing of the split connector cap around the battery terminal post by a camming action of curved connector cap sections with a coacting curved wall of the housing.

Sealed Through Connector

A process described by *L.K.W. Ching, Jr. and C.E. Coleman; U.S. Patent 3,964,934; June 22, 1976* relates to electrolytic devices, such as conventional electrochemical cells and batteries, and particularly to a sealed through-the-partition connector for conducting electrical current. The sealed through connector includes a connector post insert positioned within a similarly configured seat of an electrically nonconductive resilient partition and having a through bore. Internal aspects of the seat define a shoulder portion interposed between a reduced neck portion and an inclined ramp portion.

The connector post insert comprises a stud of electrically conductive material, at least the portion of the stud opposite the ramp portion of the seat being encased by a deformable lead or lead alloy shroud of higher electrical resistivity than the stud. Prior to engagement of the connector, the shroud has an outer dimension which is greater than the inner dimension of the seat so that upon subsequent engagement an interference fit is formed between the inclined surface of the shroud and the inclined ramp portion of the seat to form a liquid-tight seal.

In another connector design, the surface position of the lead shroud and the internal surfaces of the seat need not be inclined. In addition to the stud having an upper portion extending through the bore and outside the partition, and a body portion connected to this upper portion and housed within the seat, there is also provided extension means, such as a leg, protruding from the body portion, both the body portion and extension portions encased within the lead shroud to provide a very low impedance connector capable of withstanding extremely high discharge currents for limited periods of time. The sealed through connector may be employed in various electrolytic devices for conveying current across an electrically nonconductive partition where a liquid-tight seal is required. The process has particular benefit in electrochemical cells and batteries as an electrode-to-terminal connector, or alternatively as an intercell connector between partitioned cells of a battery, particularly where a compact, relatively lightweight connector is desirable.

Quick Disconnect System

A. Cella; U.S. Patent 4,042,759; August 16, 1977 describes a quick disconnect system for removably connecting an electrical cable to a battery terminal post. The system comprises either a pin which is engaged by an angulated slot in a spring loaded connector which slot includes a notch for locking the connector onto the pin, the connector being quickly disengageable from the terminal post by exerting a force opposite to the spring biasing force while sliding the connector off the pin; or a threaded type in which a threaded sleeve is secured to an existing battery terminal post which interfaces with a threaded connector for locking thereon by 90 degree rotation and unlocking by 90 degree rotation in the opposite direction; or a quick disconnect comprising a snap ring secured to a groove of an existing terminal post wherein a grooved resilient cover is snap-fit onto the split type snap ring secured to the existing terminal post.

Each of the arrangements provides an electrical connection between the battery cable and the terminal post with the spring providing the electrical interconnection in the first instance, either conductive threads or the protrusion of the existing terminal post through the sleeve providing the electrical interconnection in the second instance, and with a conductive inner lining for the snap fittable cover providing the electrical interconnection between the battery cable and the terminal post in the third instance. Each of the above arrangements may readily convert an existing battery terminal post into a system in which the battery cable may be quickly connected to and disconnected from the terminal post.

A process described by *C.P. Santo; U.S. Patent 3,989,544; November 2, 1976* relates to a lead acid storage battery having a case and a plurality of cells with each cell having positive and negative plates. A first cell connector strap connects negative plates of one cell to the positive plates of an adjacent cell successively and terminates in a positive terminal connector upon the interior of the case. A second cell connector strap connects the positive plates of one cell with the negative plates of an adjacent cell successively and terminates in a negative terminal connector upon the interior of the case which has an apertured closure base.

A platform is mountable upon a vehicle or other support and includes a pair of projecting terminal connectors adapted to be received and projected up into the case for respective connection with the positive and negative terminal connectors respectively. Mounted upon the platform is a connector adapted for connection to a ground and to a powerline respectively. A pair of conductors respectively interconnect the platform terminal connectors and the connector. The terminal connectors are so arranged and spaced relative to the battery connectors that upon mounting of the battery upon the platform, the platform connectors project up into and register within the battery connectors.

HANDLES

Integrally Molded Handles

According to a process described by *D.D. Hakarine; U.S. Patent 4,042,762; August 16, 1977; assigned to Gould Inc.* straps are molded integrally with and hinged to opposite sides of a plastic battery case and may be swung upwardly to form a handle for enabling easy carrying of the battery. In one form, the straps are formed with hand holes and are grabbed individually when the battery is carried.

In Figure 1.18a, there is shown an electrical storage battery **10** of the type commonly used in automobiles. The battery includes a box-like case **11** and is defined by a container **13** having a generally rectangular shape and by an overlying cover **14** which is sealed to the container. According to the process straps **15** are molded integrally with opposite upright wall **16** of the case **11** and may be hinged upwardly to form a handle which facilitates carrying of the battery. By virtue of the straps being molded integrally with the case, the battery can be provided with a handle at very little cost because the handle is formed as an incident to formation of the case and need not be attached to the case in a separate assembly step.

In the example of the process shown in Figures 1.18a through 1.18c each strap **15** is molded integrally with the container **13** and is located just below the junction of the container and the cover **14**. As initially molded, each strap projects horizontally from the container as shown in Figures 1.18a and 1.18c. With the straps being molded in this position, the design of the molding dies is simplified and the dies may be easily closed and separated. A hinge **17** is located adjacent the junction of the inner end of each strap and the associated wall **16** of the container in order to permit the straps to swing upwardly and downwardly from their horizontal positions.

Each hinge is a so-called "living" hinge and is formed during the molding operation by thinning or grooving opposite walls of the strap along lines extending transversely of the strap at the inner end portion thereof. The hinges permit the straps to be swung upwardly to substantially vertical carrying positions (see Figure 1.18b) in which the straps may be grabbed to enable easy lifting and carrying of the battery. To facilitate such lifting and carrying, a hand hole **19** is formed through the outer end portion of each strap and is sufficiently large to receive the fingers of an adult male.

The hinges also permit the straps to be swung downwardly alongside the walls of the container to an out of the way storage position as shown in phantom in Figure 1.18c. To hold the straps in the storage position, a retainer in the form of a small opening **20** is formed through the outer end portion of each strap and is adapted to interfit releasably with a coating retainer in the form of a lug **21** which is molded integrally with and projects outwardly from the associated side of the container.

FIGURE 1.18: PLASTIC HANDLE

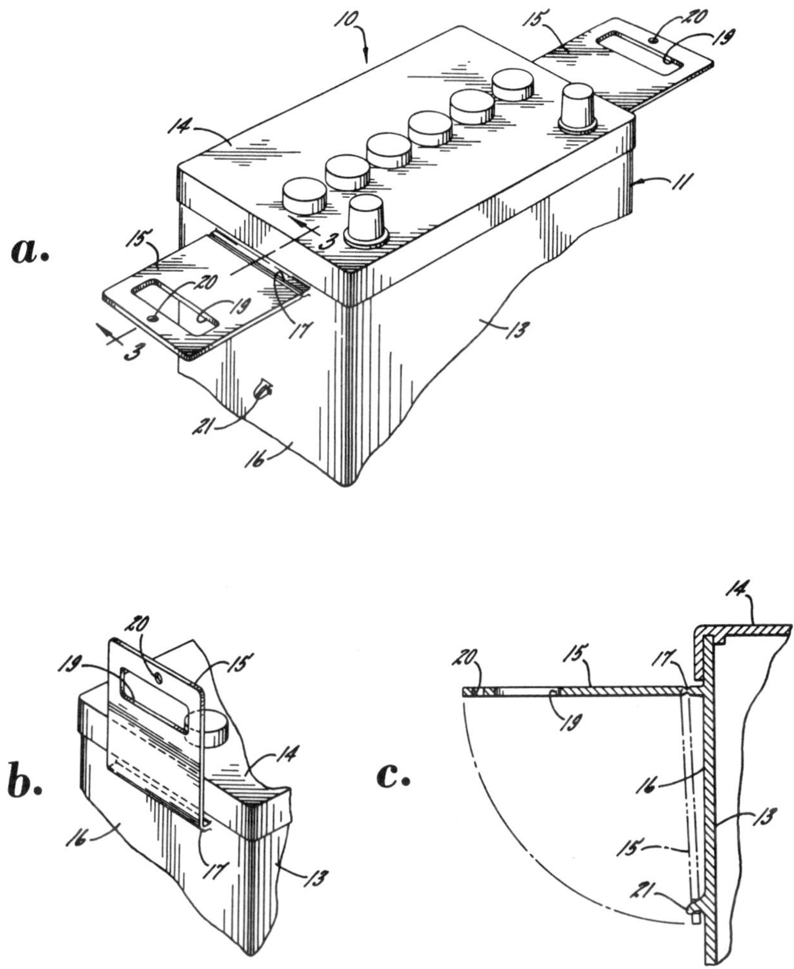

(a) Fragmentary perspective view of a battery equipped with a handle

(b) Fragmentary perspective view of the battery and showing one of the handle straps in its carry position

(c) Enlarged fragmentary cross-section taken substantially along the line **3–3** of Figure 1.18a

Source: U.S. Patent 4,042,762

The box-like case and the integrally molded straps are formed from conventionally known polypropylene polymers and copolymers of propylene with other monomers such as, for example, ethylene. Suitable molecular weights and other characteristics required to provide the straps with the necessary hinge feature are well known. The cover may be formed from the same material as in the case or from any of the other materials known in the art.

C.-C.F. Lee; U.S. Patent 4,029,248; June 14, 1977; assigned to Gould Inc. describes a detachable plastic carrying handle having a noncircular aperture at each end which is aligned and snapped over the noncircular head of a post integrally formed on each end of a plastic battery casing. The handle is then rotated upward to cause nonalignment of the aperture and head.

Rope Holder

According to a process described by *N.G. Grabb; U.S. Patent 4,013,819; Mar. 22, 1977; assigned to Varta Batteries Limited, Canada* a rope holder for a battery handle rope is integral with a side of the battery case. The holder has two separate laterally spaced apart sets of first and second passageways, the bottom opening of each passageway being unobstructed. Each passageway receives and permits passage therethrough of a length of battery handle rope having a rigid elongate lug on each end.

The area beneath the bottom opening of each passageway is sufficiently unobstructed to permit threading of the battery handle rope through the first and second passageways. The rigid elongate lug of the rope is secured in the upper portion of the second passageway. Upon failure of the means for securing the lug, the lug is frictionally bound in the lower portion of the second passageway because the cross-sectional area of the second passageway is less than the longitudinal length of the rigid lug. The battery handle rope is retained in the battery holder even with failure of the means for securing the lug.

The rope holder for the battery handle rope may be integrally molded with the side of a battery case, such as the hard rubber and other types of plastic battery cases. If, while the battery is used in the field, the battery handle rope is severed, a new battery rope may be inserted in the holder without requiring special tools. The replacement rope may be easily threaded through the passageways of the rope holder and secured by an appropriate locking means.

Handle Design

C. Groby, C.-G. Crafoord and J. Grieves; U.S. Patent 3,910,800; October 7, 1975; assigned to Aktiebolaget Tudor, Sweden describe a carrying device that is formed of a subassembly composed of an intermediate member having a flat surface that can be welded or otherwise attached to the battery casing and one or more upwardly extending legs having hinges for joining to the handles. The handles are preferably formed with a flat upper surface to minimize the space requirements and to have a size and shape that approximately conform with the cover member for the battery.

A central hand engaging portion may be provided so that one or two persons can carry the battery. The entire subassembly may be fabricated as a unit and added to batteries as desired after fabrication of the battery is completed. A further optional feature is to provide a bar which carries the filler caps that is hingedly mounted on the intermediate member. By means of the process, it is possible to carry batteries more easily and conveniently than before since the handle subassembly can be welded to batteries as an integral part of the battery as delivered to the user. Other advantages of the device are that several functions are brought together in one unit, e.g., a carrying handle which can be folded flat over the casing cover member and contain caps which have ventilation openings and are removable for electrolyte level inspection and for permitting the addition of electrolyte as needed.

HYDROGEN-OXYGEN RECOMBINATION TECHNIQUES

Sealed Battery Using Gas Phase Reaction Catalyst

A process described by *S. Sekido, S. Fukuda, M. Matsumoto and K. Murakami; U.S. Patent 3,904,434; September 9, 1975; assigned to Matsushita Electric Industrial Co., Ltd., Japan* relates to an improvement in a sealed lead storage battery and it provides a sealed lead storage battery where excess hydrogen gas generated when the battery is fresh and excess oxygen gas generated when the battery becomes old are avoided by automatic change of discharge depth at the negative electrode. A slight amount of hydrogen generated from the negative electrode is allowed to vanish together with oxygen on a gas phase reaction catalyst.

The remaining oxygen is allowed to vanish by electrochemical reaction and gas phase reaction at the negative electrode. If oxygen evolution from the positive electrode decreases by the corrosion of positive grid material, the amount of oxygen which is absorbed by the negative electrode will decrease and consequently the discharge depth at the negative electrode will be decreased. Therefore, the excess hydrogen gas is avoided.

When hydrogen is excessive, the following can be considered as simple methods for removing the excessiveness, namely:

(a) Discharge is allowed to automatically occur at the positive electrode in accordance with the excess hydrogen;
(b) Charging is allowed to automatically occur in accordance with the excess hydrogen; and
(c) Oxygen is automatically taken in from outside in accordance with the excess hydrogen, etc.

However, the method (a) is possible only when the charging current is small as in a floating charging since reaction speed is low. According to the method (b), the battery must be assembled in such a manner that discharging at the negative electrode is deeper than that at the positive electrode. In this case, since even when the positive electrode is in the complete charging state, the negative electrode is in the partial discharging state, there is the possibility that oxygen which is generated in a large amount when the positive electrode is in the complete charging state cannot be absorbed by the negative electrode which is in the partial discharging state.

According to this process, it has been found that even when the capacity of the negative electrode is made greater than that of the positive electrode by about 40% and a partial discharge corresponding to the excess 40% is carried out at the negative electrode, the negative electrode can absorb oxygen generated in charging of about 0.05 C with use of silica gel from which the electrolyte does not drop. Even in the higher charging rate, the absorption of oxygen is possible, for example, by connecting an auxiliary electrode comprising water-repellent treated active carbon to the negative electrode through a nonlinear element which biases the voltage applied to the auxiliary electrode to 0.6 to 1.5 V.

In the case of a lead storage battery, generation of hydrogen cannot be avoided even when the negative electrode is in the partial discharge state. Of course, it is necessary that the oxygen absorption rate of the negative electrode is high, but when the gas phase reaction rate is lower than the oxygen absorption rate, hydrogen generated from the negative electrode remains. Therefore, for avoiding the excessiveness of hydrogen, the following requirement must be satisfied, namely, (gas phase reaction rate)>(oxygen vanishing rate at the negative electrode)>(gas generating rate from the positive electrode).

As to the method (c) on the preceding page, the process utilizes an artificial graphite carrier to a part of which is added Pd black and to the other part of which is not added Pd black and two parts which are piled on and pressed into a mold. The artificial graphite carrier is mounted to a container by piercing there-through in such a manner that the part to which no Pd black is added is outside the container, whereby if the two parts are connected by a liquid junction, the inner part to which Pd black is added acts as a hydrogen electrode and the outer part to which no Pd black is added acts as an oxygen electrode whereby oxygen in the air is automatically supplied to overcome the excessiveness of hydrogen. However, since according to this method the oxygen is supplied from the air, it is impossible to use the battery outside the air or in water.

Next, when oxygen is in excess, the following are simple methods for overcoming the excessiveness, namely, (a) charging is allowed to automatically occur at the positive electrode depending upon excess oxygen and (b) discharging is allowed to automatically occur at the negative electrode depending upon excess oxygen. Both of the methods (a) and (b) can be employed, but it is naturally desired to use the method together with the method for removing the excess of hydrogen.

According to the process, the discharge capacity of the negative electrode is made larger than that of the positive electrode and a battery is constructed in such a state that discharge at the negative electrode is made greater by the degree corresponding to the excess of capacity than that at the positive electrode. After the excess liquid has been removed, the electrolyte is filled and a gas phase reaction catalyst is provided in the space within the battery whereby oxygen gas generated from the positive electrode is absorbed in the negative electrode and at the same time hydrogen gas generated from negative electrode is reacted with oxygen gas on the gas phase reaction catalyst to revert them to water. Especially, when the amount of oxygen gas generated is reduced due to corrosion of the plate grid or oxidation of active materials, excess of hydrogen gas is avoided by charging the negative active material in the discharge state.

Foraminous Catalytic Device

According to a process described by *E.L. Kreidl and D. Shooter; U.S. Patent 3,929,422; December 30, 1975; assigned to Koehler Manufacturing Company* hydrogen and oxygen gases are safely recombined in the presence of a catalytic device which operates in two stages in a secondary battery system. Recombination continues in a controlled manner while the battery is on discharge and charge for periods corresponding to the normal life of a conventional secondary battery.

In sealed secondary battery systems hydrogen and oxygen gases evolved during operation of the battery are releasably contained through a range of pressures during which long-term stoichiometric recombination of the gases takes place.

The catalytic device in a preferred form comprises a foraminous body which includes exposed catalytic surfaces of relatively low gas recombining capacity and an inner nucleus of enclosed catalytic material. The nucleus comprises enclosed catalytic surfaces on which gases passing through the foraminous body react initially within a thermally favorable environment whereby cumulative heating can take place.

A resulting buildup of heat at the nucleus is released to raise the temperature of the exposed catalytic surfaces to temperatures at which gases coming into contact with the exposed catalytic surfaces will start to recombine thereon, and continue to react in a sustained manner within a limited temperature range. When, as may occur, the exposed catalyst surfaces are subjected to increasing gas pressure in the battery they become self-limiting in the extent to which their temperature may be raised by exothermic reaction.

Thus the rate of recombination of gases at catalyst surfaces is controlled and limited to positively prevent the exposed catalyst surfaces from reaching temperatures at which an explosion may take place, i.e., temperatures found to be within upper limits of from about 400° to about 600°C, and the range of control may be extended downwardly to values as low as 250°C for some battery operations. Water resulting from the controlled recombination at the catalyst surface is returned to the electrolyte with catalyst surfaces being maintained in a constantly reactive condition.

Catalytic Device

A process described by *H. Nitta and M. Watanabe; U.S. Patent 4,002,496; January 11, 1977; assigned to Japan Storage Battery Company Ltd., Japan* provides a catalytic device which is not affected by overcharging current even when the overcharging current becomes large. The process aims to control the maximum level of gas reaction by introducing a rate controlling the gas reaction so that the temperature will not be increased due to occurrence of abnormal generation in the gas reaction process. The equilibrium of the gas generation and the gas reaction can be expressed as follows:

Quantity of Gas Generated = Quantity of Reacting Gas +
Quantity of Escaping Unreacted Gas

The maximum value of the quantity of the reacting gas can be controlled by making the quantity of the reacting gas and the quantity of steam generated, which is condensed, equal so that the amount of condensation becomes constant. The process involves a catalytic device for a battery containing an aqueous electrolyte into which oxygen and hydrogen gases generated as a result of an electrolysis of the aqueous electrolyte pass and are catalytically recombined into water, which returns to the electrolyte. The device includes a catalyst body; a gas-permeable catalyst vessel having the catalyst body arranged therein; a condenser vessel housing the catalyst vessel, the condenser vessel having an opening only at the lower portion; an outer vessel of the catalytic device housing the condenser vessel; and a conduit or passageway for passing a gas from the condenser vessel through a gap between the condenser vessel and the outer vessel.

Gas Reacting Electrode and Sensor

According to a process described by *J.C. Duddy; U.S. Patent 3,901,729; August 26, 1975; assigned to ESB Incorporated* a storage battery cell comprises a positive electrode group, and a negative electrode group submerged in aqueous electrolyte. The storage battery couple is such that when the cell is fully charged and receives a further charging current gas bubbles will evolve from both plate groups.

A baffle arrangement segregates a portion of the gas bubbles given off by one of the plate groups from the gas bubbles given off by the other plate group. A gas reacting electrode is so located in the electrolyte that segregated gas bubbles will impinge thereon. The gas reacting electrode produces an output current corresponding to the rate of gas bubble evolution. The gas bubble evolution increases rapidly at the time the cell reaches a fully charged state and the corresponding change in the output of the gas electrode can be used to initiate charge termination means.

The segregated gas may be hydrogen from the negative plates (best for lead acid cells) or oxygen from the positive plate (useable with nickel iron cells). The electrical circuit includes, connected in series, the gas electrode, a current sensing device and the cell electrode which is not evolving the gas bubbles that impinge on the gas electrode.

Alternately, two electrodes, one working on positive gas, e.g., oxygen, and one working on negative gas, e.g., hydrogen, may be used. In this case the circuit will include, connected in series, the first gas electrode, the current sensing device and the second gas electrode. The current sensing device may be a simple current sensing relay or it may include electronic amplifiers, etc. It is usual for the end of charge sensor to initiate an overcharge timer, the timer actually providing the charge termination and thereby giving the battery a desired overcharge.

Figure 1.19a indicates in outline form a storage battery cell 10 having two species of plates **12** and **14** therein. The plates **12, 14** are submerged in an aqueous electrolyte **16**. A gas reacting electrode **18** is placed over the top of the first electrode species **12** so that some of the gas bubbles **20** emitted by the electrode **12** impinge upon the bottom surface of electrode **14**. The gas reacts at the electrode **18** and produces an electric potential with respect to either of the electrodes **12** or **14**. A separator **22** made of porous material located between the two plates forms a baffle for and segregates the gas from the electrode **14** from the gas from electrode **12** thus preventing the former gas from inadvertently impinging on the surface of the electrode **18**.

FIGURE 1.19: STORAGE BATTERY CELL

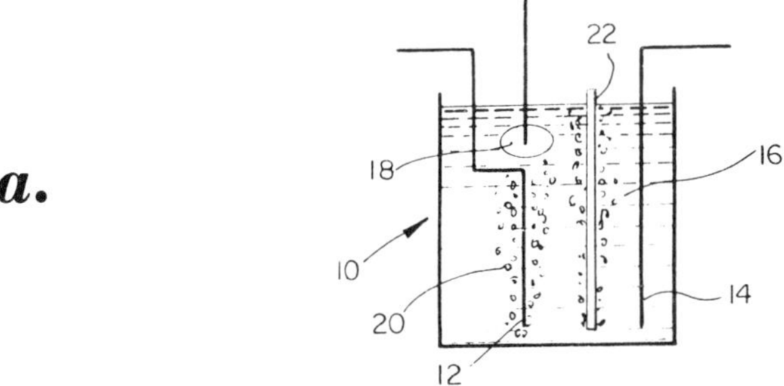

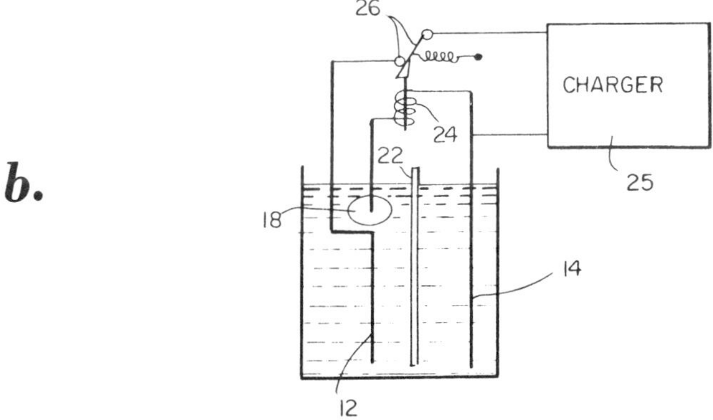

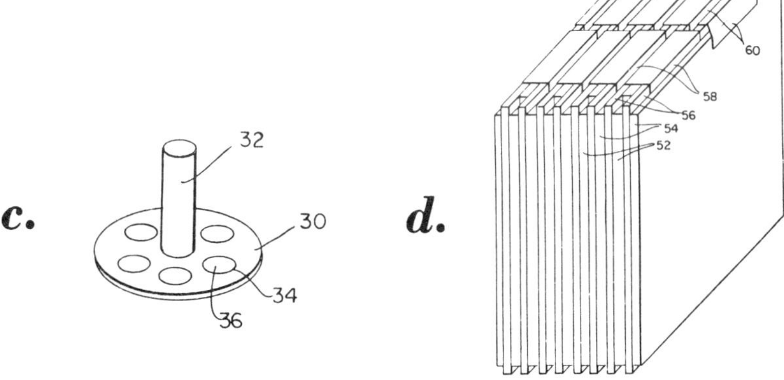

(continued)

FIGURE 1.19: (continued)

e.

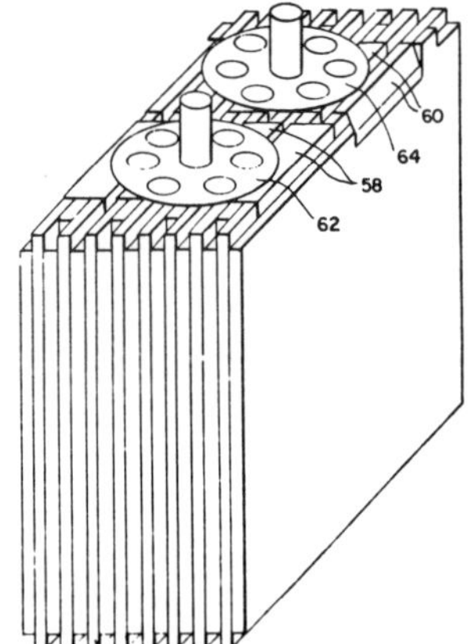

f.

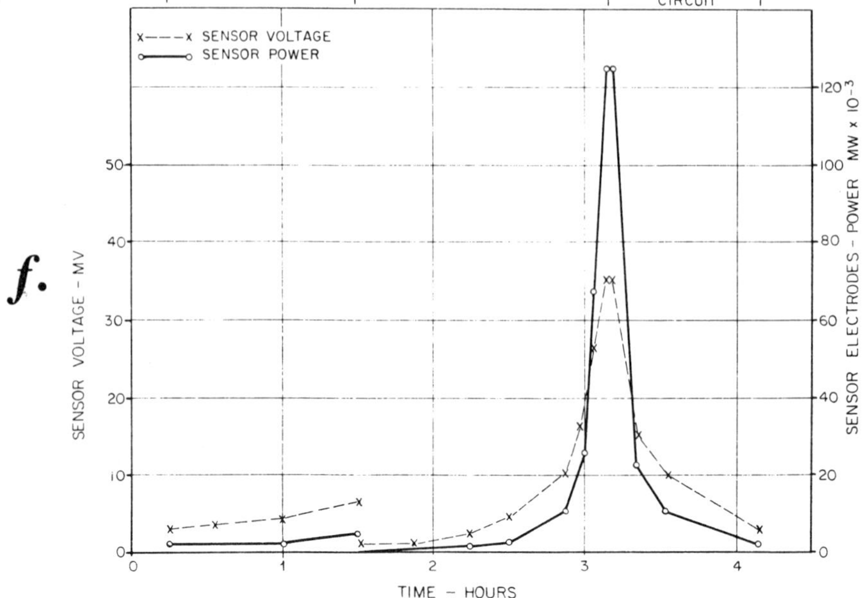

(a) Diagrammatic form of process
(b) Cell of Figure 1.19a with an associated electrical circuit
(c) Sensing electrode as used in the examples shown in Figures 1.19a and 1.19b
(d)(e) Lead acid battery elements
(f) Electrical characteristic of the cell of Figure 1.19e

Source: U.S. Patent 3,901,729

Because the electrolyte is aqueous the gasses evolved from electrodes **12** and **14** at full charge are hydrogen (from the negative plate) and oxygen (from the positive plate). If both gasses should be fed to a single electrode they could, under some circumstances, recombine on the electrode with no flow of current. When the cell is fully charged, the total gas evolved from the positive plate and the total gas evolved from the negative plate is related to the charge current by the well known stoichiometric equation:

$$2H_2O \xrightarrow{4e} 2H_2 + O_2$$

A basic circuit for making use of the process in its simplest form for charge termination is shown in Figure 1.19b. A low voltage relay **24** is connected between the gas electrode **18** and the antipode electrode **14**. The relay **24** is of the type that is normally open and can be manually closed. When a sufficient current flows greater than a preselected value in its coil, the armature will move causing the contacts to open and stay open till manually reset.

A battery charging device **25** is connected to the electrodes **12** and **14** through the contacts **26** of relay **24**. With the lead acid battery system the negative plates give off very little gas until fully charged at which time the gassing rate rises very rapidly. The gassing behavior of the positive plate is very similar. There is a reaction known as local action which occurs at the negative plate which tends to keep the negative plates in a lower state of charge than the positive plate. The local action is made apparent by a slight but continuous evolution of gas bubbles from the negative plates.

Because the negative plates tend to lag the positive plates and because the negative plates have a very sharp change in gassing rate at the time full charge is reached, it is often desirable to locate the gas electrode over the negative plates rather than over the positives. The relay **24** must be such that the gas flow from local action is not sufficient to cause it to operate, yet it must operate on the gas evolution from the plate **12** at a normal charge rate and including normal variations in the charge rate due to line fluctuations, etc. This does not impose too great a restriction on relay **24** because in a typical instance with gas electrodes as described, the current produced by the local action gas stream on a typical electrode is in the order of tens of microamperes, whereas at full charge with a normal charge current passing the gas, recombination current is in the order of milliamperes.

It should be noted that since there is roughly a 100-fold change in current between the not charged and fully charged gas evolution, the exact charge rate is not critical and can vary over a fairly large range without materially changing the determination of the point of full charge.

Many electrodes have been described as suitable for reacting hydrogen and reacting oxygen in aqueous electrolytes, both acidic and basic. Such electrodes are used in the fuel cell art. In general, these electrodes operate with one side exposed to the gaseous phase and the other side exposed to the electrolyte. Such electrodes must be waterproof in order that the electrolyte will not flood the pores and reduce their operating efficiencies.

The use of electrodes of the process differs from these electrodes in that both the reacting gas and the electrolyte are present on the same side of the electrode. Figure 1.19c illustrates a preferred electrode for the implementation of the process.

In Figure 1.19c a graphitized carbon wafer **30** in the order of one inch in diameter and about ⅛ inch thick is threaded to a waterproofed carbon rod **32**. A series of holes **34** contain small biscuits **36** of catalyst material. For a hydrogen consuming electrode, the catalyst is a mixture of activated carbon powder and polyfluorocarbon binder and a small quantity of platinum. To prepare the electrode described, a mix is made of activated carbon, polyfluorohydrocarbon binder (such as polytetrafluoroethylene) in emulsified form and a small quantity of 10% chloroplatinic acid. The ratio of dry carbon to dry binder should be on the order of 10 to 20:1.

This mix is pasted into the holes **34** drilled through the graphite disc **30**, dried and compacted. After this, the entire disc is given a light coating of polyfluorocarbon emulsion as by spraying. It is important that the bottom surface of the electrode including the biscuits of catalyst be flat. If there are depressions in the bottom surface they will trap gas bubbles and detract from the sensitivity of the electrode. A similar electrode but without the platinic acid is suitable for an oxygen sensing electrode.

Other configurations of electrodes may be used as called for by any specific design of storage battery cell. In Figure 1.19d is shown a typical lead acid storage battery cell element **50**. This element comprises positive plates **52**, negative plates **54** and separators **56**. A separator is located between each adjacent positive and negative plate. Straps, not shown, connect the positive plates together and the negative plates together. A series of gas impervious baffles **58** are arranged to cover a portion of the top of the positive plates and the adjacent separator portions but leaving the tops of the negative plates in the same area open to the top of the cell. A second series of similar baffles **60** close off the tops of the negative plates in a second area at the top of the element.

Figure 1.19e shows the same element with two of the electrodes as described above. A hydrogen consuming electrode **62** is located above baffles **58** and an oxygen consuming electrode **64** is located over baffles **60**. The element as described including the gas consuming electrodes must be located in a battery cell jar (not shown) and covered with electrolyte (sulfuric acid solution, not shown) to become operative.

Figure 1.19f shows in graphical form a typical charge cycle of a lead acid automobile battery cell having an element as depicted in Figure 1.19e. From Figure 1.19f it is readily seen that the power available from the electrode (current times voltage) varies from a value of less than 1×10^{-3} milliwatt at the start of the charge cycle to a maximum in the neighborhood of 125×10^{-3} milliwatt or a ratio of at least 120:1.

RECOVERY OF SCRAP BATTERY MATERIALS

Separation Process

G. Tremolada; U.S. Patent 4,026,477; May 31, 1977; assigned to A. Tonolli & Co., SpA, Italy describes a process for treating scrap storage batteries or their parts through which it is possible to separate: the metallic components, i.e., grids and conductors; the inactive materials, i.e., those which constitute the inactive parts of batteries, such as containers, separators, sealing compositions; and the active substance, or paste, formed by lead oxides and sulfates. Particularly favorable results are obtained when for the battery crushing is employed the drum crusher described in the U.S. Patent 3,614,003, which allows a convenient size reduction and a thorough separation of the dry paste.

The process allows one to selectively separate the metallic components from the paste and from the inactive materials so that it is possible to obtain, in the subsequent metallurgical treatment of the paste, lead containing about 1% of antimony or less and a lead-antimony alloy with an antimony content practically equal to that of the metallic parts of batteries. The process employs for the separation a liquid or a pulp having respectively a density intermediate between that of the metallic components and that of the inactive materials. The process in its general lines is as follows:

(a) batteries are crushed in sizes between 2 and 15 cm and dried so to obtain a paste with a humidity content between 2 and 7%;

(b) the crushed material is screened to remove the fines, i.e., practically all the active substance, while the remainder is fed to a separating zone;

(c) in the separating zone, which contains the pulp of a certain density, the inactive material, consisting essentially of the fragments of container separators and sealing composition, is removed as float and the metallic components consisting substantially of lead-antimony alloy, are removed as sink;

(d) the float and the sink are drained, washed and separately collected;

(e) the drainings and washings are recycled to the separating zone while maintaining the density of the pulp within the values by regulating the fresh water added to the cycle and withdrawing any excess of the paste;

(f) the pH of the aqueous suspensions is maintained within the value of 7 to 10 by continuous or discontinuous addition of sulfuric acid.

Preferably the screening of crushed batteries is so regulated that the amount of paste entering the separating zone equals the amount removed with the sink and the float. By so operating, while a certain necessary replacement of the paste in the pulp is maintained, the regulation of the process becomes very simple.

Magnetic Separation Process

The consumption of lead for storage batteries is increasing annually and, therefore, it becomes essential to process used lead storage batteries to recover the metallic lead, lead oxide and sulfate pastes from the batteries for reuse. In addition, it also is desirable to recover the plastic values from the batteries since plastic casings are used in great quantities today.

According to a process described by *J.P. Cestaro, R.K. Hebbar, and U.S. Sokolov; U.S. Patent 4,042,177; August 16, 1977; assigned to N L Industries, Inc.,* the scrap storage batteries are first crushed into small pieces, preferably to size less than 3 inches, and the metallic lead or lead-antimony alloy and the lead oxide and lead sulfate pastes are recovered from the inert constituents which include hard rubber and plastic battery casings, partitions, separators, paper and other inactive materials. In order to recover the lead and plastic values, the crushed pieces are added to an aqueous suspension of magnetic granular material having a density less than the density of metallic lead, lead oxide and lead sulfate. The lead values remain at the bottom of the vessel while the inert material including the plastic material remains in the suspension.

The suspension is prepared by adding granular magnetite ore to water and with agitation suspending the ore particles in the water. Sufficient magnetite ore is employed to produce a suspension having a density from about 1.75 to about 3.5 g/cc, and an average granular size of -100 mesh (Tyler screen). Although many types of magnetic material may be used to form the suspension, various ores, such as magnetites and ilmenites, are particularly desirable to employ since they are both inexpensive and readily available.

After the metallic lead and the lead compounds are separated from the suspension, the magnetite ore granules are separated from the inert and plastic material by magnetic separation. The remaining plastic and inert material are then separated from one another in water by floating the plastic material from the inert material which sinks to the bottom. By use of this process up to 95% of the lead values and up to 95% of the plastic material present in the batteries may be recovered.

When batteries which contain pitch as a sealer are used in the recovery process, the pitch material is recovered with the plastic material, and in most instances should be separated from the plastic material before the plastic material is recycled.

Rotating Drum Separator

According to a process described by *A.E. La Point; U.S. Patent 4,018,567; April 19, 1977* whole or shredded lead-acid batteries, together with a quantity of sodium carbonate and water, are fed continuously into a rotating drum separator containing a charge of grinding balls. Agitation of the mixture, aided by the internal drum construction features, further breaks up and degrades the battery fragments, neutralized any contaminating electrolyte, transposes the finer particles of lead sulfate into lead carbonate, and forms a heavy-medium suspension of the active material on which organic battery fragments float.

The suspension of active material constantly overflows from the drum at one end and carries the organic fragments with it into a first trommel, while sinkable fragments of grid metal and other battery parts of lower grade antimonial lead alloy are mechanically removed at the opposite end and deposited in a second trommel. A portion of the overflowing suspension is pumped back to the drum together with a regulated quantity of water, while the remainder is delivered to a thickener for further processing.

Smelting Process Using Carbon Powder

M. Liniger; U.S. Patent 4,030,916; June 21, 1977 describes a process for recovering lead from the active material of used lead batteries, this material having been separated from the other battery parts in the form of a so-called lead sludge. The process is characterized in that all the water is removed by drying from the lead sludge obtained from the active material and the dried material is ground down to the particle size of the original active material. The ground, dried material is then intimately mixed with anhydrous carbon powder and the resulting mixture heated in a closed furnace, the lead compounds constituting the active material thus being decomposed or reduced to metallic lead and the carbon powder being oxidized. The gaseous reaction products produced by this heating are cooled and filtered and the SO_2 component is subsequently removed from the gaseous reaction products.

The lead sludge is advantageously dried in completely closed vessels which are heated from outside to a temperature of 100° to 150°C. At least some of the heat necessary for this step can be taken from the hot waste gases given off when the lead sludge is heated with carbon powder as described later. It is of practical advantage if the vessels are fitted with mechanical agitating, grinding and discharging means. The steam generated during the drying stage is as a rule condensed and the condensed water used in removing the active material from the accumulator plates. Significantly less energy is required to evaporate the water from the lead sludge at a temperature of just over 100° than at about 1000°C in a smelting furnace as in prior art processes.

Carbon powder is added to the active material which has been dried and simultaneously ground down to its original particle size (usually 20 to 50 μ). The carbon powder can for instance be coke dust. The carbon powder is generally added in an amount of from 2 to 9% by weight based on the weight of the ground, dried material. The two materials are then blended–as a rule in a separator mixer. The mixture is transferred either continuously or batchwise to a preferably electric furnace.

In the furnace the mixture is heated strongly enough for the lead compounds to be split into Pb, SO_2 and oxygen, the latter oxidizing the coke dust to CO and CO_2. The reaction velocity can be controlled by choosing a particular amount of coke dust of a particular particle size to mix in with the dried active material. The lead produced collects at the bottom of the furnace from where it can be drawn off continuously or at intervals. Only insignificant amounts of lead are lost.

Now in contrast to normal coal, carbon (e.g., coke) contains practically no organic constituents which give steam on burning. Furthermore, no water is liberated by the smelting process. Thus the gaseous reaction products contain no steam. The gas leaving the furnace is cooled and filtered. It contains SO_2, and CO and a little CO_2 and O_2, the relative proportions of which are somewhat dependent on the operational conditions and on the composition of the starting material.

It can easily be seen that, apart from loss of a certain amount of radiant heat, the smelting procedure of the process only requires the heat necessary to decompose the lead compounds PbO_2 and $PbSO_4$ to be supplied from outside. Since no water at all is produced or liberated during the reaction, the amount of gas formed is practically that produced by the decomposition and reduction processes.

The minimization of the amounts of waste gases formed and the fact that the gases are dry leads to two important advantages: (1) Fabric filters can be used to filter the waste gases (dust removal). Such filters enable any desired degree of purity to be achieved but involve only relatively low expenditure per unit volume of gas for equipment. This is particularly important as far as environmental pollution is concerned. (2) The SO_2, which makes up about 30 to 40% by weight of the waste gases, can be removed from the waste gases by known and reliable means so that the waste gases ultimately consist only of CO_2, CO and O_2.

Owing to the relatively high proportion of SO_2 in the waste gases it is profitable to remove it. This can be done by compressing the waste gases and cooling to a temperature at which the SO_2 liquifies under the pressure applied. The liquid SO_2 can then be drained off and stored in a pressure tank. Any aerosols which may be present are washed out at the same time, this increasing the purity of the residual gases which are discharged to the atmosphere.

GENERAL CONSTRUCTION AND FABRICATION

Folded Plates

T.J. Dougherty and V.W. Bast; U.S. Patent 4,029,855; June 14, 1977; assigned to Globe-Union Inc. describe a storage battery and method of making the same which involves providing a single positive and negative battery plate which are folded with an interspersed nonconductive separator sheet. Each plate is provided with a protruding lug and a folded pair of plates and separators comprises a battery element. The lug of each plate is joined to the lug of an opposite polarity plate in an adjacent battery cell by forming a sealed fusion connection through an aperture in the partition wall. The endmost plates of the battery are coupled to positive and negative battery terminals by a similar welded joint through the battery case walls.

Referring to Figure 1.20a, a battery **10** is seen to comprise an outer case **12** having a cover **13** which may be provided with a vent cap **14** for venting gases formed within the battery in a conventional manner. Positive and negative terminals, **15** and **16** respectively, are provided.

FIGURE 1.20: STORAGE BATTERY

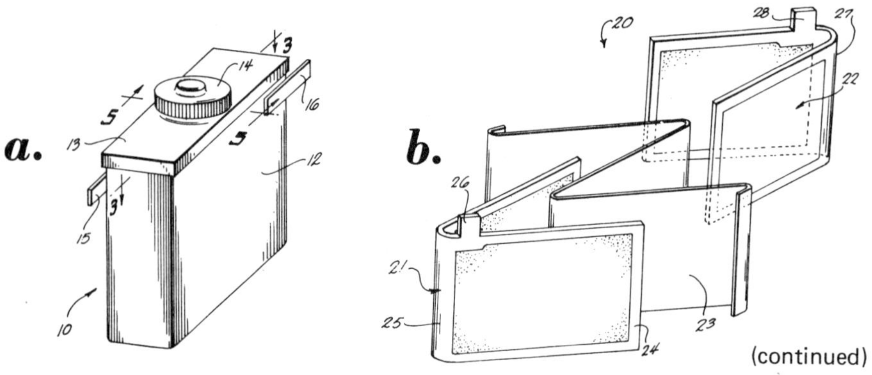

(continued)

FIGURE 1.20: (continued)

c.

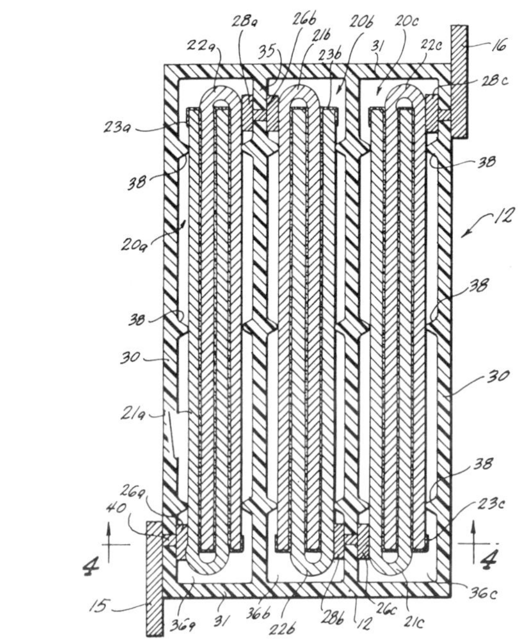

d.

e.

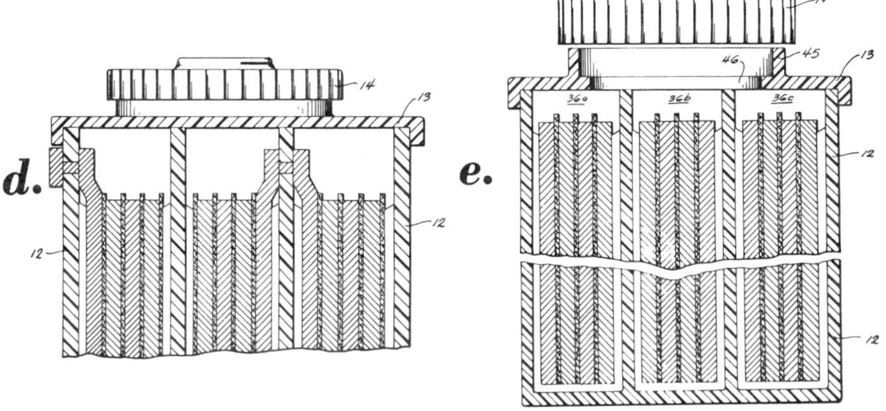

(a) Perspective view of battery
(b) Perspective exploded view of a battery element
(c) View taken along line **3–3** of Figure 1.20a
(d) View taken along line **4–4** of Figure 1.20c
(e) View taken along line **5–5** of Figure 1.20a

Source: U.S. Patent 4,029,855

Figure 1.20b shows a battery element **20** which may be provided within each cell of the battery **10**. The element **20** comprises a positive battery plate **21**, a negative battery plate **22** and a porous nonconductive separator sheet **23**. The positive plate **21** comprises a frame **24** and a solid vertically extending lead portion **25**. Plate **21** is folded along a vertical center line through portion **25** and includes a conventional conductive lead grid within frame **24** to which lead paste is adhered. An upstanding lug **26** is provided on one half of the plate **21** adjacent to the fold line.

The negative plate **22** may also comprise a similar pasted conductive lead grid. Plate **22** is folded along a vertical center line of solid portion **27** and also has an upstanding conductive lug **28** formed on one half adjacent to the fold line. The folded plates **21** and **22** are interleaved with separator **23** having corresponding folds to define a continuous nonconductive separating partition between adjacent plate surfaces. The conductive lugs **26** and **28** are positioned on diametrically opposed sides of the element assembly **20**.

With reference to Figures 1.20c and 1.20d, the internal construction of the battery **10** will now be described. The battery case **12** consists of outer side walls **30** and end walls **31** defining a hollow interior having partition walls **35** which divide the interior of the case **12** into individual battery cell compartments **36a, 36b and 36c**. A battery element **20** is disposed within each of the cells **36a** through **36c** and hereafter, the battery elements and their associated components will be designated with an a, b or c depending upon whether they are in cells **36a**, **36b** or **36c** respectively.

As is also seen in Figure 1.20c, spaced ribs **38** are provided on the interior surfaces of the side walls **30** and partitions **35** in order to hold the elements **20** in spaced relationship with the side walls **30** and partitions **35**. The first cell **36a** has an element **20a** in which the lug **26a** of the positive plate **21a** is sealingly coupled to the positive terminal **15** on the outside of the battery case **12** by any suitable welding connection through an aperture **40** provided within one of the side walls **30**. The lug **28a** of the negative plate **22a** is coupled by a similar weld through its adjacent partition wall **35** to lug **26b** of the positive plate **21a** of cell **35b**. Lug **28b** of the negative plate **22b** of the second element is coupled by welding to lug **26c** of the positive plate **21c** in the third cell compartment **36c**.

Again, this connection is made by a suitable intercell weld through the adjacent partition wall **35**. Finally, lug **28c** of the negative plate **22c** in the third cell is coupled to the negative terminal post **16** on the exterior of the battery case **12** by a suitable through-the-wall connection. It will thus be appreciated that the construction illustrates a 6 volt battery having three cells **36a** through **36c** containing elements **20a** through **20c** which are comparable to conventional four plate battery constructions but which utilize a pair of folded plates instead. Obviously more or fewer cells could be provided if desired.

The provision of the lugs **26a** through **26c** and **28a** through **28c** eliminates the need for a strap connection between plates of similar polarity within the cells and the intercell connection is simply made by joining a single lug within each cell with a corresponding single lug in the adjacent cell or with an external terminal **15** or **16**. The folded design thus permits construction of battery elements **20** having three pieces which perform the same as conventional elements consisting of four plates and three separators thus reducing assembly time and material costs.

Because the battery design is adapted for reduced capacity batteries, the case **12** has reduced external thickness with corresponding reduced individual cell sizes. Accordingly, as seen in Figure 1.20e, a single vent plug **14** is provided which engages an annular boss **45** formed in the cover **13** and defining an aperture **46** which opens into the interior of the battery case **12**. Aperture **46** extends over the three cells **36a, 36b** and **36c** to permit filling of the cells with electrolyte or water through the single aperture and also allow venting of gases from the cells.

Plates Embedded in Sealing Unit Cover

N. Sanekata and O. Hamada; U.S. Patent 3,915,751; October 28, 1975; assigned to Yuasa Battery Company Limited, Japan describe a process for producing intercell connector type storage batteries. The process for manufacturing storage batteries comprises inserting anode and cathode plates having lugs into a hollow container therefore in such a manner that the lugs project outwardly from the container. A cover for the container having means presenting at least one concavity adapted to open toward the container after assembly is provided and the process includes the step of positioning the cover with the concavity opening upwardly and then filling the concavity with a liquid, hardenable sealing agent.

The container is positioned above the cover with the lugs projecting toward the liquid sealing agent in the concavity and then the cover and the container are moved into contact with one another whereby the lugs are thrust into and embedded in the liquid sealing agent. The cover is attached to the container and the sealing agent is allowed to solidify. Thereafter a portion of the cover and of the sealing agent are removed to present a cavity which exposes the embedded lugs. The process then includes the step of filling such cavity with a molten conductive material and allowing the molten material to solidify to present connectors for the lugs.

In Figures 1.21a and 1.21b, the reference numeral **1** designates a battery container constructed of a synthetic resin such as polypropylene, an acrylonitrile-butadiene-styrene copolymer or an acrylonitrile-styrene copolymer. The container is preferably a monoblock container having a plurality of partition walls **2** presenting the required number of cells.

FIGURE 1.21: STORAGE BATTERY

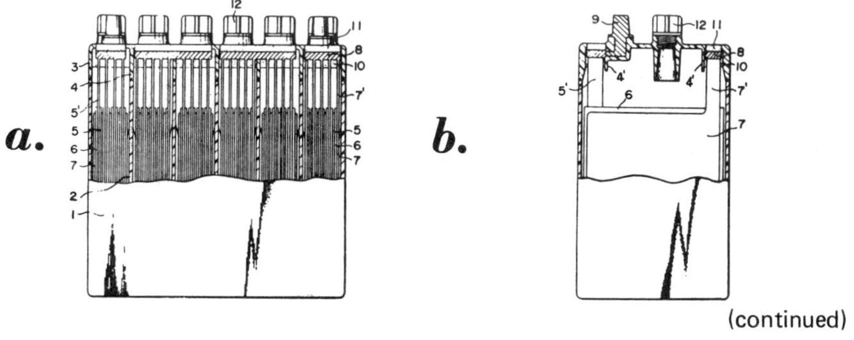

(continued)

FIGURE 1.21: (continued)

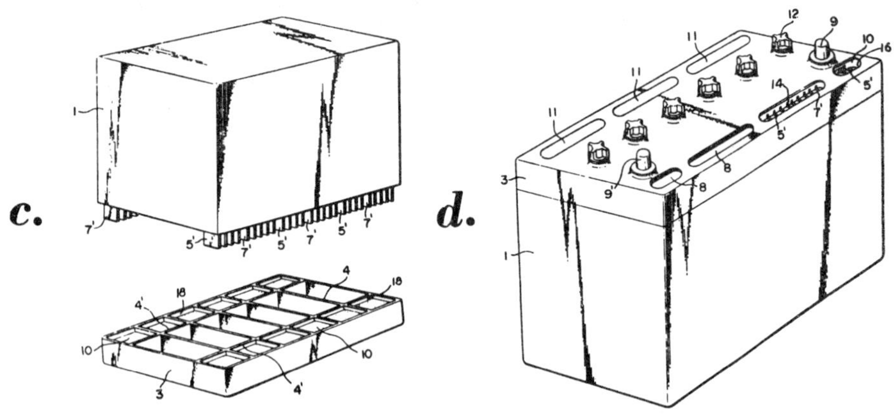

(a) Partially cross-sectional, side elevational view of a storage battery
(b) Partially cross-sectional, side elevational view of the battery of Figure 1.21a
(c) Perspective view illustrating one step of the process
(d) Perspective view illustrating steps which follow the step of Figure 1.21c

Source: U.S. Patent 3,915,751

The reference numeral **3** designates a cover constructed of the same material as container **1** and cover **3** has partition walls **4** which correspond with partition walls 2. Cover **3** is also provided with a pair of partition walls **4'** (see Figure 1.21c) which cross partition walls **4**. These walls in cover **3** present a series of concavities **18** which are illustrated particularly in Figure 1.21c where it can be seen that concavities **18** open toward the interior of container **1**.

The required number of battery elements, each consisting of a cathode plate **5**, a separator **6** and an anode plate **7** arranged in the order mentioned, are pushed into each of the cells presented by container **1**. A connector **8** constructed of a conductive material such as lead or a lead alloy connects the lugs of the similar plates in one cell with the lugs of the different plates in the adjacent cell.

The lugs of the cathode plates **5'** of one end cell are connected with the terminal **9** through a connector **8** while the lugs of the anode plates **7'** of the outer end cell are connected with the terminal **9'** through another connector **8**. (See particularly Figure 1.21d.) A sealing agent for the plates is designated by the reference numeral **10** while the reference numeral **11** designates a filler. It is desirable that the sealing agent and the filler each consist of a thermosetting synthetic resin such as, for example, an epoxy resin. An electrolyte port plug is designated by the reference numeral **12**. The process for producing storage batteries is explained in detail in the following paragraphs.

As illustrated in Figure 1.21c, the required number of battery elements, each consisting of a cathode plate **5**, a separator **6** (invisible) and an anode plate **7** arranged in the order mentioned, are pushed into each cell of container **1** in a manner such that the same are forced against each other. It is to be noted in this regard, that one of the most important features of the process is the omission of the conventional steps of accurately arranging the positions of the respective lugs of the anode and cathode plates thus pushed into the container and the connection of the lugs using a burning jig. Thus, the clearances between the respective lugs of the anode and cathode plates do not need to be uniform and the lugs may be deformed and may be positioned at different heights whereby the storage battery of the process is very simple to construct.

Moreover, it is not necessary to push the anode and cathode plates and the separators into the battery container so forcefully that the lugs are damaged and the performance of the storage battery is impaired.

The concavities **18** of cover **3** which are formed by partition walls **4** and **4'** are filled with a sealing agent **10** as can be seen viewing Figure 1.21c. Lugs **5'** and **7'** of cathode plates **5** and anode plates **7** which project from container **1** are then pushed into and embedded in the sealing agent **10** in concavities **18**. For this purpose, the lugs of the anode and cathode plates are adjusted in advance so that the same project outwardly of the upper edge of the battery container.

Battery container **1** and cover **3** are then joined and sealed together through the use of an appropriate sealing agent. For this purpose, it is desirable to form a groove in the edge of cover **3** and to fit a tongue of the edge of battery container **1** into the groove, all as illustrated in Figures 1.21a and 1.21b. Then, cover **3** and container **1** may be joined through the use of a sealing agent so that the same are completely sealed together. It should also be noted that if container **1** and cover **3** are made of polypropylene or the like, the same may be heat-sealed.

After the solidification of the sealing agent **10** disposed in cavities **18** and in which lugs **5'** and **7'** of cathode plates **5** and anode plates **7** are embedded, the storage battery may be turned upright. Then, respective portions of cover **3** are removed with an end mill or the like. During this operation, corresponding portions of the sealing agent **10** are also removed. Thus, cavities **14** which expose respective lugs **5'** and **7'** of cathode plates **5** and anode plates **7** are presented. In the end cells, a portion of terminal **9** or **9'**, as the case may be, which has been molded in the cover in advance, will also be exposed by cavities **16**. Terminals **9** and **9'** are illustrated in Figure 1.21d.

During the foregoing material removal operations, if the upper ends of the lugs **5'** and **7'** are cut slightly to remove the film of oxidized metal which is often disposed thereon, the fusion between the connector and the lugs will be improved such that electrical resistance is reduced and a storage battery having comparatively favorable performance characteristics is obtained.

After the material removal steps referred to above, the respective cavities **14** and **16** are filled with a molten conductive material, such as lead or a lead alloy, up to a level lower than the upper surface of cover **3**. Thus, exposed lugs **5'** and **7'** and/or the exposed portions of terminals **9** and **9'** are contacted directly by the molten metal in the cavities **14** and **16**. The molten metal is then permitted to solidify whereby to present connectors **8** which are illustrated in Figure 1.21d. The upper surfaces of connectors **8** are filled with fillers **11** which are also shown in Figure 1.21d. It is to be noted that insulative covering plates may be used instead of fillers **11**.

Positioning Device for Plates

J. Klein; U.S. Patent 4,016,638; April 12, 1977; assigned to Accumulatorenwerk Hoppecke Carl Zoellner & Sohn, Germany describes an assembly device which makes it possible to immediately obtain an exact longitudinal positioning of the storage plates and insulating plates during assembly of the cells for lead-acid batteries.

The process utilizes a device having longitudinally moveable positioning members arranged outside the lateral edges of the plates, the positioning members defining opposing abutment faces oriented at right angles to the bottom edges of the plates. The two positioning members each carry at least one positioning knife defining a sharp edge at a parallel distance inwardly of the abutment faces and running transversely to the edges of the storage plates. The distance of the knife edges from the abutment faces of the positioning members is equal to or greater than the nominal overhang of the insulating plates.

The proposed assembly device makes it possible to obtain the accurate positioning of the plates in a single assembly operation, thereby affording considerable savings in assembly time and cost over the previously employed assembly method.

Resilient Support for Plates

According to a process described by *G. Trippe and J. Brinkman; U.S. Patent 3,996,065; December 7, 1976; assigned to Varta Batterie AG, Germany* the plates of lead storage battery cells are supported at least from the top by resilient means attached to the cell cover. They may also be supported by resilient means from the cell bottom. The respective resilient means are placed diagonally relative to each other. At least the bottom support means has variable resiliency depending on the external forces applied to the cell.

Figure 1.22 shows a cell housing **13**, within which oppositely poled plate stacks are positioned, one such plate **12** being visible in Figure 1.22. These plates have plate bolts **1** which protrude partially from cell housing **13**. This bolt has a necked-down portion **1a** which is surrounded by a split bushing **7**. This is held in place by a sleeve **6**. Sleeve **6** is made of electrolyte resistant elastic plastic or rubber having relatively low elasticity. Above sleeve **6** there is positioned a ring **5** of rubber or plastic which has relatively high elasticity. This soft ring **5** provides a seal for the lower surface **9** of cell cover **8** and also for the shaft of plate bolt **1**.

Above cell cover **8** there are also one or more rings **4** of electrolyte resistant, elastic plastic or rubber having soft resilient properties. Above ring **4** there is a washer **3** which can be tightened more or less tightly by nut **2**. Through tightening of nut **2** pressure can be applied against cell cover **8** from both top and bottom via bolt **1**, pole bridge **10**, the two bushing elements **7**, sleeve **6** and rings **4** and **5**. By more or less forceful tightening of nut **2**, controllable compression of elements **4, 5** and **6** can be produced.

FIGURE 1.22: LEAD STORAGE BATTERY

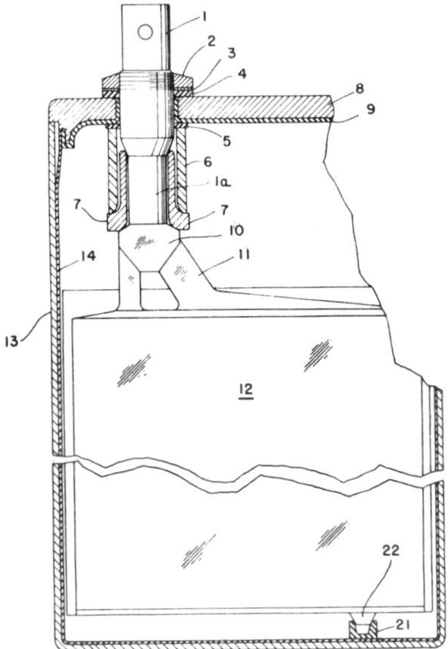

Source: U.S. Patent 3,996,065

Before inserting plates **12** into cell housing **13**, the pedestals **21** are engaged in the rests **22** of both the positive and the negative plates. Pedestals **21** are made of electrolyte resistant material, such as rubber or plastic, and are designed so as to have softly resilient properties. The sides of rests **22** and of pedestals **21** slope at different angles.

Specifically the inner sides of the recess in pedestal **21** converge at a more acute angle than the sides of the inserted rest **22**. As a result, the resiliency decreases for heavier impacts because the plate rest then increasingly stretches the pedestal and ultimately hits bottom. The diagonal positioning of this pedestal relative to the suspending arrangement for the plate further allows distortion of the grid frame and displacement of individual plates within the plate stack to take place.

It has proven particularly desirable to make resilient elements **4, 5** and **21** of rubber, and resilient elements **6** and **22** of plastic, preferably polyethylene. The advantage of this arrangement is that low, medium and high frequency vibrations of small displacement and low acceleration originating outside the cell housing are, to a large extent, kept from the plate stack and damped out. Such vibrations are produced in motor vehicle batteries by the characteristics of the path traveled by the vehicle, and by the vibrations of the engine.

The arrangement has the further advantage that occasional heavy blows at high acceleration applied from outside the cell are also largely isolated from the plate stack and damped out. These advantages are attributable to the following:

(1) The means supporting the plate stack from the bottom and suspending it from the cell cover, collectively exhibit a progressively varying elasticity.

(2) The suspension of the plate stack from the cell cover and the support of the plate stack from the bottom are disposed diagonally relative to each other, so that the forces which operate generally vertically upon the cell housing act upon the plate stack with their lines of force spaced as far apart as possible.

(3) The means suspending the plate stack from the cover in effect integrate into the damping system, as an additional resilient element thereof, the entire cell housing with its slightly elastic vertical walls and slightly elastic horizontal cell cover.

(4) Through elastic suspension and support of the plate stack relative to the cell housing, accelerations applied from outside lead to movement of the plate stack relative to the cell housing.

Because of this relative movement the electrolyte present within the cell housing is forced, at relatively high velocity, through the closely packed plate stack with its quite large surfaces. The large frictional surface and the relatively high flow resistance exhibited by this tightly packed plate stack also create considerable damping of relative movements.

Such damping, which includes the braking effect upon relative movement attributable to the high flow resistance, has the advantage that these braking forces are distributed over the entire surface of all the positive and negative plates and their intermediate separators. Consequently, even a very high total braking force places only minimal mechanical stresses on the individual elements of the plate stack.

Rib Structure for Electrolyte Circulation

A process described by *M.S. Tsygankov, N.A. Bitjutskaya, V.N. Fateeva, V.N. Kosholkin, B.D. Karev, N.A. Kudinov, M.M. Kholkin, N.N. Moiseev, and O.G. Malandin; U.S. Patent 3,972,736; August 3, 1976* relates to storage batteries intended for use in transport facilities in the capacity of a standby source of electric power and for nonassisted starting of engines. The storage battery according to the process can be used, for example, in aircraft, e.g., airplanes and helicopters.

The main object of the process is to provide a storage battery with a considerably facilitated circulation of electrolyte in its cells, where provision is made to prevent the throw-out of electrolyte in case of physical effects on the battery operating at a buffer duty on an aircraft.

This object is accomplished by providing a battery comprising a container accommodating cells interconnected in series by intercell connectors, the cell jars containing plate groups assembled with the aid of elements which fasten the terminal posts to the plates wherein, according to the process, each battery cell is provided with vertical ribs on the internal surface of two opposite walls facing the surfaces of the end plates of the plate group, the ribs ensuring circulation of electrolyte in the electrolyte layer around the plate groups and where each cell is provided with a dielectric plate freely resting in the cell jar on the elements which fasten the terminal posts to the plates. It is practicable that the relation of the distance between any two adjacent vertical ribs to the width of the plate be selected from 0.15 to 0.25 and the relation of the rib width to the distance between the two adjacent ribs be from 0.10 to 0.25, in which case it is practicable that the relation of the area of the dielectric plate to the cross-sectional area of the cell limited by the internal dimensions of the jar and lying in the same plane as the plate should range from 0.85 to 0.95.

The storage battery is guaranteed against throw-out of electrolyte under any service conditions and has a better performance at starter duties.

Thermoplastic Frames

W.L. McDowall; U.S. Patent 4,022,951; May 10, 1977 and 3,941,615; March 2, 1976; assigned to Dunlop Australia Limited, Australia describes a multicell battery comprising a plurality of frames each divided into a number of side by side active paste support areas. The frames are assembled and secured together in a stacked formation so that the perimeter portions of the frames form the top, bottom and two opposite sides of the battery, and the division in the frames form cell partitions.

Each frame is pasted with active material to form plates with adjacent plates in each frame being of opposite polarity and adjacent plates in adjoining frames also being of opposite polarity. Electrolyte-porous separator material is provided between adjacent plates in adjoining frames.

The frames of the battery are formed from a moldable material which is electrically insulating at the intended operating voltage of the battery and is inert to the active materials of the battery and any material produced during operation of the battery. Suitable thermoplastic materials for use in the construction of frames are high impact polystyrene, ABS, and polypropylene.

Conveniently each frame is of a rectangular form having a continuous perimeter member and a number of division portions parallel to two opposite sides of the frame to define the plurality of support areas. The perimeter member and division portions of adjacent frames are sealed together, and may also interfit, so that the perimeter member forms two opposition walls, a top, and a bottom of the battery and the division portions form the plurality of cell partitions of the battery. The adjacent frames may be arranged so that the frames are secured together by ultrasonic weld which will also provide the required seal between the perimeter member and division portion of adjacent frames.

There is thus provided a plurality of side by side column-like spaces in the frame to receive active material or separator material. Each column-like space may be divided into smaller areas by a plurality of transverse elements of lesser thickness than the division portions to provide additional support, which may be particularly desirable for the areas which receive active material. The frames which support the separator material may be overall thicker than the frames supporting active material, and as the separator material is conveniently made of thin sheet material, the thicker frames provide a greater electrolyte capacity. The frame may be molded in situ about the separator material so that the margin of the latter is embedded in the frame.

Some of the advantages of the battery constructed in accordance with the process are (1) the reduction of battery weight and size by elimination of bulky lead alloy grids; (2) elimination of forming intercell connections during assembly, with the avoidance of consequent sealing problems; (3) the possibility of eliminating a separate battery case; (4) additional support for active material to eliminate shedding of active material; and (5) increased capacity per unit weight of battery.

Referring to Figure 1.23a, the battery **10** comprises a housing **11** enclosing an assembly of frame members **12** arranged and bonded together so as to form a plurality of cells **14** separated one from the other by a partition wall **13** composed of the abutting vertical elements of the respective frame members **12.**

Referring to Figure 1.23b, each frame member **12** comprises a perimeter member **15** having top and bottom elements **16** and **17** and opposite side elements **18** and **19.** Extending between the top and bottom elements **16** and **17** are division elements **20** which are parallel to the side elements **18** and **19.** The frame and the vertical division elements together define three column-like areas **22, 23** and **24** which will, in the finished battery, be pasted with appropriate active battery materials.

FIGURE 1.23: MULTICELL BATTERY CONSTRUCTION

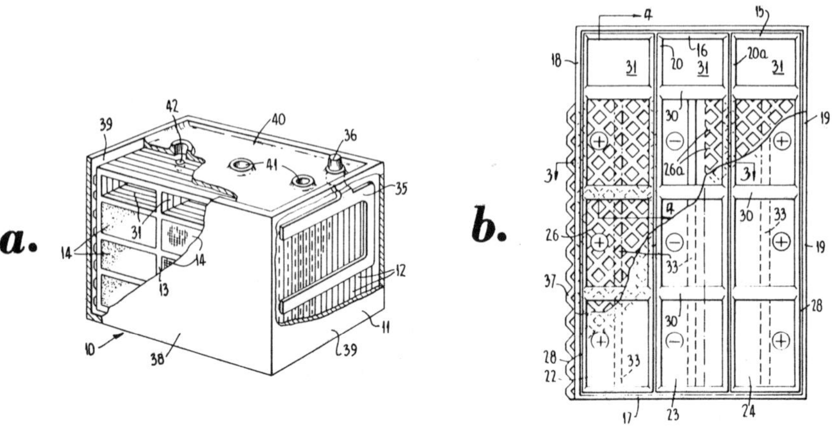

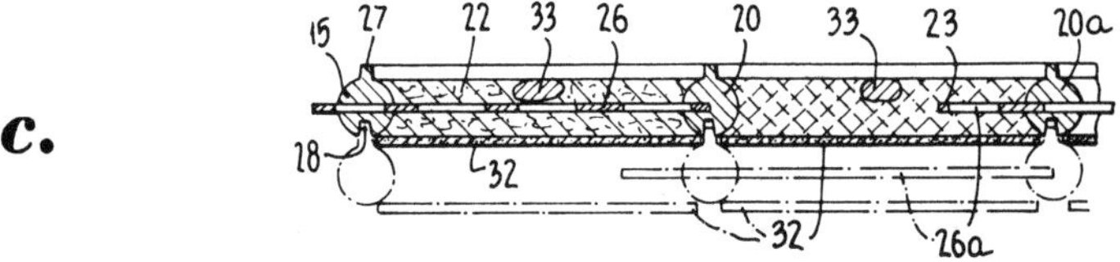

(continued)

FIGURE 1.23: (continued)

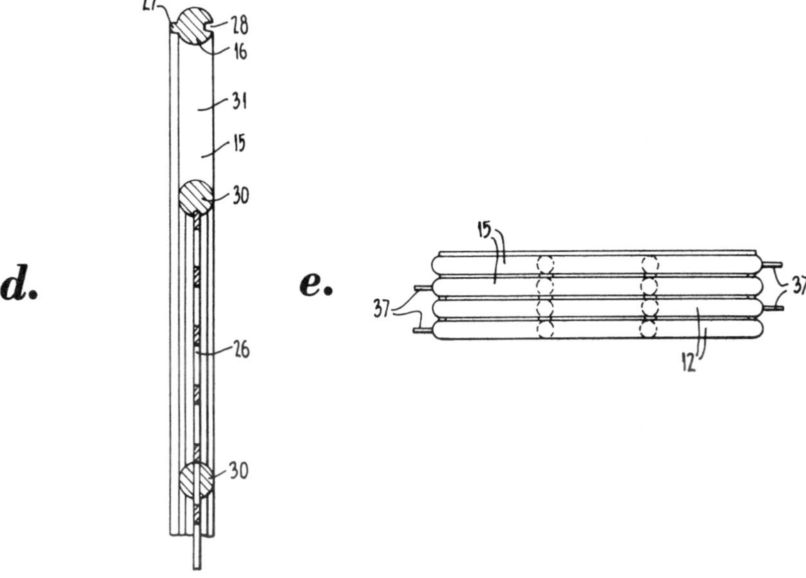

(a) General perspective view, partly in section, of a battery
(b) Side elevation of one grid assembly suitable for use in the battery shown in Figure 1.23a
(c) Enlarged cross-sectional view along line **3–3** in Figure 1.23b
(d) Enlarged sectional view along the line **4–4** in Figure 1.23b
(e) Plan view of an assembly of grids as shown in Figure 1.23b

Source: U.S. Patent 4,022,951

The top and bottom elements and the side and division elements are provided on one face with a continuous tongue **27** at right angles to the general plane of the frame, and on the opposite face a continuous groove **28**, so that when a plurality of frames are assembled in side by side relationship as shown in Figures 1.23c and 1.23d, the tongue on one frame interfits with the groove on the adjacent frame.

The interfitting tongue and groove are adhered or otherwise secured together, with or without the use of additional sealing compound, so that the connection between the top and bottom elements, the side elements and division elements in respective adjacent frames will not permit the leakage of electrolyte between the cells in the finished battery.

An assembly of a plurality of frames of this construction provides an open-ended box-like structure with internal partitions, the sides and top and bottom of the box being formed by the interfitting top and side bottom elements and the partitions being formed by the interfitting division elements.

The frames also include a plurality of vertically spaced support elements **30** extending between the opposite side elements **18, 19** and integral with the division elements **20.** In the frame shown in Figure 1.23b there are three support elements and these elements are of a thickness less than the thickness of the side and division elements so that when a number of frames are assembled in a side by side relationship, the support elements in adjacent frames are spaced one from the other. The support elements **30** divide the area between the respective division elements **20** so that the material subsequently located between the division elements is given additional support and will not become dislodged during service.

In the preferred form, grid-like structures **26, 26a** are provided to span part or all of the area between the respective division elements to provide even further support for the active battery material in those areas which form active material areas in the finished battery. The grids may be formed of the same material as the remainder of the frame and molded as an integral part thereof, but preferably are made of an electrically conductive material which is not adversely affected by the materials of the battery, such as lead alloy, and are embedded in the elements of the frame during the molding of the frame.

As shown in Figure 1.23b the grid **26** extends through the side element **18** of the frame and is embedded in, but does not extend through the adjacent division element **20.** The portion of the grid **26** external of the frame provides a terminal for electrically connecting the cells constituted by an assembly of frames. The grid **26a** extends through the other division element **20a** and is embedded in but does not extend through the side element **19.** The grid **26a** thus forms an intercell connector between plates **23** and **24.** The grids **26** and **26a** thus provide support for the active battery material, act as a current collector for the respective plates, and form intercell connectors and/or terminals as required in respect to cells formed by the assembly of frames.

During the pasting of the frames the area above the upper support element **30** in each frame is not pasted so that when the frames are assembled together electrolyte reservoirs **31** are formed as shown in Figure 1.23a. Also during assembly, separator strips **32** are inserted between the plates of active material in adjacent frames. The separators perform their normal function in a battery, and are made of conventional material.

Battery Jar

J.S. Hardigg; U.S. Patent 3,993,507; November 23, 1976 describes a battery jar for supporting a plurality of elongated battery plates together with an electrolyte. The battery jar comprises an elongated casing formed of a plastic material such as polycarbonate plastic or ABS plastic wherein the plastic material provides both an acid barrier and a means for supporting the weight of the electrolyte and the battery plates.

The casing includes a plurality of transversely disposed pressure-supporting ribs or hoops which are positioned about the periphery of the casing for providing support against the radial pressure exerted against the casing. In one example the transverse ribs are formed integral with the casing and permit longitudinal expansion of the casing when the casing is lifted.

Molded Plastic Container

B.N. Spiegelberg; U.S. Patent 3,995,008; November 30, 1976; assigned to Gould Inc. describes a battery container made of molded plastic and preferably formed in a single unitary structure. The container comprises a generally rectangular hollow container with a closed bottom and an open top. A plurality of internal partitions formed as integral parts of the container bottom and side walls divide the interior space into a plurality of battery cells. These partitions are formed by a plurality of spaced mold cores disposed within the mold that forms the outside shell of the container.

A plurality of ribs are formed on the mold surface that forms the bottom of the container, the ribs being in alignment with the cavities between adjacent mold cores and forming a plurality of parallel grooves in the lower surface of the container bottom. The ribs are at least as wide as the width of the cavities between adjacent cores at the top of the container so that the upper ends of the partitions formed in the cavities fit into the resulting grooves formed by the ribs when the containers are stacked on top of each other, thereby preventing distortion and warping of the partitions while they are stacked.

The ribs also affect the flow of plastic through the mold, preventing lateral displacement of the mold cores so that the resulting partitions are produced with a uniform thickness, providing a more reliable battery container and a lower reject rate. The mold gates are formed in the centers of the ribs, and the feed nozzles extend all the way through the ribs to avoid any reduction in the groove depth due to buildup of plastic at the gates. To maintain a substantially uniform thickness in the bottom of the container in the gate regions, the corners of the mold cores are provided with recesses adjacent the gate regions, thereby increasing the thickness of the lower ends of the partitions in those regions.

The process provides a battery container which prevents distortion of the internal partitions when the containers are stacked on top of each other, due to the nesting of the grooved bottom surfaces of the upper containers with the top ends of the partitions and end walls of the lower partitions. Furthermore, the same ribs which form the bottom grooves also improve the flow characteristics of molten plastic within the mold during the molding operation, producing partitions with a more uniform thickness and density.

The reduced throats formed between the ribs and the adjacent mold cores apparently reduce displacement of the mold cores due to differential feed rates at adjacent gates in the mold wall which forms the bottom of the container. Consequently, the resulting battery container greatly reduces the possibility of battery failure due to defective partitions due to cold shorts and voids or incomplete welds resulting in fluid leakage between adjacent battery cells.

Another major effect of the ribbed bottoms is a reduction of the incidence of partition warpage, which condition precipitates problems in subsequent manufacturing operations, notably the inability to effect a good cover seal along the line of a warped partition, resulting in leakage potential from adjacent cells.

Fixed Installation Case

A process described by *K.H.M. Koch and R. Schmechtig; U.S. Patent 3,988,170;*

October 26, 1976; assigned to Aktiebolaget Tudor, Sweden relates to cases for electric storage batteries which are to be fixed on their installation site. The sites may be either stationary or mobile, located, for example, in a building or in a vehicle or boat. In a preferred form, the process provides a battery case which comprises side wall members and a bottom member, the side wall members having extensions which extend below the bottom member of the case and which are adapted to be fixed to a substructure or support member.

In Figure 1.24a, two storage batteries in cases **10** and **12**, having usual terminals **13**, filler opening caps **13a** and plates **13b**, are shown installed together on a support member **14.**

FIGURE 1.24: BATTERY CASE

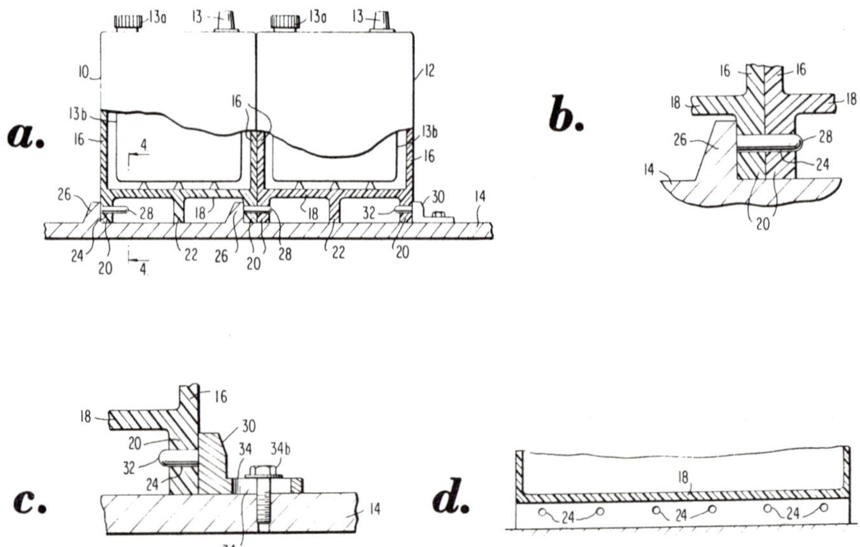

(a) View partly in cross section of two storage batteries with cases installed in operating position on a support member
(b) Sectional view to an enlarged scale of one arrangement used in Figure 1.24a for fastening the batteries to a support member
(c) Sectional view to an enlarged scale of another fastening arrangement used in connection with Figure 1.24a
(d) Partial sectional view taken along the line **4–4** of Figure 1.24a

Source: U.S. Patent 3,988,170

The side walls of the cases are designated by reference numeral **16** and the bottoms of the cases by reference numeral **18**. The side walls have downwardly projecting portions or extensions designated by the reference numeral **20** and a support rib in the bottom is designated **22**. As shown in Figure 1.24a, for example, these downwardly projecting portions **20** are continuations of and in substantial alignment with the other surfaces of the side walls themselves. By making the walls **16** in one piece with the extensions **20**, forces directed in the battery wall are distributed over the whole battery wall.

In order to obtain external reinforcement, the bottom plate or wall **18** may have downwardly extending edges corresponding to the lower part of the battery wall. As shown, the extensions **20** are thicker than the side walls. In this way external strengthening of the walls is achieved, which strengthening may be important in the event of the whole battery having to stand on these edges.

Extensions **20** in Figure 1.24a have space holes **24** therein as also shown in Figure 1.24d. Two groups of attaching members **26**, shown also in Figure 1.24b project from support member **14**. These attaching members **26** include pins **28**. The pins **28** of one group of attaching members extend into the openings **24** in the extension **20** of the left side wall of battery **10**. The pins **28** of the other group of attaching members extend into the openings in both of the adjoining extensions of the adjacent side walls of batteries **10** and **12** as shown in the enlarged view of Figure 1.24b.

Modified attaching members **30** are shown engaging the extension **20** of the right hand side wall **16** of battery **12** in Figure 1.24a. The enlarged view of Figure 1.24c shows each member **30** as having an upright portion carrying a pin **32** extending into the hole **24** of the extension **20**. Member **30** is mounted so as to be moved into an adjustable position on the support member **14** in any suitable manner. For example, arm **34** of member **30** may include a narrow slot **34a** which receives bolt **34b** threaded into support member **14** to allow removal of pin **32** from the hole **24** in the battery casing wall.

In providing the assembly of Figure 1.24a, battery **10** may be placed on the support member **14** and slidingly moved to bring the pins **28** on the two groups' attaching members **26** into the holes **24** in the extensions **20** on opposed side walls. Battery **12** may then be moved into position adjacent battery **10** with the extension on its left side wall receiving pins **28** of the intermediate group of attaching members **26**. Thereafter, attaching members **34** are adjusted into positions with their pins **32** in holes **14** and fixed in place by tightening bolts **34b**.

While Figure 1.24a represents an arrangement where holes for attaching means are required only in the extensions **20** of two opposed side walls of a battery, it is possible to attach a battery or batteries to a support member by means of holes in extensions provided on three or four sides. Various combinations of fixed attaching members **26** and adjustable attaching members **30** for doing this are described.

Injection Molding Technique for Intercell Connection

G. Bergh, K. Brass, J. Hessner and T. Varberg; U.S. Patent 3,900,343; August 19, 1975; assigned to Sonnak Batterier A/S, Norway describe a process for simplifying the conditions relating to the assembly of batteries disposed in a case with a cover of thermoplastic material. It is particularly useful in the small battery factory which has a small production series and modest investment per battery type. The process is characterized in that the partition walls of the case are provided with recesses allowing space for the connectors when the element-unit is disposed in the case, and in that all elements and connectors of each battery are molded integrally in one operation to a finished unit prior to disposal in the battery case.

The connector between two elements is disposed directly onto two adjacent bridges each connecting the plates of the same polarity in one element. An injection molding tool constructed as a simple pipe tool, together with the connector and partition wall, forms an adequately sealed injection mold which, during injection molding with liquid thermoplastic material injected at correctly determined pressure during the injection cycle, can be held together by the modest resistance provided by the base support of the case through two relatively weak adjacent elements of the connection, and through a weak, supported case wall with recess.

By the injection molding, a creep-current resistant restoration of the partition wall is achieved with sealing around the connector, and the edge of the partition wall is thereby suitable for mirror welding or adhesion between cover and case. The process further provides an injection molding tool for carrying out the method, which is characterized by being a single cavity tool, which restores one partition wall while simultaneously sealing the corresponding connection, or as a multicavity tool which restores a plurality of partition walls and simultaneously seals the corresponding connections.

Battery Cover

A process described by *A.H. Wolf; U.S. Patent 4,009,322; February 22, 1977; assigned to Gould Inc.* concerns an improved cover for such batteries which facilitates the removal of electrolyte or other liquid from the battery when this is required. In the process, a guide means is provided on the inside of a battery cover for directing liquid to the vent openings when the battery case is inverted and rocked from side to side thereby facilitating substantially complete draining of the liquid from the battery.

Referring to Figure 1.25a, there is shown an inverted battery **10** including a case **11** and a cover **12** having fill and vent holes **13** therein from which liquid, such as forming electrolyte or rinse solution, is being drained into a trough **14** as the battery is rocked from side to side by a suitable rocking dumper, shown schematically here at **15**.

The battery case **11** and cover **12** may be made of hard rubber or thin-wall plastic material such as, for example, polypropylene or ethylene-propylene copolymers and in the illustrated case the vent holes **13** are disposed in a recessed, channel-like trough **16** formed in the cover **12**. The battery **10** also includes positive and negative terminals **17** and **18**, respectively, extending through the cover **12** in recessed areas **19** and **20** adjacent the rear corners of the battery. The battery **10** is preferably of the lead-acid secondary type capable of activation by the addition of electrolyte or water after being stored in substantially dry condition.

FIGURE 1.25: DRY CHARGE BATTERY

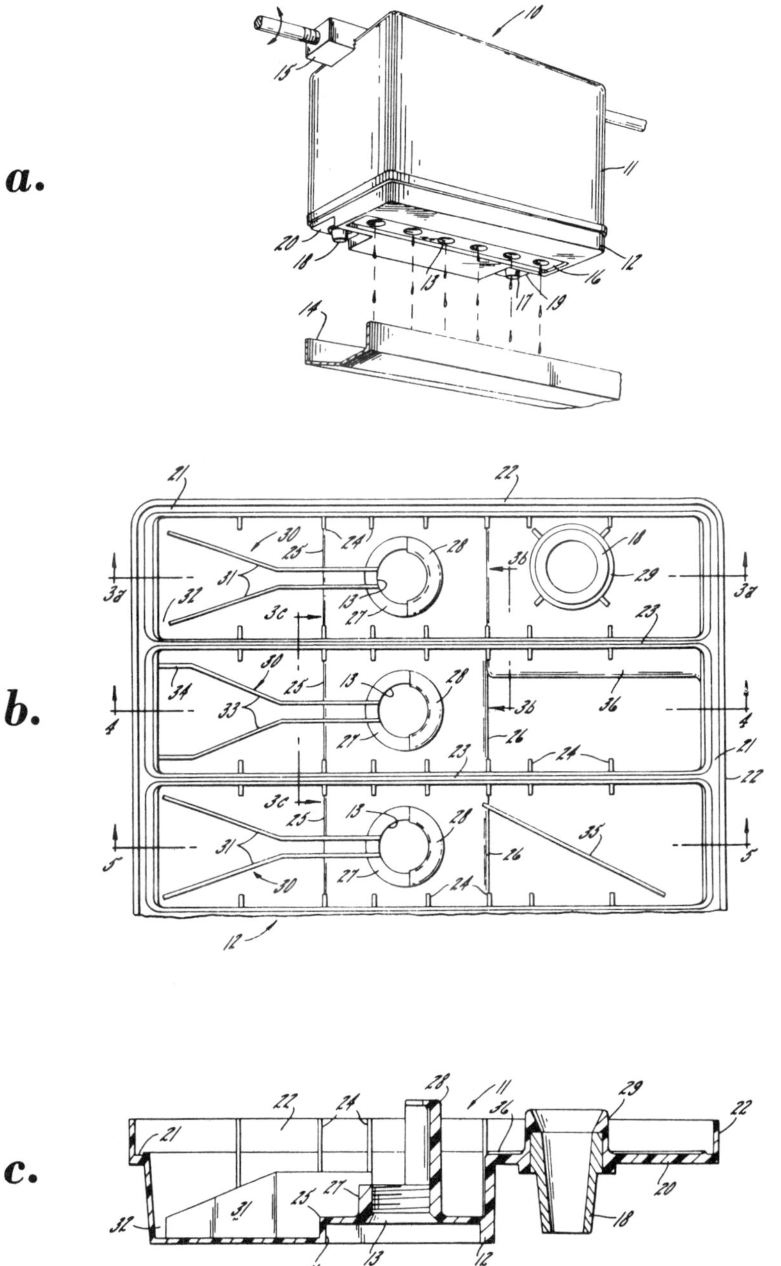

(continued)

FIGURE 1.25: (continued)

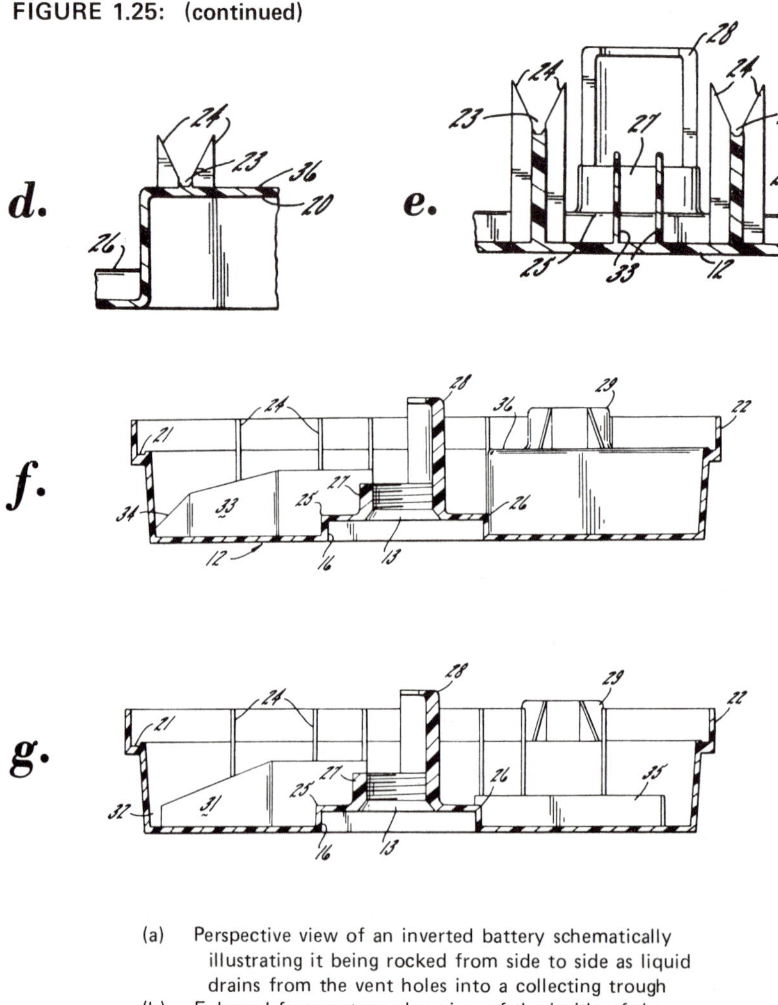

(a) Perspective view of an inverted battery schematically illustrating it being rocked from side to side as liquid drains from the vent holes into a collecting trough

(b) Enlarged fragmentary plan view of the inside of the battery cover illustrating three different examples

(c)(f)(g)(d)(e) Sections as seen substantially along lines **3a–3a, 4–4, 5–5, 3b–3b** and **3c–3c**, respectively, in Figure 1.25b

Source: U.S. Patent 4,009,322

Referring to Figures 1.25b and 1.25c it will be seen that the inverted cover **12** includes a shallow peripheral groove **21** into which the side walls of the case **11** are adapted to fit and be sealed and an upstanding flange **22** overlaps the side walls of the case. Only half of the cover **12** has been shown in Figure 1.25b and it will be understood that this portion of the cover **12** includes intermediate grooves **23** which cooperate with separator walls in the case **11** to divide the battery **10** into separate cells. Three such cell areas are shown in Figure 1.25b. As also shown in Figures 1.25b, 1.25c, 1.25d, and 1.25e the underside of the cover **12** also includes a plurality of tapered fingers **24** for guiding the walls of the case **11** into the grooves **21** and **23**. When the cover **12** is inverted, as shown, the recess or channel **16** in which the vent holes **13** are located forms two raised step-like shoulders **25** and **26** in each of the cell areas. In the cover **12** illustrated, the vent holes **13** also include integral collars **27** and vent hole extensions **28** which project into the battery case **11**. The vent hole extensions **28** permit more accurate visual inspection of the level of electrolyte in the battery cells and the collars **27**, in addition to reinforcing the cover **12** at the vent holes **13**, may be internally threaded for receiving the vent caps (not shown).

Alternatively, the inner surface of the collars may be smooth to receive push-in type vent caps. As shown in Figure 1.25c, the terminal post **18** is molded into a reinforcing integral boss **29** in the cover and, prior to placement of the cover on the case **11**, the posts **17** and **18** are hollow for reception of upstanding elements connected to the respective positive and negative plates of the battery.

Means are provided on the inside of the cover **12** to facilitate removal of the forming electrolyte or other liquid from the battery **10** when it is inverted and rocked from side to side as in Figure 1.25a. Simply for orientation purposes, the side of the battery **10** and cover **12** containing the posts **17** and **18** will be referred to as the back side and, of course, the other side is the front of the battery.

Referring now to Figure 1.25b, it will be seen that the cover **12** is formed with a plurality of integral guide vanes or ribs **30** for concentrating and directing the flow of liquid toward the vent holes **13**. As more particularly shown in Figures 1.25c and 1.25e, the ribs **30** for the uppermost cell area (as seen in Figure 1.25b) comprise a pair of upstanding guide vanes **31** extending from adjacent one side of the cover **12**, over the step **25** and into the collar **27**, opposite the vent hole extension **28**.

As the front of the inverted battery **10** (left side of cover **12** in Figure 1.25b) is tilted down, the fluid flows into the V-shaped area between the vanes **31**. Then as the front of the battery is rocked up, a substantial portion of the fluid is channeled between the vanes **31**, over the step **25** and collar **27** and into the vent hole **13**. During subsequent rocking cycles, the electrolyte or other liquid in the cell flows back and forth over the inside of the cover **12** defining the cell area. Each time the front is lowered, most of the liquid collects at the V-shaped entrance between the vanes **31** and then it is concentrated and directed up over the step **25** and collar **27** when the front is raised. After 15 cycles of alternating 15° tilts in about 4 minutes all but a very small amount (less than 5 cc) of the electrolyte is drained from the cell.

As shown in Figure 1.25c, the vanes **31** do not extend completely to the front wall of the cover **12**. This leaves a small channel indicated at **32** for the liquid to flow into the area between the vanes **31**. An alternative example is shown in Figure 1.25f where the vanes **33** do extend all the way to the front edge of the cover but at this point, indicated by **34**, they have a very low profile.

In this case it will be understood that as liquid flows to the left (as seen in the center cell area of Figure 1.25b) it is first concentrated outside the vanes **33** and then spills over at **34** into the area between the vanes. Pursuant to a further feature of the process, means are also provided for concentrating and directing the reverse flow of liquid over the step **26** and around the vent hole extension **28**.

As shown in Figure 1.25g and the lowest cell area of Figure 1.25b, the cover **12** includes an integral guide vane or rib **35** which extends substantially diagonally across the back side of the cell area from adjacent the back edge of the cover **12** to the step **26**. When the back of the battery **10** is tilted down, fluid flows around the rib **35** and then as the back is raised the liquid is directed by the rib over the step **26**, and around the forward directing means **30**, illustrated here, in the form of ribs **31** as in Figure 1.25c. It will be appreciated that a rib similar to the rib **35** could also be formed in the back side of the central cell area shown in Figure 1.25b. Such a rib is not necessary, however, in the uppermost cell area of Figure 1.25b since the recessed portion **20** containing the post **18** forms a platform **36** higher than the shoulder **26** when the cover is inverted (see Figure 1.25d).

The process provides a simple yet effective means for removing substantially all of the liquid from a battery **10** when it is inverted and rocked from side to side. The liquid concentrating means **30** in the form of one or mores ribs directs liquid over the step **25** and collar **27** and out the vent hole **13**. An additional rib **35** may be employed in the back side of the cell area to direct the reverse flow of liquid over the step **26** and around the vent extension **28**. A plurality of ribs may be provided to funnel the liquid into smaller and smaller channels leading to the vent hole **13**.

Cover Assembly

W.B. Hayes, Jr., and R.L. Kreutzfeldt; U.S. Patent 3,957,539; May 18, 1976; assigned to Gould Inc. describe a battery cover assembly that insures a substantial sealing region between the sealant and both the cover and the jar through a relatively wide range of vertical cover positions.

Referring to Figure 1.26, there is shown a storage battery cover **10** intended for a battery cell cover assembly, shown in Figure 1.26c, wherein the cover **10** is fitted into the walls **11** of an open-topped jar **12** defining a cell **13**, and a flowable, settable sealant **15** joins the cover **10** and the jar **12**.

FIGURE 1.26: BATTERY COVER

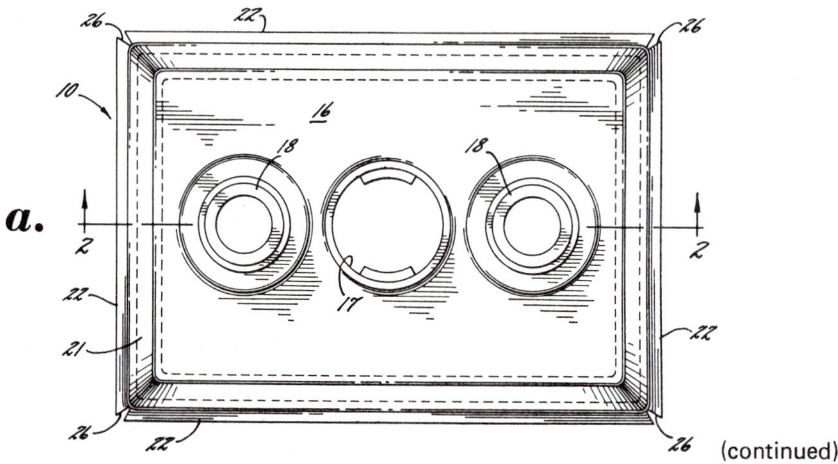

(continued)

FIGURE 1.26: (continued)

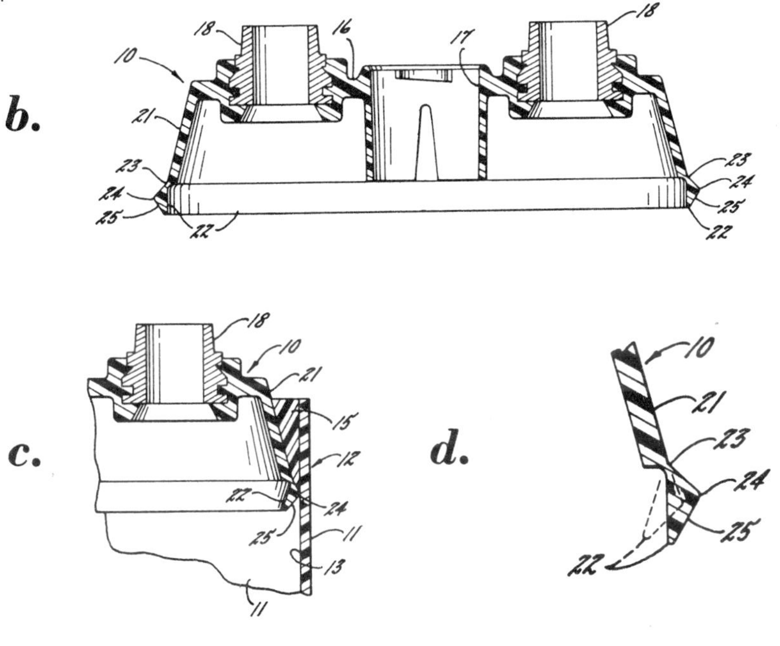

(a) Top plan of cover
(b) Section taken along the line **2–2** in Figure 1.26a
(c) Fragmentary section of a cover assembly embodying the cover of Figure 1.26a
(d) Enlarged fragmentary section showing alternate positions of the parts assumed in the making of a cover assembly

Source: U.S. Patent 3,957,539

As is conventional, the cover **10** and the jar **12** are formed of hard rubber or, preferably, resilient plastic, and the sealant **12** is a bituminous compound poured hot and molten into place where it hardens in sealing contact with both the cover and the jar.

The cover **10** includes a planar main body **16** in which is formed the usual cap-receiving, vent collar **17** and a pair of lead terminal inserts **18** forming parts of terminal assemblies. The cover body **16** is surrounded by a depending, outwardly sloping skirt **21** whose lowest and greatest periphery is well within the opening of the walls **11**, and having a plurality of straight, integral sealing strips **22** extending down from straight line, reduced thickness, hinge sections **23**. The strips **22** define jar wall contacting lines **24** as well as camming surfaces **25** for swinging the strips **22** down and in upon insertion of the cover **10** into the top of the jar cell **13**.

The relaxed, unstressed positions of the hinge sections **23** cause the contacting lines **24** to define a periphery well outside of the nominal dimension of the opening defined by the cell walls **11**. Thus, considerable variation in tolerances between the cover and the jar can be accommodated by the sealing strips **22** swinging in (see Figure 1.26d) when the cover is inserted down into the jar. The ends of the strips **22** are flared but spaced to define notches **26** (see Figure 1.26a) so as to provide for inward swinging movement about the straight line hinge sections **23** without distortion of either the cover or the jar.

It will be appreciated that the engagement between the contacting lines **24** and the walls **11** is not, in itself, relied upon for sealing of the cell **13**. This engagement need only be continuous enough to contain the relatively viscous hot sealing compound until the latter cools and sets. The true seal of the cell **13** is provided by the intimate contact between the hardened sealant **15** and both the cover **10** and the walls **11**, and it can be seen that the cover **10** defines a deep groove for the sealant having a range of possible groove depths substantially equal to the depth of the cover skirt **21**. Thus, the top of the cover **10** can be disposed well above the tops of the walls **11** (see Figure 1.26c) and there is still provided substantial sealing area defined by the sides of the sealant groove.

It can thus be seen that the cover **10**, in cooperation with the simply formed jar **12**, provides a battery cover assembly that insures substantial sealing regions between the sealant and both the cover and the jar through a rather wide range of vertical positions of the cover. Insertion of the cover **10** into the jar **12** is relatively simple because of the hinged strips **22** with their camming surface **25**, and the permissible tolerances between the cover and the jar can be relatively great without altering the eventual cooperation of the parts.

R.L. Schenk, Jr.; U.S. Patent 4,011,364; March 8, 1977; assigned to Gould Inc. describes a cover arrangement for a battery having symmetrically disposed posts of opposite polarity including abutments on at least some of the posts and opposing projections on the underside of the cover which prevent the posts from fully extending through apertures in the cover when the cover is misoriented. Thus, the so-designed cover arrangement prevents misorientation of the cover with respect to the polarity of the posts and polarity indicia on the cover.

Control of Electrolyte Level

A process described by *W. Perkams; U.S. Patent 4,008,355; February 15, 1977; assigned to AS-Motor GmbH KG, Germany* relates to an electrical storage battery having a filler and a storage space which is common to all the cells of the storage battery, and more particularly to such a storage battery which is so constructed that it is essentially immune to changes in electrolyte level differentials in the various cells.

Briefly, in the battery, the inlet connection between a common filler space and any one of the individual cells is U-shaped; this inlet connection is formed with flow resistance means located between the expansion space and the individual cell in order to increase the resistance of fluid flow. The flow resistance portion is so dimensioned that gas pressure arising in the individual cell tends to drive liquid from the cell through an overflow tube into the expansion space before the gas can drive electrolyte through the inlet tube into the expansion space.

The overflow tube, typically, extends for some distance above the bottom wall of the expansion space so that galvanic connection between liquid from the overflow tube and a liquid pool in the expansion space is inhibited. Typically, the flow resistance means is a column of liquid itself.

In a preferred form of the process, the housing or box for the battery is formed with a side pocket at the side wall of any one of the cells in which the respective inlet tubes terminate. The side pocket is open at the top and communicates with a gas expansion space of the respective cell. A tube stub extends from the common expansion space into the gas space, the lower end of which is spaced from the bottom of the pocket. The stub forms a gap, open at the top with a pocket. The flow resistance in the inlet tube is increased by forming the inlet tube as a stub pipe as well, having a length which is greater than the height of the overflow tube.

Multicellular Sealing Member

A process described by *O. Jache; U.S. Patent 3,963,521; June 15, 1976; assigned to Accumulatorenfabrik Sonnenschein GmbH, Germany and 3,919,371; Nov. 11, 1975* relates to the leakproof bonding of components of an electrical storage battery, and particularly to the leakproof bonding of storage battery components made of thermoplastic material either to other components made of substantially the same material or to current-carrying components made of metal such as lead.

According to the process components of an electrical storage battery are bonded together in leakproof relation by assembling the components into relative positions where a cavity is provided between those regions of the components at which the components are to be bonded together, injecting into the cavity a foamable thermoplastic resin composition rendered expandable by addition of a foaming agent, and allowing the injected thermoplastic resin composition to expand to a multicellular mass having closed wall cells and to solidify while maintaining the components in their relative positions.

As only a low pressure is exerted on the walls of the cavity by the thermoplastic resin composition during injection and the following expansion within the cavity there is no need for reinforcing the components surrounding the cavity. Although the foaming or swelling pressure of the expanding thermoplastic resin composition is relatively low, the resulting multicellular mass penetrates all voids within the cavity and firmly adheres or is fused to the walls of the components which are preferably made of thermoplastic material or metal. Thus the multicellular sealing member produced from the foamed up thermoplastic resin composition and its subsequent solidification ensures a tight joint between the components.

If desired, solidification of the expanded thermoplastic resin composition may be accelerated by cooling. The volume of the closed-wall cells within the foamed up mass is variable within a wide range and depends on the operational characteristics (amount of foaming agent added to the thermoplastic resin composition, viscosity of the plastified thermoplastic resin composition, injection rate, etc.). It is possible to obtain multicellular sealing members having a specific gravity as low as 0.45 g/cc. The bonding method according to the process is superior to the known methods of bonding storage battery components made of thermoplastic material either to other thermoplastic components or to metallic components. This superiority is due to the fact that the resulting one piece molded multicellular sealing member between the components has thermal properties which are by far better than those of compact thermoplastics, i.e., thermoplastics which do not contain cells. The multicellular sealing member is flexible and can easily take up volume changes of embedded metal components when subjected to sudden temperature changes.

The multicellular sealing member is able to counteract and absorb the volume shrinkage occurring during the transition from the plastic into the solid phase and upon further cooling so that no stresses are transmitted to the components adhering and/or fused to the solidifying foamed thermoplastic resin composition.

Conventional injection molding machines may be used for plastifying and injecting the thermoplastic resin composition capable of expansion. It is preferred to use a screw extruder or an extruder feeding a shooting cylinder. To avoid long feeding distances, the plastified thermoplastic material may be injected into the cavity through two or more injection nozzles. It is also possible to use several shooting cylinders which are fed by a common screw extruder to fill several cavities simultaneously and independently of each other.

Positive Displacement Bonding

R.L. Schenk, Jr.; U.S. Patent 3,909,301; September 30, 1975; assigned to Gould Inc.; J.A. Bruzas and W.E. Coville; U.S. Patent 3,908,738; September 30, 1975; assigned to Gould Inc. and R.L. Schenk, Jr. and W.B. Hayes; U.S. Patent 3,909,300; September 30, 1975; assigned to Gould Inc. describe methods and apparatus for automatically forming, on a continuous, reproducible basis, fusion bonds devoid of structural, electrical and cosmetic defects between two or more workpieces.

The process involves (1) moving a heated electrode into the area to be bonded so as to uniformly heat and melt the portions of the workpieces to be bonded; while, at the same time, (2) displacing substantially all of the molten material from the area to be bonded into a storage area or reservoir surrounding the heated electrode where such molten material is maintained in its uniformly heated molten state, and (3) maintaining the area of the bond substantially free of oxidants and (4) then retracting the electrode so as to permit the molten material to return to the cavity formed by the electrode in the workpieces where such molten material is allowed to cool and solidify—thus forming a flawless bond between the workpieces.

Thermal or fusion bonds are made in accordance with the process and by a combination of (1) elevated temperature levels sufficient to melt the material to be bonded, and (2) displacement of the molten material; as contrasted with more conventional techniques and/or apparatus which combine elevated temperature levels and pressure.

For battery construction, the flexibility and versatility of the bonding system of the process permits use of the system to bond intercell connectors either externally of or internally of the battery casing and, additionally, the system can be utilized to bond straps, plates and/or terminal posts to one another irrespective of the particular configurations of such battery components.

Vibration-Proof Battery

T. Kosuge and I. Sano; U.S. Patent 3,909,294; September 30, 1975; assigned to Furukawa Denchi Kabushiki Kaisha, Japan describe a vibration-proof battery of the type including an electrode plate group holding member adapted to compress from above an electrode plate group positioned on a rest and contained in a battery casing. The holding member is formed so as to have along its lower edge comb-tooth-shaped projections and is arranged whereby each of the projections compresses the upper edge of a corresponding one of positive and negative electrode plates constituting the electrode plate group.

Referring to Figure 1.27, numeral **1** denotes a synthetic resin battery casing comprising a battery container portion **2** and a battery cover portion **3** adapted to be secured in an air- and liquid-tight manner to the upper edge surface of the container portion through a suitable sealing agent. The interior of the battery container **2** is divided by partition walls **4** into three cell containers **2a**, and an electrode plate group **5** is mounted in each cell container **2a** and positioned on a rest **6**.

FIGURE 1.27: VIBRATION-PROOF BATTERY

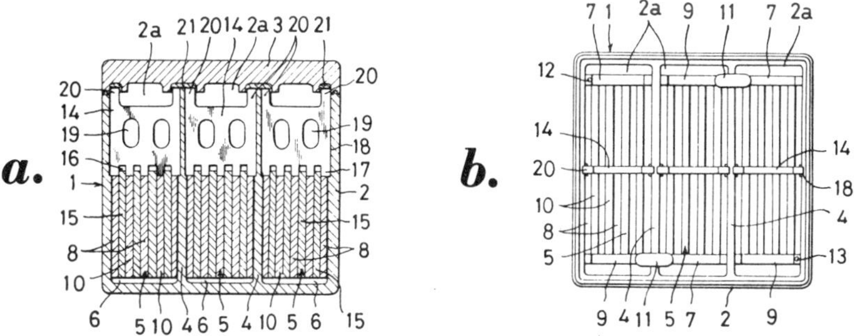

(a) Elevational side view, in section, of one form of a battery
(b) Top plan view of the battery of Figure 1.27a, illustrated with its cover removed for purposes of clarity

Source: U.S. Patent 3,909,294

A negative strap **7** interconnects negative electrode plates **8** at their ear portions, while a positive strap **9** interconnects positive electrode plates **10** at their ear portions. The negative strap **7** and the positive strap **9** in adjacent cell containers **2a, 2a** are interconnected by a U-shaped cell-bridging strap **11** which extends over the partition wall **4**. The three electrode plate groups **5**, which are thus connected in series, are electrically connected to the exterior of the casing through negative and positive terminal posts **12, 13** provided on the oppositely located endmost straps.

According to the process, in order to completely immovably fasten each of the electrode plate groups **5** against vibration, a holding member **14** for imparting pressure from above and retaining the plate groups is located on the upper end thereof within each cell container **2a**. In effect, each holding member **14** consists of, for example, a plate-shaped synthetic-resinous molded member formed independently of the battery casing **1** and being almost equal in width to the width of the space for insertion of the electrode plate group into each of the cell containers **2a**, having comb-tooth-shaped projections **17** in a quantity corresponding to the number of the negative and positive electrode plates **8, 10** constituting the electrode plate group **5**, and having recessed space portions **16**, each adapted to receive a separator **15**, previously formed along one edge.

The plate-shaped holding member **14** is inserted in press-fit relationship into each cell container **2a** until the comb-tooth-shaped lower edge is in contact with the upper edge surface of the electrode plate group **5**. In this instance, the holding member **14** may have applied to its plate group contacting edge portions an adhesive agent, a sealing agent or the like, so as to be rigidly fastened to the cell container **2a**. Thus, the projections **17** on the lower edge of the plate-shaped holding member **14** are adapted to press from above against the upper edges of all of the negative and positive electrode plates **8, 10** of the electrode plate group **5**, whereby the fastening thereof is effected firmly and strongly.

The holding or retaining position for the electrode plate upper edges by the holding member **14** lies between the positive and negative straps **7, 9** in each cell container **2a**, in effect, at the intermediate portion of the electrode plate upper edges, in view of which the pressing and fastening of the plate groups can be effected extremely effectively and stably. It is ordinarily preferable that, as illustrated, each holding member **14** is inserted so as to extend along the longitudinal center line of the battery casing **1**, whereby each electrode plate is pressed along its upper edge center portion.

Under this compressive condition, the upper projecting edge of each separator **15** is received by the corresponding recessed portion **16** between the projections **17, 17** so that the separator **15** is not damaged thereby and is safely retained. A pair of vertical guide grooves **18** for insertion of the holding member **14** are formed along the inner surfaces of the opposite side walls of each cell container **2a**, so as to assure the insertion of the holding member **14** at a predetermined position. A liquid passage opening **19** is made in the holding member **14**, so as to permit battery electrolyte to flow to the opposite sides of the member **14** in order to maintain the battery in a good operating condition.

The battery cover portion **3** is provided at its rear surface with a number of recesses **21** for receiving the upper projecting edges **20** of the holding member **14** upon closing of the cover portion **3**, each upper edge **20** being embedded in a sealing agent applied to the recess **21** so as to obtain thereby an air- and liquid-tight covering connection between the cell containers **2, 2** in that region thereof.

The holding member **14** is in compressive contact or connected with the opposite inner side walls of the cell container so as to concurrently form a reinforcement member for the battery cell container, which may be of comparatively small thickness. Thus, each of the comb-tooth-shaped projections formed along the lower side edge of the holding member inserted on the upper surface of the electrode plate group, which is located on the rest and contained in the battery casing, compresses and fixes the upper edge of the corresponding one of the positive and negative electrode plates of the electrode plate group, so that each electrode plate is directly compressed along its upper edge by each projection, so as to positively eliminate or prevent vibration of the electrode plate group, and the vibration-proof characteristic of the battery is greatly enhanced.

Inverted Construction

A process described by *E. Voss and H. Rabenstein; U.S. Patent 4,031,293; June 21, 1977; assigned to Varta Batterie AG, Germany* relates to a maintenance-free lead storage battery whose electrode grid is made of pure lead or antimony-free lead alloys and which includes a sulfuric acid electrolyte which is absorbed in the pores of the separator and of the active mass.

The process involves keeping the weight of the negative active mass at less than 80% of the weight of the positive active mass. The electrolyte quantity and concentration are so proportioned that, after 20-hour current discharge the acid density is at most 1.07 g/cm^3. The proportioning of the electrode mass is preferably achieved by departing from the construction which is customary in the lead storage battery art and by building, for example, a cell which has four positive and three negative electrodes (inverted construction). This produces a capacity increase which corresponds to the positive mass increase.

The greater acid dilution accompanying the increased capacity discharge is offset by the lower surface loading of the positive electrode. In this construction, the capacity is also limited by the positive electrode. The positive mass utilization has its customary values. On the other hand, the apparent mass utilization of the negative electrode is increased by this inverted construction. Such proportioning results in considerable enhancement of the cycling capability. This is attributable to the fact that the recharging takes place at considerably diluted acidity, and this promotes the reformation of α-PbO_2.

Pancake Battery

R.R. Aronson; U.S. Patent 3,928,080; December 23, 1975; assigned to Electric Fuel Propulsion Incorporated describes a battery preferably for use as a power source in an electrical vehicle in which the batteries can be disposed across the bottom of the car to form a structural part of the vehicle. Each cell is preferably divided into three chambers–an upper chamber, a middle chamber containing the horizontally mounted battery plates and a lower chamber–so that the electrolyte flows by gravity from the upper chamber through the plates, flushing waste products, such as water, which result from the production of electricity by the battery and which eventually insulate the electrolyte from the plates and prevent further electricity from being produced.

After all of the electrolyte has flowed from the upper chamber into the lower two chambers, it can be pumped back to the upper chamber to repeat the journey, either while the vehicle is stopped, by applying a source of compressed air to the lower chamber, or continuously while the vehicle is moving, for example, by pumping the liquid during regenerative braking. The plates can be of any shape or size including circular, and preferably the horizontal plates are disposed at a slight angle to prevent gas produced during charging from building up between the plates.

In another example the plates are formed as cones and disposed in a stack in the battery so that the electrolyte flows downward through the cones.

Deep Sea Battery

A process described by *D.W. Kraft; U.S. Patent 4,012,234; March 15, 1977; assigned to Bunker Ramo Corporation* relates to electrochemical storage battery systems for deep ocean applications, and more particularly to systems with devices for use at some depth for automatically activating electrochemical storage batteries when dropped into an ocean, and for preventing the batteries from crushing due to the high external pressure at great ocean depths.

In the process, a pack of electrochemical batteries is adapted to be stored for an extended period and then put into use in a body of water below some predetermined depth by removing all electrolyte from the batteries after they have been fully charged. The pack is provided with means for restoring the electrolyte in the batteries in response to water pressure at the selected depth. The electrolyte restoring means may be comprised of a reservoir filled with fully constituted electrolyte.

Each battery is connected to the reservoir by a nonconductive tube having suitable means for blocking passage of electrolyte from the reservoir into the battery until pressure in the reservoir exceeds water pressure at the selected depth. The reservoir is provided with suitable means for transferring the pressure of the body of water to the electrolyte in the reservoir. Alternatively, the electrolyte restoring means may be comprised of a frangible vessel filled with concentrated electrolyte inside the reservoir. Water is admitted into the reservoir through suitable means.

At some pressure depth below the selected pressure, the vessel breaks allowing concentrated electrolyte to be diluted by the admitted water. The reconstituted electrolyte is then forced into the batteries as before. Another alternative is to fill the frangible vessel with an electrolyte forming substance, such as a salt or a gel, which forms fully constituted electrolyte upon being mixed with water. Still another possibility is to fill the reservoir with just water, instead of fully constituted electrolyte, and to impregnate the battery plates with electrolyte forming substance.

In each case, the battery pack is encased and spaces inside the encasement are filled with inert fluid to equalize pressure on the outside of the batteries with pressurized electrolyte inside. The compliance of the reservoir is preferably greater than of the walls of the batteries. The encasement may be a rigid canister with a compliant diaphragm on at least one end, or the equivalent in the form of a closed sock filled with inert fluid around the battery pack and an open ended rigid canister around the stock to protect it from tears in handling.

Referring to Figure 1.28a, a canister **10** of suitable plastic or metal, such as stainless steel or brass, is provided to contain banks of batteries **11** and **12** to power an electronic device, such as a sonobuoy, in an extension of the canister. Only two batteries are shown for each bank, namely batteries **11a** and **11b** for bank **11** and batteries **12a** and **12b** for bank **12**. However, it is to be understood that as many as 22 D-size batteries may be supported in one bank with a canister of 7.5 inch diameter, as shown in Figure 1.28b, for the bank **11**.

FIGURE 1.28: WATER ACTIVATED, DEEP SEA BATTERY

a.

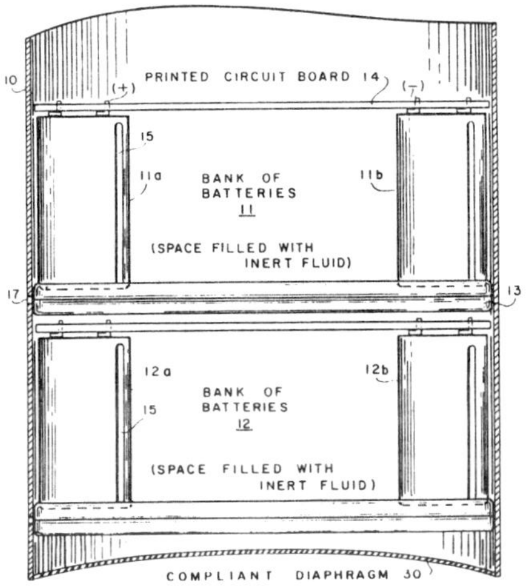

b.

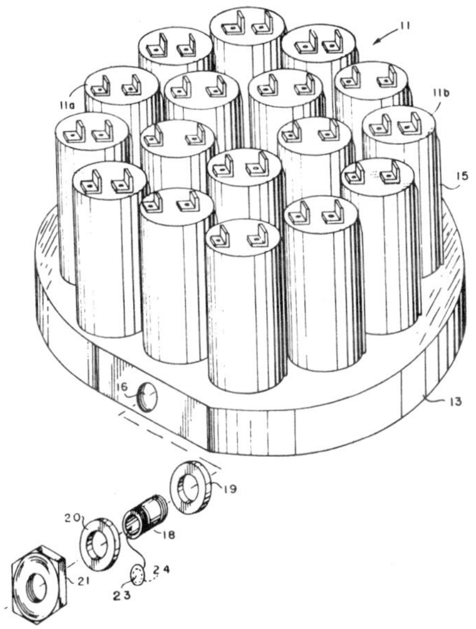

(continued)

FIGURE 1.28: (continued)

c.

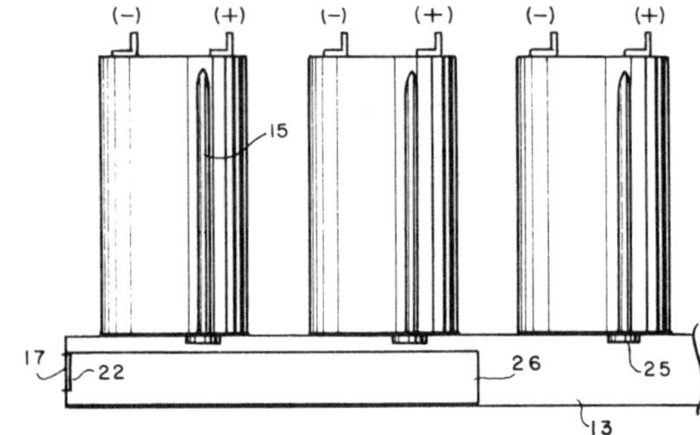

(a) View of a canister in cross section showing two banks of batteries mounted on reservoirs for electrolyte, or electrolyte forming substance

(b) Isometric view of one of the banks of dry cell batteries mounted on a reservoir with an exploded view of parts for coupling a reservoir inlet with water outside the canister of Figure 1.28a

(c) Schematic diagram of a portion of a bank of dry cell batteries mounted on a reservoir

Source: U.S. Patent 4,012,234

The base of each battery in the bank is seated in a depression molded into a reservoir **13**. In an illustrative example, each battery is a fully charged lead-acid battery modified by removing the electrolyte, either during manufacture or in a separate operation as described previously. These dry-charged batteries are connected to the electronic device through printed circuit boards, such as a printed circuit board **14** (Figure 1.28a), and cables extending along the inside wall of the canister. Positive and negative tabs (+ and -) connect the batteries to the printed circuit board. The device is not turned on until it is about to be dropped into the ocean, but even then the batteries do not power the device for lack of the electrolyte.

The electrolyte for each bank of batteries is stored in the supporting reservoir, such as the reservoir **13** for the bank **11**. Each battery is connected to its supporting reservoir by a hollow nonconductive tube, such as tube **15** for the battery **11a**. For lead-acid batteries, the electrolyte stored in the reservoir could be a fully constituted solution of sulfuric acid ready to be injected directly into the batteries, or an electrolyte-forming substance, such as concentrated sulfuric acid to be diluted with water in the reservoir prior to injection, or just water injected into the batteries for reconstituting electrolyte from substance impregnated into the battery plates.

A frangible diaphragm in each tube isolates the reservoir from the battery, as will be more fully described hereinafter. When the canister is dropped into the ocean, it will sink to at least some predetermined level. As it sinks, the pressure on the canister increases, but the canister filled with an inert fluid, such as Dow-

Corning 200 Silicon fluid, will not collapse. Each reservoir has a water inlet communicating with the ocean through a port in the canister, such as an inlet **16** shown in Figure 1.28b, communicating with the ocean through a port **17** (Figure 1.28a). A threaded sleeve **18** (Figure 1.28b) and rubber washer **19**, seal the communication passage between the canister port and the reservoir inlet. The sleeve is threaded into the reservoir through the port **17** (Figure 1.28a) with the washer **19** over the sleeve before it is so inserted. A second washer **20** is then placed over the sleeve outside the canister and a nut **21** is tightened over the washer **20**. When a sufficient depth pressure has been reached, a frangible diaphragm **22** (shown schematically in Figure 1.28c) blocking the water inlet ruptures and ocean water enters the reservoir.

The diaphragm indicated schematically may take any one of a number of different forms that may occur to one skilled in the art, but it is preferably a sheet of elastic material **23** stretched over a ring **24** press fitted into the sleeve **18**. When the sheet ruptures, water rushes in, but no part of the ruptured diaphragm is carried into the reservoir to block any of the tubes leading to the batteries. Alternatively, a plug of silicone grease may be used as the diaphragm in the sleeve.

When the external pressure rises above a few psi, the ocean water forces the plug into the reservoir. A net inside the reservoir could be used to assure that the loose plug in the canister does not block any of the battery tubes. Still another alternative for the water inlet is a spring-loaded one-way valve which opens into the reservoir when the external pressure exceeds the force of the spring. Frangible diaphragms **25** at the inlets of the battery tubes (shown schematically in Figure 1.28c) may also take any one of a number of different forms, such as a stretched membrane over the inlet, or a plug of silicone grease.

Assuming the reservoir is charged with concentrated sulfuric acid as an electrolyte forming substance, the ocean water mixing with acid forms a hot mixture that flows into the cells because the reaction of water and acid is exothermic. For example, the mixing of concentrated H_2SO_4 with water to form a 35% solution by weight produces a reaction that warms the solution about 40°C. The heat produced thus significantly warms the batteries. This produces enhanced performance over a cold battery.

The concentrated sulfuric acid would not, of course, fill all of the reservoir. Instead it fills only a polyethylene or thin glass vessel **26** in the reservoir. This vessel would be strong enough not to prematurely break in handling, but fragile enough to break at 50 to 100 psi pressure. As ocean water enters the vessel **26**, and pressure mounts, the vessel breaks, releasing the concentrated acid as it continues to mix with water for a 36% acid mix. The frangible diaphragms **25** are designed not to break until an even higher pressure is reached, thus giving the mixture of water and acid time to stabilize.

Plugs of silicone grease would inherently provide this delay due to the small diameter of the tubes and the distance the plugs must travel in the tubes. Each battery eventually becomes very nearly filled with fluid. Trapped air is harmlessly compressed in the batteries. If the reservoir is charged with fully constituted electrolyte, it would fill the entire reservoir, and the vessel **26** would not be required. Instead a compliant diaphragm would be provided to isolate the electrolyte from the incoming ocean water. Otherwise the electrolyte would be further diluted.

The vessel **26** could be charged with some other electrolyte forming substance as well, such as sodium sulfate (Na_2SO_4) salt or sodium hydrogen sulfate ($NaHSO_4$) salt. It could also be filled with an acid gel, such as an acid rich mixture of sulfuric acid immobilized in a boron-phosphate gel. Other possibilities are acid anhydrides, or sulfur trioxide (SO_3). The fully constituted electrolyte has the disadvantage of not yielding any heat to warm the batteries. The salts yield very little heat, but would need a smaller vessel **26**. The table below summarizes important characteristics of these electrolyte forming substances for comparison with fully constituted (mixed) H_2SO_4 acid.

Substance	Physical Form	Needed Volume per D-Size Battery	Temp. Rise Entire Cell
Mixed acid	Liquid	30 ml	None
Conc. acid	Liquid	7 ml	42°C
Normal acid salts	Solid	6.5 ml	1.2°C
Hydrogen salts	Solid	6.5 ml	3°C
Gelled acid	Gel	8-9 ml	≅40°C

Cooling System

J. Brinkmann and H. Franke; U.S. Patent 4,007,315; February 8, 1977; assigned to Varta Batterie AG, Germany describe a method of dissipating the heat developed in the cells of a battery, and particularly a vehicle propulsion battery, utilizing cooling elements immersed in the electrolyte above the plates. Preferably, the cooling medium is ordinary water, which has not even been demineralized. Of course, other cooling media may be used, whose specific electric conductivity is about the same or even greater than that of water.

As shown in both Figures 1.29a and 1.29b, the cooling elements **3** are so arranged in the individual storage battery cells **2** that they are permanently submerged in the electrolyte. By known means this can be assured under all operating conditions.

FIGURE 1.29: BATTERY CELL COOLING SYSTEM

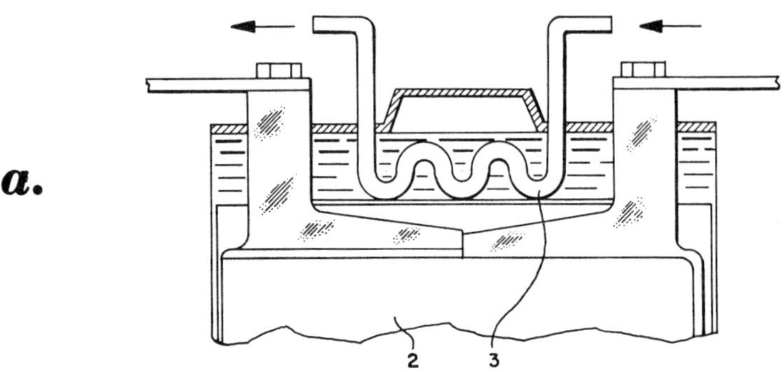

(continued)

FIGURE 1.29: (continued)

b.

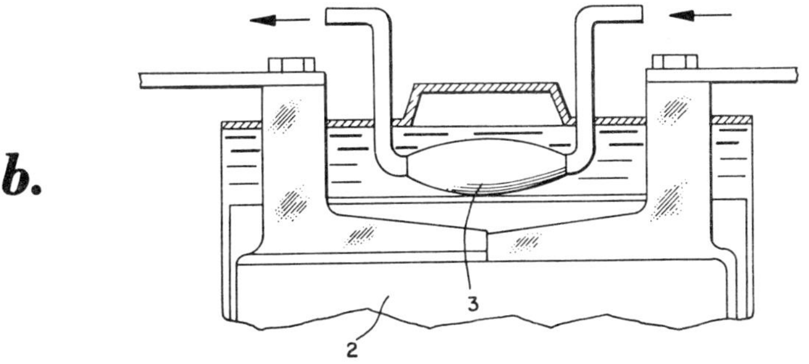

c.

d.

e.

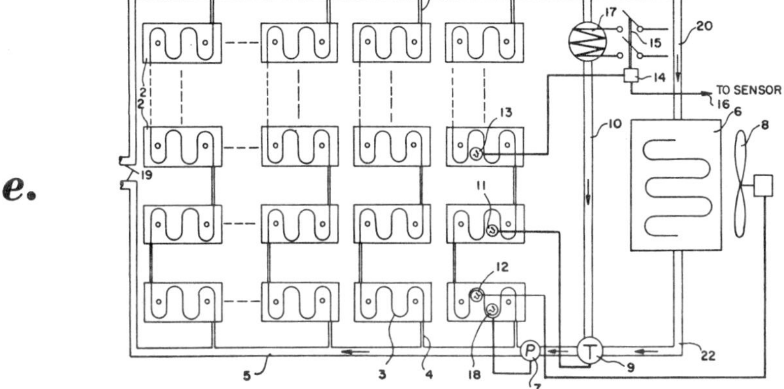

(a)(b) Schematic illustration of battery cells provided with cooling arrangements
(c)(d) Schematic diagrams for two types of cooling arrangements for a multiplicity of battery cells
(e) Cooling arrangement for complete storage battery assembly

Source: U.S. Patent 4,007,315

As shown in Figure 1.29a these cooling elements **3** are made of cooling coils fabricated from pipes made of materials having suitably high heat conductivity and specific electric resistivity. As shown in Figure 1.29b, flat, large-surface cooling bladders **3** can also be used as the cooling elements. Appropriate materials for these cooling coils or cooling elements are polypropylene, polyethylene and PVC. Glass may also be used.

The individual cooling elements **3** within cells **2** may be connected in series, as shown in Figure 1.29d, or in parallel as shown in Figure 1.29c. Combinations of these connections may also be utilized. A series-parallel connection of a cooling system is shown in Figure 1.29e. In the Figure 1.29e a predetermined number of individual cell cooling coils **3** are connected in series. The inlets and outlets **4** of such cell series terminate at manifolds **5**, to which is connected a heat exchanger **6** with inlet **20** and outlet **22**. A circulating pump **7** causes a sufficient quantity of cooling liquid to flow continuously through the entire closed cooling system. The cooler or heat exchanger **6** can be forced-air cooled in conventional manner by a blower **8**.

The broken line connections appearing in Figure 1.29e between rows and columns of cells **2** indicate the possible presence in the battery as a whole of rows and columns of such cells in addition to the ones specifically diagrammed. For example, in a typical vehicle propulsion battery, there might be 180 such cells.

Reverting to Figure 1.29e, elements **11, 12, 13** and **18** are temperature sensors, by means of which the circulatory cooling system is controlled. A switching valve **9** permits the cooling medium to flow either through cooler **6** or through bypass conduit **10**. Elements **14, 15** and **17** are the components of a heater system, by means of which the cooling medium may be warmed.

The driving motors for pump **7** and blower **8**, and also the heating element **17** are supplied with the potential of the battery via switches. Likewise, the operating voltage for temperature sensors **11, 12, 13** and **18**, and for the operating control elements of the cooling circulatory system is derived from the battery.

Pump **7**, cooler **6**, blower **8**, valve **9**, bypass conduit **10** and the components **14, 15** and **17** of the heating system are preferably assembled in a compact unit which can be mounted on the battery in various positions. Manifolds **5** then flexibly interconnect cells **2** with that unit.

In accordance with the process, the cooling coils or cooling bladder **3** inside the cells are made of plastic–preferably polyethylene because of its relatively good heat conductivity accompanied by relatively high specific resistance. Therefore, the necessary electrical potential separation between cooling medium and cooled cell electrolyte is automatically obtained, inasmuch as the plastic serves as electrical insulator between the cell electrolyte and the cooling medium. At the same time, there is good heat transfer through the walls of the cooling elements.

Preferably, ordinary water can be used as the cooling medium. No substantial leakage currents arise, because there is no difference in electrical potential between the cooling medium of the individual cooling coils in the cells of the battery assembly and the circulatory cooling system. Special precautions for insulating the cooling coils inside the cells to prevent electrode short circuiting are therefore unnecessary. During start-up of batteries from a very low temperature condition, the internal losses of the battery are often not sufficient to bring the cells quickly to their optimum operating temperature.

Heating element **17** in bypass conduit **10** makes it possible to heat the cooling medium. In such operation of the cooling system via bypass **10**, the appropriate elevated temperature of all cells can be reached. The energy for heating the cooling medium by means of heating element **17** is preferably taken from a stationary power source, in order to avoid drain on the stored energy of the battery. To that end, the heating element **17** is connected via switch **15** to the electrical power supply for the battery. A sensor **16** permits closing of switch **15** only when energy supply to heating element **17** can be made from a stationary power source.

This would be the case, for example, during charging of the battery. In principle, however, heating may also be accomplished using the stored energy of the battery, and doing so is advantageous even under those conditions. Heating element **17** may be a conventional electrical heating coil surrounding bypass conduit **10**. One or more temperature sensors **11** through **13** inside the battery cells switch blower **8**, as well as two-way valve **9** and heating element **17** at the appropriate predetermined temperatures.

On the other hand, the system can be simplified through omission of bypass conduit **10** even during start-up, when conditions are such that the ambient air has little effect and the main purpose of the system is to equalize internal temperature. While this simplifies the system, it entails a concomitant increase in its thermal lag, due to the thermal inertia of the cooler **6** through which the cooling medium must then circulate. Referring to Figure 1.29e, the operation of the battery cooling system will be described, based on the assumption that the battery initially starts from a low temperature level.

At battery start-up, pump **7** is first set into operation by means of a main switch and the cooling medium is caused to circulate through the circulatory cooling system. This main switch may be operated by hand or controlled automatically, if desired, in response to activation of the main switch of the vehicle in which the battery is located. Because of the low temperature level in the cells, the temperature sensor **11** causes valve **9** to switch into that position in which the cooling medium flows through bypass conduit **10**. This mode of operation initially equalizes the temperature in the cells.

If the temperature in the cells at start-up is below a predetermined value, then a pulse is supplied from temperature sensor **13** to operating element **14** of switch **15**. This pulse switches on heating element **17**, provided sensor **16** permits this. This takes place when the energy supplied for the heating element **17** is provided by an external power source.

The simplest way is to have the sensor **16** measure the potential of the battery, since this is known to be considerably higher during charging than when the battery is disconnected or is discharging. Alternatively, sensor **16** can be a mechanically actuated element which responds to physical placement of the battery into connection with a charging element in a charging station.

Upon activation of operating element **14** by sensor **16** due to the measured higher potential, the heating element **17** is supplied with energy from the charging circuit. Operating element **14** may be the coil of a solenoid which is energized by the pulse from sensor **13** to attract the armature of solenoid operated switch **15**.

When a predetermined temperature is reached in the cells (namely the lower limit of their optimum operating temperature range) due to heating by heating element **17** or due to the losses occurring in the battery during charging or discharging, then the heating element **17** is first disconnected by temperature sensor **13**. When the further rising temperature of the electrolyte in the cells reaches a predetermined value in the upper half of the optimum temperature range, then temperature sensor **11** switches valve **9** in such a manner that the cooling medium flows through cooler **6** and is there cooled down. The cooler is so arranged that the air flow produced by forward movement of the vehicle flows through the cooler.

If the heat dissipation by means of the cooler is insufficient and the temperature of the electrolyte rises further, then, upon exceeding of the upper limit of the optimum temperature range, the temperature sensor **12** turns on blower **8**. This considerably increases the heat dissipation from the cooling medium via cooler **6** to the ambient air. The cooling system is so designed that the permissible maximum value of cell temperature is not exceeded, even at extreme loads in driving and charging and at extremely high ambient air temperatures.

Upon decline in temperature of the battery during use of the cells, the blower **8** is first disconnected; upon further decrease in temperature the valve **9** is then switched over to cause the cooling medium to flow through bypass conduit **10**. When a warm battery is taken out of service, the previously mentioned switch disables the entire cooling system and the battery then cools off slowly solely through its outer surfaces. On the other hand, if the cooling system remains operative, then the battery continues to be cooled until it reaches its lower predetermined temperature value. This is sensed by temperature sensor **18** and circulating pump **7** is then the final component of the cooling system to be turned off.

During propulsion service, i.e., when the battery is being discharged, the energy for driving the pump **7** and blower **8** must be delivered by the battery. To conserve this energy, the circulatory cooling system is so controlled that during battery discharge either the entire cooling system or at least portions thereof, preferably blower **8**, are disconnected. The heat which is not removed during propulsion service is then stored in the battery and is dissipated on a suitable occasion, such as during charging when the energy for operating the blower is supplied by the charging equipment, through turning on of blower **8**. Control of the switching of blower **8** and, if necessary, of pump **7**, can be accomplished in that case in known manner, e.g., in response to the potential level or the current direction.

Another way to save energy during discharge of the battery involves controlling the motor of blower **8** in such a manner that it rotates more rapidly with increasing potential. This causes the battery to be cooled intensively during charging due to the higher potential supplied to blower **8**. During discharging of the battery it is cooled less because of the lower blower speed attributable to the lower potential. However, energy is simultaneously conserved because the motor then requires less current due to its current voltage characteristics.

The cooling system shown in Figure 1.29e further affords the possibility of internal cooling. For that purpose, connector plugs **19** of manifold **5** are connected in known manner to an external cooling system or, in the simplest case, to a water pipe supply. By introducing into the circulatory cooling system a cooling medium (water) at low temperature, intensive cooling of the battery is accomplished. This type of external cooling is particularly suitable when the battery is in exchange service and is, from time to time, in a battery exchange station where it is being charged for its next service period. Coupling of the battery circulatory system through plugs **19** to an external circulatory cooling system can then be accomplished automatically during battery exchange.

ZINC ELECTRODES AND BATTERIES

ZINC ELECTRODES

Strips of Polytetrafluoroethylene

A process described by *R. Gadessaud and C. Audry; U.S. Patent 3,967,976; July 6, 1976; assigned to Compagnie Generale d'Electricite, France* concerns a method for preparing electrodes for an electrochemical storage cell, these electrodes comprising a collector part constituted by an electronically conductive and chemically inert material coated on an electrochemically active part comprising a metal or a compound of the metal, as well as a partly reticulated polymer.

In the process, a substantially saturated solution of the polymer in water, to which is added an element supplying an aldehyde in a quantity sufficient for partly reticulating the polymer in contact with at least a catalyst, in a proportion of 5 to 20% by weight, is prepared. A salt of the metal is added to the aqueous solution of the polymer in a proportion of 20 to 50 parts by weight for one part of polymer, the salt being, on the one hand, soluble in water and, on the other hand, compatible with the polymer, such an addition being effected at a temperature between 50° and 100°C.

The solution thus prepared is run into a mold in which the collector part of the electrode has previously been placed until a complete immersion of the latter occurs. The mold is cooled to a temperature that is no more than -20°C. The cake constituted in the mold is stripped, and it is immersed in an alkaline solution saturated with an oxide of the metal. The cake is then washed in cold water and dried at a temperature between 20° and 60°C, finally being compressed at a pressure between 2 and 10 bars/sq cm.

The method is characterized in that longitudinal strips of a material having, with respect to the electrochemically active part of the electrode, nonadhesive properties are deposited on each of the faces of the collector part so that a space thus remains between the electrochemically active part and each of the strips.

According to Figure 2.1a, the electrode comprises a part or collector web **1** formed by a thin sheet of silver-plated nickel or of another suitable conductive material. Thin longitudinal strips **2** of a water-repellent material, such as polytetrafluorethylene (PTFE) or polydichlorodifluoroethylene, have been arranged on each of the faces of that sheet, the width of such strips being in the order of one-half or one-eighth of the width of the residual metallic surfaces **3**.

FIGURE 2.1: PREPARATION OF ZINC ELECTRODE

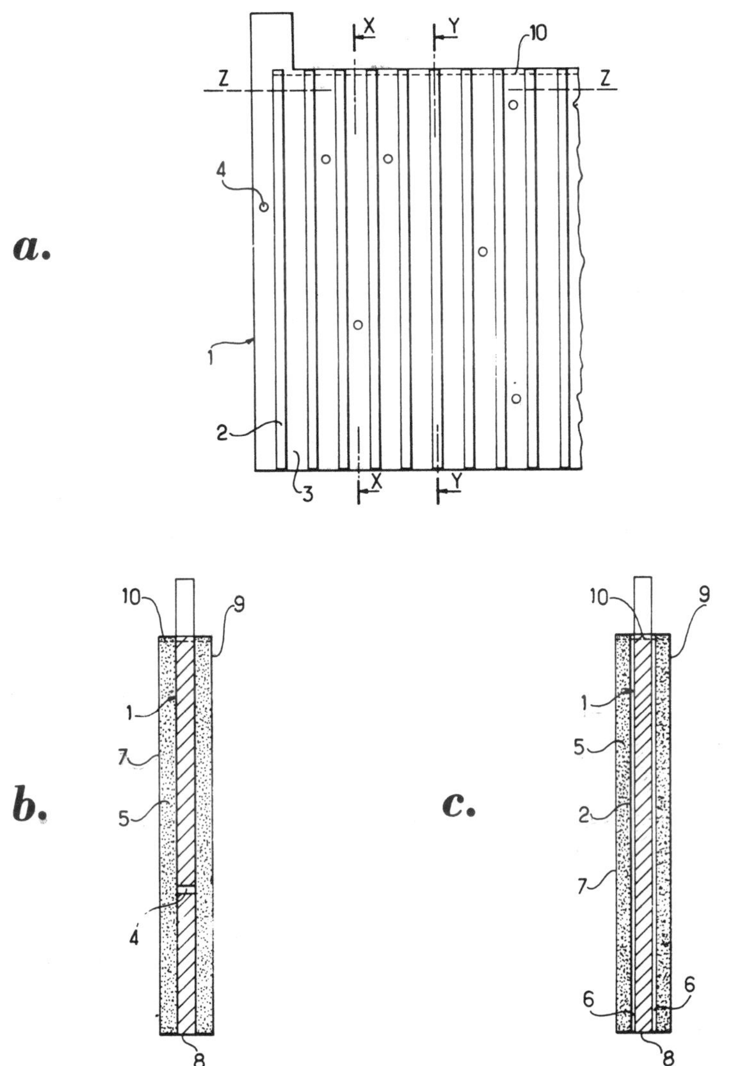

(continued)

FIGURE 2.1: (continued)

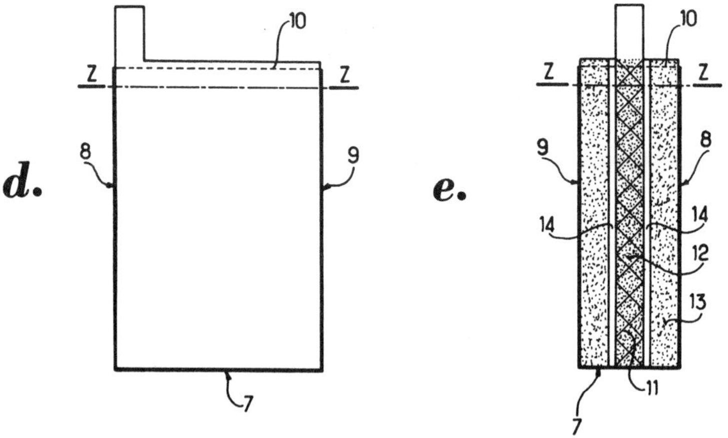

(a) Elevation view of an electrode
(b) Side view of a cross section through the line **XX** in Figure 2.1a
(c) Side view of a cross section through the line **YY** in Figure 2.1a
(d) Elevation view of a second example of an electrode
(e) Side view of a cross section of Figure 2.1d

Source: U.S. Patent 3,967,976

Openings or drillings such as **4** are provided right through the residual surfaces **3**, such openings being intended to ensure effective fixing of the electrochemically active part deposited subsequently. The producing of such longitudinal strips is effected as follows: An even layer of PTFE is deposited on each face of the sheet **1** by spraying by means of a pneumatic spray gun. This layer is sintered at about 320° to 380°C so as to obtain good adherence. Then, by means of a suitable tool having teeth whose width is equal to the width of the metallic surfaces **3** and whose spacing is equal to the width of the strips **2** of PTFE, the layer thus deposited is scraped off, so as to remove the PTFE only from the metallic surfaces **3**, on each of the faces of the sheet **1**.

One variation comprises depositing the layer of PTFE by immersion of the sheet **1** in a PTFE emulsion, such as Soreflon, then in sintering and producing the strips **2** as described below. Another variation comprises providing, on the surfaces of the sheet **1**, masks corresponding to the dimensions of the metallic surfaces **3** and in spraying a PTFE aerosol. The masks are then removed so as to leave the metallic surfaces **3** clear, then the PTFE is sintered. Of course, polydichlorodifluoroethylene may be used instead of PTFE. The collector web **1**, therefore, has the shape shown in Figure 2.1a.

According to Figures 2.1d and 2.1e, the electrode comprises, in a second example, a part or collector web **11** formed by a silver-coated nickel grating or a grating made of any other suitable conductive material. A paste **12** is spread on that grating **11** so that the openings in the grating **11** are stopped up. Then, the excess paste is removed by any suitable means from the surface of the grating, leaving only a thin film on the grating **11**. This is then dried to remove the diluting agent of the paste. The composition of that paste may, by way of an example, have the following: zinc oxide, 100 g and polytetrafluorethylene, 2 g. That mixture is diluted in ethanol in quantity sufficient for obtaining a paste suitable for being spread as mentioned above. To great advantage, 1 g of mercuric acetate can be added to that mixture with a view to increasing the overpotential. Then, 10 g of polyvinyl alcohol, designated by the abbreviation PVA, are inserted cold in about 200 cc of water.

The mixture is heated almost up to the boiling point until complete dissolution. The following are mixed into the solution thus obtained: 1 g of dimethylolurea, 0.5 g of ammonium chloride and 0.3 g of sodium sulfate. Dimethylolurea in contact with NH_4Cl and Na_2SO_4 partly reticulates PVA by formation of a polyvinyl acetal. A zinc salt, which is very soluble in water and compatible with the PVA [in this instance, zinc acetate $(CH_3CO_2)_2Zn{\cdot}2H_2O$] in a proportion of about 340 g, is then added to the solution, while shaking it, at a temperature of substantially 70°.

To great advantage, 3 g of mercuric acetate may also be mixed in, in order to increase the overpotential. In this way, a viscous liquid, which is run into a mold in which, according to the process, the collector web **1** such as shown in Figure 2.1a or the grating illustrated in Figure 2.1e has previously been inserted, is obtained with the running off being effected until the assembly is immersed within the liquid. The mold is then cooled suddenly to about -20°C so as to form zinc acetate crystals of very small size.

The cake thus obtained is then immersed in a potassium hydroxide solution having a concentration of 8 to 12 N saturated with zinc oxide ZnO, for substantially 24 hours. During that operation, the potassium hydroxide diffuses into the mass of the cake and transforms the zinc acetate into oxide while coprecipitating the PVA.

At the end of that treatment, the cake is thoroughly washed with water in order to draw away the residual potassium hydroxide, then dried in an oven at about 40°C; during such a drying operation, the assembly is held slightly clamped between two gratings with a view, on the one hand, to promoting the removal of the water and, on the other hand, to prevent any deformation or cracking by thermal stress. The cake is compressed at a pressure of about 5 bars/sq cm so as to form the electrode.

It will be observed that in all cases, the structure of the electrodes is in the form of very small metallic grains coated with a thin film of PVA, which wets the grains completely and hence prevents their migration towards the lower part, while forming an obstacle for zincate ions.

On referring to Figures 2.1b and 2.1c, the sheet **1** will be seen to be immersed in a mass **5** of zinc oxide and of PVA particles precipitated by the potassium hydroxide. Figure 2.1b shows that this mass adheres to the metallic parts **3** of the sheet **1** which are not covered with PTFE. On the other hand, on referring to Figure 2.1c, it will be seen that subsequent to the mutual nonadhesive properties between the mass **5** and the strips **2** of PTFE, spaces **6** are thus provided between the strips and the mass **5**. On referring to Figure 2.1e, it will be seen that the grating **11** is surrounded with a mass **13** of zinc oxide and of PVA particles precipitated by the potassium hydroxide. It will be seen in this figure, that subsequent to the mutual nonadhesive properties between the mass **13** and the material within the grating **11**, a space is thus provided between the grating **11** and the mass **13** on either side of the grating. Moreover, in both examples, the faces as well as the thick edges of the electrode are coated, preferably by painting, with a solution of reticulated PVA substantially similar to the solution used for forming the electrode itself. Thus, fluid-tight layers such as **7, 8** and **9** preventing any diffusion of zincate ions are formed.

After drying, the deposit is removed from the upper thick edge of the cell simply by cutting out the upper end along the line **10** materially illustrated by interrupted lines with scissors. Of course, during the use of the electrode, the latter is not completely immersed in the electrolyte. In the case illustrated by the figures, the level of the electrolyte is materially illustrated substantially by the line **ZZ**. The electrodes thus formed have improved mechanical and electrical characteristics, thus imparting excellent operational characteristics to the electrochemical storage cells in which they are incorporated.

Such advantages result more particularly from the fact that the hydrogen evolved during the electrochemical process fills the spaces **6** or **14** and flows out freely into the atmosphere through the upper thick edge of the electrode cut out as previously mentioned, thus avoiding any swelling within the electrode. It will be observed, moreover, that the existence of zinc oxide in the grating **11** as shown in Figure 2.1e imparts to the latter improved conductivity, thus resulting in a further increase in the performances of the storage cell in which the electrode is used.

Perforated Metal Strips

According to a process described by *R.F. Enters; U.S. Patent 3,918,990; Nov. 11, 1975; assigned to McGraw-Edison Company*, an alkaline-electrolyte storage battery, having nickel and zinc active materials for the positive and negative electrodes, is characterized by the use of perforated metal strip pockets for the negative electrodes containing a powdered zinc active material under pressure to hold the active material in situ and in close electrical contact with the metal strip. By preventing loss of active zinc material during cycling, a greatly extended cycle life is obtained at nearly the initial electrical capacity of the battery.

In fabricating the zinc electrode according to the process, strip pockets **10** as shown in Figures 2.2d and 2.2f are first formed, filled with zinc active material **11**, and then seamed along their edges to form the plate structure **12** shown in Figure 2.2c. Each strip pocket is made from two metal strips **13** (Figure 2.2e) having a central perforated band portion **14** lengthwise thereof and having imperforate border portions **15** along the sides. These strips are made of nickel-plated steel of 2 to 5 mil thickness, having 3 to 15 mil perforations.

One of these strips **13** is formed into a channel **16** (Figures 2.2d and 2.2f) by bending the solid border portions **15** at right angles and the edges thereof then outwardly so that the perforated band **14** forms the bottom wall of the channel and the solid border portions **15** form the side walls with outwardly extending flanges **17**. The other metal strip **13** is a cover strip having the solid border portions **15** curled back on themselves to form guideways **18** lengthwise of the strip.

FIGURE 2.2: ZINC ELECTRODE

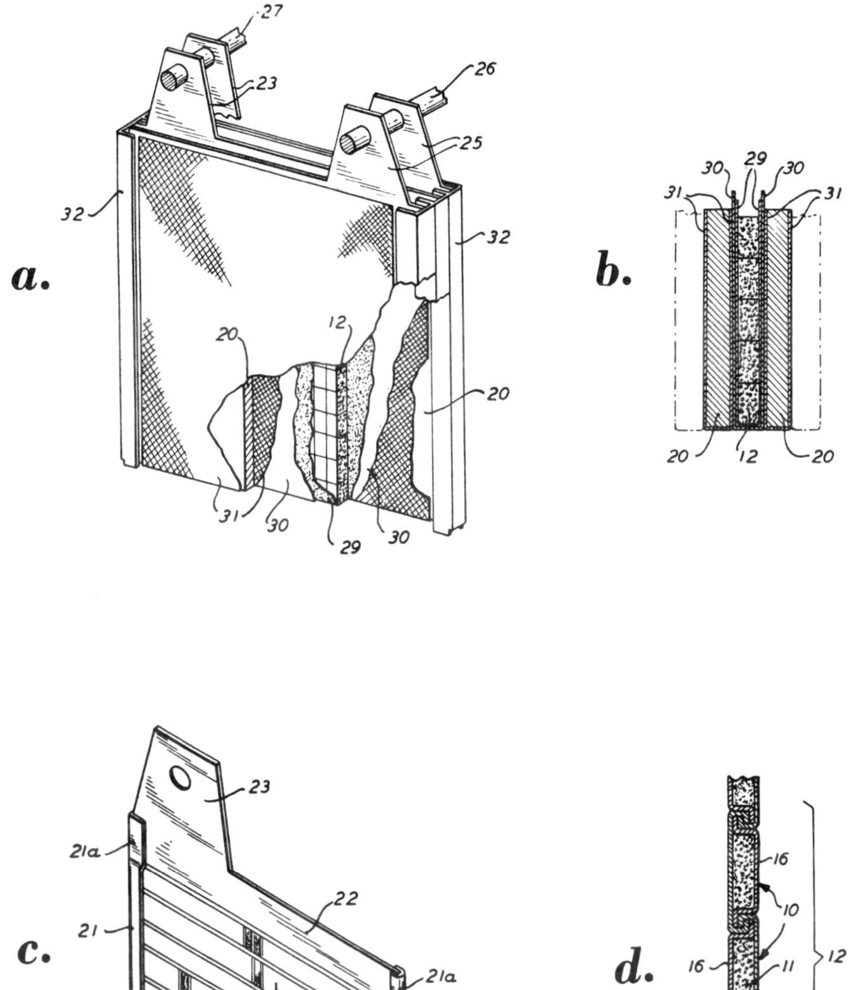

(continued)

FIGURE 2.2: (continued)

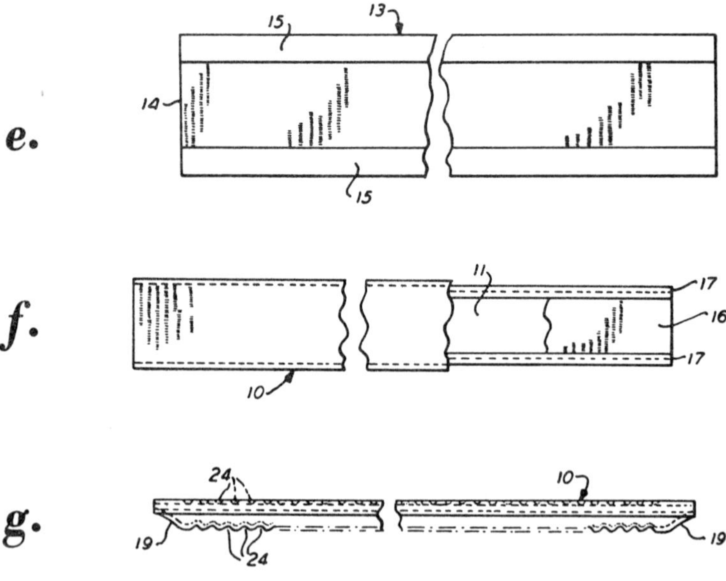

(a) Perspective view, partly broken away, of a nickel-zinc storage battery
(b) Fractional side section of Figure 2.2a
(c) Perspective view of a zinc negative electrode of the battery comprising a plurality of parallel strip pockets seamed together along their edges into a plate structure
(d) Fractional sectional view to enlarged scale through the negative plate structure taken on the line **3–3** of Figure 2.2c
(e) Perforated metal strips from which the strip pockets are made
(f) View of a partially assembled strip pocket comprising a channel strip containing a briquette of compressed zinc active material and a cover strip in slidable telescoping arrangement
(g) Side view of a finished strip pocket ready for being seamed with a plurality of such pockets into a plate structure

Source: U.S. Patent 3,918,990

These two strips are telescoped together as shown in Figures 2.2d and 2.2f with the side flanges **17** of the channel members engaging the guideways **18** of the cover strip. Before the channel members and cover strips are telescoped together, the channel members are filled with zinc active material **11** preferably comprising a mixture of powdered zinc and zinc oxide including a small amount up to 5% by weight of mercuric oxide and a binding agent up to 3% by weight such as polyvinylpyrrolidone. This active material may be in powdered form or it may be compressed into a rectangular briquette form to fit the channels **16** as shown in Figure 2.2f. After the channel members and cover strips are telescoped together to form a strip pocket containing the zinc active material **11**, the ends **19** are crimped closed to form a finished strip pocket **10** as appears in Figure 2.2g.

A plurality of finished strip pockets **10** are seamed together along their edges to form the plate structure **12** shown in Figure 2.2c. The seaming is done by interengaging the guideways **18** of successive strip pockets edgewise in a hooked arrangement as shown by the sectional view of Figure 2.2d. After a group of such finished strip pockets are so engaged with each other in a planar arrangement, U-shaped channel members **21** of nickel-plated steel are crimped tightly onto the end portions of the finished pockets within the thickness dimension of the pockets. Further, extensions **21a** of the U-shaped channel members **21** are crimped onto and welded to relatively heavy metal strips **22** completing the frames for the zinc negative electrodes.

Each metal strip **22** has an extending tab or leg **23**. However, after the group of finished strip pockets are secured in plate form to the U-channel members **21** and metal strips **22**, the strip pockets are subjected to a compressive operation to form the plate structure with transversely corrugated walls **24** so as to tightly compress the zinc active material in the pockets into electrical contact with the perforated metal strips and so as to confine the successive portions of the active material against displacement or migration during the cycling of the battery.

A typical nickel-zinc pocket-type battery, according to the process, comprises a succession of alternate nickel positive plates **20** and intervening zinc negative plates **12** in a parallel side-by-side arrangement as shown in Figures 2.2a and 2.2b. As is common, the successive plates are oriented so that the terminal lugs **25** of the positive plates are aligned along one end of the battery for connection via a strap or rod **26** to a positive terminal, and the terminal lugs **23** of the negative plates are aligned along the opposite end of the battery for connection via a strap or rod **27** to a negative terminal.

Each negative plate **12** is covered by a single-layer felt separator **29** of about 25 mils thickness applied down across one side of the plate, around the bottom edge, and then up across the other side. Next, an ionic membrane separator, typically a cellophane tape **30** of about 1½ mil thickness, is wrapped vertically around the felt-covered zinc negative plate with an overlap of the successive convolutions to constitute, in effect, a three-wrap covering. Each positive nickel plate is covered across one side, and around the bottom edge, and up the other side by a semi-flexible plastic mesh insulator **31** typically of polyethylene.

Prior to placing these plates in a pack against one another face to face, molded plastic U-channel members **32**, typically of polystyrene, are pressed onto the opposite vertical edges thereof in clamping engagement therewith. The plates are packed together with the end U-channel members applied side by side against each other and with these channel members forming a slight spacing between the plates for better flow of electrolyte to all surfaces of the plates.

In this battery construction, the active zinc material is retained in situ in perforated metal strip pockets without loss of electrical contact with the perforated metal walls, with the end result that the voltage and ampere-hour capacity of the battery are maintained at approximately their starting values after repeated cycling of the battery. Although one cannot be certain as to the exact way in which the zinc active material is so efficiently retained in the battery system, it is believed to be due (1) to starting with a zinc active material in a powder form, (2) packing this powdered material in long, thin metal strip pockets without inclusion of any substantial inert material, (3) using such strip pockets that when compressed will maintain the active material under pressure, (4) corrugating the walls of the strip pockets in the compressive operation so as to effectively isolate successive portions of the active material against settling or migration, and (5) forming the metal strips constituting the pocket walls with perforations of a size compared to that of the active zinc particles which permits ingress of electrolyte but no substantial loss of zinc active material.

Calcium Hydroxide

A process described by *L. Kandler; U.S. Patent 3,930,883; January 6, 1976; assigned to Rheinisch-Westfalisches Elektrizitalswerk AG, Germany* provides a negative zinc-containing plate for alkaline accumulators (i.e., alkaline storage batteries or cells) which permits an increased number of charge/discharge cycles and yet restricts dendrite formation and the tendency to form short-circuit bridges between the opposite-polarity plates or electrodes. This is accomplished using a negative electrode for alkaline accumulators in which the active mass (i.e., the mass participating in a principal electrochemical reaction at the electrode) comprises zinc or a zinc compound and is present in association with at least one substance capable of forming corresponding zincates of low solubility in the electrolyte during discharge of the plate.

The substance, which is capable of forming the low-solubility zincates, is preferably an alkaline-earth-metal hydroxide with the compound formed being the corresponding low-solubility alkaline-earth-metal zincate. Of particular suitability, in this connection, is calcium hydroxide although magnesium hydroxide may also be used. In the first case, the product is calcium zincate, while in the second case it is magnesium zincate.

The alkaline-earth zincates, resulting from the substantially instantaneous reaction of the alkaline-earth hydroxide and the solubilizing zinc during discharge, does not irreversibly remove the reacted zinc from the system in terms of the active mass; moreover, it has been found that the zinc redeposits or replates from the low-solubility zincate, in the form of a porous spongy mass. While it is doubtful that the spongy porous mass is formed directly from the precipitated alkaline-earth zincate, this hypothesis cannot be excluded although it is assumed that the small quantity of this low-solubility compound, which is in the solution, is readily depleted by electrodeposition and galvanic techniques and is replenished from the low-solubility salt which is precipitated from the solution.

According to the process, the zinc or zinc compounds incorporated in the negative electrode plate for an alkaline accumulator are mixed with an auxiliary substance tending to form low-solubility salts with the zinc as it goes into solution during discharge of the cell, i.e., low-solubility zincate salts. From the low-solubility compounds thus formed, sponge zinc precipitates on the plate during charging of the cell without dendrite formation.

Example 1: A zinc-plated wire mesh of rectangular configuration and having dimensions of 31 x 51 mm is provided with a metal tab by spot-welding and is formed with spacer ribs along the edges on both sides of the support. The spacers are composed of Plexiglas and are cemented to the wire grid. The electrode thus prepared is immersed in an electrolyte containing 600 grams per liter of zinc chloride and 200 grams per liter of calcium nitrate in water, the bottom of the vessel being provided with a layer of zinc oxide. Using the electrode as a cathode against a zinc anode, electrolysis is carried out for two hours at a current of 0.5 ampere. Thereafter, the electrode is cathodically electrolyzed in an electrolyte containing 800 grams per liter of calcium nitrate over a period of one-half hour with a current of 0.3 ampere. The electrode is then immersed in an aqueous potassium hydroxide solution (37%) and electrolyzed for 20 hours with a current of 0.1 ampere. In this last step, zinc compounds previously deposited are converted into metallic zinc. The electrode is then dipped, after rinsing, in a mercury chloride solution to amalgamate the zinc. Under light pressure, a nonwoven synthetic-resin layer (Viledon) is wrapped around the electrode and cemented to the spacers.

The electrode is then dipped in a calcium chloride solution and then again in a potassium hydroxide solution to precipitate calcium hydroxide in the pores of the wrap. The electrode is then used in a storage battery in which nickel(III) oxide constituted the counterelectrode or in combination with an air electrode in a zinc/air battery. In each case, potassium hydroxide served as the electrolyte. The plate had a capacity of 1.4 ampere-hours corresponding to 10.5 ampere-hours/decimeter^2 of apparent surface. Repeated charge/discharge cycling showed no evidence of dendrite formation.

Example 2: Upon the grid prepared as in Example 1, a viscous paste was coated flush with the spacers, the paste consisting of equal parts by weight of zinc oxide and calcium hydroxide in the presence of sufficient water to impart a pasty consistency to the mass. After drying, a zincate resin layer of nonwoven fabric, as described in Example 1, was supplied and the plate thereafter electrolyzed in an electrolyte containing 800 grams per liter of calcium nitrate as described in Example 1. After a cathodic treatment in potassium hydroxide solution, the electrode was found to have a capacity of 1.1 ampere-hours.

Alkaline Fluoride

J. Sandera, M. Calabek, O. Kouril, M. Cenek, J. Vanacek, J. Malik and V. Koudelka; U.S. Patent 4,017,665; April 12, 1977; assigned to Prazska akumulatorka, narodni podnik, Czechoslovakia describes an electrode composition which includes (a) powdered zinc or a powdered compound of zinc, (b) a binding agent, (c) a porosity-increasing agent, and (d) an ingredient that forms a negligibly soluble solution with zinc, such as sodium fluoride, potassium fluoride, sodium borate, sodium carbonate, sodium acetate or their mixtures. The latter ingredient is present in a range of 5 to 55% by weight of the composition.

The improved composition is applied in a conventional manner to the electrode collector and is exposed to a conventional heat treatment. The inhibiting effect of the resulting composition is found to be effective to (1) limit the sedimentation of zinc in the accumulator tank, (2) prevent the formation of undesired dendrites, and (3) prevent the shifting of the active composition of the electrode during the charge and discharge processes, thereby increasing its efficiency of utilization. Additionally, the incorporation of the composition in the electrode substantially reduces the stringency of control during the charging process.

A plurality of zinc electrode compositions were formulated which included (1) zinc powder, (2) dispersion employed as a binding agent, (3) oxalic acid powder employed as a porosity-increasing agent, and (4) various types of alkaline fluorides, and the remainder water and a Teflon dispersion employed as a binder.

Several alkaline fluorides were employed, specifically sodium fluoride, potassium fluoride and their mixtures. Each of the resulting compositions was applied to a collector of the electrode to be manufactured, illustratively. It was found that all of the so-constructed zinc electrodes formed, using any of the compositions within the ranges specified in these examples, performed at at least 60% of their theoretical capacity after the 50th charging cycle of the accumulator, and performed at 30% of their theoretical capacity at the end of 200 charging cycles. Even after the attainment of the 200th cycle, no significant dendrite or sedimentation formation was observed.

Cadmium Additive

A process described by *A. Charkey; U.S. Patent 4,022,953; May 10, 1977; assigned to Energy Research Corporation* provides an electrode structure containing zinc active material and having cadmium compounds dispersed therein, the surface area of the metallic cadmium particles originally present or present on conversion of the cadmium compound being not less than 1.0 square meter per gram and the particle size thereof being not greater than 10 microns. The weight of the cadmium particles is from 1.0 to 10% of the weight of the zinc active material.

Alkaline secondary batteries employing electrodes in accordance with the process exhibit deep discharge-recharge cycling capacity in excess of 300 cycles. It is believed that the cadmium additive greatly improves electrode surface morphology to retain zinc electroreduced from the solid zinc compound state or deposited thereon from solution and provides the basis for the improved secondary battery performance.

Fluoride Resin and Calcium Oxide Sheet Electrode

T. Takamura, T. Shirogami, H. Niki and K. Aizawa; U.S. Patent 4,037,033; July 19, 1977; assigned to Tokyo Shibaura Electric Co. Ltd., Japan describe a rechargeable nickel-zinc battery, which is capable of undergoing many charge-discharge cycles (cycle life), with deep discharge, and of high performance with heavy drain discharge during use. The increased cycle life is accomplished by a suitable combination of electrochemical generating elements. The high performance with heavy drain discharge service is obtained by an improved zinc electrode construction. The battery has a sheet-like kneaded zinc electrode, a sheet-like nickel oxide electrode with limited capacity ratio, a separator, an electrolyte absorber, and a concentrated alkaline electrolyte in limited amounts.

The negative electrode is prepared by fixing the sheet of negative electrode mixture to a current collector. The sheet of negative electrode mixture is made by kneading a mixture of 2 to 13 weight percent of a fluoride resin which is dispersed in an alkaline solution, 3 to 20 weight percent of calcium oxide and/or calcium hydroxide, 1 to 20 weight percent of metallic compound which comprises at least one material selected from the group consisting of bismuth oxide, bismuth hydroxide, cadmium oxide and cadmium hydroxide, 2 to 10 weight percent of zinc powder and the balance being zinc oxide and/or zinc hydroxide, and laminating several layers of such sheet.

The positive electrode is provided with a sintered-type or a plastic bonded-type hydrated nickel oxide electrode and has a theoretical capacity limited to one-quarter to one-half that of the negative electrode. The separator is desirably produced by coating a paste of polyvinyl alcohol and a boric compound on nonwoven fabric which is durable in the concentrated alkaline solution and difficult to oxidize. The electrolyte absorber is provided with a thin nonwoven cloth which is also durable in the alkaline solution. It is soaked with a solution of a surfactant and dried before use so as to be easily wet by the alkaline electrolyte during the process of manufacturing the cells. If the separator or both electrodes absorb the alkaline electrolyte, the absorber is not necessary. The amount of the concentrated alkaline electrolyte is preferably within the range of 1.0 to 1.7 ml per 1 Ah of the theoretical capacity of the negative electrode.

Example: The negative electrode, which includes the various components as the active materials tabulated in the table below, was produced. The sheet-like plate had a thickness of 0.8 to 1 mm and was pressed into the current collector of expanded silver-plated copper to adhere the mixture sheet of the active materials thereon.

	type of the zinc negative electrode (weight per cent)						
Contents	A_1	A_2	A_3	B_1	B_2	B_3	C
$Ca(OH)_2$	0	10	20	5	15	25	10
Bi_2O_3	10	10	10	0	0	0	0
CdO	0	0	0	10	10	10	0
ZnO	77	67	57	65	62	52	77
Zn	5	5	5	5	5	5	0
HgO	0	0	0	0	0	0	5
Fluorine-containing resin	8	8	8	15	8	8	8

For the comparison, the conventional-type zinc electrodes, which contain mercury oxide in the negative electrode mixture, were tested, indicated as type-C. All test batteries were made in C-size shape having 1.2 Ah nominal capacity. The results are shown in the following table, along with the combination of the electrodes, the amount of electrolyte and the cycle life of the batteries.

All the batteries were prepared with use of a separator of nonwoven fabric of a copolymer of acrylonitrile and polyvinyl chloride which was coated with the mixture of polyvinyl alcohol (PVA) and boric acid aqueous solution after pretreatment with a wetting agent. A mixture of PVA and TiO_2 powder was soaked into a nonwoven fabric material after treatment with a nonionic sufactant to prepare the electrolyte absorber.

The battery tests of cyclic life were accomplished under the conditions of 0.5 C charge for 2.5 hr and 0.1 C discharge fully to the end voltage of 0.9 V every time. The tabulated number represents the number of charge-discharge cycles the battery has undergone when the capacity goes down to the half value of the nominal capacity.

As is apparent in the following table, the batteries, which had the longer cycle life and thus have the most utilities, have a negative electrode made of the mixture of calcium hydroxide of 10 weight percent, bismuth oxide 10 weight percent, zinc oxide 67 weight percent, zinc powder 5 weight percent and PTFE 8 weight percent, an amount of electrolyte of which is from 1 to 1.5 ml per 1 Ah of the negative electrode capacity, and a positive electrode which has a capacity of from ¼ to ½ the theoretical capacity of the negative electrode.

Number of tested battery	The type of zinc electrode	The capacity of zinc electrode	The liquid quantity of electrolyte	The capacity of nickel electrode	Cyclic life
1	A_1	4 Ah	6 ml	1.5 Ah	90
2	A_1	4 Ah	6 ml	1.5 Ah	90
3	A_1	5 Ah	5 ml	1.5 Ah	100
4	A_1	3 Ah	4 ml	1.5 Ah	60
5	A_1	2 Ah	3 ml	1.5 Ah	50
6	A_2	4 Ah	6 ml	1.5 Ah	320
7	A_2	5 Ah	5 ml	1.5 Ah	350
8	A_2	4 Ah	4 ml	1.5 Ah	280
9	A_2	6 Ah	6 ml	1.5 Ah	220
10	A_2	3 Ah	5 ml	1.5 Ah	120
11	A_2	3 Ah	6 ml	1.5 Ah	60
12	A_3	4 Ah	6 ml	1.5 Ah	180
13	A_3	4 Ah	4 ml	1.5 Ah	105
14	B_1	3 Ah	5 ml	1.5 Ah	20
15	B_2	4 Ah	6 ml	1.5 Ah	280
16	B_2	4 Ah	5 ml	1.5 Ah	320
17	B_2	4 Ah	4 ml	1.5 Ah	270
18	B_2	3 Ah	4 ml	1.5 Ah	210
19	B_3	3 Ah	5 ml	1.5 Ah	120
20	C	4 Ah	5 ml	1.5 Ah	38

In related work, *T. Takamura, T. Shirogami, Y. Sato, K. Murata and H. Niki; U.S. Patent 3,951,687; April 20, 1976; assigned to Tokyo Shibaura Electric Co., Ltd., Japan* describe a nickel-zinc storage battery which is prepared by winding into a whirlpool a sheet-like zinc electrode, a sheet-like nickel electrode and a separator with the separator interposed between the zinc and nickel electrodes. The zinc electrode is obtained by bonding under pressure to a collector a mixture sheet consisting of zinc oxide, zinc, bismuth oxide, calcium hydroxide and fluoroplastic. The nickel electrode is obtained by bonding under pressure an oxide of nickel to a collector together with a conductive material. The separator is obtained by coating on an alkali-resisting nonwoven fabric a mixture of polyvinyl alcohol and at least one material selected from the group consisting of boric acids and metal oxides having low solubility to alkali solution and drying the nonwoven fabric thus coated.

SEPARATORS

Nonmembrane Separator and Cellulosic Gelling Agent

L.M. Gillman and R.E. Stark; U.S. Patent 3,894,889; July 15, 1975; and U.S. Patent 3,980,497; September 14, 1976; both assigned to Gates Rubber Co. describe a separator construction for alkaline cells having at least two bibulous, nonmembranous separator layers laminated together with a thin layer of a gelling agent which provides mechanical integrity to the separator layer and functions as a semipermeable membrane. Alternatively, the separator is comprised of bibulous, nonmembranous separator layers sandwiching and laminated to a membranous layer utilizing a gelling agent to provide an integral separator of mechanical integrity. These separators have particular utility as interelectrode spacers in alkaline batteries in which a separator resistant to alkaline electrolyte, oxidation, dendrite growth and other degrading cell environment factors are required.

The gelling agents serve the dual purpose of providing a barrier film or semipermeable membrane through which ions permeate, but through which electrode materials are impermeable, and as a binding agent to render mechanical integrity and protection to the separator. In general, the gelling agent must be compatible with and substantially insoluble in alkaline electrolyte within the cell, retentive of electrolyte, resistant to oxidation, and capable of adhering to the aforementioned bibulous layers, or in another aspect of the process, it must be additionally adherent to a semipermeable membrane.

Preferred materials include the cellulosic materials. Compatible mixtures of the above may be utilized, e.g., methyl cellulose and sodium alginate.

The following working examples illustrate certain preferred aspects of the process as applied to the nickel-zinc type cell.

Example 1: A number of sealed pillbox cells are constructed, having one anode and two cathodes arranged in a horizontal parallel stack compressed to ⅔ psi. The cathodes are conventional nickel-sintered plates. The anode is of the pasted zinc variety, composed of 70 weight percent zinc powder, 25 weight percent zinc oxide as a charge reserve, and 5 weight percent mercuric oxide. In conventional manner, the anode ingredients are made pliant by addition of a suitable binder such as aqueous methyl cellulose and plasticizer. The resulting smooth paste is applied to a metallically conductive, expanded mesh, allowed to dry and made smooth.

Two types of separators are assembled. Type A consists of two layers of high-grade, roll-coated cotton-fiber-based filter paper enclosing a single layer of cellophane. Type B differs from Type A in that the three layers are laminated together with a gelling agent according to the process; a bibulous layer of filter paper is roll-coated on one side with a viscous solution of methyl cellulose; a presoaked (in water) layer of cellophane is placed on the sticky bibulous layer, carefully smoothing the cellophane; onto this composite layer is placed an additional bibulous layer coated with sticky methyl cellulose to form a composite separator; this separator is hot-pressed to form the laminated separator. Three Type A and Type B cells are repeatedly discharged to approximately 50 percent of nominal capacity and then recharged. In both cells, the mid-life voltage is about 1.6 volts. However, the average cell life of Type A is 223 cycles, while the average cell life of Type B is 320 cycles.

Example 2: The same procedure of assembling pillbox cells as in Example 1 is employed. In this example, two additional separators are tested, both of which lack a cellophane layer. Type C is composed of two layers of high-grade filter paper without any gel layer. Type D is the same as Type C except that the two bibulous layers are bonded together with a gelling agent composed of 5 weight percent methylcellulose according to the procedure of Example 1. Three cells of Type C had a mid-life voltage of 1.70 volts and a discharge/charge life of 29 cycles. In contrast, three cells of Type D had a mid-life voltage of 1.55 volts and a cycle life of 180.

Reticular Flexible Envelope Using Thin Interlacing Rods

S.S. Axelrod, V.G. Belozerov, M.B. Gershman, E.G. Margolin and M.B. Shapot; U.S. Patent 4,041,218; August 9, 1977 describe a separator for alkaline storage batteries which is a reticular flexible envelope arranged about each electrode of the same polarity and formed by interlacing rods, some of the rods being arranged in parallel along the longitudinal axis of the electrode and other rods being inclined with respect to the first rods. According to the process, a thickness ratio of the rods arranged in parallel to the longitudinal axis of the electrode to the inclined rods is within the range of about 2 to 4. The described separator prevents current leakages and shorts, thereby insuring a reliable operation. It is achieved in the following manner.

The cause of current leakages and shorts between electrodes of opposite polarity resides in the separation of active mass particles from the electrode's surface and their disposition on the inclined rods. Consequently, the thinner the inclined rods the smaller the amount of active mass particles that will be deposited thereon, which are conducive to current leakages and shorts. Therefore, the inclined rods should be as thin as possible whereas the vertical rods should be as thick as possible, with the gap remaining optimal.

The minimal thickness of the inclined rods manufactured by the extrusion method is within 0.2 to 0.35 mm. Any further decrease of the rod thickness results in their breakage both in the process of manufacturing and of assembling the storage battery and also in operation. The maximum thickness of the vertical rods is determined, on the one hand, by the high specific volume characteristics and, consequently, by the optimal interelectrode gap. On the other hand, it is determined by the optimal interelectrode gap for the purpose of avoiding shorts in the storage batteries. As a rule, the gap is about 1.0 mm.

Thus, if the thickness of the inclined rods is 0.2 to 0.35 mm, the thickness of the vertical rods is accordingly 0.8 to 0.7 mm, and the thickness ratio of the vertical to the inclined rods is within the range of about 2 to 4. This is the optimal ratio. Any further decrease of the lower limit will result in shorts owing to the disposition of the active mass particles on the inclined rods.

Any further increase of the upper limit is either technologically impossible because of problems in manufacturing very thin and unreliable rods, or results in poor specific electrical volume characteristics of the storage battery due to the increased interelectrode gap. The gap is increased because of the increased thickness of the vertical rods. The separator, as described above, simplifies battery operation, since it does not require frequent washing of the battery owing to the disposition of the active mass particles on the inclined rods.

Auxiliary Electrode and Barrier

A process described by *G. Benczúr-Ürmössy, K. von Benda and F. Haschka; U.S. Patent 4,039,729; August 2, 1977; assigned to Deutsche Automobil GmbH, Firma, Germany* relates to a rechargeable galvanic cell with at least one negative, completely soluble zinc electrode (i.e., an anode), at least one positive metal oxide or oxygen electrode (i.e., a cathode), an alkaline electrolyte, as well as at least one electrically conductive auxiliary structure of low gas over-voltage, which is not permanently connected conductively with the electrodes for cathodic hydrogen evolution.

According to the process, it is important that (1) the auxiliary structure is segregated from the positive and/or from the negative electrode by at least one, preferably microporous, separator and serves simultaneously as an auxiliary electrode for after-discharging the zinc electrode as well as dendrite barrier; and (2) the auxiliary structure is arranged between, respectively, a positive electrode and a negative electrode so that it is close to the positive electrode but electrically insulated while an interspace is provided as an electrolyte chamber between the auxiliary structure and the current collector grid of the negative electrode, wherein the zinc deposition takes place during the charging of the cell and wherein, during discharging, the oxidation products of the negative electrode are completely dissolved.

The process for operating this cell resides in that the auxiliary structure (arranged between, respectively, each zinc electrode and each positive electrode), which is metallically conductive and is not constantly conductively connected with either the zinc electrode or the positive electrode, is utilized as an auxiliary electrode for an after-discharge of the zinc electrode and that the after-discharge takes place by short-circuiting the zinc electrode with the auxiliary electrode after the normal discharging (i.e., useful discharge). The after-discharge can take place after each cycle or after several cycles, depending on the operating conditions, especially depending on the intensity of discharge. If a battery is normally discharged only partially, an after-discharge will be conducted only after several cycles. In any event, such after-discharging should be effected if a drop in the capacity of the cell is noticed.

The interspace between the auxiliary electrode and the current collector of the negative electrode is dimensioned so that it offers sufficient space for the amount of zinc to be deposited. Because of varying capacities per unit area of the usable positive electrodes and because of different charging methods, no fixed value can be indicated for the interspace. However, when using nickel oxide electrodes, interspacings of from 0.5 to 8 mm and, when using air electrodes, interspacings of 1 to 5 mm are preferred between the current-discharge structure of the zinc electrode and the auxiliary electrode.

The distance between the current collector of the zinc electrode and the adjacent separator may be determined by insulating electrolyte permeable elements, which can also serve for the fixation of the electrodes, such as webbings, ribs, corrugated elements or dimensionally stable nettings made of synthetic resin or another insulating material. Preferred spacers are pins extending perpendicular to the plane of the electrodes and which are fixed to the cell wall and/or to the current collector grid of the zinc electrode.

The auxiliary electrode fulfills a dual task in accordance with the process; on the one hand, the prevention of short circuits during charging and, on the other hand, the removal of residual zinc from the current-discharge structure of the zinc electrode after the useful discharge by means of after-discharge. Residual zinc remains on the current conductor, i.e., current-discharge element, of the zinc electrode after useful discharge by passivation under very high current densities; due to local nonuniformities in the distribution and utilization of the active masses; and/or by a higher charging efficiency of the zinc electrode. The after-discharge takes place voluntarily without an external current source because of the small hydrogen overvoltage of the auxiliary electrode and produces a quantity of hydrogen at the auxiliary electrode equivalent to the amount of zinc being dissolved, until the reaction stops due to zinc consumption. The velocity of the reaction and/or the duration of the after-discharge can be influenced by the resistance of the short-circuiting bar or wire; the cutoff can be time-, current- or voltage-controlled. Normally, the after-discharge is terminated when the cell voltage is about 0.25 to 0.3 volts lower than before the after-discharge.

Separator Vibration Technique

M.D. Kocherginsky, S.L. Kalachev, V.A. Naumenko and L.F. Penkova; U.S. Patent 3,907,603; September 23, 1975 describe storage cells with moving zinc electrodes, having a long service life, simple structure, high specific electrical characteristics and small charging time. This is attained in a storage cell according to the process, wherein plastic separators with holes are provided between the positive and dissoluble zinc electrodes with current collectors, the current collectors of the dissoluble zinc electrodes being connected to a means for imparting vibration to the dissoluble zinc electrodes so as to ensure mechanical interaction between the whole surface of the zinc electrodes and the plastic separators and compaction of the zinc-sponge deposit forming on the electrodes in the process of charging.

To provide intensive stirring and accelerate the dissolution of zinc oxide accumulating on the bottom of the jar in the course of discharging, it is preferred that the ends of the electrodes disposed near the bottom of the jar be bent at a certain angle or connected by a perforated plate. The plastic separators in the storage cell may be in the form of perforated plastic sheets of abrasion-resistant materials, such as hard polyvinyl chloride or polyamide. The separators may also be fabricated from woven or pressed plastic grates or nets resistant to abrasion.

It is also expedient to dispose in the interelectrode clearance of the storage cell, according to the process, supports to which the positive electrodes and plastic separators are clamped, thereby providing a space for deposition of the sponge-zinc deposit. The supports prevent deformation of the positive electrodes in the course of cycling. The supports may be provided with slots in which the side edges of the zinc electrodes move. The slotted supports prevent deformation of both the positive and the zinc electrodes in the course of cycling.

It is further expedient that the dissoluble zinc electrode be composed of at least two layers, the layer adjacent to the plastic separator preferably being less porous. The porosity of the zinc electrode layer adjacent to the plastic separator should preferably be from 30 to 70 percent, while that of the inner layer contacting the current collector plate of the zinc electrode should be within 70 to 90 percent.

As a means for imparting vibration to the zinc electrodes, it is expedient to use a cam drive connected to the current collectors from the negative electrodes, thus providing the most favorable oscillating circular motion of the zinc electrodes. In case the area of the storage cell base is limited (for example, in transport vehicles), the current collectors of the zinc electrodes and the electrodes connected thereto are vibrated in the direction of the vertical axis of the storage cell.

To ensure reliable operation of the storage cell without short-circuiting when perforated separators are used, the amplitude of vibration of the zinc electrodes should exceed the diameter of the separator holes. It is preferred to dispose at the bottom of the electrolyte jar an electrochemical filter in the form, for example, of a porous nickel plate which is vibrated when the storage cell is charged. To ensure sealing of the storage cell, the moving current collector of the zinc electrodes is provided with an elastic compensator with a corrugated surface at least twice as large in total length as the distance between the points of securing the compensator to the current collector and cover.

Nickel-zinc storage batteries comprised of the storage cells, according to the process, have a specific energy from 45 to 50 Wh/kg and are completely charged within 45 minutes. Air-zinc and silver-zinc storage batteries, according to the process, have a specific energy from 100 to 110 Wh/kg and are charged within 2 hours. All these types of storage batteries have long service life determined by the service life of the positive electrodes. The high specific energy and power characteristics, fast charging and long service life make the storage batteries ideal for use on electromobiles intended for urban electrical passenger transport, thereby contributing to the cleanliness of the ambient air in large cities.

CONTROL OF DENDRITE FORMATION

Quaternary Ammonium Compounds

T.S. Lee; U.S. Patent 3,944,430, March 16, 1976; assigned to Union Carbide Corporation describes a cell having a zinc anode and a zinc-containing electrolyte which comprises an aqueous alkaline or aqueous acidic medium containing dissolved zinc ions and a quaternary ammonium compound for suppressing zinc dendrite formation during the charging cycle.

Specific illustrative quaternary ammonium compounds that can be used in the process include alkyl ammonium compounds, such as tetramethylammonium hydroxide, tetraethylammonium hydroxide, tetraethylammonium perchlorate, and tetraethylammonium acetate.

Example: A galvanic cell was constructed employing an epoxy-lined 30 ml beaker as the container. The electrolyte employed was 9 N potassium hydroxide containing 5 weight percent zinc oxide and was made from reagent grade chemicals and distilled water. Both the anode and the cathode of the cell were made from pure zinc discs. Only one side of each disc, which side had an active face area of about 2 cm^2, was exposed to the electrolyte. The remaining electrode surfaces were electrically insulated with epoxy. The exposed surfaces of the electrodes were prepared by first abrading them with abrasive paper down through 00 grade and then etching them in an 18 weight percent hydrochloric acid solution containing about 1 to 3 weight percent nitric acid. Finally, the electrodes were rinsed thoroughly in distilled water and placed in the cell.

The test cell was charged and discharged using direct current from an external power supply in series with a variable resistor. The charging current was varied from a few milliamperes to about 30 milliamperes per cm^2 of electrode surface. During charging, metallic zinc was deposited onto that zinc electrode which was intended to be the anode in a rechargeable cell system.

The deposit was nonadherent and ranged in form from mossy zinc to crystalline zinc dendrites in the charging current density range employed. As tetraethylammonium hydroxide or its salts were added to the electrolyte in a final concentration of about 5×10^{-4} to 1×10^{-2} molar, the deposit obtained at the zinc electrode during charging became adherent and smooth. After more than ten continuous charge-discharge cycles, the final deposit obtained after the last charging cycle was still smooth and adherent. It is expected that, even after several hundred continuous charge-discharge cycles, the final deposit would still be smooth and adherent. Based upon this fact, the additives of this process would be very beneficial in a rechargeable cell employing a zinc anode.

Another good feature of this process is that the effectiveness of that additive was not destroyed by electrolysis at both electrodes during charge-discharge cycling. The following experiment illustrates this point. A pair of pure nickel electrodes were placed in a cell containing only 9 N potassium hydroxide solution and 1×10^{-3} molar tetraethylammonium hydroxide dissolved therein. Continuous electrolysis was carried out for 76 hours by supplying to the pair of nickel electrodes a direct current of 30 mA/cm^2. The nickel electrodes were then removed and 5 weight percent zinc oxide was dissolved in the electrolyzed solution. Finally, zinc electrodes were placed in the electrolyzed solution, and charging and discharging from an outside dc source was begun. The deposit obtained during the charging cycle was still smooth and adherent at a current density ranging from a few to about 30 mA/cm^2 as indicated above.

Other examples of additives in alkaline media which produced smooth zinc deposits during the charging cycle are the following. The electrolyte used was 9N KOH, 5 weight percent ZnO.

(1) Additive, trimethylphenylammonium hydroxide (or chloride), 1×10^{-3} M; charging current density, 30 mA/cm^2
(2) Additive, methyltri-n-butylammonium iodide, 1×10^{-3} M; charging current density, 20 mA/cm^2
(3) Additive, triethyl(2-hydroxyethyl)ammonium iodide, 5×10^{-4} M; charging current density, 30 mA/cm^2

Film Barrier

J. Steffensen; U.S. Patent 3,970,472; July 20, 1976; assigned to McGraw-Edison Company describes a rechargeable battery having a zinc negative electrode, and a positive electrode preferably of nickel hydroxide. The battery is provided with a special dendrite barrier to prevent zinc spurs growing out of the zinc electrode during charge and shorting the battery. A separator pocket, which includes the barrier and is adapted to confine the zinc negative electrode against shape change, comprises an inner wrap of an absorbent film, an intermediate wrap of a microporous film, a next adjacent wrap of the flexible dendrite barrier, and an outer wrap of a cellulosic film.

This improved barrier is a porous, flexible sheet material comprising a substrate of a flexible, porous cloth woven from random fiber polypropylene in and through which is deposited a porous metal having a low hydrogen over-voltage to oxidize the zinc dendrites on contact in an alkaline electrolyte. The barrier in this multiple wrap allows free flow of ions and free escape of evolved hydrogen, and prevents any dendrites from passing through and penetrating the outer cellulosic film.

Vibratory Action

According to a process described by *O. von Krusenstierna; U.S. Patent 3,923,550; December 2, 1975; assigned to AGA Aktiebolaget, Sweden* in order to avoid dendrite formation when charging an alkaline accumulator battery cell having a zinc anode, a cathode and an ion-permeable separator arranged in the electrolyte between the anode and cathode, either the separator or the anode is subjected to a vibratory movement during the charging process.

The ion-permeable separator may consist of a separating wall made of a nonconducting material, for example, a plastic such as cellophane, polyvinyl chloride, polythene, polyvinyl alcohol or polyamide or of an inorganic material such as oxides which are insoluble in the alkaline electrolyte, for example, zirconium or titanium dioxide. The electrolyte preferably consists of potassium hydroxide dissolved in water to a solution containing 20 to 45 percent by weight of the hydroxide. The vibratory movement is suitably carried out with a frequency of 0.01 to 1,000 Hz, preferably 1 to 500 Hz, and an amplitude of 0.1 to 10 mm.

In related work, *O. von Krusenstierna; U.S. Patents 4,025,698; May 24, 1977; and 4,015,053; March 29, 1977; both assigned to Aktiebolaget Tudor, Sweden* describes a rechargeable accumulator battery apparatus in which either the zinc anode part or the porous separator part of each cell is vibrated in its own plane to produce macro and micro turbulence effects in the surrounding electrolyte to inhibit the occurrence of irregular zinc deposits on the anode in the course of repeated charge cycles. In one form, the amplitude of vibration appropriate during the charging of the particular battery is fixed mechanically by an eccentric element rotatably mounted in a permanent fashion on the battery case and adapted to be coupled to externally located motor means having a rotary power output.

SILVER-ZINC BATTERIES

Zinc Metal Strips and Grid

A.S. Berchielli; U.S. Patent 3,923,544; December 2, 1975; assigned to Electrochem, Inc. describes an electrode for electrolytic cells which comprises, in combination, a porous metal grid, at least one solid strip of zinc metal disposed on the grid along an edge thereof, and a zinc oxide-containing shape-retaining particulate mixture covering at least one side of the grid and overlying the strip.

The mixture includes a binder. A portion of the mixture overlying the grid has a density, for example, in excess of 2.5 grams per cubic centimeter, which is higher than that of the remainder of the mixture. Preferably, a pair of solid zinc metal strips are disposed along the opposite edges of the grid and the mixture overlies both of the strips. Most preferably, the strips are wrapped around the edges and overlie opposite grid sides adjacent the edges. The strips may be about 2 to 15 mil thick each and have the advantage of preventing cell short circuiting, because of separator perforation, by the grid. They also obviate overcharging of the electrode and prevent electrode shape loss during use of the electrode. Such shape loss causes changes in electrical characteristics of the electrode.

As shown schematically in Figures 2.3a and 2.3b, zinc electrode **10** in accordance with the process comprises a grid **12** which is fabricated of porous noncorrodible metal, such as copper or silver, i.e., contains uniformly spaced apertures **14** defined by a metal lattice **16**. The grid **12** may be, for example, about 2⅞ inches high and about 1⅞ inches wide with a thickness of about 0.010 inch with apertures of about 1/16 inch. The electrode **10** also includes a pair of strips **18** of solid zinc metal, each of which may be, for example, approximately 5 mils in thickness.

FIGURE 2.3: ZINC ELECTRODE

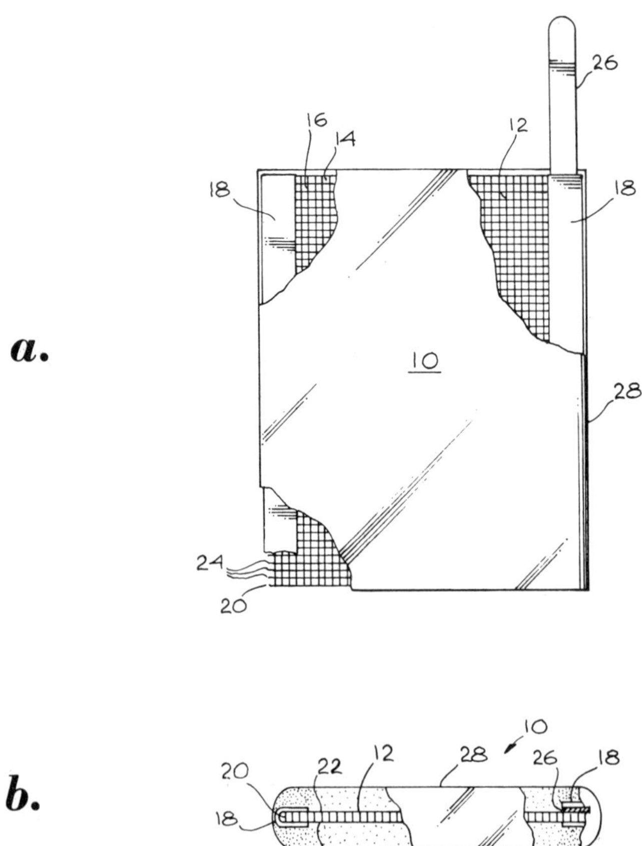

(a) Side elevation of electrode
(b) Top plan view of the electrode of Figure 2.3a

Source: U.S. Patent 3,923,544

Such thickness can vary from less than about 2 to over 15 mils or more depending on the particular application. As shown in Figure 2.3b, each strip **18** can be curved around a long edge **20** of the grid **12** so as to adjoin opposite sides **22** of the grid adjacent edge **20** and enclose edge **20** and prevent short circuiting in the electrode **10** by the sharp prongs **24** of edge **20** shown in Figure 2.3a.

A metal electrode lead **26** of noncorrodible metal, such as copper or silver, is disposed between one of the strips **18** and the grid **12**, abutting both, as shown in Figure 2.3b. Such lead **26** can be of any suitable size and shape. The electrode **10** also includes a continuous outer shell **28** adherent to the grid **12** and strips **18** and wholly enclosing the same. The shell **28** comprises a set shape-retaining mixture of particulate zinc oxide and binder. The binder may be, for example, tetrafluoroethylene, polyvinyl alcohol, polypropylene, polyethylene or the like plastic material, or carboxymethylcellulose (CMC) or the like. Such material may be thermoplastic or thermosetting, and may be initially mixed into the zinc oxide in dry or in wet form, i.e., dissolved or dispersed in a solvent. The zinc oxide particles can have an average particle size of, for example, about 0.1 μ to about 0.5 μ (diameter), and the zinc oxide is usually present in a concentration by weight in the mixture of between about 85 and about 99.5%, depending on the type of binder and the end use of the electrode.

The binder and zinc oxide are uniformly and thoroughly mixed together so that an essentially homogenous blend is obtained. Other materials, in addition to solvent or dispersant, can be added to the mixture before it is applied to the grid. Such materials may, for example, modify the physical, chemical and/or electrical properties of the electrode **10**. Typically, rayon fibers, metallic powders (Ni, Zn, Ag, Cu) or finely chopped fibers and expanders can be added in amounts of from about 1.0 to about 10% by weight.

In accordance with the process, the unset mixture of particulate zinc oxide and binder is applied around the grid **12** and strips **18** and is then pressed in place, fully shaped and then set to the form of the hard shape-retaining shell **28**, as shown. The shell **28** tightly adheres to and holds the grid **12** and strips **18** in place. The pressing operation can be carried out in any suitable manner such that the portion of the mixture which overlies the strips reaches a higher density than the remainder of the mixture. This result is easily effected, for example, by applying a uniformly thick layer of the mixture over the grid and strip assembly and then pressing this layer down onto the grid **12** and strips **18** to form a flat surface parallel with the grid **12**.

With such a technique, the portion of the layer overlying the strips **18** is thinner than the portion of the layer overlying the grid by an amount equal to the thickness of the strip **18** overlying the grid **12**. Thus, the density of the finished set mixture is increased, for example, to about 2.5 or more in the area overlying the strips **18** while the density of the mixture only overlying the grid remains at, for example, about 1.8 grams per cubic centimeter. The higher density mixture adjacent the long edges **20** is more durable and less subject to shape change than undensified material during use of the electrode. Moreover, the higher concentration of zinc-containing material in this area, due to the mixture and strips **18**, eliminates the necessity of overcharging the electrode in order to have it function properly. Accordingly, electrode **10** has improved properties.

Example: An improved electrode having the characteristics set forth in the table below is provided.

Component	
Grid:	Copper metal
Size:	1-7/8" × 2-7/8"
Type:	Perforated Beckley No. 2 Pattern
Lead:	Silver
Size:	1/8"w × 3"l
Type:	Solid Silver sheet 005" thick
Strips:	Zinc
Shape:	"U" strip

(continued)

Component	
Size: 1/4"w × 2-7/8"	
Number: 2-	
Particulate Mixture:	10 Grams of Mix per Electrode
ZnO Particle Size:	0.15 μ
Concentration:	95% ZnO
Binder	
Type:	CMC
Concentration:	0.5%
Solvent	
Type:	Water used to disperse CMC
Other Materials:	1.0% Mercuric Oxide (HgO) as amalgamating agent
	0.1% Rayon Fibers

The electrode is typically fabricated as follows:

(1) Prepared a mixture of ZnO powder, HgO powder, rayon fibers, CMC binder and distilled water. Blend until uniform mixture is obtained.

(2) Cast slurry to form thin strips of electrode material. Dry, and cut to size.

(3) Prepare grid with welded tab and edge strips attached.

(4) Make a sandwich consisting of a grid between two sheets of electrode material. Press composite at 2 tons per square inch to a thickness of 0.040 inch.

The electrode exhibits the following electrical characteristics: An electrode formed as described above is assembled into a Ag/Zn cell by positioning it between a pair of positive silver electrodes. The separator system for the cell is four layers of cellophane in the form of a U-wrap. The cell is charged at 0.2 amps equivalent to a current density of about 18 milliamps per square inch. The total input is 4.8 ampere-hours. The cell is then discharged at 1.0 ampere to 1.0 volt cut-off. The average voltage is 1.52, under load, and the capacity of the cell (and therefore of the zinc electrode since the cell is zinc limited) is 4.0 ampere-hours.

Inorganic Titanate Fibers for Electrode

A process described by *A.S. Berchielli and R.F. Chireau; U.S. Patent 4,041,221; August 9, 1977; assigned to Yardney Electric Corporation* comprises the inclusion in negative zinc electrodes in rechargeable alkaline electrochemical cells of about 0.2 to about 1.8% by weight of the weight of the zinc oxide of an inorganic titanate compound.

The inclusion of such a titanate in the negative zinc electrode improves the cell maintenance capacity while, at the same time, decreasing negative electrode shape change. Additionally, if the titanate is employed in fiber form, it provides improvements in electrode mechanical strength. The benefits are obtained when using the titanate compound in the stated concentration range while maintaining a commercially acceptable limiting charging current density.

As used in the examples, the term parts means parts by weight

Example 1: A slurry was prepared consisting of 100 parts of zinc oxide, 2.2 parts of Fibex L (approximately 57% by weight potassium titanate fibers), 40 parts of 1% solution of carboxymethylcellulose and 33 parts of distilled water. The slurry constituents were blended until a uniform, thixotropic blend was obtained. The pasting operation was performed by a pasting machine which spread the material uniformly between two layers of Aldex paper by means of two oscillating doctor blades.

The resulting zinc oxide strips were dried at about 93°C. The weight per unit area of each strip was 0.9 gram per square inch. The strips were cut to size by means of a die, one layer of the Aldex paper was removed, the current collector (2 mil perforated copper sheet) was placed between two strips on the paper-free side and the assembly pressed to a thickness of 42 mils. The total active mix weight per electrode was 10.4 grams.

Test cells were fabricated using the above-described electrodes. The cells consisted of four positive (silver) electrodes, each 1.975" wide x 3.00" high x 0.021" thick and five negative (zinc) electrodes measuring 1.975" wide x 3.00" high x 0.042" thick, made into an LR10-5 cell assembly. The cell pack was placed in a plastic cell case which was sealed with a cover terminal assembly and the cell was filled with a 45 weight percent solution of potassium hydroxide.

Exact replicates of the test cells just described were built except that the negative electrodes in these cells contained no additive. These cells were designated Controls. The test regime consisted of charging at 0.6 amp to a cut-off voltage of 2.05 V and discharging at 4 amps to 1.0 V per cell (100% DOD). The results are shown in Table 1.

TABLE 1

	CONTROLS (Average of 3 Cells)	TEST CELLS (Average of 3 Cells)
Cycle No.	Capacity (amp-hr.)	Capacity (amp-hr.)
1	15.5	15.0
5	12.5	14.5
10	12.0	14.0
20	11.2	14.0
50	9.0	11.0

Example 2: The procedure of Example 1 was repeated except that potassium titanate fibers were employed which differed dimensionally from the Fibex L fibers employed in Example 1, as shown in Table 2.

TABLE 2

Trademark	Avg. Fiber Dia. (microns)	Avg. Fiber Length (microns)	Density (gm/cc)
"PKT"	0.1 - 0.2	3 - 8	3.3
"Tipersul"	1.0	100	3.58
"Fibex D"	0.1 - 0.15	5 - 10	3.3
"Fibex L" (hydrated)	0.1 - 0.15	5 - 10	3.3

Substantially no difference in results was found among the fibers listed in Table 1.

Heat-Resistant Plastic Separator

A. Langer, L.C. Scala and C.R. Ruffing; U.S. Patent 3,953,241; April 27, 1976; assigned to Westinghouse Electric Corporation have found that battery separators able to withstand sterilizing cycles involving heating at 135°C in 40 percent KOH and which essentially prevent silver ion flow and resist concentrated, highly branched zinc dendrite penetration can be fabricated using a continuous dip process.

This is accomplished by coating at least one side of a flexible, porous support, resistant to caustic solutions, with a mixture of an alkali-resistant, water-insoluble polymer, inert inorganic filler particles and a water-soluble organic solvent. Once the mixture is applied to the support, the solvent still present in the composite is removed, preferably leached out by immersing the composite in a water-acid or water-organic solvent extracting solution.

This action results in the formation of very small pores throughout the polymeric matrix which will allow K^+ and OH^- electrolyte ions to pass through; simultaneously the aqueous treatment precipitates out and solidifies the polymer from the deposited mixture onto the porous support. The removal of the residual water and solvent from the treated composite by means of a modest heat treatment finally results in a heat-resistant porous substrate, consisting of a central support layer coated over with a rubbery, solid, porous mixture of KOH-resistant polymer and inorganic filler particles.

Example: A filler-coating formulation mixture, containing 720 grams of vacuum oven-dried 20 to 50 mesh hydrated zirconia filler particles, 1,840 grams of a polysulfone-dimethylacetamide coating solution (having a weight ratio of 15 parts resin binder to 100 parts solvent), and 9.6 grams (1%) of a cationic low molecular weight substituted oxazoline wetting agent (Alkaterge E) were ball-milled for 5 days to insure homogeneous mixture. The filler:resin binder weight ratio was 3:1. The polysulfone was a linear aromatic thermoplastic resin consisting of phenylene units linked by isopropylidene, ether and sulfone groups (Union Carbide P1700 Bakelite polysulfone extrusion and molding compound).

A 1' wide, nonwoven 3.5 mil thick polypropylene support weighing 30 grams per square yard (Webril tape EM476) was fed from a payoff roll into a dip-coating bath containing the above-described filler-coating formulation mixture. The coated support tape was then passed into an extracting bath containing by volume 2 parts (gallons) water:4 parts (gallons) 1,4 dioxane (diethylene dioxide), where the water-insoluble polysulfone resin solidified into a rubbery mass and the aqueous extracting solution leached out most of the water-soluble dimethylacetamide solvent.

The coated support tape was then fed through a 12' high vertical oven operating at 75° to 80°C. The line speed of the tape was 6' per minute. The resulting dry composite tape, about 7 mils thick, was wound onto a takeup roll. This substrate was porous and pinhole-free. Several 2" x 1.5" samples of the substrates were cut from the takeup roll. They were placed in stainless steel cups lined with Teflon and containing 8 M (40 weight percent) aqueous KOH solution. The cups were sealed with Teflon washers and steel covers held by C clamps. The cups containing the samples in KOH were then sterilized in a metal box containing sand in a 135°C oven for 60 hours. The battery separator material withstood sterilization without losing its physical or electrical properties. Weight loss was about 10 percent and the thickness of the samples seemed relatively unchanged. The specific resistivity of the separator material was of the order of 40 ohms-inch before sterilization cycling and about 20 ohms-inch after sterilization. Acceptable resistivity values are considered to be below 60 ohms-inch.

These porous battery separator substrates were then tested to polarographically measure the flow of silver ions through them as a function of time. The instrument used was a Heath polarograph Model EUW-402 M, employing a dropping mercury electrode as the working electrode and a massive external mercury-mercuric oxide electrode (in 40% KOH) as the reference and counter electrodes. A saturated silver solution was prepared by dissolving silver nitrate in 40% KOH. Measurements were carried out continuously over a period of about 30 to 60 hours. Each membrane sample was placed between two Teflon cells, one containing the known Ag solution in 40% KOH and the other containing only 40% KOH. The runs were made under a blanket of N_2 to prevent formation of potassium carbonate. Silver ion flow through the battery separator was almost negligible, the worst case being less than 4 parts per million after about 50 hours' exposure to the saturated KOH-Ag solution. In addition, the samples did not appear to be especially physically degraded.

Zincate ion diffusion through the separators was measured using flame absorption spectroscopy. The apparatus consisted of two Teflon cups (2" i.d., 100 cc vol) each with a flat face. A hole (½" diameter) was drilled in each flat face, a square of separator to be tested was placed between the holes, and the cups clamped together. One cup was filled with 40% KOH and the other cup filled with 40% KOH containing 40 ppm zinc ions. Samples of 1 cc were taken at regular intervals and the zinc concentration measured by flame absorption spectroscopy. The samples tested had previously been tested for silver ion diffusion. The results showed that the rate of zincate ion diffusion was of the same order as silver ion diffusion, with a maximum of 4.2 ppm after 50 hours.

The electrolyte flow through the battery separator was very fast; this was indicated by the rapid passage of KOH molecules through it as shown by pH measurements as functions of time of exposure to 40% KOH at 26°C by the separator. The results showed an increase in pH from about 6.5 to 12 in about two minutes. The substrates produced were flexible, prevented rapid transfer of silver oxides through the material, were not easily degraded, allowed transport of K^+ and OH^- electrolyte ions easily, were structurally sound and had a low electrical resistance. They were uniquely suitable as high-temperature separators for use in electrochemical cells since they withstood very vigorous sterilization cycles at 135°C in 40% KOH.

These substrates have been used in silver oxide-zinc cells with very high initial efficiency. A comparison of zinc dendrite growth was made between the polysulfone-refractory oxide battery separator and a multiple-layer cellophane battery separator. It could be observed microscopically that the zinc dendrites did not pierce the separator but grew into and inside of the structure. In the cellophane separator, the growth was at many points and the dendrites were relatively broad and highly branched. In the porous filled membrane although there was zinc dendrite penetration, it was usually only at a few points, and the dendrites were thin and hardly noticeable.

Electrolyte-Bonded Oriented Cellophane Foil Separator

According to a process described by *H.T. Mote, L. Hajdu and B. Ronay; U.S. Patent 3,945,851; March 23, 1976; assigned to Medicharge Limited, England and Medicor Muvek, Hungary,* in order to increase the cyclability life of rechargeable silver-zinc electric storage cells, the semipermeable membrane separator between the silver and zinc electrodes comprises a laminate of plies of semipermeable oriented cellophane foil bonded together with the electrolyte with some of the plies arranged crosswise, as regards their direction of orientation, relative to another ply of the membrane. In a preferred case a further single layer of cellophane is disposed between the silver electrode and the absorbent paper layer which is normally positioned between the silver electrode and the membrane.

Titanate-Containing Interseparator

R.F. Chireau; U.S. Patent 4,034,144; July 5, 1977; assigned to Yardney Electric Corporation describes a negative interseparator for use as a component in a separator system for alkaline rechargeable batteries which comprises about 50 to 95% by weight of a titanate material such as potassium titanate, a matrix-forming fibrous material, and a thickening agent for use in forming the interseparator. The negative interseparator is interposed between the negative electrode and main separator in the batteries.

The titanate is the principal or active material in the interseparator in that it is primarily responsible for the improvements produced by the interseparator. The titanate may be, for example, potassium titanate, sodium titanate, magnesium titanate, calcium titanate, cerium titanate, barium titanate, complex titanates such as magnesium-calcium titanate, and mixtures of the foregoing titanates. These titanates are usually fibrous themselves. A typical size of available potassium titanate is 0.2 micron in diameter by 10 microns in length. The following example illustrates the process.

Example: Silver zinc cells were constructed with each cell including two silver positive electrodes (measuring 1.62" x 1.5" x 0.034") and three zinc negative electrodes (measuring 1.62" x 1.5" x 0.042") with the zinc electrodes being disposed on each side of the silver electrodes. The negative electrodes consisted of 95% by weight zinc oxide and 5% by weight mercuric oxide.

A separator system was employed in each cell utilizing the "U" wrap technique with each separator system comprising: a positive nylon (Pellon) interseparator; a silverized cellophane main separator; and a negative interseparator comprising 89% by weight potassium titanate; 9% by weight chrysotile asbestos fiber, and 2% by weight carboxymethylcellulose.

The structure was disposed within a plastic casing to which there was added an aqueous potassium hydroxide solution comprising 40% by weight potassium hydroxide. Each cell had the above structure and differed only in that different thicknesses of the negative interseparator were utilized as shown in the table below and with the further exception that one cell included a 5 mil thick Pellon interseparator in place of the negative titanate-containing interseparator. Each of the cells was cycled on a 100% depth of discharge level with a charging procedure consisting of overcharging the cells by 50% on each cycle. The data derived from such tests are shown in the table.

- - - - Negative Interseparator - - - -			
Type	Thickness, mils	Cycles to 50% of Original Capacity	Cycles to Short
Titanate	6.5	105	120
Titanate	13	113	115
Titanate	20	98	105
Pellon	5	20	25

As will be seen from the table, the presence of a titanate-containing negative interseparator provides a substantial advantage over cells which do not contain such an interseparator. It will also be noted from the table that thickness of the titanate interseparator appears to have little effect on the working life of a cell.

ZINC-HALOGEN BATTERIES

Alkylammonium Perchlorate Electrolyte Additive

A process described by *M. DeRossi; U.S. Patent 3,915,744; October 28, 1975; assigned to Consiglio Nazionalle delle Ricerche, Italy* is directed toward a battery including a plurality of cells arranged in series. Each cell comprises: a bipolar electrode which serves to separate neighboring cells; an aqueous electrolyte of zinc bromide which is continuously circulated through the cells during the charging and discharging processes, and stored in an autonomous tank when the battery is not in use. The bipolar electrode has deposited on one side a zinc anode and on the other side a bromine cathode including an active cathodic mass. The active cathodic mass contains an active cathodic substance which is substantially insoluble in water and which is capable of combining with cathodic bromine to form solid addition products.

The active cathodic substance is selected from at least one of the group comprising alkylammonium perchlorate, diamine bromides, diamine perchlorates, triamine bromides and triamine perchlorates. The additive substances are dissolved in the electrolyte to reduce the solubility of the active cathodic substance, to increase the conductivity and the acidity of the electrolyte and to encourage the zinc to deposit in a thin and uniform layer.

Electrolyte Circulation System

A process described by *T.G. Hart; U.S. Patent 4,025,697; May 24, 1977; assigned to Energy Development Associates* relates to an apparatus for circulating and distributing an electrolyte through a multisection battery. Battery systems are well known which utilize a circulating electrolyte such as a metallic halogen, zinc-chloride, for example. Such a battery system is illustrated in U.S. Patent 3,713,888. The electrolyte is generally pumped through the system and to electrodes at which an ionic exchange occurs and a potential difference is created across the electrodes. This source of energy can be used to electrically power motor vehicles or serve as a standby power reserve and in many other situations where electrical energy is needed for a long-term use.

The circulating system described in this process provides a good solution to the various problems associated with the design of such high-flow, multiple-section, circulating electrolyte battery system. In particular, electrolyte is distributed in a two-stage system in which a large pump (first stage) distributes the electrolyte through hydraulically driven circulators (second stage) to individual electrode compartments, which are electrically isolated from each other. This results in minimizing intercell leakage. No rotating mechanical seals are used; only a single magnetic coupling is provided to circulate the electrolyte. This is achieved in the two-stage pumping system where an electric motor has the only magnetic coupling to drive the pump. The first stage generates a relatively high-pressure, low-flow rate of electrolyte fluid, but the second stage is designed to generate a relatively low-pressure, high-flow rate of the electrolyte.

This conversion of relatively low-flow, relatively high-pressure to relatively high-flow, relatively low-pressure in the circulators feeding the individual cells allows for higher overall efficiency and power use. Furthermore, the system lends itself to the pump and circulators being constructed of noncorrosive parts, mainly of polyvinyl chloride or polypropylene, except where bearing surfaces are required, and then these can be manufactured from ceramic and carbon. The connections of the various conduits can also be made of polyvinyl chloride or polypropylene with Viton rubber gaskets for sealing the couplings.

Ruthenium Dioxide Coated Silicon Cathode

According to a process described by *H.H. Hoekje and P.L. Dietz; U.S. Patent 3,982,960; September 28, 1976; assigned to PPG Industries, Inc.* a secondary electrochemical cell is provided having a consumable anode collector, a nonconsumable cathode collector, an electrolyte, and provisions for recovering the electrical current generated in the cell. The cathode collector has a silicon substrate with a suitable electroconductive surface on the substrate.

In a zinc bromide electrochemical cell prepared according to this process, it has been found that the discharge voltage is from about 0.02 to 0.20 volt higher than is the discharge voltage of a control cell having a graphite cathode collector, and the charging voltage was from about 0.1 to 0.30 volt lower, at the same current densities, electrode gap, normality, and electrolyte temperature as a cell having a graphite cathode collector. Under substantially similar conditions, a zinc bromide secondary cell having a ruthenium dioxide coated silicon cathode appears to have a voltage efficiency approximately 17% better than the voltage efficiency of a zinc bromide electrochemical cell having a graphite cathode.

High Temperature Dendrite Control

A process described by *C. England; U.S. Patent 3,912,999; October 14, 1975; assigned to California Institute of Technology* relates to an electrochemical cell which uses zinc and either chlorine or bromine as the primary reactants, with a molten, nonaqueous zinc halide electrolyte. During charging of the cell, growth of dendrites on the zinc electrode is controlled by operating the cell continuously or intermittently at a temperature above the melting point of zinc, thereby positively preventing such growth; or by operating at such temperature that incipient dendrites are melted by heat generated by the charging current.

That current may be intermittently increased to insure such dendrite control. In a large battery of cells utilizing chlorine as the halogen, excess halogen gas developed during charging of the battery is compressed and liquefied to facilitate external storage, and a large portion of the energy required for compression is recovered upon expansion of the gas during battery discharge.

Ion-Exchange Diaphragm

A process described by *J.J. Leddy and G. Gritzner; U.S. Patent 3,929,506; December 30, 1975; assigned to The Dow Chemical Company* provides a zinc-bromine secondary cell suitable for use as a battery with a longer shelf-life than had been heretofore possible. The zinc-bromine secondary cell comprises in combination an at least partially cell-encasing body having therein an anode compartment and a cathode compartment. A first electrode is positioned within the anode compartment and is spaced apart from a second electrode positioned in the cathode compartment by an ion exchange diaphragm. The ion exchange diaphragm is of the cationic type suitable to minimize passage of anions such as a bromide ion from the cathode compartment into the anode compartment. The ion exchange diaphragm further substantially entirely separates the anode compartment from the cathode compartment.

The secondary cell further includes an electrolyte system which at least at the time of initial operation of the cell consists essentially of an aqueous solution of a bromide ion disposed in the cathode compartment and an essentially bromide ion free, aqueous solution of a zinc cation disposed in the anode compartment.

Operation of the cell includes adding the electrolyte system to the respective anode and cathode compartments to at least partially immerse the electrodes in the respective solutions. An electromotive force is then impressed between the anode and the cathode to deposit metallic zinc on the anode and to release bromine at the cathode. During release of the bromine at the cathode substantially only cations will pass through the diaphragm into the anode compartment to thereby maintain the solution in the anode compartment in an essentially bromine free condition.

At least partially as a result of the absence of bromide ion in the anode compartment the described secondary cell has an improved shelf-life in comparison to nondiaphragm zinc-bromine secondary cells having a single bromide ion containing electrolyte.

Vented Electrode

P. Carr; U.S. Patent 3,909,298; September 30, 1975; assigned to Energy Development Associates describes an electrode for use in an electrical energy storage device which comprises a first side and a second side, thereby forming a channel for flow of an electrolyte with gas dissolved therein. The first side has holes for venting of undissolved gas.

The electrolytic cell of Figure 2.4a has two bipolar electrodes **13** and **15** with an area **11** between the electrodes for electrolyte flow. The electrodes are held together in a frame and are communicating with electrolyte inlet duct **17** and outlet manifold **19**. The electrodes each have a gas-impervious and electrolyte-impervious wall **21** of graphite, which extends vertically and on its outer surface, after charging or fuelling of the secondary battery of which the cell is a part, has a coating or plated layer **23** of highly electropositive metal, e.g., zinc. The inner surface of the impervious wall is cemented to porous electrode base member **25**, by an electrically conductive resinous polymeric cement layer **27**.

FIGURE 2.4: VENTED ELECTRODE STRUCTURE

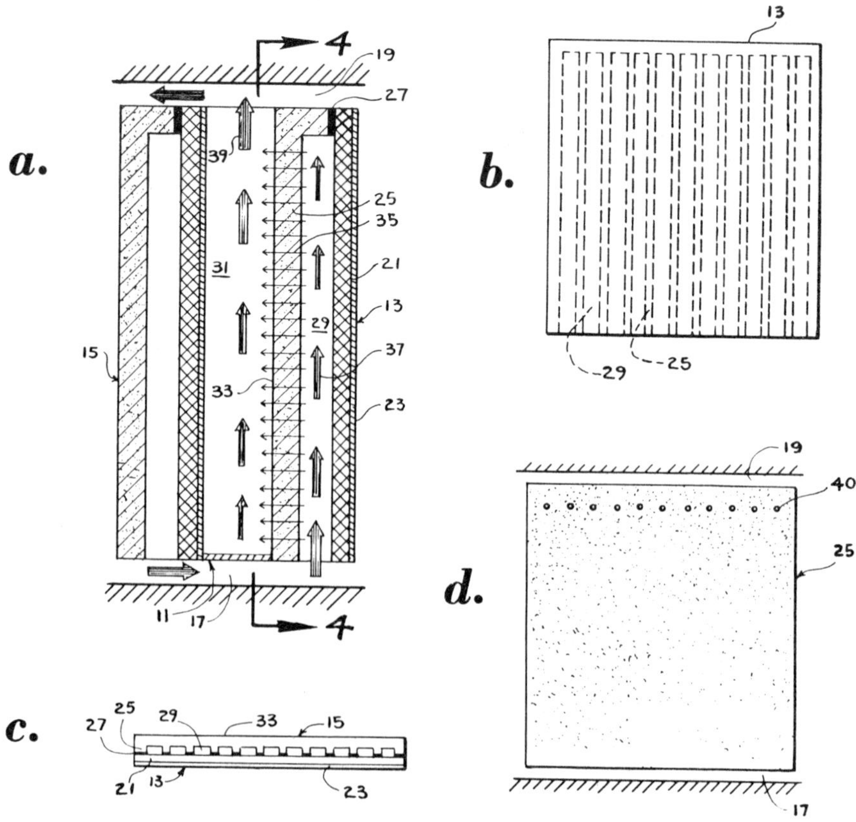

(a) Central vertical section of a pair of bipolar electrodes.
(b) Side elevation of the electrode, seen from the metal coated side.
(c) Bottom plan of the electrode of Figure 2.4b showing the flatness of the electrode and illustrating the passageways.
(d) Frontal view of Figure 2.4a along lines **4–4** showing the face of the porous electrode with the holes drawn in the upper portion.

Source: U.S. Patent 3,909,298

The porous member has a plurality of vertical passageways **29**, also well illustrated in Figures 2.4b and 2.4c, through which electrolyte may pass, usually from the lower to the upper portion of the electrode, as shown. Because the porous carbon member has pores or passages in it extending from the inner portion near the vertical passageways to the outer portion, fronting on the reactive zone of the cell, electrolyte pumped into duct **17** by pumping means, not illustrated, penetrates the porous body of member **25** and enters reaction zone **31**. Due to the pumping pressure, the flow of the electrolyte is vertically upward and out manifold duct **19** wherein it mixes with other electrolytes from other cells and, after enrichment with chlorine, is recirculated through the electrodes.

It is to be appreciated that the impervious wall is relatively impervious when compared to the porous electrode. The electrolyte may weep through the impervious wall with the passage of time but the rate of flow through the impervious wall is substantially less than through the porous side.

An important feature of the process is that the inner surface **33** of porous carbon member **25** is maintained in continuing contact with dissolved chlorine that is in the electrolyte passed through the porous member to the reaction zone. No boundary layer of stagnant electrolyte insulates the electrode surface from chlorine, as could be the case if it were to enter the cell only at the bottom of the reaction zone.

By causing an excess of electrolyte to enter the passageways **29**, one maintains them full at all times, prevents the porous carbon member from having a stagnant electrolyte and the deficiency of halogen at an upper surface portion, maintains the desired direction of electrolyte flow, and prevents undesirable backflows. The proportion of electrolyte may be regulated by suitable means, such as by adjustments of valves, not illustrated, or pumping pressures or capacities.

A number of cells of the type illustrated may be joined together in series to form cell banks and these may be further joined in series to increase the voltages developed, in parallel to increase current capacity or in mixed series-parallel, to do both.

The batteries made according to the illustration and descriptions given are of an improved power-weight ratio, usually over 50, and preferably over 100 watt-hours per pound, utilizing a zinc chloride solution in which the molar ratio of $ZnCl_2$ to H_2O is about 1:8. Such batteries are strong and are suitable for use in automobiles and trucks, where they withstand the ordinary shocks attending uses of such vehicles.

They are also long lasting, comparatively easy to manufacture, utilize readily available materials, recharge well or are easily refuelled, and are efficient and economical to operate. Among the most important advantage of the batteries, cells and electrodes made from the particular materials of construction and electrolyte, is the passage of the electrolyte so readily and evenly through one of the walls of the bipolar electrode.

As indicated by arrows **35** passage of the electrolyte through the porous carbon is substantially even from the bottom to the top while arrows **37**, which diminish in size as they move upwardly, indicate the decreasing of the volume of electrolyte flowing in passageways **29**.

Arrows **39** show the correspondingly increasing volume of electrolyte through reaction zone **31**. It will be evident that without the particular mechanism for contacting surface **33** with chlorine-enriched electrolyte, the efficiency of electricity generation at the upper portion of the electrode would be diminished, due to loss of chlorine in the electrolyte as it moves upwardly.

Such uneven generation of electric potential at different locations on the electrodes would tend to lead to inefficient operations, due in part to internal short-circuiting.

AIR-ZINC BATTERIES

Thin Platelike Electrode

B. Warszawski; U.S. Patent 3,902,916; Sept. 2, 1975; assigned to Societe Generale de Constructions Electriques et Mechaniques (ALSTHOM), France describes an electrochemical generator rechargeable electrochemically with liquid electrolyte having negative electrodes of zinc and positive nonporous electrodes that are contacted on the same face by the electrolyte and an oxygen-containing gas.

The generator or battery comprises a repetitive unit of groups of thin, platelike components, each group having a total thickness that is preferably smaller than about 1 millimeter and comprising an impervious bipolar electrode, a microporous membrane, and means for maintaining an essentially constant spacing between the bipolar electrode and the membrane. These generators are capable of providing long service lives and/or a large specific power or output.

In this process, the forced convection movement of the electrolyte prevents in particular the passivation of the zinc electrodes due to accumulation of zinc oxide and reduces considerably the formation of dendrites (i.e., zinc dendrites) at the time of the recharge. Moreover, the air electrodes used in fuel cells, which have a very thin, nonporous structure and in which the electrolyte and the gas are supplied to the same side of the electrode do not undergo any appreciable deterioration during the recharging of the generator.

In Figures 2.5a and 2.5b, reference numeral **1** identifies a bipolar electrode. This electrode comprises a frame or housing **2** made from insulating plastic material having a thickness that is essentially equal to the mean thickness of the central portion of the electrode and defining an essentially rectangular shape or configuration for the active portion of the electrode. In Figure 2.5a, the visible face **3** of the bipolar electrode **1** constitutes the positive electrode and the hidden or opposite face constitutes the negative electrode.

The bipolar electrode **1** is made up of a thin embossed conductive plate or foil **4** of stainless steel or plastic material made conductive by means of an appropriate charge, for example, a carbonaceous charge, having a thickness of several tens of microns, for example about 50 microns. The thin foil or plate **4** is equipped, on that side of face **3** which is supplied with a gas containing oxygen, with a coating **5** of a catalyst for effecting reduction of the oxygen, e.g., carbon, fixed or secured by a plastic binder, preferably a hydrophobic binder such as polytetrafluoroethylene. Similar electrodes are described particularly in Belgian Patents 754,335 and 776,160.

Reference numeral **6** designates a zinc deposit carried by the other face of the plate or foil **4**. This deposit may be produced electrochemically and may have a mean thickness of about 100 microns. An even number of openings **7** on one edge of the electrode, and opening **8**, on the opposite edge, allow for forming, by superposition with opposite adjacent opening in the other components, channels for supplying and removing the electrolyte from the assembled generator. Grooves or microchannels are made or obtained by impression in the frame or housing of the electrode. The microchannels **9** and **10**, on the side of the positive electrode with the oxidizer, and microchannels **11** and **12** on the side of the negative electrode assure equal distribution in the supply and removal of the electrolyte for the two faces of the bipolar electrode. This arrangement is similar to the general structure defined in French Patent 1,522,305. The length and the section thereof are such that they render negligible the shunting currents of the electrodes and allow for obtaining an equal loss of charge. These microchannels assure a uniform supply of fluid over each electrode, as well as a uniform supply from one electrode to the other.

FIGURE 2.5: ZINC-OXYGEN CELL

a.

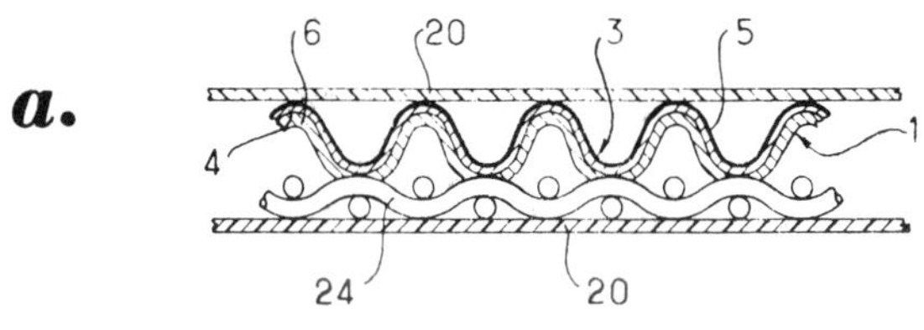

b.

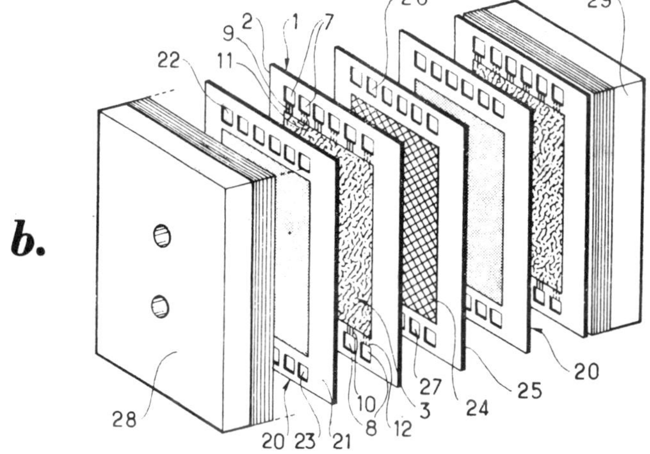

c.

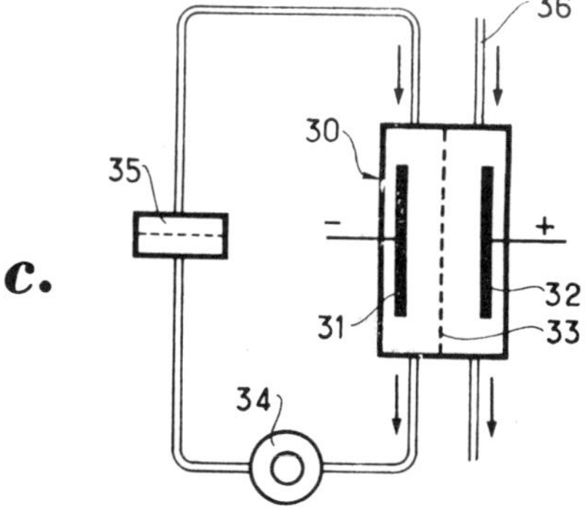

(continued)

FIGURE 2.5: (continued)

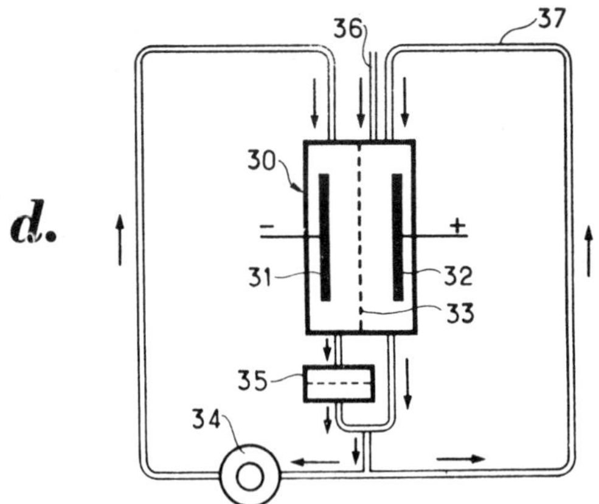

(a) Schematic partial cross-sectional view through a group of thin platelike components used in the makeup of an electrochemical generator.
(b) Perspective view at an enlarged scale representing schematically a generator.
(c)(d) Schematically illustrated circuits of the fluids used in two types of generators.

Source: U.S. Patent 3,902,916

The microchannels which assume the same supply or removal of one and the same fluid also provide an equal loss of charge. These microchannels may diverge from the openings in the frames toward the edge of the electrode plate. The cross sections thereof may be proportional to the lengths thereof. By way of example, the length of these microchannels may be in the order of 1 centimeter, and the width and depth thereof may be of several tenths of a millimeter. For instance, in the case of an equal distribution of these microchannels on the edge of the electrode, the electrode may have one channel spaced about every 5 millimeters.

Projections may be disposed on an insulating band between the edge of the electrode and the outlets of the microchannels in order to avoid a deformation of the adjacent microporous membrane. They are preferably cylindrical in shape and distributed in a staggered or offset fashion so as to assure a uniform distribution of the electrolyte flow.

Reference numeral **20** designates a microporous membrane. It is equipped at the periphery thereof with a frame or housing **21** made of insulating plastic material perforated by openings such as **22** and **23** on opposite edges so as to form the channels for feeding and removing the electrolyte. As shown, one microporous membrane **20** is in contact with the face **3** of the bipolar electrode **1** supplied with the oxygen-containing gas. On the other hand, the face of the bipolar electrode **1** provided with the zinc coating or deposition is spaced from another membrane **20** by means of a screen or web **24** made of insulating material such as, for example, a polyamide. This screen or web **24** is also equipped at its periphery with a frame or housing **25** made of insulating plastic material perforated by openings such as **26** and **27** on the opposite edges thereof which also form channels for feeding and removing the electrolyte.

The channels formed or constituted by the superposition of the openings placed on one of the edges of each of the above-described components are alternately used for feeding or removing fluid from the positive face and for feeding or removing fluid from the negative face of the bipolar electrode. The streams of the fluids flowing along the faces of the bipolar electrode are directed parallel with respect to each other and in the same direction.

The sum total of the thicknesses of the peripheral parts or portions, i.e., the frames, of each of the platelike thin components is essentially equal to the sum of the thicknesses of the central parts or portions thereof, which assures that the components be kept in position with respect to each other. This arrangement is a necessary condition for obtaining a uniform distribution of the fluids.

The equal distribution of the charge losses inside the various compartments defined by the adjacent components, as well as in the microchannels in communicating with the channels formed by the superposition of the openings provided in the edges of the thin components with these compartments, conjugated to the rectangular shape of these compartments, allows for obtaining uniform currents without turbulence, or eddy currents, and without any stationary zones. This has the effect of avoiding the formation of dendrites therein.

The electrochemically rechargeable generator according to the process is made up by assembling a plurality of the thin plate-like components such as those shown in Figure 2.5a. This assembly may be obtained in a manner similar to those of the fuel cells which have been described particularly in the following French Patents: 1,379,800; 1,399,765; 1,522,304; 1,522,305; 1,522,306; 1,564,864; 1,584,577; and 1,604,897.

In Figure 2.5b reference numerals **28** and **29** designate the tightening plates providing the ends of the assemblage **3** of the thin generator components. These plates are equipped with means not shown in Figure 2.5b which allow for feeding and for removing fluids therefrom. The entire unit of the components and the tightening plate may be connected by tie beams or the like securing means (not shown) which extend through appropriate openings in the frames or housings made of plastic material. This consolidation or integral structure may also be achieved by molding a casting from a resin.

In Figure 2.5c and Figure 2.5d, reference numeral **30** designates a generator in accordance with the process. A negative electrode **31** of zinc, and a positive electrode **32** supplied with a gas containing oxygen, separated by a microporous membrane **33** are schematically shown in these figures. Figure 2.5c shows that case in which the positive electrode **32** is supplied directly with a gas containing oxygen, for example air, and the process of operation is then called "process with gliding layer," and the arrow shows the direction of gas introduced through supply means **36**. Reference numeral **34** designates a pump, and reference numeral **35** represents a device or means for selectively precipitating and filtering zinc oxide from the electrolyte being recirculated past negative electrode **31**.

Figure 2.5d is an example in which the positive electrode **32** is supplied by an emulsion of a gas containing oxygen, for example air, in a liquid electrolyte, particularly an alkaline electrolyte, for example, of the same type as that circulating in contact with the zinc. Reference numeral **37** designates the supply of electrolyte allowing for making up or producing this emulsion and **36** again designates the air supply. The electrolyte that has been supplied to the negative electrode is mixed, after filtering, with that electrolyte that has been supplied to the positive electrode, and the resultant electrolyte is divided and recycled.

The operation of a generator is as follows: In the course of the discharge, the pump **34** effects circulation of a potash solution having at least one normal concentration along the zinc electrode. At the contact of the hydroxyl ions of the potash solution, the zinc passes in solution in the form of zincate which precipitates at the beginning of the discharge as soon as it has reached its saturation concentration. The zinc oxide formed is taken along by the electrolyte circulation.

The fact that the thickness of the space between the zinc layer and the microporous membrane remains at several tens of millimeters provides a very rapid circulation which eliminates any oxide layer which might have the tendency to form on the zinc layer and to passivate it. The zinc oxide entrained or taken along by this circulation must be separated from the potash solution in the filtering device **35** prior to the recycling of the solution in contact with the zinc layer. With the discharge continuing, the zinc oxide precipitates at the level provided by the device **35**.

In the course of the recharge of the generator, the zinc oxide stored in the device **35** is redissolved in the potash solution placed in circulation and zinc will be deposited on the face of the bipolar electrode opposite the side forming the electrode with the oxidizer gas. The active and regular circulation of the potash in the capillary interval between the bipolar electrode and the membrane reduces to a very large extent the formation of zinc dendrites which could take place if the electrolyte layer were thicker and assures a zinc deposit having an essentially uniform thickness.

Of course, an electrode with oxidizer will be used which is susceptible to undergoing the recharge without appreciable deterioration during a significant number of cycles by oxidation of the hydroxyl ions of a potash solution circulating at its contact, with freeing or the emission of oxygen. The process thus provides for electrochemical generators having a capacity and an energy density per unit of volume or by weight being in the same order as those of the fuel cells, and very superior to those of the known accumulators with a solid oxidized compound.

Funnel Design for Air Supply

A process described by *G. Gerbier; U.S. Patent 4,009,320; February 22, 1977; assigned to Saft-Societe des Accumulateurs Fixes et de Traction and Compagnie Industrielle des Piles Electriques Cipel, France* relates to air depolarization cells or batteries and, more particularly, to air-zinc batteries, whose positive electrode is fed with air through at least one cavity provided in the latter. An aim of the process is to ensure a circulation of the air in the cavity so as to renew the oxygen consumed during the discharge of the battery. The process provides an air depolarization battery comprising a negative electrode, an electrolyte and a positive electrode fed with air through at least one cavity. The cavity is formed by at least two funnels extending substantially from top to bottom of the positive electrode, each communicating with the outside air at their upper parts and separated from each other by a portion of the mass of the positive electrode and communicating with each other at their lower parts or ends. The requisite ratio between the area of the surface over which air flows through each funnel of the electrode and the volume of air therein is different for the two funnels.

FIGURE 2.6: AIR DEPOLARIZATION CELL

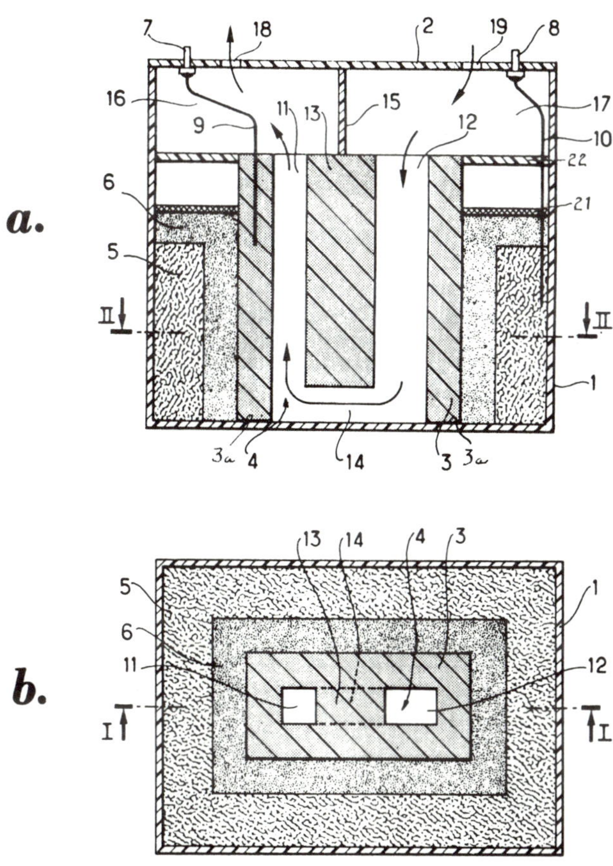

(a) Vertical cross-sectional view of a cell or battery taken along the line I—I of Figure 2.6b.

(b) Cross-sectional view taken along the line II—II of Figure 2.6a.

Source: U.S. Patent 4,009,320

The directional references are relative to the normal operating position of the battery. The speed of oxygen depletion of the air contained in the funnels being a function of its area-to-volume ratio, different concentrations of oxygen and hence different densities of gas tend to be established in the two funnels, this causing a draught effect from one funnel to the other and thus promoting the circulation of air.

The air depolarization cell or battery shown in Figures 2.6a and 2.6b comprises a casing formed by a can **1** and a cover **2** both made of plastic material, a positive electrode **3** basically containing active carbon, provided with a cavity **4**, a negative electrode **5** basically containing zinc powder, arranged along the lateral wall of the can **1** and surrounding the positive electrode **3** and an alkaline gelled electrolyte **6** arranged between the positive electrode and the negative electrode and also above the latter. A positive terminal **7** and a negative terminal **8** are connected to the electrodes **3** and **5** respectively by metallic wires **9** and **10.**

In the battery or cell in Figures 2.6a and 2.6b, the cavity is formed directly in the mass of the positive electrode by two vertical funnels **11** and **12** separated from each other by the mass of the positive electrode in the zone **13**. These funnels are connected together at their bases by a tunnel **14** also formed directly in the positive electrode.

In the example illustrated, the funnel **11** has a square cross-section whose dimensions are a x a and the funnel **12** has a rectangular cross-section whose dimensions are a and b = 1.5a. If h is the height of the funnels above the tunnel **14**, the lateral surface of the funnel **11**, that is, the surface by which the air contained in that funnel is in contact with the positive electrode, has an area S = 4 ah and the corresponding volume of air is $V = a^2h$.

The ratio between these two dimensions is R = S/V = 4/a. For the funnel **12**, the corresponding values are S = (2a + 2b) h = 5 ah, V = abh = $1.5a^2h$ and R = 10/3a. The ratio R regulates the speed of oxygen depletion of the air for a given discharge rate of the cell or battery. That speed is, therefore, higher for the funnel **11** than for the funnel **12.**

It ensues that the gaseous mixture contained in the funnel **11** is more depleted in oxygen, hence less dense, than that in the funnel **12** and a circulating flow is established in the cavity **4**, the air entering through the upper end of the funnel **12**, passing through the tunnel **14** and leaving through the upper end of the funnel **11**.

The two funnels communicate by their upper parts with a free space in the top part of the battery. To improve the circulating flow, that space is divided by a partition **15** into a chamber **16** communicating with the funnel **11** and a chamber **17** communicating with the funnel **12.** The air enters the chamber **17** through an orifice **19** in the cover **2** and leaves the chamber **16** through an orifice **18** in the cover **2.** The positive electrode **3** provided with the cavity **4** according to the process can be made very simply by agglomerating the catalytic mass round a core mating the shape of the cavity and withdrawing the core from below. To prevent leakages, in one direction or another, between the cavity **4** and the electrolyte **6** under the positive electrode, it is necessary to provide fluid-tight sealing means, for example, by glueing the electrode **3** at **3a** onto the bottom of the can **1.** If it were required to prevent the cavity from reaching the lower face of the positive electrode, it would be possible to cut funnels obliquely, after the forming of the electrode, in such a way that they are joined together at the lower part without opening out. An appropriate layer of pitch **21** overlies the upper surface of electrolyte **6**. A separating disc **22** of suitable plastic material closes off the bottoms of respective chambers **16** and **17** to prevent access of air therein to the electrolyte **6**.

Connecting Elements for Forced-Flow Systems

According to a process described by *J.-P. Pompon; U.S. Patent 3,979,222; September 7, 1976; assigned to Compagnie Generale d'Electricite, France* an electrochemical cell of the air-zinc type is formed by connecting up elements in hydraulic and electric series and, in each of the elements, in electrically insulating the inside of the negative end fitting situated hydraulically the nearest to the following element. In this way, the interference transfer of charges resulting from the electric series connection of two neighboring elements can be effected only by ionic conduction in the electrolyte flowing between the two elements.

Figure 2.7 illustrates two elements **1** and **1'** which are substantially identical to each other, fed, in series, with a potassium hydroxide solution containing zinc powder in suspension, such feeding being effected through electrically insulating pipes **2, 2', 2''** and being materially shown by the arrows **F**. Such elements are, for example, of the type which is described in U.S. Patents 2,173,637 and 2,227,651.

The element **1** comprises a negative collector grid **a**, a porous separator **b**, a porous catalytic active layer **c**, a positive collector grid **d** coated in a porous water-repellent layer **e**. Moreover, conductive end fittings or terminals such as **f** and **g**, connected to the negative and positive collector grids **a** and **d** respectively ensure the electric and hydraulic connection of the elements together.

The element **1'** comprises, of course, the same parts referenced by the same letters but bearing the "prime" index. In the elements **1** and **1'** the oxidation of the zinc by the outside air at the level of the catalytic layers **c** and **c'** generates an electromotive force collected between the end fittings **f** and **g** on the one hand and **f'** and **g'** on the other hand. The elements are connected in electrical series by means of a conductor wire **3** consequently connecting together the end fittings g and **f'** of the elements **1** and **1'** respectively.

It will therefore be seen that the electric circuit formed successively by the end fitting **g**, the layers **c** and **b**, the electrolyte flowing in the pipe **2'**, the layer **a'**, the end fitting **f'** and the conductor wire **3** constitute an interference circuit discharging a leakage current generated by the two half-elements thus short-circuited. Of course, that leakage current can be reduced by providing a sufficiently great length of pipe **2'** with a view to increasing the ohmic resistance of the interference circuit.

However, at the level of the end limited by the zones **A** and **B** of the element **1**, two paths are open simultaneously to the transfer of the electric charges, that is, to the current flowing in the interference circuit. The first method of transfer consists in the moving of the OH^- ions from the layer **c** where they are formed, to the electrolyte, crossing the layers **b** and **a**, then moving under the effect of the electric field in the pipe **2'**.

FIGURE 2.7: AIR-ZINC CELL

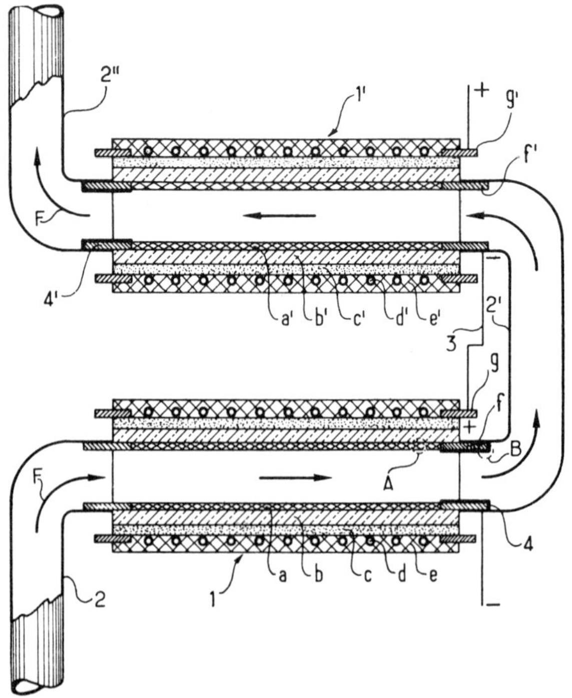

Source: U.S. Patent 3,979,222

The second method of transfer consists in the moving of the OH^- ions from the layer **c** where they are formed to the metallic collector **a**, for example in the zone A; at that level, each OH^- ion can supply, in contact with a particle of zinc in the suspension, a zincate ion and an electron; that last charge transits, in the end piece **f** and, having reached the zone **B**, substantially supplies, with a zincate ion of the solution, a particle of zinc on the one hand and an OH^- ion on the other hand, which, moving along in the pipe **2'**, will become discharged on coming into contact with the end fitting **f'**.

It can therefore be seen that progressively, a zinc deposit will be effected on the internal wall of the end fitting and will grow according to the electric field in the pipe **2'** and, when such a deposit reaches the end fitting **f'**, the element **1** is clearly short-circuited. To avoid such a detrimental effect, according to the process, an insulating coating **4** is provided on the internal face of the end fitting **f**, so that the transformation described in the zone **B** may not take place, thus preventing the second transfer method.

Such a coating **4** can be a plastic tube or cap fitted or cemented into the end fitting **f** or else a layer of insulating varnish or the like. It must be noted that such a coating **4'** will also be arranged at the output of the element **1'**, itself connected to an element, but it will not be necessary to provide it inside the end piece **f'**, on the input side of the element **1'**.

J. Jacquelin; U.S. Patent 3,977,903; August 31, 1976; assigned to Compagnie Generale d'Electricite, France describes the construction of an electrochemical battery of the type comprising several cells fed in series by an electrolytic solution comprising an active material in suspension and more particularly zinc powder. The solution is conveyed by forced flow in the cells from a storage tank and made to flow back into the tank after having passed through the last cell. The cells are grouped together in modules and are electrically connected in series within the modules (m), the modules each giving a voltage V and a current I, themselves being connected electrically in series to constitute a battery.

They are fed in series by the solution. The battery is suitable for supplying an electromotive force mV and a rated current I. The modules themselves are grouped into p assemblies and are electrically connected in series in each assembly and are characterized in that the number p of assemblies is an even number, the assemblies each comprising the same number m of modules each capable of supplying an electromotive force V and a rated current I/p. The assemblies are connected together in parallel, in such a way that the potential of the solution conveyed from tank towards the first cell of the battery is equal to the potential of the solution leaving the last cell of the battery before being made to flow back into the tank.

Circulating Electrolyte

According to a process described by *Z. Stachurski and M.N. Yardney; U.S. Patent 3,985,581; October 12, 1976; assigned to Yardney Electric Corporation* to prevent the formation of internal short circuits between electrodes of a rechargeable electrochemical cell, the electrolyte is maintained in continuous motion at least during charging.

In the case of an air-depolarized cell with a positive oxygen electrode and a negative zinc electrode, an air stream is introduced into a cylindrical cell casing through a multiplicity of orifices in a generally tangential direction and at increasing distances from the housing axis to impart rotation about that axis to the liquid. During the subsequent discharge, a valve redirects the air stream into a compartment separated from the electrolyte by the oxygen electrode.

In Figures 2.8a and 2.8b, there is shown a cell **16** comprising a cylindrical housing **1**, zinc electrode **14** and air electrode **15**. The electrolyte in the cell is designated **2**, and is alkaline and may consist of sodium, potassium or lithium hydroxide, with or without special-purpose additives. The air feed channel **3** herein shown as centrally located serves for the conduction of air coming from gas compressor **10** via channel **13** and valve **11** and is discharged through apertures **4**. The compressed gas previously entrapped in channel **3** is driven out through the apertures **4** into the electrolyte **2** in the form of multiple rapidly moving gas jets which break up into bubbles **5**.

The impingement of the jets of gas on the body of electrolyte causes the latter to move in a circular path. Under the continuing gas flow the electrolyte assumes the characteristics of a rapidly rotating body of liquid. The plates **6** are held to guide and/or direct the flow of the liquid electrolyte along the established path. The gas escapes from the electrolyte through the apertures **7** arranged in the housing **1** into a gas disentrainment chamber **8** which vents through a gas outlet **9**.

FIGURE 2.8: AIR-DEPOLARIZED CELL

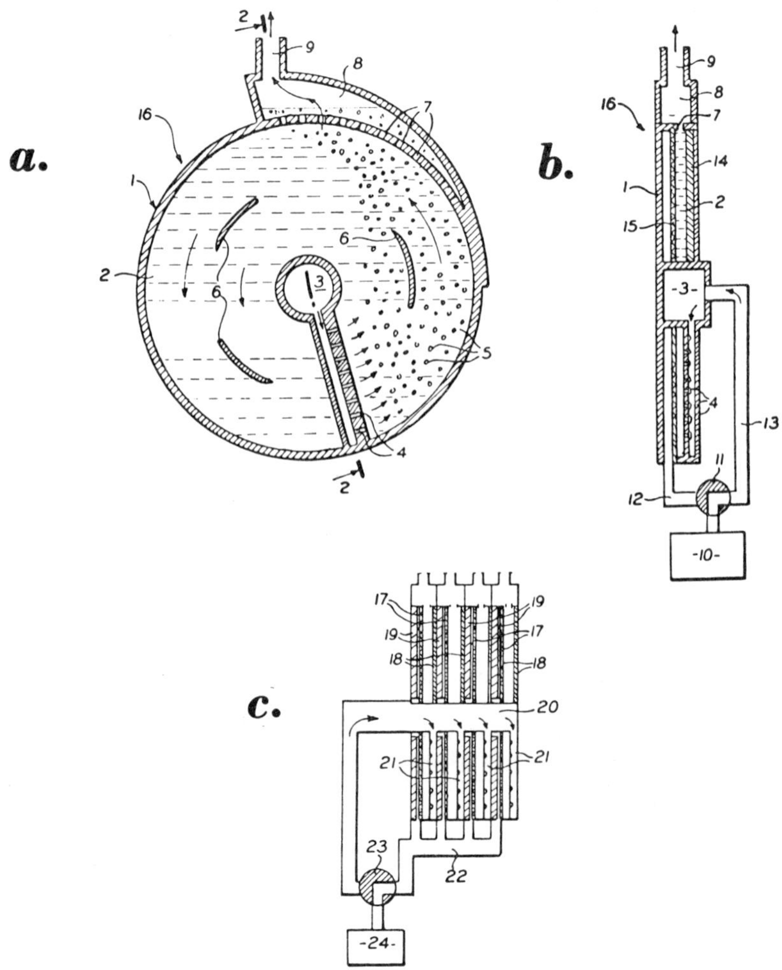

(a) Section through electrochemical cell.
(b) Section taken along line **2–2** of Figure 2.8a.
(c) Partial cross-sectional view of a cell assembly.

Source: U.S. Patent 3,985,581

In operation, on charge, compressed gas having been released from compressor **10** is directed by valve **11** into channel **13** and is passed by the continuing gas pressure into channel **3**. The gas is discharged from channel **3** via jet-like apertures **4** streaming into the electrolyte and imparting thereto a rapidly rotary motion. The gas is removed from the electrolyte via apertures **7** in the housing passing into a chamber **8** from which it is discharged at **9**.

The example shown consists of an air cell wherein it is desirable to use air as the circulation-imparting agent. On discharge in this case, the air leaving the compressor **10** is shunted by valve **11** into channel **12** leading directly to the oxygen electrode **15** where the oxygen in the air is reduced, the excess of air flowing across the oxygen electrode into the electrolyte whence it is discharged via apertures **7** into air disentrainment chamber **8** and out through outlet **9**.

The air is pumped from the compressor and is introduced via valve **11** into channel **13** and from there discharged through apertures **4** into the body of electrolyte. The size of the apertures **4** is approximately 0.1 to 1 mm. The air discharges as a result in the form of multiple finely divided streams, the latter being rapidly divided into bubbles on introducing into the electrolyte. The electrolyte assumes a rotary motion in this case because of the general shape of the cell, manner of gas introduction and guide plate, the speed of rotation amounting to approximately 2 inches per second to 30 inches per second.

Figure 2.8c illustrates a cell assembly or zinc/air battery utilizing a pile construction. As shown in the figure, four cells are assembled in a series arrangement wherein the air electrodes **17** are electrically connected to the electrodes **18** of opposite polarity of the next succeeding cell across metallic ribs **19** spacing the air compartment. Air under pressure is passed from compressor **24** via valve **23** into channel **20**. From channel **20** the air is passed into channel **21** and out through the jet openings into the electrolyte as set out in detail in connection with Figures 2.8a and 2.8b.

Channel **22** serves for conducting air to the oxygen electrodes. The valve **23** serves for directing the air either into the oxygen electrodes or into the electrolyte and is regulated according to the phase of operation, i.e., charge or discharge. Reference numeral **24** designates the air compressor or pump.

Electrolyte Circulation System

H. Ikeda, M. Inaba and M. Ide; U.S. Patent 3,915,745; October 28, 1975; assigned to Agency of Industrial Science and Technology, Japan describe the operation of a metal-air secondary battery of an electrolyte circulation type. The method comprises a charging circulation system including an electrolyte tank with a large capacity detachably connectable by a pair of pipe connectors to the apparatus and a discharging circulation system with a small capacity, these two circulation systems being alternatively changed over by means of a two-circuit three-way cock.

In a charging mode, the body of a metal-air secondary battery is connected to the tank which is usually installed in a charging station so that an active metal may be efficiently electrodeposited using an ordinary circulation method, and in a discharging mode, the tank is detached and the battery is connected to the discharging circulation system so that a minimum quantity of electrolyte, necessary for discharging, may be circulated. Therefore, the system of the process is advantageous in that the weight and capacity size of a power supply are reduced by half as compared with those of conventional systems, with the result of an improved ratio of energy to weight and size, and is extremely useful when used in a vehicle-borne metal air battery, for example, an air zinc battery for an electric automobile.

Polytetrafluoroethylene Diaphragm

A process described by *D. Groppel and D. Kuhl; U.S. Patent 3,922,177; Nov. 25, 1975; assigned to Siemens AG, Germany* relates to rechargeable metal-air cells and batteries of the type having an aqueous electrolyte in general, and more particularly to a method for controlling the water budget of such cells and batteries. This is achieved by using air electrodes having a hydrophobic layer on the gas side, and by at least partially covering the air electrodes on the gas side with water during the charging process. When this is done, water vapor diffuses toward the electrolyte through the gas-filled pores due to the water vapor pressure gradient.

The arrangement permits making up water losses that occur during charging and discharging in an extremely simple manner. With regard to single cells, the cell need only be placed, during the charging phase, at least partially in a container of water. Alternatively, the cell may be placed in a container and water added in such an amount so that the layer of the air electrode on the gas side is at least partially covered with water.

Figure 2.9a illustrates a metal air cell **12** which is partially immersed in a container **10** filled with water **11**. The cell **12** comprises a cell frame **13** which may be made, for example, of plastic, within which is contained a metal electrode **14** immersed in an electrolyte such as 6 M KOH in the electrolyte chamber **15**. The metal electrode **14** is attached to the frame **13** with a holder **16** which also provides for carrying current out of the cell.

At each side of the electrolyte chamber **15** are air electrodes designated **17** and **18**. These are also inserted into the frame **13** and held therein in a liquid-tight manner. The hydrophobic layer on the gas side of electrode **17** is designated with the reference numeral **19** and that of electrode **18** with **20**. As shown, the electrodes **17** and **18** are not flush with the frame **13** but are instead recessed to form an air space **21**. If a plurality of single cells such as cell **12** are joined to form a battery, the water may be placed in these air spaces. The contact leads of the air electrodes **17** and **18** are designated by the reference numerals **22**.

Figure 2.9b is an enlarged section showing the hydrophobic layer **19** of the air electrode **17**. On the gas side, the hydrophobic layer is covered with water (H_2O). On the opposite side it is covered with the electrolyte, for example, KOH. The H_2O and KOH each form a meniscus of the liquid with the pore **23** being filled with air. The pore is of such a size so that the H_2O or KOH will not pass therethrough. However, diffusion of water vapor takes place within the pore from the water to the electrolyte because of the existing vapor pressure gradient. Figure 2.9c is a schematic illustration of a possible electrode arrangement in a rechargeable metal-air cell having an auxiliary charging electrode. During discharge, the air electrode **30**, which permits only oxygen dissolution, is connected along with the rechargeable metal electrode **31** to the load. For recharging, the rechargeable metal electrode and the auxiliary charging electrode, which causes oxygen precipitation, are connected across the source. The auxiliary electrode **32** may, for example, be a nickel screen.

FIGURE 2.9: METAL-AIR CELL

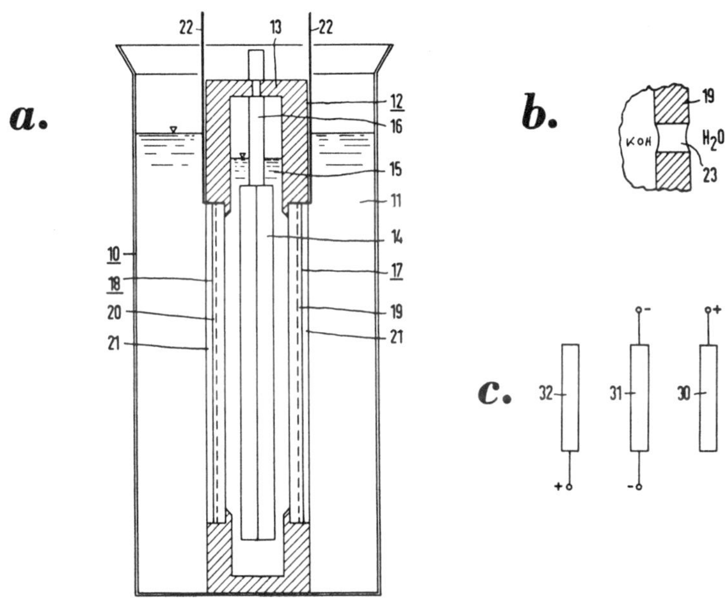

(a) Cross-sectional view illustrating a cell immersed in water.
(b) Sectional view of a portion of the air electrode of the cell of Figure 2.9a showing a pore.
(c) Schematic diagram illustrating the electrode arrangement for a cell which uses an auxiliary charging electrode.

Source: U.S. Patent 3,922,177

As noted above, both three-layer and two-layer electrodes may be used as the air electrode **17** and **18**. A three-layer electrode may be manufactured using carbon powder, which may contain a binder such as polytetrafluoroethylene or polyethylene and a filler such as sodium sulfate or ammonium oxalate and possibly an additional catalyst such as silver, as the middle portion and arranged between a polytetrafluoroethylene diaphragm (PTFE diaphragm) and a porous nickel plate, the arrangement being pressed together with a pressure of about 1,000 to 2,000 N/cm^2 (100 to 200 kg/cm^2). After pressing the arrangement will subsequently be sintered for one hour at a temperature of about 380°C. The electrode will be made in a thickness of about 1.2 mm. Preparation of the porous nickel plate can be accomplished by pressing nickel powder having a grain size of about 8 to 10 μ at a pressure of about 8,000 N/cm^2 and sintering for half an hour at about 800°C in a hydrogen atmosphere. The nickel plate should be made in a thickness of about 0.3 mm, the area density should be about 0.15 g/cm² and the volume porosity about 50%. To make the porous PTFE diaphragm, polytetrafluoroethylene in powder form and having a grain size of about 30 to 50 μ is suspended in n-propanol to form a fine dispersion after which it is drawn off, dried and sintered for half an hour at about 380°C.

The resulting PTFE diaphragm is about 0.5 mm thick with an area density of the dry diaphragm of about 0.04 g/cm² and a volume porosity of about 65%. It is possible, however, to make diaphragms of smaller thickness. A PTFE diaphragm of this nature will exhibit essentially uniform microporosity. The micropores are such a size that the electrolyte is prevented from passing through the pores of the diaphragm due to the hydrophobic nature of the diaphragm. Through using PTFE powder with a grain size in the range of between 30 and 50 μ a microporous PTFE diaphragm which meets the above-mentioned requirements is easily obtained.

In the operation of a metal-air cell of the kind described above and having an air electrode area of 100 cm², water losses take place as follows: The water consumption due to electrolysis during the charging phase and including water vapor carried along by the gases escaping during charging amounts to approximately 1 ml/hr. Water losses due to evaporation of water through the porous air electrodes during discharge amounts to approximately 2 ml/hr. Assuming a discharge time of eight hours and a charging time of 16 hours, the total amount of water given out by a cell of this type is 32 ml. Thus, this is the quantity which must be made up or added during charging.

The amount of water which will be transferred to the interior of the cell per hour is a function of the thickness and porosity of the PTFE diaphragm along with the water temperature. It has been found that between 1 and 4 ml of water per hour can be brought into the interior of the cell where a total air electrode area of 200 cm² is provided, i.e., an air electrode on each side of 100 cm².

Thus, a maximum of 64 ml of water can be diffused during 16 hours. In addition to control through water temperature the amount of water transferred can be controlled by the degree of coverage of the air electrodes with water along with the amount of time the water is present outside the electrodes.

With a three layer electrode such as that described above, and having a total electrode area of 200 cm² and complete coverage of the air electrodes with water at a temperature of about 20°C, approximately 32 ml of water will be brought into the cell during a charging period of 16 hours. Thus, an improved manner of managing the water budget of a rechargeable metal-air cell is provided.

Hydrophilic Air Electrode Composition

A process described by *E.S. Buzzelli; U.S. Patent 3,925,100; December 9, 1975; assigned to Westinghouse Electric Corporation* relates to an air/oxygen electrode for use in metal/air batteries, and, in particular, an air electrode for use in metal/air rechargeable batteries. Generally, the air electrode includes a hydrophilic layer which comprises a current collector of metal fibers and a hydrophilic composition containing electrochemically active materials into which the current collector is pressed or mold formed, and a hydrophobic layer laminated to the molded hydrophilic layer. The hydrophilic composition includes a high surface area carbon, preferably a silver-mercury catalyst, a wet-proofing agent such as polytetrafluoroethylene, and manganese dioxide. It has been found that by the incorporation of manganese dioxide into the hydrophilic layer in combination with the high surface area carbon, substantial improvement in the voltage performances as well as life characteristics results.

The high surface area carbon is utilized in the hydrophilic composite for the absorption of oxygen which can be electrochemically converted to active carbon carrier. Preferably, a silver-mercury catalyst is used to decompose secondary products that build up in the electrode as a result of the oxygen reduction mechanism.

Thus, while the electrode performs well without a catalyst, the life characteristics are not as good, because the secondary products, typically perhydroxyl groups, create severe electropolarization results if allowed to build up in significant amounts. The Ag-Hg catalyst, on the other hand, has been found to decompose peroxide in a manner substantially equivalent to formerly used platinum and other noble metal catalysts. Accordingly, the use of a catalyst is preferred.

Most importantly, however, is the utilization of manganese dioxide in combination with the Ag-Hg catalyst and high surface area carbon. It is believed that the manganese dioxide decomposes peroxides as well as reduces the oxygen. For example, manganese dioxide participates in the oxygen reduction by giving up oxygen ions to the system, and then, because of its intimate contact with the high surface area carbon, can be reoxidized to a higher oxide form, where it again participates in the electrochemical reduction. Furthermore, it is believed that the manganese dioxide, along with the Ag-Hg catalyst, catalytically decomposes the peroxide groups formed within the electrode.

Preferably, the components of the hydrophilic composite are mixed together in powder form to which deionized water is added to form a pastelike consistency. The fiber metal mesh current collector, preferably nickel or nickel plated steel, is then integrated within the pastelike composition and press molded into the hydrophilic layer.

To this molded composite layer is laminated a hydrophobic layer of porous, fibrillated, unsintered polytetrafluoroethylene having a total porosity of from about 35 to 60% by volume and a thickness from 0.005 to 0.20 inch. The hydrophobic layer is capable of permitting rapid diffusion of gas, such as oxygen and air, but prohibits the passage therethrough of electrolytes such as alkali hydroxides used in the metal/air batteries of the process.

The electrodes have been found to operate for longer than 4,000 hours. Moreover, air electrodes of the process were found to have a performance level of 50 to 200 mV better than other known useful air electrodes systems at 50 milliamps per cm^2.

Hydrophobic Polymer Cathode

According to a process by *H.P. Louie; U.S. Patent 3,963,519; June 15, 1976; assigned to Leesona Corporation* primary and secondary lighweight AA, C, D, and the like metal/air cells comprising an anode, a cathode, and an electrolyte in which the anode comprises a consumable porous metal and the cathode comprises a hydrophobic member in contact at one surface with an electrocatalyst. The components of the cell are retained in a substantially rigid protective shield permitting access of air to the cathode over substantially the entire cathode surface area.

The structural elements which include the protective shield spacers to space the shield from the cathode comprise lightweight materials such as lightweight metals or a thermosetting or thermoplastic resin or the like. The electrolyte can be a solution or paste or an ion-conductive material or a water-activatable solid, such as dry potassium hydroxide adhering to or absorbed in the anode and/or cathode and/or a hydrophilic separator between the anode and cathode.

Referring to Figure 2.10a, the exploded perspective view permits the anode **1** and accompanying structural elements, the cathode **11** and accompanying structural elements, and the outer or protective shield **31** to be considered separately. Spacer element or casing **15** and the cathode **11** are partially cut away to reveal the water reservoir **25** and puncturing means **23**. Considering first the anode assembly, it will be seen that the anode **1** comprises a cylindrical metallic screen, i.e., gauze or mesh filled with active metal particles.

FIGURE 2.10: METAL-AIR CELL

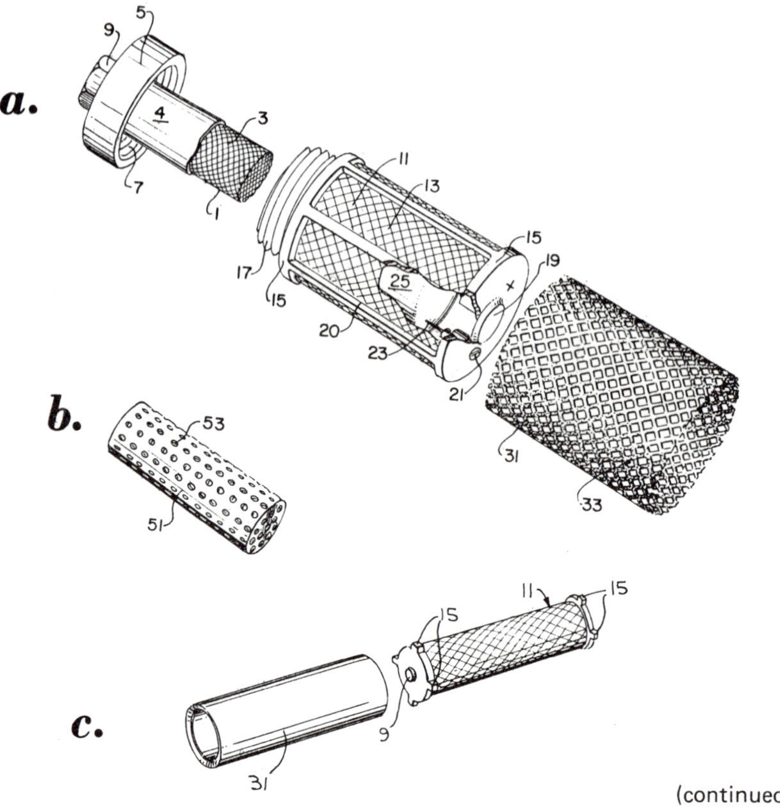

(continued)

FIGURE 2.10: (continued)

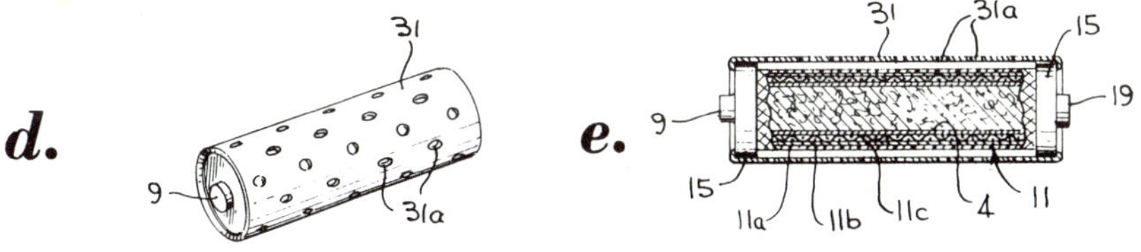

(a) Exploded, perspective view of a metal/air cell.
(b) Perspective view of an alternative form of cylindrical, porous metallic element for use as the anode of the cell of Figure 2.10a.
(c) Exploded, perspective view of an alternative cell design.
(d) Perspective view of another design.
(e) Cross-sectional view of Figure 2.10d.

Source: U.S. Patent 3,963,519

The spaces in the mesh **3** permit good contact between the electrolyte and the anode which can be a liquid or, alternatively, the anode can include an anhydrous or dry electrolyte material which is activated upon contact with water. A separator material, **4**, circumscribes the anode. This material can be any suitable hydrophilic material such as Visking, a copolymer of vinyl chloride and acrylonitrile, or the like.

A closure means comprising a cap **5** is attached at the upper end, and an external terminal **9** protruding through the cap is electrically connected to the anode **1**. The inside of the cap **5** is threaded and the female threads **7** form a liquid-tight seal with the male threads **17** of the spacer **15**.

The cathode **11** as shown comprises a cylindrical hydrophobic polymer member **11a** in contact on its inner surface with a catalytic **11b** such as a metal of Groups VIII or I-b of the Periodic Table, mixtures or alloys thereof. A conductive screen **11c** is adjacent to the catalyst layer. It is seen from the outside as a result of the thinness of the polymer member.

Further, as shown, the cathode **11** is bonded to spacer **15** to form a liquid-tight seal as a result of the hydrophobic, gas-permeable member which can be a polymer such as a polytetrafluoroethylene. Spacer **15** comprises a lightweight synthetic resin such as polyethylene. An external positive terminal **19** protrudes through the case **15** and is electrically connected to the cathode **11**.

The cell is activated by rupturing, in this case puncturing, a thin, flexible water reservoir **25**. This reservoir is preferably a hydrophobic plastic membrane. The puncturing means is a needle **23** attached to a threaded member which protrudes through the outer case **15** and has a slotted head **21** which, when turned, advances the needle into the bag **15** and punctures it, releasing the water or electrolyte solution. The water or electrolyte floods the interior of the cell and, in the case of water, dissolves the dried electrolyte material, e.g., a powdered alkali metal hydroxide, retained in the anode.

Additional structural strength and protection is provided by an outer or protective shield **31** comprising metal or plastic. The shield **31**, as shown, is gauze-like in structure, and the spaces **33** in this structure permit ingress of air to the cathode **11**. Alternatively, the shield can be solid and since it is positioned away from the cathode by spacer **15**, air access is permitted through the end or ends of the shield as seen more clearly from the embodiment of Figure 2.10c.

Figure 2.10b shows an alternative form of cylindrical metal anode structure **51**. It can be seen that the plurality of holes **53** in the anode **51** provide a large surface area available for reaction. As in Figure 2.10a, the anode can be impregnated with an electrolyte material capable of water-activation, e.g., a solid, anhydrous alkali metal hydroxide. Figures 2.10c, 2.10d and 2.10e show examples of the process which are more closely related to conventional AA, C, or D cells and do not utilize electrolyte activating means for readying the cell for use.

The cell of Figure 2.10c comprises spacer elements **15** at either end of the cell which, together with the cathode **11** and closure means, provide an electrolyte-tight cell between the anode and cathode and electrical contact means. Outer shield **31** fits over the cell and is spaced from the cathode by means of the spacer element **15**, the outer periphery of which is notched to permit air access into the cathode. Figure 2.10d is substantially similar to Figure 2.10c except in this instance the outer or protective shield **15** is integral with the closure means of the cell with air access being permitted to the cathode **11** through holes or openings **31a**.

Figure 2.10e is a cross-sectional view of the cell of Figure 2.10d and depicts the structural relationship of the components of the cell. It is seen that the pressure-deformable cathode **11** which is made up of the hydrophobic polymer **11a**, catalyst layer **11b**, and conductive metal grid **11c** is spaced from protective shield **31** by means of spacer elements **15** at either end of the cell. The anode **1** comprises a pastelike gel of potassium hydroxide electrolyte and powdered zinc. A porous hydrophilic separator **4** electrically separates the anode material from the cathode. The cathode is in electrical contact with terminal **19** and the anode material is in direct contact with terminal **9**.

OTHER PROCESSES

Profiling of Conductivity Path

A process described by *E.M. Cohn; U.S. Patent 3,972,727; August 3, 1976; assigned to the U.S. National Aeronautics and Space Administration* is related to secondary or rechargeable cells or batteries and concerns an improved technique for combatting electrode shape change.

The process is thus concerned with combatting shape change in the zinc electrodes of a rechargeable battery by profiling the conductivity of the path between the electrodes so that it is easier for ions to travel at and near the edges of the electrodes and more difficult for the ions to travel in and near the center of the electrodes. Thus, rather than accepting substantial migration of the zinc from the edges to the center of the electrodes and compensating for this migration by increasing the amount of the density of the zinc at the edges, the process involves preventing or reducing this migration of zinc ions toward the center by controlling the ionic conductivity of the path between electrodes.

According to one form of the process, the amount or thickness of the separator material is profiled so that the density or thickness of the separator is greatest at the center of the electrodes and least at the edges, so that the flow path for the ions in the electrolyte is easier and more direct at the edges of the electrodes. In a second case, a porous separator is used which is constructed so that the pores therein are concentrated at the edges and that there are relatively few pores at the center. A similar effect is produced in a third example by the use of a porous separator constructed so as to include relatively large pores at the edges and relatively small pores at the center.

In another design, a separator arrangement is provided which defines a series of chambers containing electrolytes of different strengths, the strengths of the electrolytes being greatest at the edges and becoming progressively weaker towards the center, so that the ionic conductivity of the paths to the edges of the electrodes is substantially greater than that of the paths to the central regions of the electrodes.

Zinc-Lead Dioxide

According to a process described by *E. Villarreal-Dominguez; U.S. Patent 3,964,927; June 22, 1976* a lead dioxide-zinc rechargeable-type cell comprises at least one positive lead dioxide plate and at least one negative amalgamated zinc metal plate spaced from the positive plate. Both plates are permanently submerged in an aqueous acid electrolyte containing soluble zinc salts, a zinc surface-controlling agent, an electrolyte conductivity-correcting agent and an amalgamating agent to prevent formation of localized cells or electrolytic pairs on the zinc plate during discharge.

An electrolyte suitable to be used in a lead dioxide-zinc rechargeable-type cell comprises an aqueous acid solution of a zinc salt, aluminum sulfate as the zinc surface-controlling agent, sodium sulfate as the electrolyte conductivity-correcting agent, mercuric sulfate as the amalgamating agent and sulfuric acid as an aggressive agent which also acts to minimize the treeing of the anodic surfaces.

More particularly an electrolyte to be used in a lead dioxide-zinc rechargeable-type cell comprises from 1 to 5 grams per liter of zinc sulfate, 25 to 75 grams per liter of sodium sulfate, 25 to 75 grams per liter of aluminum sulfate, 2 to 8 grams per liter of mercuric sulfate and 250 to 500 grams per liter of sulfuric acid, the remainder being water. Preferably the electrolyte may additionally contain 10 to 30 grams per liter of a suitable gum such as dextrin to improve the physical structure of the zinc deposited, and 5 to 15 grams per liter of dibutyl amine as a corrosion inhibitor.

Pressure Sensitive Switch in Housing

According to a process described by *M.H. Rackin; U.S. Patent 3,933,526; January 20, 1976; assigned to Motorola, Inc.* a housing for a battery cell includes a built-in switch for opening the circuit to the battery in response to excess pressure and/or temperature. The housing has a first conducting contact connected to the first battery electrode, and a second contact connected to the second electrode and connected through a wall of the housing to provide a circuit to the battery cell. The second contact may be a porous wall to allow gases which form in the battery cell to pass to the outer housing wall, and when the pressure exceeds a predetermined value the outer wall will flex and open the electrical connection to the battery. The outer wall can be a bimetallic disc which flexes with rise in temperature of the battery cell to open the connection when the temperature exceeds a predetermined value. The pressure responsive action can be provided alone, the temperature responsive action can be provided alone, or the two features can be used in combination to provide a housing wherein the connection to the battery is opened in response to excessive pressure or temperature, or a combination of the two conditions.

Perovskite Compounds as Cathodes

J.M. Longo and L.R. Clavenna; U.S. Patent 3,939,008; February 17, 1976; assigned to Exxon Research and Engineering Company describe an electrical energy storage device which has a cathode-active material, an oxide having the perovskite or perovskite related structure and the general formula ABO_3. In the formula A is an element from the group consisting of nontransition metals selected from Group II-a of the Periodic Table, and includes Ca, Ba, and Sr and B in a first row transition metal selected from Groups VII-b and VIII of the Periodic Table, and includes Mn, Fe, Co, and Ni. Specific examples of the cathode-active materials include, among others, $CaMnO_3$, $BaNiO_3$, $SrCoO_3$, $SrFeO_3$.

The anode in the electrical energy storage device of the process is a metal selected from a group consisting of cadmium, zinc, lead, lithium, sodium, and potassium. The electrolyte useful in the process includes aqueous electrolytes such as aqueous solutions of potassium hydroxide, sodium hydroxide, and ammonium chloride and nonaqueous or organic electrolytes such as propylene carbonate solutions of alkali metal salts. The following examples illustrate the process.

Example 1: In this example, a number of tests were carried out on half-cell cathodes using an electrolyte of 9 M KOH. In these tests the cathode-active materials were mixed with acetylene black typically in a ratio of 2:1. To this mixture was added enough electrolyte, 57 ± 3 weight percent, to form a thick paste. The paste was compressed into a polytetrafluoroethylene cell between a gold current collector and a separator supported by a gold screen. The cell was designed such that it could contain from 0.25 to 0.5 gram of cathode-active material and that the effective working area of the cathode was 5 cm^2. A Permion membrane or a glass fiber filter paper was used as the separator, and the results appeared to be independent of these separators.

The polytetrafluoroethylene cell containing the cathode material was immersed into 400 ml of electrolyte. The cathode was examined at room temperature using a standard half-cell arrangement with a graphite counter-electrode and a saturated calomel reference electrode, SCE. The electrolyte was purged with argon to remove dissolved oxygen. The half-cell was discharged or recharged by driving it at constant current.

Example 2: Following the general procedures of Example 1, a half-cell containing 0.27 gram of $CaMnO_3$ was discharged to completion by repeated closed circuiting at a current density of 1.0 mA/cm^2 for 0.5 hour and opened circuiting (zero current density) for 1.0 hour. Figure 2.11a shows the half-cell potentials at closed circuit and opened circuit versus the depth of discharge. The depth of discharge is indicated as the number of electrons supplied to the cathode per manganese atom in the cathode active material. The initial state of manganese in $CaMnO_3$ is Mn^{4+}.

FIGURE 2.11: PEROVSKITES AS BATTERY CATHODES

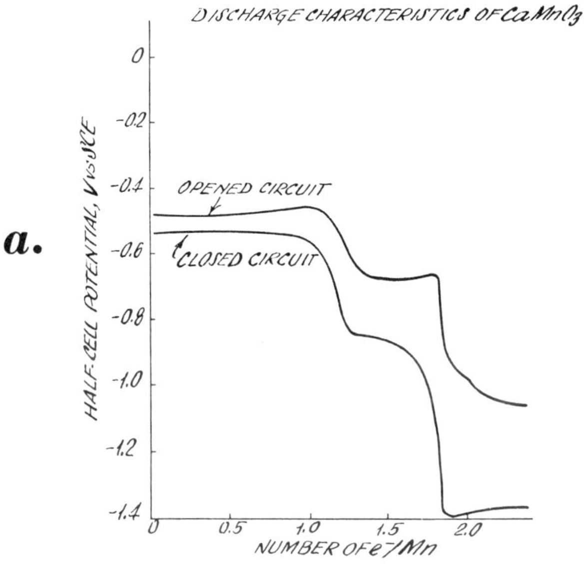

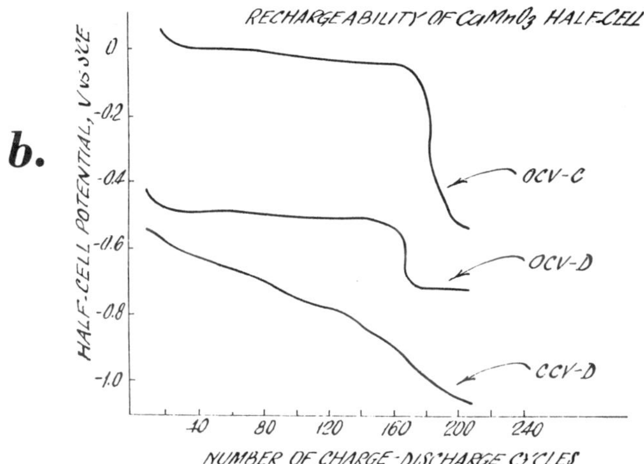

(a) Graph showing the half-cell potential, both at closed circuit and at open circuit, when $CaMnO_3$ was employed as a cathode-active material.

(b) Graph showing the charge-discharge characteristics for a half-cell having $CaMnO_3$ as the cathode-active material.

Source: U.S. Patent 3,939,008

If the cathode material was fully utilized, complete discharge would correspond to conversion of all Mn^{4+} to Mn^{2+}, or supplying $2e^-/Mn$. The results in Figure 2.11a show that initially the half-cell potential is approximately constant but that there is a rapid decrease in potential at a discharge depth of $\sim 1e^-/Mn$ and another at $\sim 1.8e^-/Mn$. The first decrease is associated with the conversion of Mn^{4+} to Mn^{3+} and the second with the conversion of Mn^{3+} to Mn^{2+}.

Example 3: In another test, the rechargeability of the half-cell cathode was examined by running the cathode through many charge/discharge cycles. A charge/discharge cycle consisted of discharging a fully charged half-cell cathode to a depth of $1.8e^-/Mn$ over a period of one hour at constant current, running at open circuit for fifteen minutes, recharging with $1.8e^-/Mn$ over a period of one hour at constant current and then running at open circuit again for 15 minutes.

The results for a half-cell cathode that was charged/discharged 205 cycles with a current density of 2.56 mA/cm^2 are presented in Figure 2.11b which shows the half-cell potential at various stages in a cycle versus the number of charge/discharge cycles. The half-cell potential is given at the closed circuit voltage during discharging, **CCV-D**; the open circuit voltage after discharging, **OCV-D**; and the open circuit voltage after recharging, **OCV-C**. The results show that this cathode material does have recharge capabilities for at least 170 cycles for a discharge depth of $1.8e^-/Mn$ per cycle.

Manganese-Treated Carbon Plates

L. Silva and G. Spector; U.S. Patent 4,025,699; May 24, 1977 describe a manganese storage battery which when being charged, deposits manganese dioxide from the electrolyte on to the positive carbon plate so to increase the power and efficiency of the positive electrodes.

Referring to Figures 2.12a and 2.12b, the reference numeral **10** represents a manganese storage battery according to the process that includes a battery case **11** divided into cells **12**, each cell containing positive carbon plates **13** and negative zinc plates **14** having separators. The battery contains a liquid electrolyte **15** which, in the charged battery, comprises manganous hydroxide and water. When the battery is being charged, a manganese dioxide **16** from the electrolyte is deposited on the carbon plate **13** to form a thick deposit thereof, as shown in Figure 2.12a.

In the process electrolysis is used to thus build up the manganese dioxide on the carbon plate. This results in increased power and efficiency of the positive electrodes. It is to be noted that in a primary cell using manganese dioxide as a depolarizer, the manganese dioxide is placed in contact with the positive carbon electrode. The reason for treating the positive carbon electrode is that new carbon is difficult to oxidize. Once a deposit of manganese dioxide is built up on the positive carbon electrode, the electrode is then ready for operative use in the battery.

Figure 2.12c shows one of the cells **12** of the battery **10** with details of an alternative mode of suspending the electrodes in the cell, **20** is a thickened portion of cell wall **11** (see also Figure 2.12d) and is formed to pivot a pin **22** riveted to an arm **21** made of any metal alloy impervious to the electrolyte, or of suitable plastics.

FIGURE 2.12: MANGANESE BATTERY

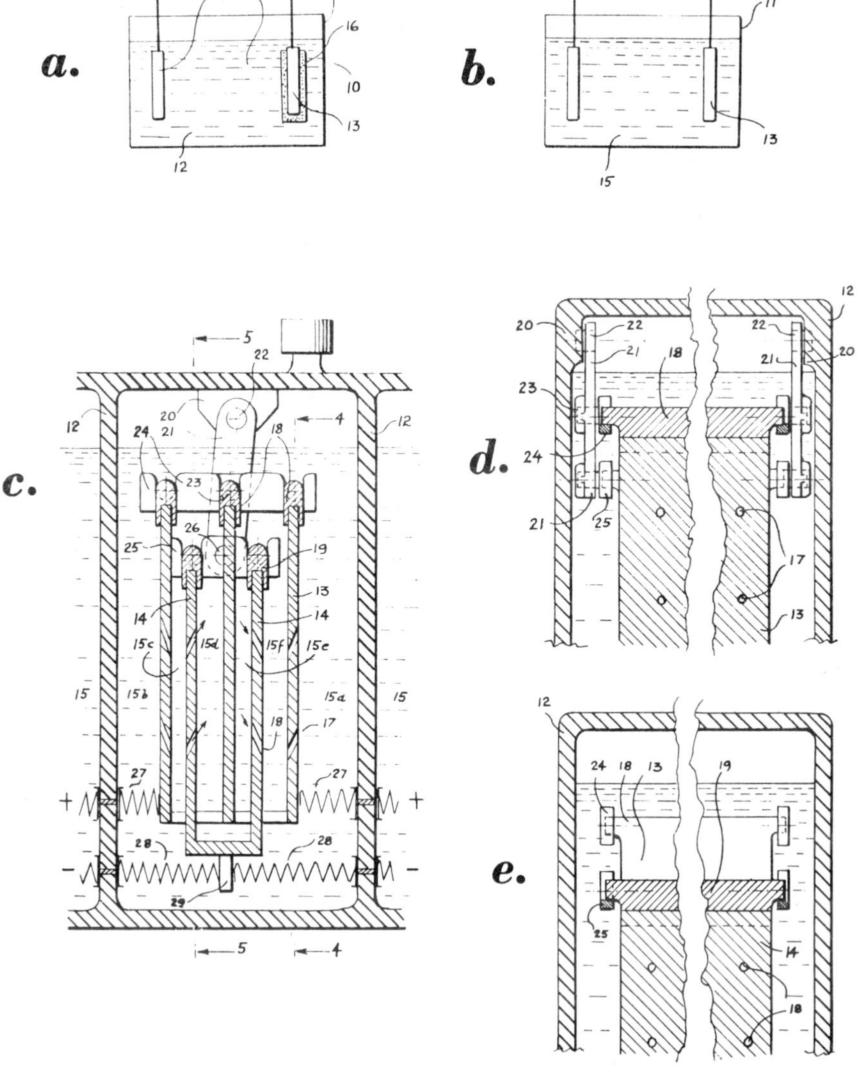

(a) Diagram of charged condition.

(b) Diagram of discharged condition.

(c) Modified design which includes the features illustrated in Figures 2.12a and 2.12b.

(d) Partial sectional view of the battery through **4–4** of Figure 2.12c.

(e) Similar section through **5–5** of Figure 2.12c.

Source: U.S. Patent 4,025,699

This suspension arm carries two levers, **24** and **25**, in such a manner that they may swing independently, around pivots **23** and **26** respectively. The levers **24** and **25** have slots equidistant with the space between electrode plates of the same polarity. For example the electrode plate **13** is fastened to a holding bar **18** whose extensions on each side fit into the slots of levers **24**. Similarly, holding bar **19** (Figures 2.12c and 2.12e) fits at both ends into slots of levers **25**. It is evident therefore that each electrode group within a cell is independently suspended from a pair of levers (**24–24, 25–25**) which in turn pivot independently in suspension arm **21**.

The lower portions of the electrodes are prevented from swinging at large amplitudes by buffer springs **27–27**, and **28–28** respectively; the same may simultaneously be used as electronic conductors from the duct to the electrodes. Actual connections depend on whether cells work in parallel or series. In Figure 2.12c it is assumed that cells work in parallel.

The position of lever **21** in Figure 2.12c is off the normal perpendicular to illustrate what happens when the vehicle moving from right to left applies the brakes. The electrodes swing towards the left and in the process alter the previously equidistant spacing between the neighboring electrode plates. This change in volume between the plates causes a forceful displacement of the electrolyte; some of this displacement passes through holes **17** and **18** respectively in the electrode plates.

Holes may be at an angle so as to further add to the circulative effect. The springs **27, 28** will introduce displacements as well but at a higher frequency. The combined resulting movement of the electrodes will tend to disperse local concentrations of the electrolyte, and thereby eliminate one of the causes of battery inefficiency. By making the holes **17, 18** in the battery electrodes conical (instead of cylindrical) a pumping action in one direction would be obtained, which may still further improve the circulation of the electrolyte within the cell and thereby its full utilization.

NICKEL-CADMIUM AND OTHER ALKALINE BATTERIES

NICKEL ELECTRODES

Cutting Technique from Continuous Band

A process described by *F. Sperandio and M. Leturque; U.S. Patent 3,928,069; December 23, 1975; assigned to Saft-Societe des Accumulateurs Fixes et de Traction, France* relates to a method for cutting out electrodes having a sintered support from a continuous strip. It is an already known method, described in U.S. Patent 2,819,962 to prepare strips of sintered nickel by a continuous method by sintering nickel on perforated bands, the strips being thereafter impregnated with active materials by any suitable method and then cut up into electrodes to be used in alkaline electrolyte electrochemical cells.

As an alternative, the cutting up may be effected on the nonimpregnated sintered band, impregnation with active materials being effected subsequently on the already cut out elements to complete the electrodes. In the current work, an improved method for cutting out substantially rectangular electrodes from a continuous strip comprising at least one zone not coated with active material disposed parallel to its edges is described.

The process is characterized in that the edges of the electrodes are cut out obliquely in relation to the longitudinal axis of the strip, with triangular corner portions only cut out in the zones free of active material. According to a preferred case, the direction of the cut is at 45° to the longitudinal axis of the strip. In this way, it is possible to cut out from the strip rectangles or squares one only of whose angular corners comprises a wiped or uncoated area, that angular corner free of active material subsequently being used as a lug for the connection of such electrodes to the corresponding output terminal of a cell.

In Figure 3.1, the metal strip **1** prepared, for example, according to the process of U.S. Patent 2,819,962 comprises, on each side, a wiped or active material free edge **2**. Four electrodes **3** separated from that strip by cutting transversely at right angles to its longitudinal axis according to the usual prior art method, with plate lugs **4** which are located in the wiped active material free edge **2**, are shown.

FIGURE 3.1: SINTERED ELECTRODES–CUTTING TECHNIQUE

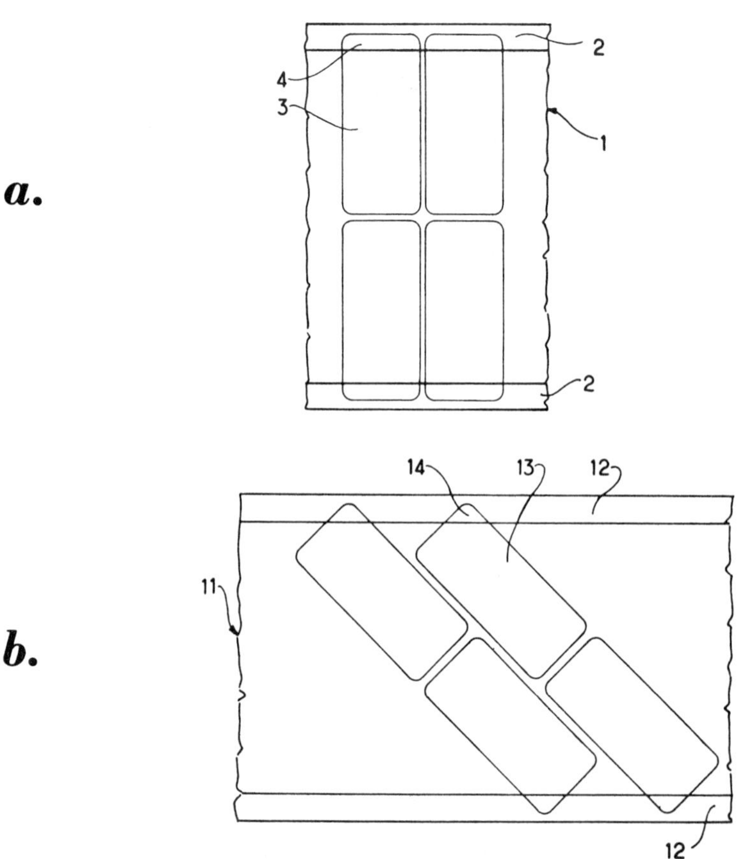

(a) View of a strip from which four electrodes according to the usual prior art technique are cut out

(b) View of a similar strip from which four electrodes having the same dimension as the previous ones, but cut out according to this process are cut out

Source: U.S. Patent 3,928,069

It is possible to calculate, for example, the ratio between the total surface and the available active material impregnated surface of one electrode **3** assuming that each electrode has a total area of 25 x 60 mm; the height of its terminal lug **4** being 5 mm. Then the ratio between the active material impregnated surface and the total surface is:

$$\frac{(25 \times 60) - (25 \times 5)}{25 \times 60} \times 100 = 91.67\%$$

On the other hand, if the electrodes **13** of the same dimensions are cut out on a bias to the longitudinal axis in the manner shown in Figure 3.1b, which shows a strip **11** having two edges **12** wiped free of active material from which strip are cut out, at an inclination of 45° to the longitudinal axis of strip **11**, the four plates **13**, the wiped portion **14** free of active material of each such plate **13** is an isosceles right-angled triangle **14**. If one assumes that the altitude or height of this triangle **14** is 5 mm and the dimensions of each electrode as to length and width are the same as those of electrodes **3** cut in Figure 3.1, then the area of the triangle **14** will be 25 mm^2. Therefore, the ratio between the active area of the electrode **13** and the total surface area will be:

$$\frac{(25 \times 60) - 25}{25 \times 60} \times 100 = 98.33\%$$

It may be seen, therefore, that 6.66% of active surface area is gained. However, in actual fact, the gain in electrode capacity will be even greater, for there is at the juncture of the wiped portion **12** and active material portion, due to that wiping, an active material border where the overall thickness is less than the overall thickness of the active material-bearing portion of the electrode in the middle of the strip and where, consequently, the capacity per unit of surface is less than the capacity per unit of surface in the middle of the strip.

Now, in the electrodes **3** of the assumed dimensions cut out by the conventional method of Figure 3.1a, the length of that border is 25 mm whereas in the electrode **13** of the same dimensions cut out by the bias method according to the process, the length of the border is now only 10 mm. It will be seen, therefore, that in actual fact, the gain in capacity of electrodes of the same dimensions prepared by the method of Figure 3.1b will, therefore, be approximately 7%. An alkaline storage cell comprising a sintered positive electrode prepared according to the process is also described.

Metal Hydroxides and Graphite Particles

A process described by *J. McBreen; U.S. Patent 4,000,005; December 28, 1976; assigned to General Motors Corporation* relates to the making of nickel positive electrodes for secondary, alkaline storage batteries, and more particularly to a method of distributing and supporting fine particles of nickel hydroxide and graphite throughout the electrode so as to achieve maximum utilization of the nickel hydroxide, and produce an active mass particularly useful with a lightweight grid/conductive support.

The process involves mechanically suspending (e.g., in a Waring blender) comminuted nickel hydroxide and graphite in a suspension medium (e.g., isopropyl alcohol) which is not a solvent for polyvinylidene fluoride (i.e., Kynar), adding thereto a solution of polyvinylidene fluoride in a solvent (e.g., acetone) which is slowly soluble in the suspension medium, blending for a sufficient time for the polyvinylidene fluoride to coagulate, drawing the blend through a conductive grid which separates the coagulum from the medium-solvent, wet pressing the coagulum-impregnated grid and finally drying the electrode.

The dissolution rate of the polyvinylidene fluoride solvent in the suspension medium is sufficiently slow that, as the polyvinylidene fluoride solution is added to the suspension in the blender, it rapidly and substantially uniformly mixes therewith before any substantial dissolution of the solvent occurs. Thereafter the solvent can dissolve in the suspension medium at its own rate and the polyvinylidene fluoride precipitates and coagulates substantially uniformly throughout the solution.

As coagulation proceeds, the nickel hydroxide and graphite are entrained in the coagulated polyvinylidene fluoride. Medium-solvent combinations other than isopropanol-acetone are useful so long as the dissolution rate of the solvent in the medium does not substantially exceed that of the isopropanol-acetone combination. Likewise, small amounts (i.e., up to about 5 wt %) of cobalt hydroxide may be added to the nickel hydroxide to improve the charge efficiency of the electrode as is already known to those skilled in the art.

After the formation of the coagulum in the blender, as above, a porous conductive support is filled with and covered by the coagulum, by passing the blender mix through the support and separating the medium-solvated solvent from the coagulum. While it is preferred to use a conventional papermaking technique (i.e., vacuum table), a centrifuge could also be used to effect the separation and impregnation. According to preferred practice, a 20 mesh, 5 mil thick expanded metal screen is laid atop a vacuum table which is covered with newsprint (i.e., porous paper). A frame is placed over the screen and has a mold/cavity therein for receiving and containing the mixture from the blender. The frame is about 2 cm thick and has a 12 cm x 12 cm central cavity into which the blender mix is poured over the screen.

A vacuum (e.g., about 20 to 25 inches of Hg) is drawn and the medium-solvated solvent sucked through the porous support leaving the nickel hydroxide, cobalt hydroxide, graphite and polyvinylidene fluoride coagulum within the interstices and built up on one side of the support. By turning the support over during this operation an additional thickness of coagulum may be built up on the other side of the screen if desired. While virtually any thickness electrode can be made by the process, it is preferred to cast the coagulum to a thickness of about 0.180 cm, and this is later reduced to about 0.076 cm by wet pressing at about 40 kg/cm^2 to about 80 kg/cm^2 (preferably about 73 kg/cm^2) after the medium-solvated solvent is removed.

Following wet pressing the electrode is dried in an oven at about 100°C for about 3 minutes and finally pressed again, at about the same pressure as before, to ensure uniform thickness, sizing, etc. of the electrode. The pressed and dried coagulum thus produced is not only conductive and porous, but also quite self-supporting, so that very thin screens (i.e., conductive supports) may be used in relation to the overall thickness of the finished electrode. Hence lightweight, materials-efficient electrodes are producible by the process.

Raney Metal Alloy

A process described by *A. Winsel and E. Buder; U.S. Patent 4,003,754; January 18, 1977; assigned to Varta Batterie AG, Germany* provides a positive electrode which is discharge limiting, and a negative electrode to which is added a Raney metal alloy of one or more metals in Group VIII of the Periodic Table, in a quantity sufficient to conduct the maximum permissible polarity reversal current of the cell. In such a cell construction, in which the positive electrode is charge limiting as well as discharge limiting, i.e., the negative electrode has a charge reserve as well as a discharge reserve, oxygen evolves during protracted charging. The known oxygen cycle takes place during protracted discharge, i.e., during polarity reversal hydrogen evolves. This has the significant advantage that little heat is evolved during polarity reversal, because the polarization of H_2 electrodes is considerably smaller than of O_2 electrodes, in both directions of the reaction.

As hydrogen catalyst the negative electrode contains a Raney metal catalyst, preferably Raney nickel. It is desirable that hydrogen evolution at the positive electrode begins at a time at which the hydrogen catalyst of the negative electrode is not yet covered with a hydroxide layer. When the Raney nickel catalyst is present with reduced surface, the hydrogen produced through polarity reversal is reduced at the catalyst in the negative electrode.

This catalyst must be proportioned for the maximum polarity reversal current intensity, but not for the expected duration of the polarity reversal. The total quantity of catalyst can therefore be much less than the conventional quantity of antipolar mass, which ordinarily amounts to 20 to 30% of the total capacity. The lower limit of the Raney nickel additive is determined by the performance requirement of the H_2 cycle during polarity reversal and is preferably in excess of 0.25 g/A, for example, approximately 1 g/A. Use of a Raney metal catalyst, or rather a Raney nickel catalyst, therefore does not lead to substantial increase in the weight of the electrode.

The Raney nickel in its reduced form is a good conductor which can be mixed in with the powder forming the negative electrode. It then performs simultaneously the function of conductive material and hydrogen consuming catalyst. For mass electrodes there can be admixed to the negative electrode, in addition to this catalyst a binder of hydrophobic substances, such as polyethylene or polytetrafluoroethylene powder. These additives produce hydrophobic regions and facilitate gas penetration.

For sinter electrodes, the active mass is customarily combined in a sinter frame. In this case, the Raney metal catalyst can be produced, for example, by forming a double skeleton catalyst structure as described in U.S. Patent 3,150,011. In that case, powdery support frame material and powdery Raney alloy are applied, pressed or rolled onto the surface of a metallic form. The body is sintered at elevated temperatures and then treated in lye or acid to dissolve out the soluble components of the Raney alloy. It is also possible to produce the Raney structure by application of an aluminum or zinc layer to the nickel carrier. This material is then alloyed and subsequently dissolved out by treatment with potassium hydrate.

Suitable activating additives, particularly for Raney nickel, are platinum, palladium, copper, aluminum, zinc oxide, and lithium oxide. To enhance the catalytic activity of the Raney catalyst, known techniques may be used, such as the process described in U.S. Patent 3,235,213. There, organic compounds are added during treatment in concentrated potassium hydrate to form soluble complex substances with the catalytically inactive element of the alloy. For example, tartrates are suitable for complex substance formation.

To provide the oxygen cycle described during overcharging and the hydrogen cycle during deep discharge and polarity reversal, it is important in what manner the active masses are assembled or treated within the cell. For example, the negative electrode may be introduced into the cell casing in reduced form, together with the fully oxidized, i.e., fully charged positive electrode. The casing is then hermetically sealed. The negative electrode should contain cadmium hydroxide, in order to provide a reliable start for the oxygen cycle during overcharging. This condition of the negative electrode can be achieved by electrochemical hydrogen separation in an alkaline medium at temperatures in excess of about 40°C.

The process proceeds more rapidly as the temperature goes up. Preferred are temperatures of about 60° to about 80°C. Not only the cadmium but also the Raney nickel surface and the surface of the remaining nickel components of the electrode thereby become reduced, so that it becomes electrochemically active relative to the hydrogen. Renewed oxidation of the Raney nickel can then not take place, because the Raney nickel present in the moist electrode is protected from oxidation by the metallic cadmium.

Raney metal catalysts that are particularly useful are Raney nickel, Raney nickel-iron, Raney nickel-cobalt and Raney nickel with palladium or platinum alloy additives. The process is particularly suitable for hermetic nickel cadmium storage batteries, but can also be utilized for other alkaline storage batteries, such as nickel-iron batteries.

Solid Nickel Hydrated Oxides

J.F. Jackovitz and E.A. Pantier; U.S. Patent 4,016,091; April 5, 1977; assigned to Westinghouse Electric Corporation describe a process that will provide an improved activated battery material mixture, by chemically reacting NiO, which may also have added to it about 2 to 12 wt % Co based on NiO plus Co content as a material selected from Co, CoO, Co_2O_3, Co_3O_4, or their mixtures, with effective amounts of Na_2O_2, generally within a weight ratio of $NiO:Na_2O_2$ of 1:1.35 to 1:2.1. This nickel oxide-sodium peroxide mixture is reacted at temperatures between about 800° to 1150°C, for a period of time, generally about ½ to 8 hours, effective to form $NaNiO_2$ or $NaNiO_2$ plus $NaCoO_2$ melted reaction product.

The reaction product, comprising $NaNiO_2$ is then hydrolyzed. If the cobalt oxide or elemental cobalt additive was not added initially, before fusion, as is preferred, it will be added generally as cobalt hydroxide after hydrolysis, or as a water-soluble cobalt salt such as cobalt chloride or cobalt nitrate during hydrolysis, after hydrolysis or after plaque loading.

This process will provide a final solid active battery material containing over about 95 wt % solid Ni hydrated oxides and hydroxide forms and Co hydroxide forms, the remainder being interlaminar sodium. It is important that about 0.5 to 5 wt % but preferably 0.5 to 3 wt % unreacted $NaNiO_2$ be present after hydrolysis and drying. The unreacted $NaNiO_2$ is present in the active material as interlaminar sodium in the nickel oxy-hydroxide layers and helps prevent swelling of the active material in the plate during the life of the battery. This activated battery material is washed and generally dried after which it can then be loaded into a supporting porous plaque to provide an electrode plate, which may then be electrochemically cycled or formed (electrically charged and discharged in an alkaline electrolyte) prior to use in a battery opposite a suitable negative electrode. The drying step is generally carried out at temperatures below about 65°C, or at a suitable temperature in a high moisture atmosphere so that water present in the active material structure is not eliminated to an extent to cause the material to lose activity.

This process thus involves conversion of nickel oxide, or nickel oxide with added cobalt as elemental cobalt or cobalt oxide to an active battery material powder without tedious filtering or washing steps and without use of expensive electrical equipment. The starting materials cost is significantly reduced, since nickel oxide is the least expensive nickel containing material commercially available. Starting with nickel oxide makes the process useful and commercially feasible, since it eliminates a prolonged oxidation step at high temperatures which is sure to degrade the reaction container.

Starting materials cost relative to the chemical, electrochemical and ozone processes is drastically reduced by at least 50%. In addition, a by-product of this process is an aqueous alkali metal hydroxide solution which may be further used as a battery electrolyte by suitable processing, or used as a basic material for neutralizing mine acid pools and the like.

Lanthanum Nickel

P.A. Boter; U.S. Patent 4,004,943; January 25, 1977; assigned to U.S. Philips Corporation describes an electrochemical cell having an electrode which consists of $La_{1-y}R_yNi_{5-z}M_zH_x$, where y = 0 to 1, z = 0 to 1, x = 0 to 1, R is a metal of the group comprising the rare earth metals, calcium and thorium, and M is a metal of the group comprising cobalt, copper and iron, with the understanding that in the presence of iron $z \leqslant 0.2$, a counterelectrode and an electrolyte. The cell is characterized in that the counterelectrode is made of a material which is capable of reversibly taking up and giving off a proton and an electron.

The electrolyte preferably is alkaline or mildly acid. Electrodes capable of reversibly taking up a proton and an electron consist, for example, of nickel hydroxide or manganese dioxide. The electrolyte may consist of an aqueous solution of KOH. When for intermetallic compounds, such as $LaNi_5$, the amount of hydrogen absorbed at a given temperature is plotted against the hydrogen pressure, it is found that with these compounds for each temperature the isothermal includes a substantially horizontal part. This substantially horizontal part indicates that with a small change of the hydrogen pressure much hydrogen is absorbed or given up by the intermetallic compound. This pressure is frequently referred to as the plateau pressure.

As the electrode material $La_{1-y}R_yNi_{5-z}M_zH_x$ for the electrochemical cell according to the process preferably a material is used which at 20°C has a plateau pressure which is comparatively low and preferably less than one atmosphere. An example of such a material is $LaNi_4CuH_x$ which at 20°C has a plateau pressure of 0.7 to 0.8 atmosphere. A rechargeable electrochemical cell according to the process has the advantage that during use high hydrogen pressure can be avoided, which simplifies the construction of the sealing-off. The electrochemical cell can be used as a rechargeable battery owing to the fact that an electrode consisting of $La_{1-y}R_yNi_{5-z}M_zH_x$ has proved to be a very good reversible electrode having fast reaction kinetics. Hence such an electrode is found to have also good load capability both when being charged and when being discharged. This was found for an electrochemical cell according to the process (size R14), an electrode which consisted of $LaNi_4CuH_X$ (where, depending on the charge condition, x can reach a value between 0 and 4.8). The charging current in a 5 N solution of potassium hydroxide was 0.5 C/h (C = capacity of the electrode in Ah). Starting from $LaNi_4CuH_X$, where x = 0, the theoretical amount of hydrogen was absorbed by the electrode in 2 hours. The counterelectrode was made on the basis of $Ni(OH)_2/NiOOH$ as is commonly used in alkaline Ni/Cd cells. Emf of the cell: about 1,380 mV. When the charged cell is discharged the following reactions occur roughly at the electrodes:

$$\text{anode: } LaNi_4CuH \rightarrow LaNi_4Cu + H^+ + e$$

$$\text{cathode: } NiOOH + H^+ + e \rightarrow Ni(OH)_2$$

When charging, these reactions are reversed. It was found that a fully charged cell after being exposed to the air at room temperature had lost less than 10% of its charge in 340 hours.

Electrochemical Impregnation

H.H. Kroger, S.F. Acree and A.P. Akridge; U.S. Patent 3,979,223; September 7, 1976; assigned to General Electric Company describe a system for the electrochemical impregnation of active material into an electrode for a galvanic cell. The process comprises utilizing a constant potential during the impregnation as well as a screen or barrier interposed, in the bath, around the electrode while it is impregnated to mitigate the deposition of active material on the outside of the electrode surfaces. A more densely packed electrode having a higher charge rating per unit volume results.

In the formation of positive electrodes for nickel-cadmium cells, an aqueous solution of about 4 molar $Ni(NO_3)_2$ is used. In a preferred example, $Co(NO_3)_2$ is also present in the solution in a weight ratio of 10 parts by weight $Co(NO_3)_2$ per 90 parts by weight $Ni(NO_3)_2$. The concentration and ratios of the nickel ions and cobalt ions in the solution are maintained using consumable electrodes on each side of a support comprising a nickel-cobalt alloy of about 90% by weight nickel and 10% by weight cobalt.

The nickel and cobalt are deposited within the pores of the support as hydroxides. As the nickel and cobalt ions are deposited within the electrode the concentration is maintained by the dissolving of like numbers of nickel and cobalt ions at the electrodes thereby causing the concentration of metal ions in the bath to remain relatively constant. The use of a nickel-cobalt alloy as consumable electrodes not only assures a constant supply of ions to replace those deposited in the support, but it insures maintenance of the desired ratio of nickel to cobalt ions in the bath.

In accordance with the process, a source of constant potential of between about 1.5 to 3.0 V, dc, preferably about 2.5 to 3.0 V is applied between the electrodes with one electrode being connected to the negative terminal of a constant potential source. The deposition is carried out in cycles for a time period of about 10 to 30 minutes each, preferably about 30 minutes.

After each cycle, the negative electrode is removed from the bath and immersed in caustic for about 30 to 60 minutes to dehydrate the hydroxide active material. In a preferred embodiment, the electrode is immersed in hot caustic comprising KOH of about 20 to 40% by weight at a temperature of about 50° to 90°C.

The electrode is thoroughly washed in distilled water, preferably hot, to remove all traces of the caustic and then dried. Preferably, the drying is carried out at an elevated temperature of, for example, about 60° to 120°C for about 10 to 30 minutes. Any deposit of active material on the external surface of the electrode is then removed and the cycle repeated. Preferably about 5 to 15 such cycles are made.

Electrolytic Activation

T.S. Turner and J.E. Whittle; U.S. Patent 3,894,885; July 15, 1975; assigned to The International Nickel Company, Inc. describe an electrolytic process for activating essentially solid nickel surfaces, e.g., nickel foil, for use as the positive electrodes in alkaline storage batteries. The electrolyte contains in solution a nickel salt, ammonia and an ammonium salt having the same anion as the nickel salt.

Examples of electrolytes according to the process, the concentrations of the salts and ammonia being given in gram-molecules, are set forth below.

Electrolyte	Nickel Nitrate	Ammonium Nitrate	Ammonia	Ammonium Nitrate: Nickel Nitrate	Weight Gain, mg
1	1	0.75	3	0.75:1	18
2	1.5	1.25	4.5	0.83:1	19
3	2	1.5	6	0.75:1	25
4	1	1.25	3	1.25:1	9
5	1.5	1.87	4.5	1.25:1	15
6	2	2.5	6	1.25:1	17
7	1	2.0	3	2:1	6
8	1.5	3.0	4.5	2:1	9
9	2	4.0	6	2:1	20

Each electrolyte was kept for two months and then examined. Electrolytes 1 to 3 showed a slight precipitate, amounting to 1% by volume of the solution in Electrolyte 1, less than 1% in Electrolyte 2 and very slight in Electrolyte 3. Electrolytes 4 to 6 showed very slight turbidity. Electrolytes 7 to 9 did not even show turbidity, but in use tended to corrode nickel anodes slightly. In contrast, in an electrolyte without the stabilizing addition which contained 825 g/l of $Ni(NO_3)_2 \cdot 6H_2O$ and sufficient 0.880 ammonia to give a ratio of NH_3/Ni^{++} equal to 2.75, 30% of the solution by volume had precipitated after 2 months.

The capacity of each of the 9 electrolytes for the production of active mass was evaluated by ascertaining the gain in weight of a circular nickel sheet 3.2 mm thick and of total area 6 cm^2, when immersed as the anode in the electrolyte through which a current was passed for 2½ hours at an anode current density of 18 mA/cm, the temperature being 20°C. The gains in weight are shown in the figures above.

Passivating Treatment

R.L. Beauchamp, D.W. Maurer and T.D. O'Sullivan; U.S. Patent 4,032,697; June 28, 1977; assigned to Bell Telephone Laboratories, Incorporated describe a process for the manufacture of alkaline batteries in which the plaques used for the electrodes are subjected to a specific pretreatment procedure prior to impregnation. This pretreatment is carried out under carefully controlled conditions so as to produce on the nickel plaque a passivating layer which is uniform in thickness and sufficiently thick to prevent corrosion under the conditions of electrolyte impregnation.

However, the thickness of the passivating layer is limited so as to avoid reduction in electrode capacity. The pretreatment procedure involves the production of an oxide layer. A procedure is described for adjusting the parameters of the oxidation procedure (reaction time, temperature, etc.) so as to obtain adequate protection against corrosion under conditions (pH, temperature, time, etc.) used later in the impregnation. Controlled plaque oxidation permits greater variation in the operating parameters of the impregnation procedure (as, for example, lower bath pH and longer exposure time of the plaque to the impregnation bath) and leads to higher and more uniform impregnation under commercial manufacturing conditions.

Particular benefits accrue from use of the procedure where continuous electrolyte impregnation procedures are used since the plaque must suffer exposure to the acid electrolyte for a longer time before the beginning of electrolysis than in a batch process. Controlled plaque oxidation also improves the surface condition of the plaque possibly by removing organic matter, such as tars and oil, so as to improve the wettability of the plaque surface and improve the electrical contact between the plaque and the active material used to impregnate the plaque.

Various pretreatment procedures may be used to produce the oxide layer including, for example, reaction with aqueous peroxide solution. Anodic oxidation and air oxidation at elevated temperatures are preferred because these methods yield uniformly high electrode capacities. In some cases, it is advantageous to wet the plaque, usually with water or water solution, prior to air oxidation.

Conditioning Technique

D.W. Maurer and L.L. Schull; U.S. Patent 3,899,351; August 12, 1975; assigned to Bell Telephone Laboratories, Incorporated describe a process for the fabrication of aqueous alkaline batteries in which at least one of the electrodes is formed under special conditions. Formation refers to the procedure (carried out after impregnation and before final fabrication of the batteries) in which the electrode is cycled (charged and discharged) to put the electrode in condition for battery operation.

In the process the formation is carried out at unusually high rates ranging in current from 4 to 10 times the ampere-hour capacity of the battery. For example, for a capacity of 1 ampere/hour charging and discharging rates would vary from 4 to 10 amperes. These high formation rates are made possible without detrimental effects to the electrodes by the use of an aqueous electrolyte with unusually high concentrations of strong base.

The concentration of base in the electrolyte should exceed 4 molar and may range up to the saturation concentration of the solution. Further, the basicity constant of the base used in the electrolyte should exceed 0.1. Typical bases are sodium hydroxide and potassium hydroxide. Preferred concentration is from 7 to 8 molar both for convenience in making up the electrolyte and the rapid formation made possible. In order to insure both the rapid results obtained and freedom of detrimental effects to the electrode due to the high currents used, the formation procedure is carried out in the temperature range between 40°C and the boiling point of the electrolyte. The preferred temperature range is from 50° to 60°C since above 60°C no advantages are obtained and increased temperatures are disadvantageous because of the corrosive nature of the electrolyte solution and detrimental crystal changes in the electrode.

The formation procedure is applicable to a large variety of electrodes used in alkaline batteries including nickel electrodes, cadmium electrodes, zinc electrodes, silver electrodes and mercury electrodes. Nickel and cadmium electrodes are of particular interest because of the economic importance of these batteries. When the formation process is carried out as described, often only one charge/discharge cycle is required to remove surface deposits and to remove unwanted ions from the plaque.

This formation procedure can be carried out in less than 30 minutes and usually in less than 12 minutes for charge and discharge. It is also easily adaptable to a continuous formation procedure. The high current densities used in the process do not have any adverse effects on the structural integrity of the electrode and electrodes made by this process have capacities as great as electrodes made by formation processes which require much longer time periods.

Sintered Porous Plaque

G.P. Cromer, H.H. Hazen and S. Pensabene; U.S. Patent 4,029,856; June 14, 1977; assigned to General Electric Company describe a sintered porous plaque battery plate which comprises an elongated substrate and a sintered porous plaque on both major surfaces. One longitudinal edge of the substrate has a region substantially free of porous plaque for welding a current collector thereto and the other opposed longitudinal edge of the substrate has a transversely tapered porous plaque region on both major surfaces of the plate adjacent and extending longitudinally of the opposed edge for preventing the plaque from chipping along the opposed edge.

The method for forming such plates includes the steps of (1) coating a substrate with double and single plate width strips of powdered metal slurry, (2) forming a V-shaped groove longitudinally of each of the double width strips to define a tapered porous plaque region along one edge of each plate, and (3) severing the substrate into strips of plate each with a tapered edge region after the slurry has been sintered into a porous plate.

Referring to Figure 3.2a, there is shown a battery coil **2** which comprises a positive plate **10**, a first separator **16**, a negative plate **20**, and a second separator **26**. Plates **10, 20** with separators **16, 26** are spirally wound to form coil **2**. Plates **10, 20** are transversely offset from one another as well as separators **16, 26** to provide an extended longitudinal edge **12** of plate **10** at one end of coil **2** and an extended edge **22** of plate **20** at the opposite end of coil **2**.

As described in U.S. Patent 3,790,408 edges **12** and **22**, respectively of plate **10**, **20** are formed with a ruffled or nonlinear configuration as best seen in Figures 3.2a and 3.2c. This configuration (Figure 3.2b) is used to impart strength and also provides a larger cross section of the exposed edge available for welding to current collector **30** provided at one end of coil **2**. A second current collector is provided at the lower end of coil **2** (Figure 3.2b) but illustration is omitted because it is identical to collector **30.**

FIGURE 3.2: SINTERED POROUS PLAQUE BATTERY

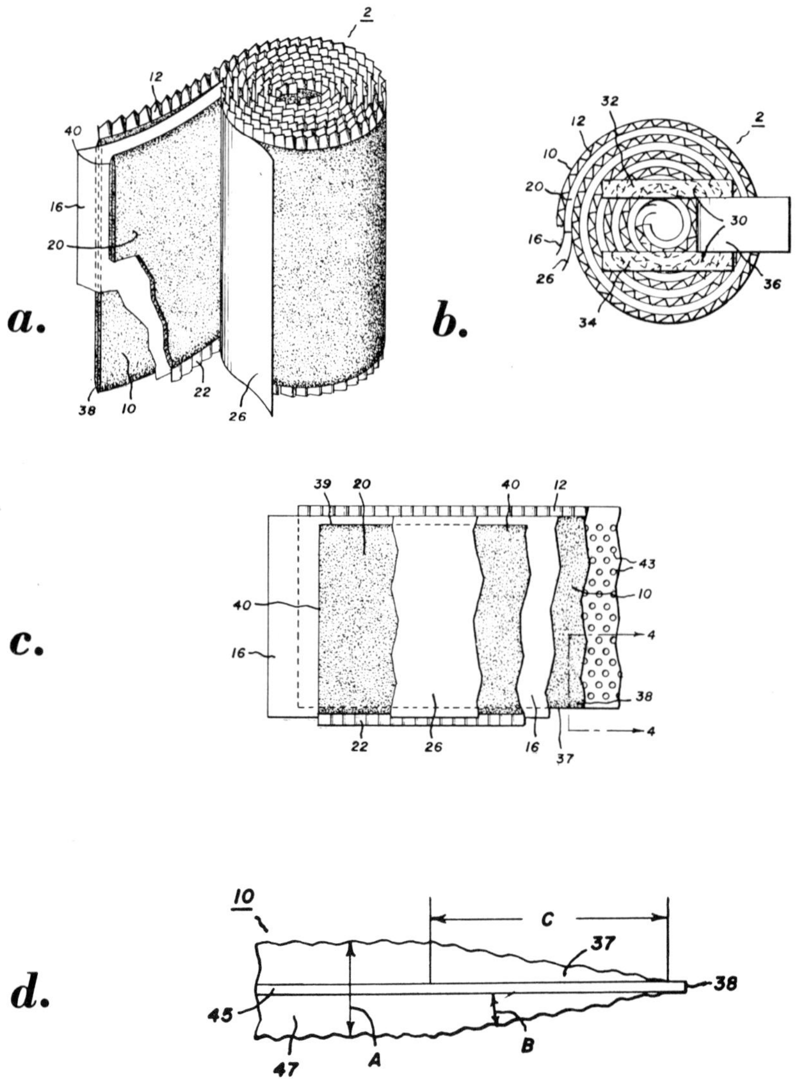

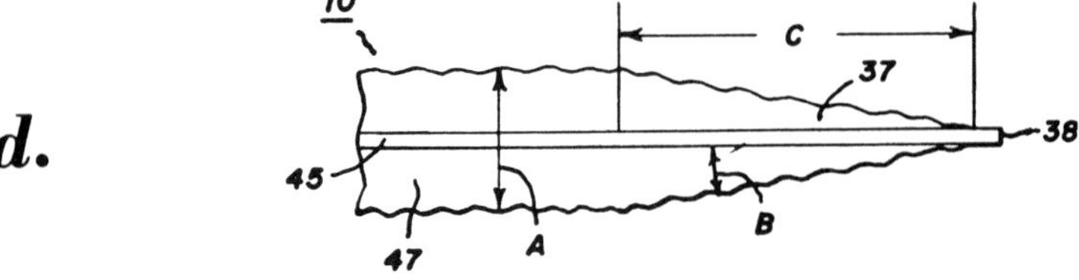

(continued)

FIGURE 3.2: (continued)

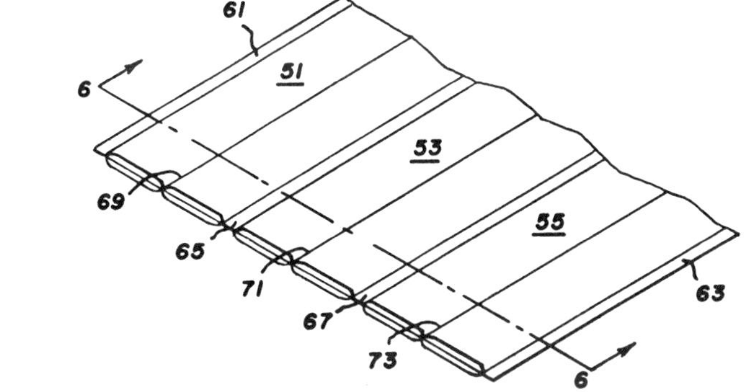

(a) Perspective view, partially broken away, of a battery coil
(b) Plan view of the battery coil of Figure 3.2a showing a current collector attached to the end of the coil
(c) Elevational view, partially broken away, of a portion of Figure 3.2a
(d) Cross-sectional view of Figure 3.2c taken along line **4–4** of Figure 3.2c
(e) Perspective view of a substrate with one arrangement of slurry coatings which view represents the plates at one stage of construction

Source: U.S. Patent 4,029,856

Current collector **30** may comprise, for example two substantially planar leg portions **32, 34** and a raised central portion **36.** The raised central portion **36** is subsequently welded to the base of a cover disc for the cell can which forms the positive terminal of the cell can. The central portion of the negative current collector is welded to the base of the cell can which forms the negative terminal of the cell.

Leg portions **32, 34** may also have downturned knife edges which are imbedded in and welded to extended edge **12** thereby to provide good electrical contact between plate **10** and current collector **30.** In accordance with the features of this process, longitudinal edges **38, 40** of plates **10, 20** are provided with a transversely tapered porous plaque region **37** on both major surfaces of the plates adjacent and extending longitudinally of the plates. This configuration is used to impart structural integrity to the porous plaque at the edge.

In Figure 3.2d a preferred example of the components of plate **10** and region **37** are shown. Plate **10** comprises a metal substrate **45** of nickel-plated steel with an array of perforations **43** therethrough and a porous plaque coating **47** of sintered nickel powder on both major surfaces of the substrate. The construction of the negative plate **20** is the same as that of the positive plate **10** except that positive active material is impregnated into plaque **45** of plate **10** and negative active material is impregnated into the plaque of the negative plate **20.**

In a preferred example of plate **10** the plate thickness **A** is generally between 0.3 and 1.0 mm. The tapering angle **B** of region **37** is generally between 5° and 30°. Tapered length **C** is between 2 and 4 mm. As will be recognized, these dimensions are intended to be exemplary only and are not necessarily critical to the process. The critical dimensional aspect of region **37** is that it be gradually tapered to terminate adjacent to edge **38** thereby to impart structural integrity to the plaque in this region. Plaque **47** is preferably terminated contiguous with edge **38**. Of course, plaque **47** can be terminated short of edge **38**, but this is not desirable because this reduces the pore volume of the plaque which is available for impregnation of active material.

As a comparison of the available pore volume of the plates constructed in accordance with the process and prior art plates with a flat edge termination, there is a theoretical loss of about 4% of pore volume caused by the practice of this process based on plate dimensions given below.

	Prior Art	Process
Plate length, cm	15.748	15.748
Plate thickness, mm	0.65	0.65
Substrate thickness, mm	0.063	0.063
Tapering angle, °	90	15
Plaque porosity, %	0.83	0.83
Pore volume, cm^3	2.3	2.2

The substrate used in the process is preferably a nickel-plated steel substrate, although other types are acceptable, and has an array of perforations in the areas which are to be covered with plaque in the finished plate.

The substrate is wide enough to accommodate a plurality of strips of battery plate being formed therefrom. For example, with a 3.25 cm wide battery plate, 6 strips of battery plate can be formed from a 20.3 cm wide substrate. In an alternative arrangement 4 strips of 3.33 cm wide battery plate and one strip of 3.33 cm wide battery plate are formed on the substrate for a total of 5 strips of battery plate.

The substrate is coated on both major surfaces with a slurry comprised of a nickel metal powder in a viscous gel carrier. The slurry can be applied by any conventional method such as, for example, calendaring. After the slurry has been applied to both major surfaces of the substrate in a continuous substantially uniform coating, the substrate is moved through a doctor blade station in which the slurry coating on the substrate is divided by doctor blades into a plurality of slurry strips running longitudinally of the substrate.

In Figure 3.2e, three double width slurry strips **51, 53, 55** are defined by slurry-free edge regions **61** and **63** and slurry-free regions or channels **65** and **67**, strip **51** being between edge region **61** and channel **65**, strip **53** being between channels **65** and **67** and strip **55** being between channel **67** and edge region **63**. Each double width strip **51, 53, 55** is divided in half by formation of V-shaped grooves **69, 71, 73**, respectively. The slurry-free edge regions **61, 63**; channels **65, 67**; and grooves **69, 71, 73** are preferably formed by the use of doctor blades of the appropriate width simultaneously at one doctor blade station. However, they could obviously be formed at separate stations if desired.

Approximately 188,000 "prior art" NiCd cells were constructed identically to 7,400 NiCd cells constructed according to features of this process with the exception that in the prior art cells (e.g., as shown by Figure 3.2e) grooves **69**, **71**, and **73** were not formed in the substrate prior to sintering. When plates were cut from the substrate after sintering, the strips of plate were formed with a flat edge termination along one longitudinal edge of each plate by severing midway between edge **61** and channel **65**. It was found that by testing the prior art cells and the cells of this process at various points during the assembly procedure that 17.1% of the prior art cells and 6.7% of the cells of this process were rejected because of shorts. Accordingly, it is concluded that this process results in a superior battery plate which upon being spirally wound into a battery coil provides a design which is less subject to shorting during assembly into a completed cell.

Coiled Electrodes

S.H. Greaser and E.T. Russell; U.S. Patent 3,900,340; August 19, 1975; assigned to Union Carbide Corporation describe a nickel-cadmium cell employing a coiled electrode assembly whereby an opening is provided in the outermost layer of the separator of the assembly so that a corresponding portion of the outer face of the outermost wound electrode will be exposed for contacting the inner surface of the outer can housing the assembly with sufficient pressure to provide an electrical connection.

Referring to Figure 3.3a, there is shown an exploded view of a prior art cell employing a coiled electrode assembly.

FIGURE 3.3: NICKEL-CADMIUM CELL

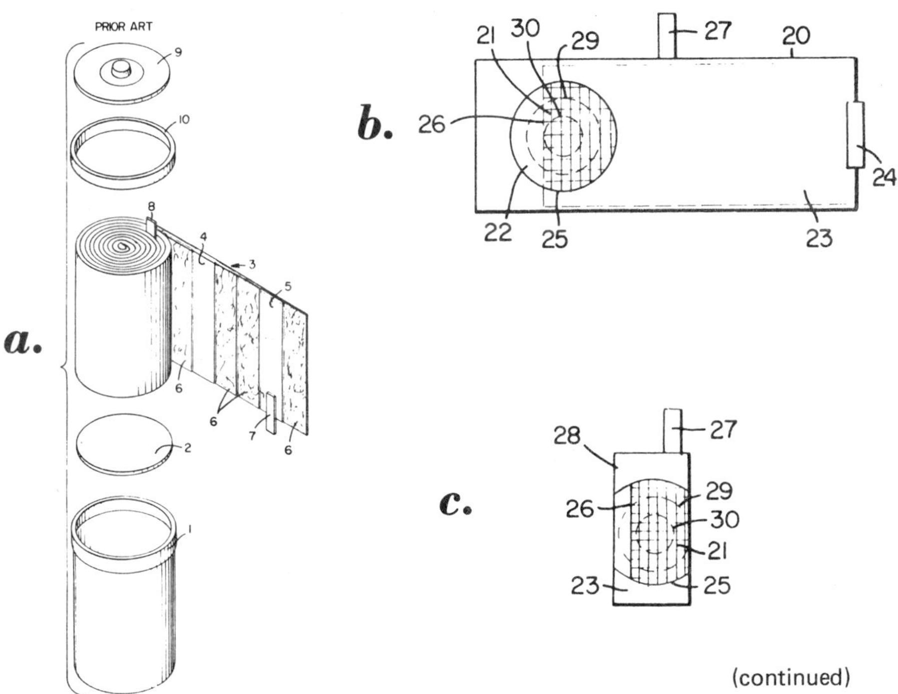

(continued)

FIGURE 3.3: (continued)

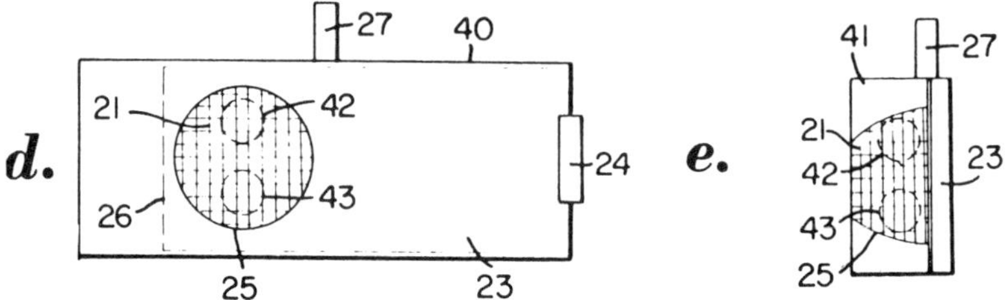

(a) Exploded view of a galvanic cell employing a coiled electrode assembly of the prior art

(b)(d) Front views of flat electrode assemblies of this process

(c)(e) Front views of coiled electrode assemblies of this process

Source: U.S. Patent 3,900,340

The cell comprises a cylindrical can **1** into which an insulating washer **2** is inserted to insulate the bottom of can **1**. A coiled electrode assembly **3** is shown with an inner wound electrode **4**, usually the positive electrode, an outer wound electrode **5**, usually the negative electrode, and separators **6** which separate the electrodes. A conductive tab **7** is shown welded to negative electrode **5** while a similar type conductive tab **8** is shown welded to the positive electrode **4**.

Conductive tab **7** is bent back upon the outermost separator **6** prior to inserting the electrode assembly **3** into the can so that when it is inserted, tab **7** will provide a pressure contact against the inner wall of the can thus adopting the can as the negative terminal. Tab **8** is welded to cover **9** of the cell thereby adopting the cover as the positive terminal of the cell. Cover gasket **10**, made of any suitable insulating material such as nylon or the like, is then interposed between the cover **9** and can **1** in a conventional manner to seal the can **1** while also insulating the can **1** from cover **9**.

In the assembled condition, it is tab **7** that provides the contact between the negative electrode **5** and can **1**. It is the welding of this tab **7** to the negative electrode **5** and the use of tab **7** as the contact member between the electrode **5** and can **1** that has resulted in poor cell performance.

Figure 3.3b shows a flat electrode assembly **20** composed of a first flat electrode member (not shown), a second flat electrode **21**, a separator **22** interposed between the first electrode and second electrode **21**, and an outer separator **23**, all superimposed and joined at one end by a conventional slotted core **24**. A circular window **25** is disposed in separator **23** thereby exposing electrode **21**. The end burrs **26** of electrode **21** are also exposed through window **25** and it is these end burrs **26** that are more easily bulged or bowed outward than is the face of electrode **21** when the flat electrode assembly is helically wound about core **24**. Conductive tab **27** is secured to the first flat electrode and provides the means for connecting the electrode to a cell cover as described in conjunction with Figure 3.3a.

Figure 3.3c shows the flat electrode assembly **20** of Figure 3.3b in a jelly roll construction **28**. It is the exposed electrode **21** through window **25** that is intended to contact the inner wall of a can when the assembly **28** is inserted in the can. The initial protrusion of electrode **21** through window **25** to contact the inner wall of a can is made possible by the fact that the jelly roll assembly **28** is very tightly wound and by the fact that when expanded metal is used as the carrier, it is capable of bulging even though coiled.

To show the variation in size of the window that can be employed according to this process, a smaller circular window **29**, shown by broken lines, along with an even smaller circular window **30**, also shown by broken lines, is illustrated in Figures 3.3b and 3.3c. The exact size of the window need only be sufficient to insure a satisfactory electrical connection between the exposed portion of the outermost electrode and the inner wall of the can.

Figures 3.3d and 3.3e show a flat electrode assembly **40** and a coiled electrode assembly **41**, respectively, having the same numbered elements as in Figures 3.3b and 3.3c except that window **25** is disposed such that it does not expose the end burrs **26** of electrode **21**. As stated above, the size of window **25** can vary and it can even be replaced by two or more smaller windows **42** and **43**, as shown by broken lines in Figures 3.3d and 3.3e. Again, all that is necessary is that a sufficient portion of the electrode is exposed for providing a satisfactory electrical connection with the inner wall of the can.

Example: Seven thousand nickel-cadmium batteries, each containing 5 AA size nickel-cadmium cells, were produced using the direct negative pressure contact of this process. Each cell comprised, in the uncharged condition, a negative electrode of cadmium hydroxide, a positive electrode of nickelous hydroxide and an electrolyte of aqueous potassium hydroxide. The outer separator of the coiled electrode assembly of each cell was fabricated with a circular window measuring approximately 0.8 square inch.

The window was disposed approximately midway on an axis parallel to the longitudinal axis of the coiled electrode assembly and positioned such that the end burrs of the negative electrode were exposed through such window as shown in Figure 3.3b.

The coiled electrode assembly was inserted in an AA size can whereupon the exposed portion of the negative electrode contacted the inner wall of the can. The limiting of the opening to a defined window rather than exposing an entire vertical strip of the electrode, substantially eliminated any of the electrode mix from adhering to the top portion of the can when the electrode assembly was inserted into the can. The positive electrode of the cell was electrically connected to the cover of the cell as described in conjunction with Figure 3.3a.

All 7,000 batteries were tested for internal shorts by the loaded voltage technique and none were found defective. In the production of similar type nickel-cadmium cells using a negative tab as discussed in conjunction with Figure 3.3a, the rejection rate due to failure of the negative contact as evidenced by high resistance shorts at the negative tab contact area, ran as high as 4% of the total cells produced. Thus with five cells in each battery, a total rejection of 20% of the batteries is possible.

VENTING AND CAPPING TECHNIQUES

Resilient Seal

A.S. Decker and R.F. Stephenson; U.S. Patent 3,994,749; November 30, 1976; assigned to Motorola, Inc. describe a vent valve for electric energy cells requiring pressure relief capability for gases formed within the cells due to overcharging or other circumstances. The valve includes a cover plate for the internal structure of the cell which contains a central relief aperture. A vent cap is affixed to the cover plate with one or more vent ports being provided. A resilient seal is provided between the vent cap and the cover plate for controlling the release of the gases, the seal being die-cut of a suitable corrosion-resistant material and having two parallel plane surfaces which contact the cover plate and the closed end of the vent cap when the cap and plate are joined.

The seal is preferably of a polygonal configuration and its vertices contact the walls of the cap for fixed positioning. The vent path for the gases is through the aperture in the cover plate, between the cover plate and the resilient seal, and out through the vent ports.

In Figure 3.4a the container can **10** of the energy cell also serves as the negative terminal as is standard practice in the art.

FIGURE 3.4: VENT VALVE

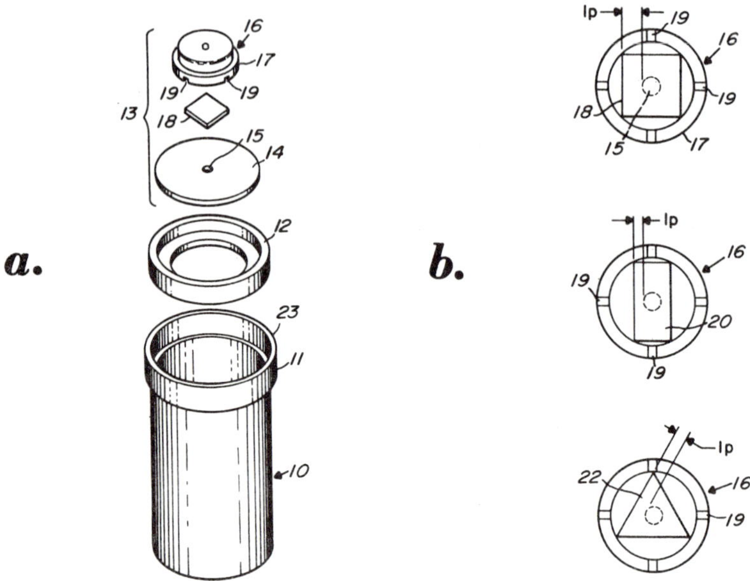

(continued)

FIGURE 3.4: (continued)

c.

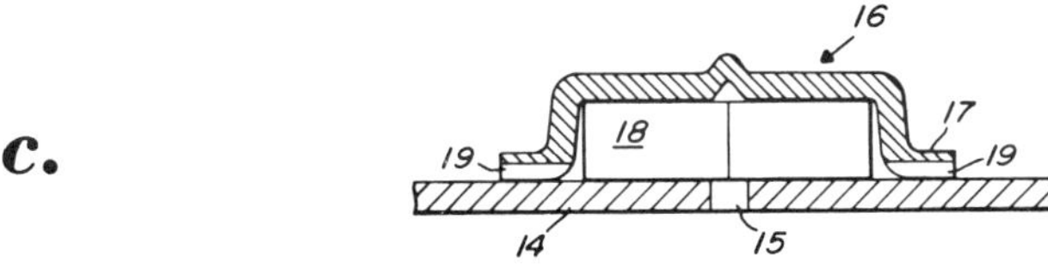

(a) Exploded view of an energy cell
(b) Plan view of three parts of the seal of the vent valve shown within the vent cap
(c) Section view

Source: U.S. Patent 3,994,749

The can **10** is generally cylindrical with one end closed (this end is generally considered the bottom of the container). The top of the can is formed with a slightly larger diameter forming a shoulder portion **11**. A ring **12** of a resilient material such as nylon having an L-shaped cross section is seated on the shoulder **11** of the can **10** and the vent valve arrangement identified generally at **13** is pressure fitted into the ring **12**.

The valve arrangement **13** comprises three parts: a flat metal cover plate **14** having a central aperture **15**, a generally cylindrical vent cap **16** having a closed end and a flange **17** on an open end, and a seal **18**. Small channels or vent ports **19** are formed in the flange **17**. The seal **18** is preferably of a polygonal configuration and which as shown is die-cut from a sheet of suitable corrosion-resistant elastomeric material. A preferred material is a rubber composition of 50 to 90 durometer hardness; 60 durometer generally being considered optimum.

Since the rubber is formed and cured in sheet form, then die-cut, it is much less expensive than small, individually molded elements, while the uniformity of physical characteristics is much greater. The thickness of the sheet rubber comprising the seal **18** is chosen to provide the precise desired pressure on the aperture **15** of the cover plate **14** when the vent cap **16** and cover plate are welded together. The choice of the other dimensions of the seal **18** is, in one respect, determined by the inner diameter of the vent cap **16**, since the central aperture in plate **14** must be covered and a sufficient number of vertices should contact the walls of the cap so as to provide fixed positioning of the seal.

The type of polygon used to a large measure is determined by the desired relief pressure. Since the pressure at which the seal will release gas from the interior of the can is related to the leakage path, i.e., the shortest distance between the edge of the aperture **15** in the plate **14** to the edge of the vent seal **18**, and designated **lp** in Figure 3.4b, the specific polygonal configuration will be chosen to provide a specific relief pressure. As seen in Figure 3.4b, the leakage path **lp** of any rectangular seal **20** other than a square will be less than that of the square seal **18**, and may be more or less than the **lp** of a triangle **22**, depending on the dimensions of each. The preferred seal is the square seal **18**, based on current energy cell and container characteristics.

In the assembly, the internal components which provide the electrical energy of the cell by chemical action are inserted in the container can **10**. (Internal electrical connections are not shown.) The seal **18** is inserted in the vent cap **16** and the cap is welded to the cover plate **14** to form the complete valve **13**. The valve is then inserted into the nylon ring **12** with a pressure fit and the ring seated in the shoulder portion **11** of the container can **10**. When the cell assembly is completed, the upper edge **23** of the can **10** is rolled or crimped down over the ring **12** to provide a permanent seal around the rim of the container can **10**.

In operation, if the pressure inside the can **10** should rise above a desirable limit as, for example, under too rapid charging conditions, the gases would be forced through the aperture **15** in the cover plate **14**, in between the vent seal **18** and the cover plate and out of the cell by way of the vent ports **19**. When the pressure is sufficiently reduced, the vent path will be self-resealed and gas will no longer leave the container.

Plastic Case and Vent

According to a process described by *D.D. Sorensen, R.A. Erickson and H.E. Iepson; U.S. Patent 4,007,060; February 8, 1977; assigned to Gould Inc.* a flat, sealed case for a secondary cell includes a flat shallow plastic base and a thin, flat plastic cover the periphery of which is sealed to the base and includes at least one internal restraint to prevent bulging of the flat cover and base upon a build-up of internal pressure within the cell case, for example, during recharging.

Referring to Figure 3.5a there is shown a rechargeable battery cell **10** having a thin, flat case **11**, and positive and negative terminals **12** and **13**. As more clearly shown in Figure 3.5b, the cell case **11** includes a shallow, substantially flat base **14** having a bottom panel **15** with a low peripheral wall **16** at the sides and ends and a thin, flat cover **17** having a peripheral edge adapted to sealingly engage the peripheral wall **16**.

FIGURE 3.5: PLASTIC CASE FOR FLAT SECONDARY CELL

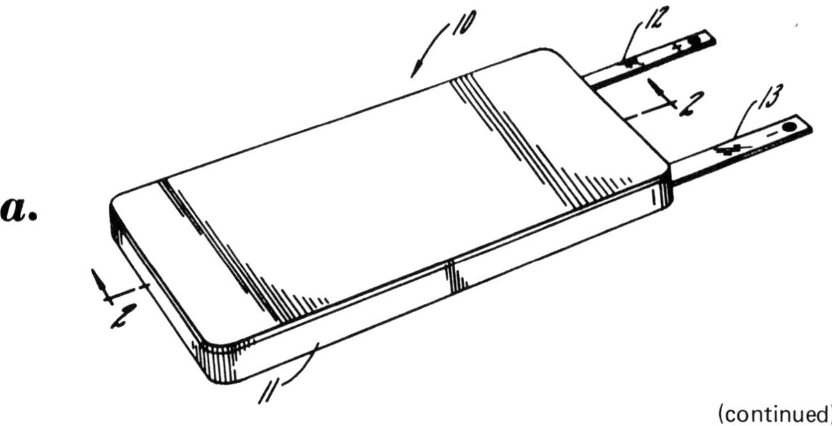

(continued)

FIGURE 3.5: (continued)

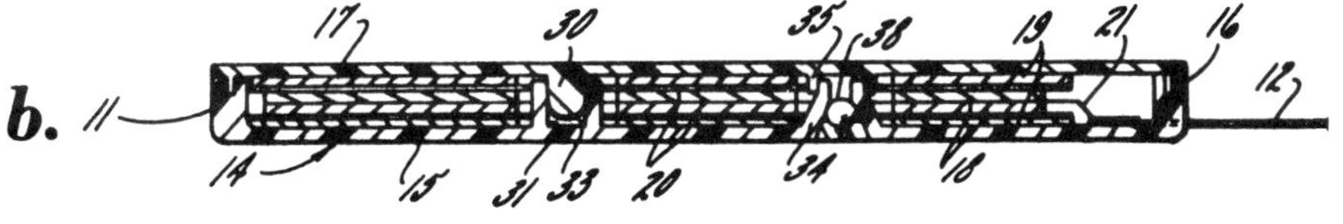

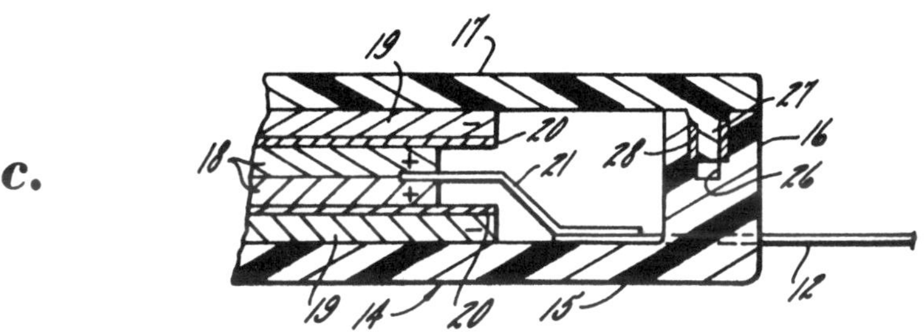

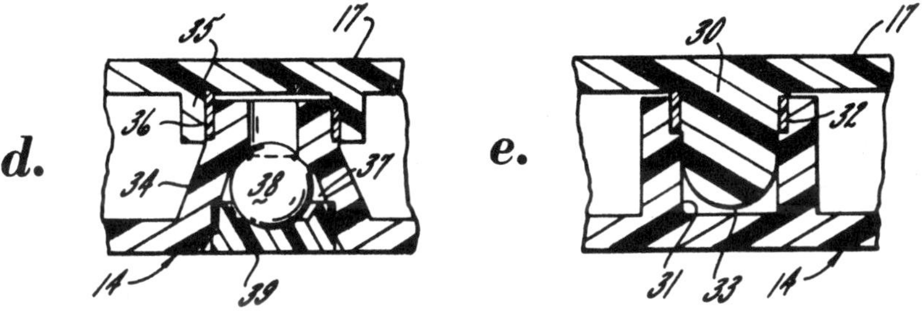

(a) Perspective view of the flat battery cell
(b) Section taken along line **2–2** in Figure 3.5a
(c)-(e) Enlarged fragmentary sections, respectively, of the right end, right center and left center portions as in Figure 3.5b

Source: U.S. Patent 4,007,060

Within the case **11** are the positive and negative electrodes **18, 19** between which are separators **20**. The positive electrodes **18** are connected to the positive terminal **12** by a connector **21** (see Figure 3.5c) and the negative electrodes **19** are similarly connected to the negative terminal **13** by a pair of connector elements. The flat base **14** and cover **17** are separately molded from plastic material and are subsequently bonded together to form the case **11** after the electrodes **18, 19** and separators **20** are in place. The plastic material used may be an ABS resin, a high impact polystyrene or a polysulfone material. It should be caustic resistant as well as being quickly and conveniently bonded such as by solvent, thermal or ultrasonic bonding techniques, the latter being preferred. As in Figure 3.5a, 3.5b and 3.5c, the terminals **12, 13** are molded in the base **14** so as to lie on the bottom panel **15** and extend through the peripheral wall **16** at the end of the base. To insure a gas-tight seal between the terminals, which may be made of nickel, and the base **14**, the terminals are preferably coated with a plastic and solvent mixture in the area where they extend through the wall **16** before the base **14** is molded.

The peripheral wall **16** of the base **14** is formed with a groove **26** and the peripheral edge of the cover **17** is formed with a mating tongue **27**. Preferably, the tongue **27** is slightly larger than the groove **26** permitting a force fit, and when ultrasonic energy is applied to the joint, molten plastic flows into any voids or irregularities between the tongue **27** and groove **26**. When this molten plastic solidifies as indicated at **28**, a gas-tight seal is formed.

In accordance with a further aspect of the process opposing means are provided on the base **14** and cover **17** internally of the peripheral wall **16** and edge and are adapted to be secured to one another for restraining the flat cover **17** and bottom panel **15** of the base **14** from bulging upon a build-up of internal pressure within the sealed case **11**. It will be appreciated that such internal pressure is generated within the case **11**, particularly during recharging of the battery cell **10** and the pressure build-up increases as the rate of charging increases. Moreover, the force exerted on the cover, for example, is also dependent on its area and the deflection of the cover varies exponentially with distance between points of support.

Referring now to Figures 3.5b, 3.5c and 3.5d two specific forms of internal restraints according to the process are illustrated. In Figure 3.5e and the left center portion of Figure 3.5b, the internal restraint is in the form of a pin **30** extending downwardly from the cover **17** and received in and secured to a socket **31** formed in the base **14**. Various forms of bonding may be employed to secure the pin **30** to the socket **31**, but ultrasonic bonding is preferred with the subsequently solidified molten material indicated at **32**. It should also be understood that the socket **31** and pin **30** may have various cross-sectional shapes such as circular, triangular, square, rectangular (in which case the terms mortise and tenon would aptly apply) or other geometric configurations. Preferably, however, the pin **30** is somewhat pointed or rounded at its tip **33** to provide for centering and ease of initial insertion in the socket **31**.

As shown in Figure 3.5d and the right center portion of Figure 3.5b, the internal restraint of this example is in the form of a hollow pin **34** which extends up from the base **14** and is received in and secured to a socket **35** formed in the cover **17**. Preferably the hollow pin **34** is ultrasonically bonded to the socket **35** with the subsequently solidified molten material indicated at **36**. In this case, the hollow pin **34** may also form the housing for a resealable vent **37** which includes a resilient ball **38** and an apertured cap **39**.

The cell **10** develops an internal pressure on the order of 20 to 30 psi during normal recharging and the vent **37** may be set to discharge gas at a pressure of about 50 psi. The flat base **14** and cover **17** with the internal restraints of the process, however, will remain dimensionally stable without bulging up to an internal pressure on the order of about 120 psi. This insures that a reasonable safety margin is maintained for the flat cell case **11**. The flat cell case **11** is particularly suited to enclose a rechargeable nickel-cadmium cell.

In related work *D.D. Sorenson and R.A. Erickson; U.S. Patent 3,980,500; September 14, 1976; assigned to Gould Inc.* describe a resealable vent for a secondary cell having a flat plastic case. The vent includes a resilient ball seated in an aperture passage with a predetermined amount of compression and adapted to unblock an opening in the passage and vent gaseous material upon an excessive build-up of internal pressure within the cell case.

Connecting Caps for Cassettes

A process described by *K. Mabuchi and T. Tsuchimochi; U.S. Patent 3,923,549; December 2, 1975; assigned to Mabuchi Motor Co. Ltd., Japan* relates to battery connection devices and, more particularly, to a battery connection device in a battery source cassette using a plurality of assembled cells, in which the electrical connection between the individual cells is insured by electrically conductive connecting caps which also serve to facilitate the discharge of gas which may emanate from the electrode regions of the cells.

Recent developments in nickel-cadmium type cells have brought forth the so-called quick-chargeable cells which have a low enough internal resistance to afford a discharge current of as high as several amperes and which may be recharged in a matter of a few minutes. This kind of cell is usually equipped with a gas discharging means such as a relief valve for facilitating the escape of the gas produced when the cell is overcharged. When such cells are used in an assembled arrangement, it is required to have the contact resistance between the cells as small as possible, although this requirement is not necessarily limited to the quick-chargeable type of cells.

In Figure 3.6a, character numeral **1** denotes generally the battery source cassette, which comprises a number of cells **2**, an insulating end plate **3** which also serves as the socket base, an opposite end plate **4**, socket apertures **5** in the plate **3**, a socket **6** for the terminal cells. Apertures **7** in the plate **3** serve the purpose of welding the terminals of the cells and also as ventilating holes. The opposite rear plate **4** has therein an aperture **8**.

FIGURE 3.6: BATTERY CONNECTION DEVICE

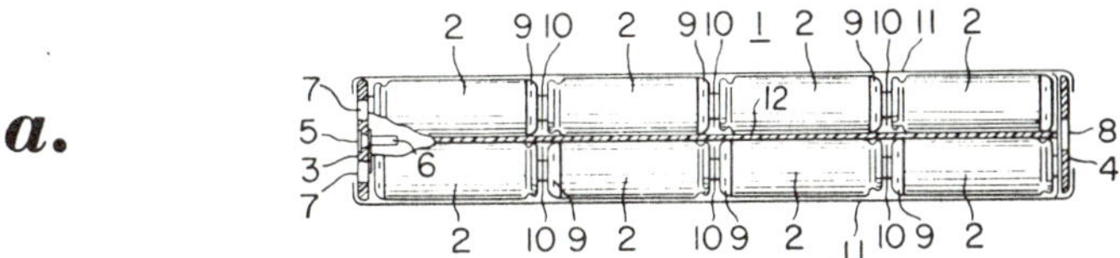

(continued)

FIGURE 3.6: (continued)

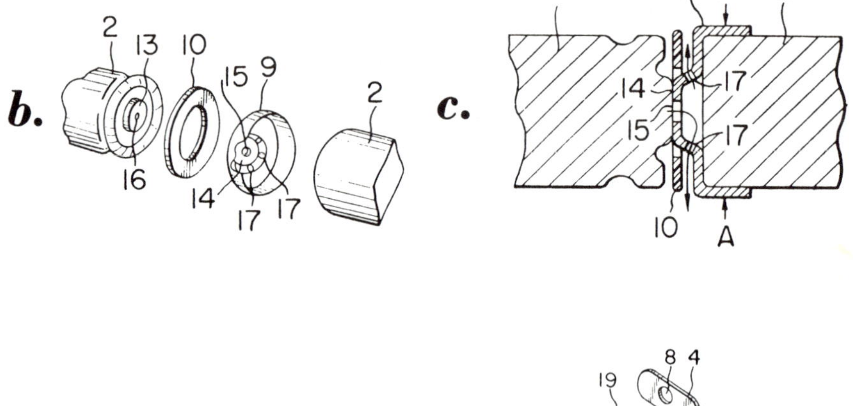

d.

(a) Plan view, in cross section, of a battery source cassette incorporating two rows of assembled cells
(b) Exploded perspective view of a battery connection means
(c) Longitudinal cross-sectional view showing how the connection between two cells is made
(d) Exploded perspective view of the battery source cassette shown in Figure 3.6a

Source: U.S. Patent 3,923,549

Cup-shaped conductive connecting caps **9** are provided for connecting the individual cells to one another and insulating washers **10** are associated therewith. A thermocontractive vinyl resin **11** covers entirely the battery source cassette, and a longitudinal cassette separator **12** separates the two rows of cells from each other. As shown in Figure 3.6a and 3.6d, the cells **2** are connected in series by welding the connecting caps **9** thereto. They are then placed against both longitudinal sides of the separator **12** and packed entirely within the thermocontractive vinyl resin **11** with end plates **3** and **4** attached in position.

The end plate **3** as stated previously is provided with socket apertures **5** for the sockets **6** providing for electrical connection and with apertures or accesses for welding the terminals. Conversely, the rear plate **4** is provided with aperture **8** for releasing the gas discharged from the cells **2**. In Figure 3.6d, reference numerals **18** show the connecting plates for connection between the sockets **6** and the battery terminals, while numeral **19** shows an electrically conductive element for the connection between the individual cells.

Figure 3.6b is an exploded perspective view showing the manner of connecting the individual cells in accordance with the process and Figure 3.6c illustrates two cells in connected condition. To the positive terminal **13** of cell **2** there is welded a cup-shaped electrically conductive connecting cap **9** at the central projecting portion **14** thereof. An insulating washer **10** is disposed therearound, while the central projecting portion **14** is provided with an aperture **15** to allow free escape of the gas from the discharge hole **16** of cell **2**, formed at the positive terminal **13** of the cell. Also in the curved region of the central projecting portion **14**, there are formed openings **17** to permit the issuance of the discharged gas as shown by the arrows.

The connecting cap **9** is welded to the positive terminal **13** of one cell and is snugly fitted over the casing of the adjacent cell which constitutes the negative terminal thereof. The peripheral portion of connecting cap **9** is then spot-welded to the casing of the adjacent cell, as shown at **A** in Figure 3.6c. The insulating washer **10** prevents the casing of one cell from accidentally contacting the casing of the adjacent cell, thus causing an inadvertent short-circuiting of the battery.

As can be seen from the foregoing description, the connecting caps **9** of the process are welded to the positive and to the negative terminals of any two adjacent cells respectively, so as to form series connection of the cells, making the contact resistance therebetween and uniting the cells with sufficient mechanical strength to facilitate the assembly of the battery source cassette as shown in Figure 3.6a. Moreover, since the connecting cap **9** is provided with a central projecting portion **14** having openings **15** and **17** therein, blocking of the gas discharge aperture provided in the positive terminal of a cell is prevented and the discharged gas is allowed to escape freely.

Intermediate Lid

A process described by *R. Lebrun; U.S. Patent 3,943,007; March 9, 1976; assigned to Saft-Societe des Accumulateurs Fixes et de Traction, France* concerns a battery of storage cells and more particularly, a semi-open storage battery, for example, a nickel-cadmium battery. The battery comprises several storage cells arranged side by side in a general casing closed by a lid with each storage cell having a bleed valve for the gases which it may emit and two terminals connected by outside electrical connections to the terminals of the adjacent storage cells. The battery is characterized in that it has a means for conveying the gases likely to be emitted through valves up to first openings formed in the lid.

The lid of the casing is provided with second openings which form a part of an air flow circuit for the cooling of the connections between storage cells. These second openings are separated from the means for conveying gases emitted during the electrochemical reaction in the cells, more particularly during the overcharge of the battery.

According to an example which is of particular advantage, the means for conveying the gases comprises an intermediate or dummy lid inserted between the lid of the casing and the storage cells. This dummy lid is perforated directly above the electrical connections, in order to allow the flow of air for their cooling; it also comprises funnels whose first ends lead out into the first openings of the lid and whose second ends are fitted about the outlets of the valves of the storage cells.

Watertight Cap

A process described by *K. Mabuchi; U.S. Patent 4,011,368; March 8, 1977; assigned to Mabuchi Motor Co. Ltd., Japan* provides a watertight cap for Ni-Cd batteries provided with gas escaping means. Referring to Figure 3.7a, a battery is shown as having a peripheral portion **1** which functions as a negative terminal on one side face or on the entire portion thereof. The positive terminal **2** is electrically insulated from the negative terminal by means of an insulating sealing ring **3**. A groove **4** is provided on the peripheral portion **1** of the battery.

FIGURE 3.7: WATERTIGHT CAP

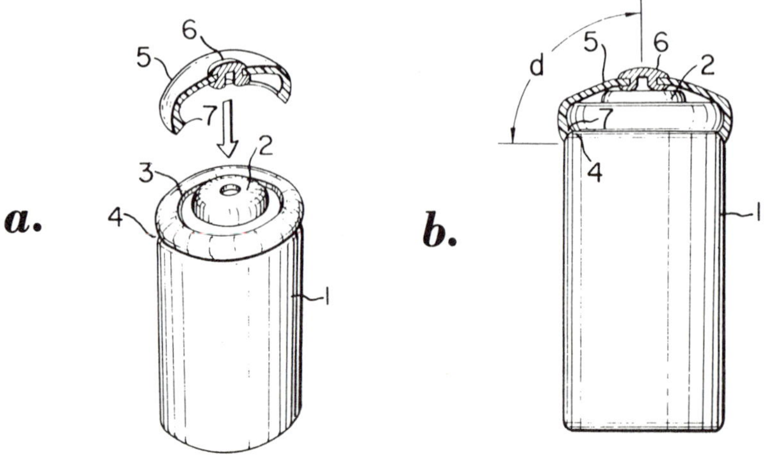

(a) Perspective view of a battery and a watertight cap, with a partially cross-sectional view of the cap

(b) Partially cross-sectional view of a watertight cap fitted onto a battery in which an electrically conductive terminal penetrating the cap and the positive terminal of the battery are in electrical contact condition

Source: U.S. Patent 4,011,368

Recently a quickly chargeable Ni-Cd battery has been developed, which provides a safety valve in order to let the gas escape during overcharging. The groove **4** serves to hold the supporting plates of the safety valve. The cap **5** is a watertight cap, and is made of resilient insulating material such as rubber. An electrically conductive terminal **6** is in electrical contact with the positive terminal **2** of the battery and serves to deliver electrical energy to the outside. The terminal **6** is inserted into the aperture of the cap **5** which is provided on the center portion thereof.

The inwardly projecting ring portion **7** which is formed at the inner edge of the periphery of cap **5**, is adapted to fit the groove **4** on the peripheral portion **1** of the battery. Namely, the inner diameter of the ring portion **7** is slightly smaller than the outside diameter of the groove **4**, so as to be snugly fitted to the groove **4** such as by stretching of the resilient cap member **5**. As shown in Figure 3.7b, when the watertight battery cap is fitted or mounted on the battery, the peripheral portion of the watertight cap can be tightly fitted to the peripheral portion **1** of the battery, so that the surface on which the positive terminal **2** of the battery is located can be protected from the penetration of water.

From the foregoing, it is to be appreciated that by using the watertight cap of the process the battery holding case need not necessarily be a completely watertight structure. Consequently, a greater cooling effect will be achieved by the ability to immerse the portion **1** of the battery directly in water. Further, the distance between the positive and the negative electrodes is greater than in a capless battery as it is shown at **d** in Figure 3.7b, so that any unexpected short-circuiting can be prevented.

Safety Resistor

B.C. Bergum; U.S. Patent 3,907,588; September 23, 1975; assigned to ESB Incorporated describes an electrochemical cell having a resistive element interposed between a cell electrode and its corresponding terminal. The resistive element acts as a safety resistor and is a high resistivity material which has a high temperature coefficient of resistance. The element significantly lessens the high flash amperage current which is characteristic of some electrochemical cells, such as nickel-cadmium cells. The resistance of the element preferably has a value as determined in accordance with the equation:

$$R_n = \frac{E}{I_p} - \frac{E}{I_i}$$

where R_n is the ohmic value of the element; E is the open circuit voltage of the cell; I_i is the initial flash current of the cell in the absence of the element; and I_p is the flash current of the cell with the element present in the cell.

Example: As an example of this process, a sub-C size cell was placed within a larger cell container such as a C or D size container and measurements were taken of cell voltages of a series of cells having no safety resistor and of cells having safety resistors which increased the cell impedance by different amounts. The cell itself has an impedance of 0.12 ohm and therefore the cells in Series No. 1 shown on the following page had no safety resistor and had an impedance of 0.12 ohm and a flash amperage current of 46 A. Series No. 2 had a safety resistor which was an alloy of 80% Ni and 20% Cr and which was cut to the proper length to raise the overall impedance to 0.19 ohm and to reduce the flash amperage current to 12 A. Series No. 3 cells had a somewhat larger piece of the same material for a safety resistor and the impedance of these cells was measured as 0.25 ohm with a flash amperage current of 8 A. In the following, the voltages of these series of cells were measured over a period of time during which the cells were connected to a load and subjected to a 100 mA drain. As the figures show, this process clearly provides a cell construction which significantly reduces the flash amperage current while not deleteriously affecting the normal operating characteristics of the cell.

Time (hr)	Voltage		
	Series No. 1	Series No. 2	Series No. 3
0	1.360	1.340	1.340
2	1.300	1.280	1.275
4	1.280	1.270	1.260
6	1.270	1.260	1.246
8	1.250	1.240	1.230
10	1.235	1.230	1.210
12	1.220	1.210	1.200

Hydrogen-Absorbing Lanthanide-Cobalt Material

D.O. Feder and D.W. Maurer; U.S. Patent 3,980,501; September 14, 1976; assigned to Bell Telephone Laboratories, Incorporated describe alkaline batteries in which a special electrode is incorporated in the positive electrode in order to give added protection against electrochemical damage due to battery reversal. This special electrode contains a hydrogen-absorbing material which also acts as a hydrogen electrode for the conversion of hydrogen ions into elemental hydrogen. Such alkaline batteries are particularly well protected against hydrogen overpressure due to battery reversal with only a small penalty in energy density. Particularly suitable for this application is the use of a hydrogen-absorbing material with nominal formula LnM_5 in which Ln represents a lanthanide metal and M is either cobalt or nickel.

GENERAL CONSTRUCTION FOR NICKEL-CADMIUM CELLS

Stacked Cells

H. Steig; U.S. Patent 3,933,522; January 20, 1976; assigned to Accumulatorenfabriken Wilhelm Hagen AG, Germany describes a battery cell arrangement where the facing topside and underside of the stacked cells are developed to complement each other in such a way that they can be built up with substantially identical cell construction into a compact stack having two fixedly arranged, electrically conductive connectors provided outside the housing, which are intended for connecting one of the poles with the corresponding pole of a similar cell disposed below or above it; and the space in each cell for the liquid electrolyte communicates with the spaces of cells disposed above or below.

Storage battery cells with optimum lengths of plates and electric characteristics, representing elementary cells, may be produced in accord with the process whereby the increase in capacity is made available by stacking the various elementary cells. Moreover, a very stable arrangement from a mechanical point of view is assured on the basis of the natural weight of the battery stack. Thus, the most diverse arrangements of storage batteries can be realized according to the mechanical assembly technique, whereby practically only a single cell type must be fabricated. The storage battery cells of the process have a housing **11** of a moldable electrolyte-resistant insulating material such as polypropylene, closed on all sides in a leak-proof manner with a leak-proof cover **19**. Inside the housing **11**, the elements **18** are disposed in a suitable manner on spaces **27**.

FIGURE 3.8: STACKED CELL BATTERY

a.

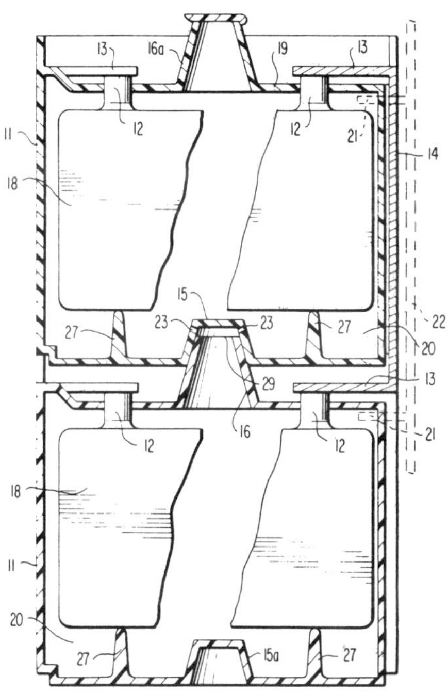

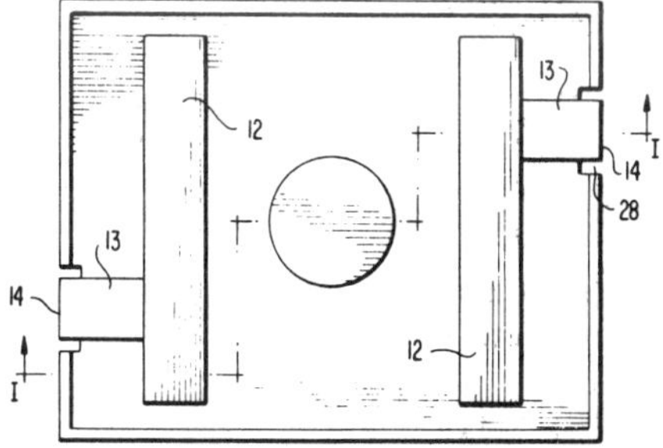

(continued)

FIGURE 3.8: (continued)

c.

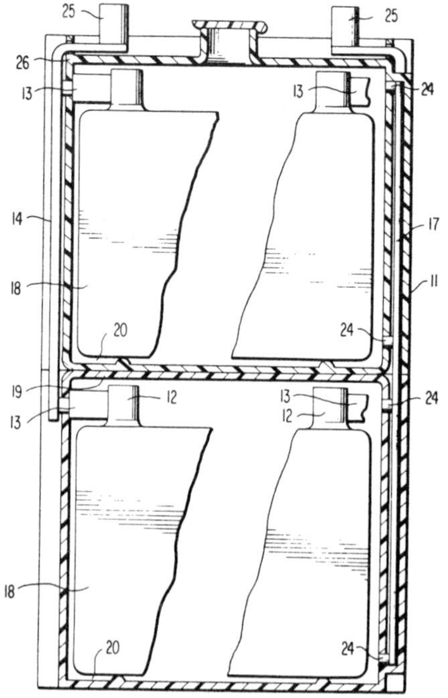

d.

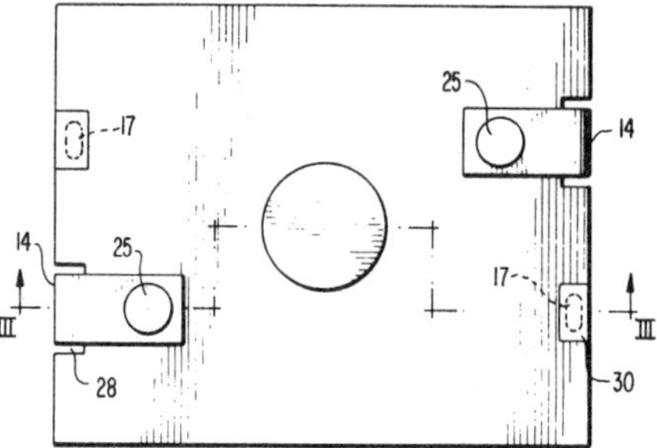

(a) Elevation in section taken along line I–I of Figure 3.8b showing two storage battery cells stacked one on top of the other
(b) Top plan view of the battery of Figure 3.8a
(c) View similar to Figure 3.8a taken along line III–III in Figure 3.8d of another arrangement of storage battery cells
(d) Top plan view of the battery of Figure 3.8c

Source: U.S. Patent 3,933,522

The plates of the same polarity, having plate lugs **12**, are connected to one another by cell connectors **21**. According to the process, lead-through conductors **13** extend from the cell connectors through the walls of housing **11**. Vertically disposed connectors **14**, are shown in Figures 3.8b and 3.8d as being embedded in recesses or indentations **28** of the sidewalls of the housing **11**. The connectors **14** are covered preferably with shrinking tubes or poured into synthetic resin.

In the area of the underside of the upper cells, the connectors **14**, with additional lead-through conductors **13**, consist of one piece, run to and are electrically connected with the underlying cell connectors of elements **18**. In this way, the elements of the elementary cells arranged on top of one another are connected with one another by electric connecting lines running outside the electrolyte chamber **20**.

The electrolyte chambers **20** of the cells that are stacked one on top of the other are in fluid communication with one another. One suitable construction involves the use of identations **15** in the bottom side or on the undersides in case of the example as in Figures 3.8a and 3.8b, while there are raised portions **16** on the top sides. Both the indentations **15**, as well as the raised portions **16**, have conical shapes and are so dimensioned that the raised portions **16** have a sealing engagement with the indentations **15** in the bottom member.

The indentations **15** of the bottom and the raised portions **16** coming into engagement with one another, have openings **23** or **29** by which the electrolyte chambers **20** are connected. The indentations **15a** or raised portions **16a**, on the contrary, are closed. In the case of the example of Figure 3.8c, the poles are brought out laterally, while the connection of the electrolyte chambers is provided by a connecting pipe **17** sunk into an indentation **30**. Transverse bores **24** branch off from the pipes **17** on the bottom or at the level of the electrolyte to the inside of the cell, so that an electrolyte replenishment and degassing can take place at the same time. The connecting pipe may be open on top while the covers and bottom may be flat.

According to Figure 3.8c, an attachment **26** is arranged on the upper cell, which has the terminal poles **25**, and makes an electrical connection with the connectors **14**. A connection is disposed in the cover of the top cell. In the case of other examples, the covers and bottoms have recesses or elevations which fit into one another and which can be suitably sealed as by means of liquid, plastic or solid rubber or plastic compounds. At the same time, it is possible to shift the lead-throughs of lead through the cover to the sides and to connect only the middle area of the covers by such elevations or indentations with bottoms of the boxes of the superposed cells.

In summary, by constructing higher storage batteries by means of elementary storage battery cells according to the process, a considerable simplification in fabrication and consequent reduction in cost results. In the case of the same height, a storage battery built from cells according to the process, has considerably reduced internal resistance conditions, as compared to the situation where only one cell of the same height was provided. As a result, the uniform discharge of all parts will be promoted.

Plastic Encapsulated Flat Cell

T.R. Beatty and H. Vourlis; U.S. Patents 3,977,906; August 31, 1976; and 3,982,966; September 28, 1976; both assigned to Union Carbide Corporation describe a flat alkaline cell where an electrode assembly including at least a pair of flat electrode elements of opposite polarity having a porous separator containing an alkaline electrolyte interposed between and a current collector disposed adjacent to and in electrical connection with one of the pair of electrode elements at one end of the electrode assembly, are enclosed within a sealed, liquid-impervious plastic film envelope.

The film has an opening in one wall which exposes at least a portion of the current collector for making external electrical connection, and a layer of an adhesive sealant which is nonwettable by the alkaline electrolyte tightly adheres and seals together the current collector and the wall of the envelope at least around the periphery of the opening. The exposed portion of the current collector constitutes a first terminal connection to which a wire lead or the like may be attached. A second and third terminal connection are provided in the sealed envelope, the former being in electrical connection with the other of the pair of electrode elements of opposite polarity and the latter being in electrical connection with an auxiliary electrode.

Figures 3.9a and 3.9b illustrate a rechargeable flat nickel-cadmium cell made in accordance with the process.

FIGURE 3.9: FLAT CELL

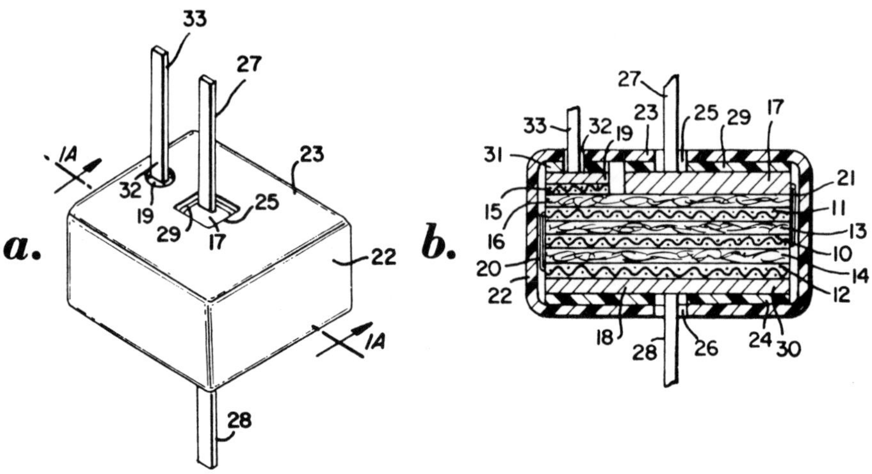

(a) Perspective view of flat alkaline cell construction
(b) Cross-sectional view of the flat cell illustrated in Figure 3.9a, taken alone the line **1A–1A**

Source: U.S. Patent 3,977,906

The cell comprises a positive electrode plate **10**, a pair of negative electrode plates **11, 12**, one of each of which is disposed on each side of the positive electrode plate **10**, porous separators **13, 14** containing an alkaline electrolyte sandwiched between and in facial contact with the positive electrode plate **10** and each of the pair of negative electrode plates **11, 12**, the arrangement of the electrodes and separators forming a conventional electrode stack assembly. An auxiliary electrode **15** which is smaller in size than the positive electrode plate **10** and the negative electrode plates **11, 12**, is positioned over the left hand side of the electrode assembly adjacent to the negative electrode plate **11** but separated therefrom by a porous separator **16**.

The positive electrode plate **10**, negative electrode plates **11, 12** and the separators **13, 14** and **16** are rectangular in shape and are of substantially the same size such that the electrode plates and separators are stacked congruently. Both the positive electrode plate **10** and negative electrode plates **11, 12** may be sintered type electrodes fabricated from a sintered metal plaque which may be made, for example, by sintering a layer of metal powder, e.g., nickel, onto both sides of an open or porous substrate such as a nickel screen, which serves as a mechanical support and electrical path.

The sintered metal plaque is impregnated with the electrochemically active material in accordance with conventional methods known in the art. The porous separators **13, 14** and **16** containing the alkaline electrolyte may be made from a conventional separator material such as a nonwoven organic fiber matte. A preferred type is made from nylon fiber under the tradename Pellon. The alkaline electrolyte used in the cell may be, for example, a 30% by weight solution of potassium hydroxide.

At each end of the electrode assembly is provided one of a pair of current collectors **17, 18**. The current collector **17** is also smaller in size than the positive electrode plate **10** and negative electrode plates **11, 12** but is larger than the auxiliary electrode **15** and overlies substantially more than half of the electrode assembly. A third current collector **19** is positioned over and in contact with the auxiliary electrode **15**. Auxiliary electrode **15** and the current collector **19** which is substantially the same size as the auxiliary electrode **15**, are spaced apart from the current collector **17** at one end of the electrode assembly.

The current collector **18** is positioned in contact with the negative electrode plate **12** at the opposite end of the electrode assembly. The pair of negative electrode plates **11, 12** are electrically interconnected by an insulated metal conductor **20** and the positive electrode plate **10** is electrically interconnected to the current collector **17** also by an insulated metal conductor **21**. The current collectors **17, 18** and **19** are made from an electrically conductive metal, preferably in the form of thin metal foil, which is inert to the alkaline electrolyte such as nickel or nickel plated steel.

All of the cell elements as described above are sealed within a liquid-impervious, electrically nonconductive, plastic film envelope **22**. The envelope **22** fits tightly around the side walls of the electrode assembly and also around the opposite ends thereof forming a pair of end walls **23, 24**. The end wall **23** completely covers the current collector **19** but does not completely overlap the current collector **17** but rather leaves an opening **25** which exposes a portion of the current collector **17**. Similarly, the end wall **24** does not completely overlap the current collector **18** but leaves the center thereof exposed forming an opening **26**. As shown in both Figures 3.9a and 3.9b, a metal terminal lead **27** is secured such as by welding to the exposed portion of the current collector **17** which constitutes the positive terminal for the cell. A metal terminal lead **28** is secured to the exposed portion of the current collector **18** which constitutes the negative terminal for the cell.

Substantially the entire outer surface of each of the pair of current collectors **17, 18** except for the exposed portion thereof, is coated with a thin layer **29, 30**, respectively, of an adhesive sealant. Similarly the entire outer surface of the current collector **19**, except for a portion at the center which is to be left exposed, is coated with a layer **31** of the same adhesive sealant. The layers **29, 30** and **31** of adhesive sealant tightly seal the interfaces between each of the pair of end walls **23, 24** of plastic film and the current collectors **17, 18** and **19** against leakage of alkaline electrolyte. Suitably, the adhesive sealant should be an organic resin which will adhesively bond to both the plastic film and metal collectors.

Preferably the adhesive sealant employs a fatty polyamide which is chemically resistant to and not readily wet by the alkaline electrolyte. The layers of adhesive sealant are first applied as a thin layer over the outer surface of each of the collectors **17, 18** and **19**.

The side wall **23** is formed with an opening **32** which coincides with the exposed center of the current collector **19**. A third terminal lead **33** pases through the opening **32** and is attached as by welding to the current collector **19** which constitutes a third terminal connection for the auxiliary electrode **15**.

The envelope **22** is made from a tubular heat shrinkable plastic film such as a vinyl film. In assembly of the cell, the positive electrode plate **10** and the pair of negative plates **11, 12**, the separators **13, 14** and **16**, auxiliary electrode **15** and current collectors **17, 18** and **19** are first stacked together in the manner as described above and then inserted inside the heat shrinkable tube with the outer ends of the tube protruding beyond the coated current collectors. The plastic film tube is then heated and is caused to shrink down tightly around the side walls of the electrode stack and at the same time, the protruding ends of the tube shrink down forming the pair of end walls **23, 24**. Application of heat and pressure to the end walls **23, 24** establish the final adhesive bond.

In cells utilizing the rechargeable nickel-cadmium electrode system, gas generation and the consequent build-up of substantial gas pressure inside the cell can occur particularly if the cell is placed on overcharge for long periods of time. On overcharge, oxygen gas initially may be liberated at the positive electrode at a faster rate than it can be recombined at the negative electrode leading to a build-up of high internal gas pressure. Hydrogen gas can for example be evolved when the cell is subjected to deep discharge. The evolution of hydrogen gas further increases the gas pressure inside the cell since it does not normally recombine within the cell as does the oxygen.

In flat cells of this process, the plastic film envelope which is made, for example, of a polypropylene or vinyl film is flexible and fairly weak. The cell can rupture, although without danger, if the internal gas pressure is allowed to build up to any significant level such as 160 psi, for example.

In order to avoid this problem, it has been found desirable to provide a mechanism for preventing the generation of hydrogen at the negative electrode while at the same time facilitating the recombination of oxygen.

Accordingly, in the preferred example of a rechargeable nickel-cadmium flat cell, the positive and negative electrodes are balanced electrochemically with respect to one another such that the capacity of the negative electrode is greater than that of the positive electrode. Preferably, the capacity of the negative electrode is at least one and one-half times greater than that of the positive electrode and may be as great as three times the positive capacity.

As a further deterrent against cell rupture due to the buildup of excessive internal gas pressure, flat cells may incorporate an auxiliary oxygen or hydrogen sensing electrode together with a third terminal connection in accordance with the process.

In the instance where the auxiliary electrode is an oxygen sensing electrode, the auxiliary electrode forms with the negative electrode a voltage differential whose value will depend on the partial oxygen pressure that is developed inside the cell. Conversely, in the case where the auxiliary electrode is a hydrogen sensing electrode, the auxiliary electrode forms with the positive electrode a voltage differential whose value will depend on the partial hydrogen pressure inside the cell.

Under conditions where the cell evolves copious quantities of oxygen gas on overcharge, for instance, the partial oxygen pressure will rise inside the cell, the voltage differential will change and this change can be utilized as a signal for actuating a control device in the charge circuit to cut off the charging current and thereby prohibit the further buildup of gas pressure inside the cell.

Commercial plastic films which may be utilized in forming the sealed envelope in flat cells of the process include those made of the following materials: vinyl polymers and copolymers, polyvinylidene chloride, polyethylene, polypropylene, nylon, polysulfone, polystyrene, and fluorocarbon polymers. For use with the preferred fatty polyamide adhesive, films made of polyethylene, polypropylene, and vinyl polymers and copolymers are preferred. Regular and shrink-type films are available in these materials

Desired film characteristics include the following: low cost, flexibility, tear and puncture resistance, chemical stability and resistance to alkaline battery electrolyte, hot-formability, low oxygen gas and water vapor transmission rates, and of course strong surface adherence with fatty polyamide or equivalent adhesive.

To reduce the gas and water vapor transmission rate of the plastic film, it may be vacuum metallized or otherwise given a surface metallic coating on one or both sides providing of course the film is not made electrically conductive enough to put a parasitic current drain on the cell. Although there are probably a number of organic compounds which exhibit a nonwetting characteristic when in contact with an alkaline electrolyte, the most preferred adhesive sealants for the use in the process are fatty polyamides.

Thin Concavo-Convex Design

A process described by *J. Coibion; U.S. Patent 4,032,695; June 28, 1977; assigned to Saft-Societe des Accumulateurs Fixes et de Traction, France* provides a cylindrical electric cell comprising a casing containing a plurality of parallel-connected elementary units, each of which is concavo-convex or trough-shaped and extends over a portion of the arc of a cylinder. The elementary units are juxtaposed within the casing to obtain a substantially entire cylinder with appropriate connections being made between electrodes of the elementary units and the terminals of the cell. Each elementary unit comprises a relatively thick concavo-convex or trough-shaped, positive electrode tightly sandwiched with interleaved layers of separator material between inner and outer relatively thin, concavo-convex, trough-shaped, negative electrode parts which are electrically interconnected.

The thick positive electrode having, for example, a thickness of 1 to 3 mm, is thus surrounded on both its faces by a thin negative electrode having a thickness of less than 1 mm, and so operates with high efficiency. In a preferred case, the elementary units are substantially half cylinders and consequently there are two of them in the casing or can.

The positive electrode can be of the compressed type and in that case will, to great advantage, be directly compressed into its concavo-convex shape, or of the plastified type, i.e., in which the active material, to which an electric conductor may be added in divided form, is mixed with a binding agent and pasted on a current collector. The plastified type of electrode will preferably be shaped to requisite concavo-convex form after manufacture as a plain electrode.

The negative electrode parts are also concavo-convex in shape and preferably of the thin sintered carrier type, and are preferably connected together by a portion of their common collector free from sintered material folded along an end edge of the arcuate positive electrode, and preferably coated with an insulating material such as insulative varnish at that portion.

In Figure 3.10a the cell has a cylindrical casing **1** made of nickel-plated steel. It is closed at its upper part by a cover **2**, also made of nickel-plated steel, which rests on a shoulder defined by an inwardly projecting groove **3** formed at the upper part of the casing **1**. An insulative sealing ring **4** surrounds the periphery of the cover **2** and serves to insulate it electrically from the casing **1**. The upper rim part of the casing is turned down onto the sealing ring **4** after assembly of the cell components therein and insures a fluid-tight seal by pressing the assembled sealing ring **4** and cover **2** against the shoulder defined by the groove **3**. The cover **2** supports a valve which comprises a metal cap **5**, welded to the cover **2**, and containing an elastomer disc **6** plugging a hole **7** formed in the center of the cover **2**. The cap **5** serves as one of the cell terminals.

Referring to Figures 3.10a and 3.10b, the casing **1** contains two elementary units electrically connected in parallel. Each of the elementary units is concavo-convex or trough-shaped, extending over an arc of substantially 180° and comprises a positive electrode **91** or **92** respectively sandwiched in the radial direction between an inner negative ultimately concavo-convex electrode part **82** or **83** and on outer ultimately concavo-convex negative electrode part **81** or **84**. The corresponding inner and outer negative electrode parts, **82** and **81** or **83**

FIGURE 3.10: CYLINDRICAL NICKEL-CADMIUM CELL

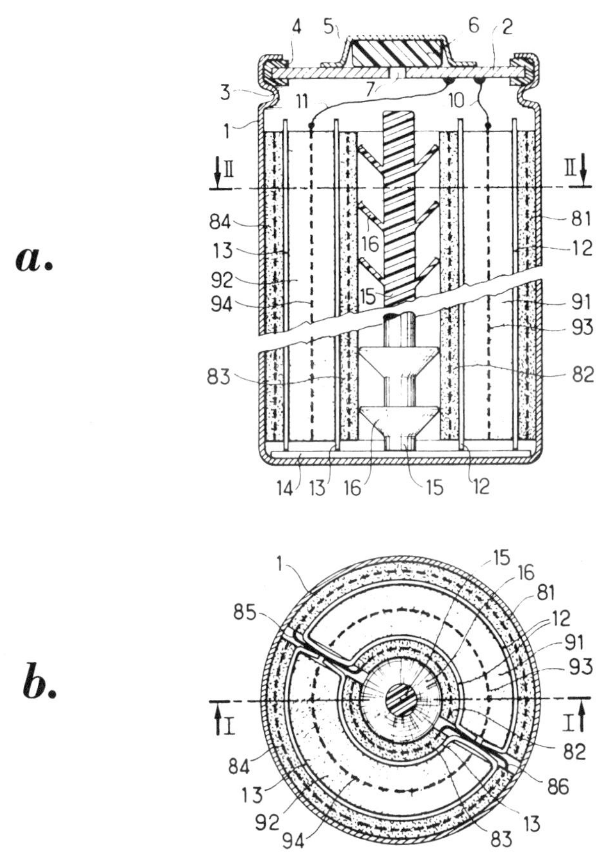

(a) Longitudinal sectional view of a cell in cross section taken along the line I—I of Figure 3.10b

(b) Horizontal cross section of the cell along the line II—II of Figure 3.10a

Source: U.S. Patent 4,032,695

and **84** are electrically interconnected by portions **85** and **86** as will be presently described. The outer convex surfaces of negative electrode parts **81** and **84** are pressed against the inner convex surface of the wall of the casing **1** by the action of radially-resilient fins **16** of insulating rod **15** which occupies the tubular axially extending space in the casing **1** defined by assembled units in the casing and which fins urge the elementary units outwardly, thus providing mechanical support for the units in the casing as well as promoting the electrical connection between the casing and the abutting convex surfaces of the negative electrode parts **81** and **84**. The positive electrodes **91** and **92** are electrically connected to the cover **2** and to the positive terminal cap **5** by respective leads **10** and **11**.

The negative electrode parts are made of pulverulent nickel layers sintered on a core of perforated metal foil and impregnated with active material. Corresponding pairs of negative electrode parts **81** and **82** or **83** and **84** share the same metal core **85** or **86** which provides the electrical connection between the parts of each pair. The base or sinter-free portion of the metal core **85** or **86** which interconnects the corresponding pairs of negative electrode parts is folded around one of the straight or end edges of the positive electrode **91** or **92**. These bare portions of the metal cores **85** and **86** are preferably and advantageously coated with an insulating varnish or the like to prevent short circuits to the positive electrodes.

The positive electrodes **91** and **92** are made of plastified material pasted on respective current collecting cores **93** and **94** which are electrically connected to the leads **10** and **11**.

The respective positive electrode and negative electrode parts of each elementary unit are separated by separators **12** and **13** which fold around the same straight end edges of the positive electrodes **91** and **92**, respectively, as the edges around which the metal cores **85** and **86** of the negative electrode parts are folded. It will be observed in Figure 3.10b that even if the ends of the separators **12** and **13** surrounding the positive electrodes are long enough to fold around the opposite edges of the positive electrodes there is nevertheless a danger of short circuiting with the bare portion of the negative electrode of the other elementary unit, i.e., the positive electrode **92** may short to the negative electrode core **85** and similarly **91** to **86**. Therefore it is preferable to coat these bare portions with an insulating varnish. Moreover, if this precaution is not taken, there is further risk of these bare portions becoming a seat of gas evolution which is to be avoided, especially in sealed storage cells.

The bottom of the casing **1** is provided with an insulating washer **14** which prevents any contact between the positive electrodes **91, 92** and the casing **1**. An insulating rod **15** is made of plastic material and is provided with deformable fins **16** which are compressed into the form of truncated cones whose generator lines are slightly curvilinear when the rod is installed in the tubular space of the casing **1** defined by the innermost concave surfaces of the innermost negative electrode, and which fins or discs **16** provide the above-mentioned radially outward resilience. The different components of an elementary unit at various stages of assembly are described in detail.

Single Common Metal Casing

A process described by *K.H. Dehmelt; U.S. Patent 3,895,959; July 22, 1975* relates to gas-tight electric batteries, and more particularly to storage batteries of the alkaline, nickel-cadmium type, constituted preferably of from 2 to 10 cells.

The process involves stacking the individual cells in a single common metal box or casing, using spacer and support rings which correspond to the height of each cell, and which are pressed essentially against each other, the individual cells being sealed tightly by means of interposed separating contact plates that engage the support rings and are insulated from each other and from the battery terminals. The single, common metal box or container is sealed so as to be gas and liquid tight, by means of a common cover and common gasket.

Only a single metal casing or box, a single common box cover, and a single common box seal are required. This results in a considerable reduction in the number of boxes and other parts, and thus a substantial cost reduction, as well as savings in weight and volume, considering the output or capacity of the battery. Despite the single, common metal box which contains all the cells, the process ensures that each cell forms a self-contained unit which is electrically insulated from its environment, except for the desired series connections. In particular, the process ensures that no ion exchange can take place between the electrolytes of adjacent or other cells.

In a preferred form of the process, the spacer and support rings consist of electrically insulating sealing material, preferably plastic. This enables the spacer and support rings to accomplish at the same time a full, reliable seal at the circumference of each cell, so that the separating contact plates need only form the front or top tight seal of the cells and can form a good insulation and seal with the spacer and support rings.

The separating contact plates can consist of metal foil, and can be electrically insulated at locations outside of their contact regions, particularly from the metal box. Such separating contact plates can be provided with embossing to improve the contact with electrodes arranged at each side of the plates. On the other hand, the electrical insulation of the separating contact plates outside of their contact regions ensures that none of the cells which are combined in the battery will be short-circuited.

Various possibilities exist for the electrical insulation of the separating contact plates beyond the contact region. For example, the plates can be separated a distance from the metal box by the spacer and support rings which receive the plates between them. In addition, the spacer and support rings can be provided, each at one end face, with an axially projecting outer edge to effect an electrically insulating centering means for the plates. Another possibility for the electrical insulation of the separating contact discs consists in providing the metal box on its inner circumferential surface with an electrically insulating coating.

Finally, it is possible within the framework of the process to make the separating contact plates of electrically insulating material, preferably plastic, and to provide in their central regions a metal contact bushing; or the separating contact plates can be made of electrically insulating material so as to be integral with one of the spacer and support rings. Still another possibility consists in making the spacer and support rings for two adjacent cells together of electrically insulating material and integral with the separating contact plate. In order to simplify the manufacture, the spacer and support ring of the topmost cell can be made integral with the sealing gasket of the container.

The storage battery can be built in any manner suitable for the intended use. If it is desired to build a storage battery with a relatively high capacity, the individual cells can consist of electrode packs containing two, three or more parallel-connected electrode plates with intermediate, separator layers. The cross-sectional form of the battery, and thus of the individual stacked cells, can be selected according to the desired use. It is important for the operating safety of the battery that the self-contained cells be constantly physically pressed against each other. The force for pressing the cells against each other can be maintained by a compression spring arranged between the common box cover and the top electrode. But it is also possible, and perhaps even preferable, to maintain the force for pressing the cells against each other by an elastic design of the common box cover which bears on the top electrode.

Plastic Bonding Using Oxalic Acid Additive

A process described by *J. Sandera, M. Calábek, M. Cenek, V. Koudelka, O. Kouril, J. Malik and J. Vanácek; U.S. Patent 4,004,944; January 25, 1977; assigned to Prazska akumulatorka, narodni podnik, Czechoslovakia* relates to compositions for a plastic-bonded or pocket-type cadmium, nickel or zinc electrode for an electrochemical source of current. The electrode composition is particularly suitable for electrodes useful in accumulators of the nickel-cadmium, silver-cadmium, nickel-zinc, silver-zinc and zinc-air type.

As an example of the composition, a powder of a base electrode active material, illustratively cadmium, nickel, zinc or zinc oxide, is mixed with a binding agent and with oxalic acid and/or a salt thereof, the oxalic acid being present in the composition in the range of 1.5 to 60% by weight.

Similar advantageous results are obtained by employing in such composition, in place of oxalic acid, a substituted ingredient such as tartaric acid, succinic acid, malonic acid, acetic acid, and/or a salt of at least one of such acids. In each case, the proportion of such last-mentioned ingredient should be 1.5 to 60% by weight, and preferably 20 to 25% by weight, of the resulting composition.

In order to form a plastic-bonded electrode for an electrochemical source of current, an electrode material having the above-mentioned composition may be applied around an expanded metal collector and then simultaneously heat and pressure treated in a conventional manner. The resulting plastic-bonded electrode exhibits a degree of active material utilization which is larger than that of prior pocket-type electrodes, and yields an internal resistance which is the same as a pocket-type electrode or which approaches that of conventional sintered electrodes. At the same time, the manufacturing cost of the improved electrode is comparable to or less expensive than that of previous conventional pocket-type electrodes. The following are illustrative examples of the process.

A plurality of electrode compositions were formulated which included (1) a powder of the base electrode active material, (2) a polymeric dispersion employed as a binding agent, and (3) various acids (i.e., oxalic acid, tartaric acid, succinic acid, and malonic acid) and/or their salts, such ingredients (3) being present in the mixture in a proportion in the range of 1.5 to 60% by weight.

Of the compositions formulated, the following were particularly noteworthy:

(a) Plastic-bonded cadmium electrodes having a current collector of ductile iron were manufactured using powdered cadmium base material, a binding agent of pulverized polyethylene and various oxalic acid contents as set forth in Table 1. During a succession of operating cycles, each comprising 1.75 hours of charging and

1.25 hours of discharging at an identical current density of 12.3 mA/cm^2, the attained capacities of each type of electrode were as follows:

TABLE 1

.... Electrode Mixture Composition				Capacity Reached at End of ...... Following Cycles					
Conventional Cadmium Material (g)	. $H_2C_2O_4{\cdot}2H_2O$. Grams	Percent	Pulverized Polyethylene (g)	5	89	90	140	225	250
				.... ampere hours........					
31	1.67	5.11	6.2	5.0	5.3	5.3	5.1	4.5	4.3
31	5.86	15.89	10.2	6.0	5.7	6.1	6.0	5.5	5.45
31	14.78	32.2	12.5	7.1	5.9	6.6	6.5	6.6	6.6
31	27.05	46.6	13.1	6.9	4.0	6.0	6.1	6.3	6.4

After the 89th cycle the electrolyte was changed.

(b) Plastic-bonded zinc electrodes having a current collector of zinc-plated ductile iron were manufactured using zinc oxide base material, a binding agent of a polytetrafluoroethylene suspension and oxalic acid as indicated in Table 2 below. During a succession of operating cycles comprising 1.75 hours of charging and 1.25 hours of discharging at an identical current density of 7.4 mA/cm^2, the obtained capacities were as follows:

TABLE 2

....Electrode Mixture Composition				Capacity Reached at End of Following Cycles						
ZnO (g)	. $H_2C_2O_4{\cdot}2H_2O$. Grams	Percent	Polytetrafluoro-ethylene Suspension (g)	23	90	160	190	227	260	354
				.... ampere hours...........						
32	48	60	9.0	8.5	7.0	5.2	9.0	5.4	4.0	1.8

After the 180th cycle the electrolyte was changed.

Each of the resulting compositions was tested in a conventional accumulator environment having a commercial, complementary nickel electrode and an alkali electrolyte. Consistently the resulting accumulators were found to have a much lower internal resistance using the electrodes formulated in accordance with the process than with similar pocket-type electrodes made by prior-art techniques.

Clamping Structure

G.J. Selinko; U.S. Patent 4,020,244; April 26, 1977; assigned to Motorola, Inc. describes a structure for clamping a plurality of battery cells to prevent deformation. The structure includes a pair of pressure plates positioned against the ends of a stack of battery cells and a plurality of bands which extend around the cells and the end plates to hold the end plates in fixed position with respect to each other.

The structure can be used with a plurality of sealed cells of rectangular configuration to prevent expansion of the cells of the type where pressure can build up therein, such as nickel-cadmium rechargeable batteries, and thereby prevent bulging of the electrodes of the cells to injure the same. The pressure plates can be made of steel to also serve as heat sinks during welding of the bands, and the bands can be formed of steel strips which are preferably cinched tight about the cells and welded together. Partitions can be positioned between adjacent cells, with edges welded to the bands to prevent expansion of the cells in a direction parallel to the pressure plates.

Sheet Metal Connector

According to a process described by *A.T. Hewitt; U.S. Patent 4,041,212; August 9, 1977; assigned to Chloride Group Limited, England* plates of each polarity of an alkaline battery are connected by a sheet metal connector which passes through a slot in the bottom of a well formed in the battery lid, and which is supported by means of a pair of ears projecting from the connector which rest on the lid. Each connector is secured against vertical movement by abutment of the ears against the upper surface of the lid and by abutment of a respective plate strap to which each connector is welded against the lower surface of the lid.

Horizontal movement in one plane is prevented by the provision of a shoulder on the lid on each side of each ear, and in the other plane by abutment of the connector, or the ears, against the edge of the well or a further shoulder respectively. The slots are sealed by means of a sealant which fills the wells. Upstanding ribs are provided on the upper surface of the lid so as to increase the length of the leakage path between neighboring cells.

Surfactant Coated Microporous Separator Film

A process described by *H.T. Taskier; U.S. Patent 3,929,509; December 30, 1975, assigned to Celanese Corporation* provides surfactant coated hydrophobic microporous polymeric film. The hydrophilic film has properties which include: a pore size sufficiently small to bar the flow of deleterious anions but large enough to permit the flow therethrough of cations; a small enough thickness to minimize the resistance caused by ionic flow across; a chemical inertness sufficient to resist chemical attack by strong acids and bases of the type used as electrolytes; and a high level of wettability which is provided almost immediately after contact with water and water base liquids. These properties make the hydrophilic film an excellent battery separator.

In accordance with the process, a hydrophilic microporous film is provided comprising a hydrophobic microporous film, characterized by having a reduced bulk density as compared to the bulk density of the corresponding nonporous precursor film from which it is formed, a crystallinity of above about 30%, an average pore size of about 100 to 12,000 Angstroms, a surface area of about 2 to 200 meters per gram, and a void volume of 20 to 45%. The battery separator also includes a coating of a silicon glycol copolymer surfactant.

Preferably the surfactant represents 2 to 20% by weight of the uncoated microporous film. In a preferred example, the microporous film is coated with a coating which comprises a silicon glycol copolymer with a second surfactant, preferably an imidazoline tertiary amine. The following example illustrates the process.

Example: Polypropylene resin having a melt index of about 0.7 and a density of about 0.92 was melt extruded at 230°C through an 8 inch slit die of the coat hanger type using a 1 inch extruder with a shallow metering screw. The length to diameter ratio of the extruder barrel was 24 to 1. The extrudate was drawn down very rapidly at a melt drawdown ratio of 150, and contacted with a rotating casting roll maintained at 50°C, 0.75 inch from the lip of the die. The nonporous precursor film produced in this fashion was found to have the following properties: thickness, 1 mil (0.001 inch); recovery from 50% elongation at 25°C, 50.3%; crystallinity, 59.6%.

A sample of this film was oven annealed in an air atmosphere with a slight tension at 140°C for about 30 minutes, removed from the oven and allowed to cool. It was found to have the following properties: recovery from 50% elongation at 25°C, 90.5% and crystallinity, 68.8%. The annealed elastic film was then cold drawn at 25°C after which the film was hot drawn at 145°C to produce a total draw (extension in length) of 100%. The extension ratio was 0.9, that is, 10% of the stretching resulted from the cold drawing step and 90% of the stretching was a result of the hot drawing step. The cold and hot drawing step resulted in the formation of micropores. The microporous film was thereafter heat set under tension, i.e., at constant length, at 145°C for 10 minutes in air to produce the polypropylene microporous substrate film of the process.

A representative length of the microporous substrate film made above was coated by the reverse roll coating method employing a 7% by weight solution of an imidazoline tertiary amine surfactant in acetone. The final coated microporous film comprised 9.7% of the surfactant measured as a percent by weight of the uncoated substrated film. The same procedure was repeated with the 7% imidazoline surfactant solution replaced with a 7% by weight solution of a nonionic water soluble silicon glycol copolymer, more specifically, a polyoxyethylene polymethyl siloxane. Again, the coating procedure comprised the reverse roll coating method and the 7% by weight solution of the silicon glycol copolymer surfactant again employed acetone as the solvent. The surfactant coating represented 8.2% by weight of the uncoated microporous film or more simply stated 8.2% add-on.

A third sample of the uncoated microporous polypropylene film was coated by the same procedure as specified for the sample employing the silicon-glycol surfactant. In this case a 14.0% add-on of the silicon-glycol copolymer resulted. The three coated microporous films were cut into shapes emulating the size of typical battery separators. They were then individually tested in a 40% potassium hydroxide solution with the following results:

Surfactant	Percent Add-on**	.....Electrical Resistance*		
		1 Hour	1 Day	13 Days
Imidazoline tertiary amine	9.7	OL***	600	6.9
Silicon glycol	8.2	12	9.1	9.0
Silicon glycol	14.0	17.4	17.2	12.8

*In 40% KOH electrolyte measured as milliohms-square inches.
**Percent by weight of the uncoated film.
***Overload, resistance beyond range of instrument.

Cadmium Formate for Cathode

G. Kramer, P. Ness, and H.-H. von Döhren; U.S. Patent 3,993,504; November 23, 1976; assigned to Varta Batterie AG, Germany describe a method for the production of a negative electrode with discharge reserve, which can be carried out easily. The method permits maintenance of the amount of the discharge reserve within narrow limits even when used in mass production. This is achieved by producing the discharge reserve by thermal dissociation of an organic cadmium compound which is introduced into the electrode structure. Cadmium compounds suitable for the process are, for instance, formates, oxalates, and acetates, of which the use of cadmium formate is preferred. At a temperature between 150° and 300°C cadmium formate dissociates as follows:

$$Cd(HCOO)_2 \rightarrow Cd + 2\,CO_2 + H_2 \qquad (1)$$

$$Cd(HCOO)_2 \rightarrow Cd + CO_2 + CO + H_2O$$

As will be noted from the above equations gases are generated producing thereby the required reducing atmosphere. The temperature to which the organic cadmium compound is heated to effect decomposition is desirably below the vaporization temperature of cadmium preferably below 500°C and within the range of 150° and 300°C. A method for the production of a discharge reserve in negative sinter electrodes is described in the following examples. Negative raw sinter foils are impregnated in the usual conventional manner by soaking with cadmium nitrate solution and subsequent precipitation of cadmium hydroxide by means of an alkali solution. This is repeated several times, as often as necessary, until the foil contains the desired amount of $Cd(OH)_2$ which is needed for the required capacity.

Thereafter, the foil containing $Cd(OH)_2$ is subjected to an additional soaking with a cadmium formate solution. The raw sinter foil is predried and then roasted at the above mentioned temperature of 150° to 300°C. There occurs as a result, dissociation of the cadmium formate according to equation (1), and also dehydration of the cadmium hydroxide according to equation (2).

$$Cd(OH)_2 \longrightarrow CdO + H_2O \qquad (2)$$

The oxidizing influence of the water vapor which could decrease the metallic fraction generated according to equation (1), can be counteracted by a flow of hydrogen during the roasting process. Alternatively, the capacity carrying negative cadmium-oxide mass may be produced by roasting in the usual manner first, then applying and impregnating it with the organic cadmium compound solution, whereby the cadmium oxide which was formed during the roasting is rehydrated by the water content. Subsequently the electrode is again roasted in order to obtain the finished electrode which contains cadmium oxide and cadmium.

Generally and favorably, the roasting takes place in the reducing atmosphere which is produced by the thermal dissociation of the organic cadmium compound. It is possible to control, within wide limits, the degree of reduction of the negative mass i.e, the content of metallic cadmium and therefore, the discharge reserve, by the amount of impregnated formates, by the dissociation or roasting temperature and by the content of the protective gas. As an aid in controlling the content of metallic cadmium, introduction of hydrogen to the reducing atmosphere generated at the dissociation as an added reduction agent is particularly advantageous. An inert gas such as nitrogen may be added to facilitate carrying away of H_2O and prevent oxidation of the cadmium. Oxygen as an oxidizing gas may be added in order to alter the ratio of the cadmium oxide to cadmium within special limits.

In a further example, a solution of an organic nickel compound, preferably nickel formate, which thermally dissociates is added to the solution of the organic cadmium compound, so that, after the dissociation process, finely distributed nickel is contained in the electrode resulting in improved conductivity. The roasting process has the additional advantages of reduction of the content of nitrogen compounds and of improvement in the ability of the gas-tight cell to maintain a charge which properties otherwise can only be achieved by an electrical preparatory treatment of the foils, the so-called preformation.

A further advantage is increased flexibility of the foils caused by the roasting process. As a result, the usual wetting of the electrodes with potassium hydroxide at the assembly of cylindrical cells can be eliminated and the assembly process made easier. Thus, by the process there results a good sinter foil with a definite discharge reserve which can be rolled or shaped and which is practically free of nitrate. The foil can be installed dry, without electrochemical "in between" treatment and after quantitatively measured addition of the electrolyte and sealing of the cell, makes the starting of operation in the gas-tight sealed condition possible.

Adaptor for Nonstandard-Sized Battery

K. Mabuchi; U.S. Patent 4,037,026; July 19, 1977; assigned to Mabuchi Motor Co. Ltd., Japan describes an adapter which is designed to enable a battery, whose shape and dimensions are nonstandard, to be housed in an electric appliance such as a model, toy, or portable electric appliance, which has a battery compartment designed to house a standard battery such as type U1, U2 or U3. In view of the fact that electric appliances which can house nonstandard batteries such as the Ni-Cd batteries and chargers for such batteries have been successfully developed, it is desirable to permanently contain such batteries in a dummy container of standard shape and dimensions.

In related work *K. Mabuchi and Y. Tsuchimochi; U.S. Patent 4,020,245; April 26, 1977; assigned to Mabuchi Motor Co. Ltd., Japan* describes a cell adapter which comprises a hollow cylindrical body having an interior space for the insertion of a cell and having an exterior diametrical size agreeing with the size of the cell-receiving space of a cell powered electric appliance in which the cell is to be placed. The hollow cylindrical body can be formed as a monoblock molding unit, that is integral unit, and has resilient holding portions at its top for elastically holding the cell inserted into the interior space. Specifically, the cell adapter in accordance with the process uses a nickel-cadmium type cell of a longitudinal dimension equal to that of the conventional UM3 type inserted in the adapter body so as to assume the standard size of a UM2 cell.

NICKEL-IRON CELLS

Surface Active Compound for Iron Electrode

V.E. Kononenko, N.S. Kononenko, V.N. Tamazina, V.N. Baranova, V.A. Gaintsev, A.J. Kuzin, G.I. Pankov, B.I. Fishman and T.K. Teplinskaya; U.S. Patent 4,021,911; May 10, 1977 describe a method for producing a pasted iron electrode for alkaline accumulators. This process involves adding, into an active mass comprising black iron ore concentrate, iron oxide, synthetic fiber, ferrous sulfide and an aqueous solution of nickel sulfate an aqueous solution of surface active compounds in an amount of 3 to 20 parts by weight of the dry components of the active mass, after which the mixture is pressed into a steel-band grid and dried for about 5 to 10 minutes at a temperature range of between about 150° and 300°C, which is followed by treating the electrode thus produced with a binder which is a 15 to 20% suspension of LiOH in a 20% epoxide resin solution with an addition of a hardener, for example, polyethylenepolyamine, after which the electrode is dried for about 10 to 30 minutes at 70° to 160°C.

According to the method for producing a negative pasted iron electrode, a substrate is first produced by notching and stretching a band-like steel material in longitudinal rows, the notching and stretching of all the longitudinal rows being done simultaneously with 2 to 3 notches in each row and a stepped displacement of notches in adjacent rows. An active mass for a negative pasted electrode is prepared from a mixture of black iron ore concentrate in an amount of 95 weight parts, iron oxide (Fe_2O_3) in an amount of 5 weight parts, synthetic fiber in an amount of 0.1 to 0.2 weight parts, ferrous sulfate (FeS) in an amount of 0.86 to 1.25 weight parts, aqueous solutions of nickel sulfate in an amount of 15 weight parts (concentration of 135 g/l) and surface-active compounds in an amount of 3 to 20 weight parts (concentration of 3 to 4 g/l).

The introduction of surface-active compounds reduces the surface tension of the system and thus facilitates the process of homogenizing the active mass, which, in turn, improves its pasting properties. In the presence of the surface-active compounds, the active sulfur content is practically unchanged after the drying, which accounts for the increased storage capacity of the electrode. The active mass is spread on a grid and rolled. As a result, the active mass becomes denser and excessive moisture is removed. After the rolling, the electrode is dried at a temperature range from about 150° to 300°C for about 5 to 10 minutes to reduce the moisture content in the active mass to no more than 0.2%.

The electrode is then treated with a binder which is a suspension of lithium hydroxide (LiOH) in an epoxide resin solution with an addition of a hardener of the polyethylenepolyamine type. Ths suspension is prepared from 150 parts by weight of a 20% solution of epoxide resin in an organic solvent by adding thereto 10 to 15 weight parts of lithium hydroxide and 2.5 to 3 weight parts of polyethylenepolyamine. In the course of treatment the suspension (the binder) penetrates into the surface layers of the active mass and binds into particles. As a result, a solid reinforcing film-like layer is produced on the electrode's surface.

After being treated with the binder, the electrode is dried for about 10 to 30 minutes at a temperature range of from about 70° to 160°C. In the course of drying the binder is hardened, and volatilization of the solvent takes place. As a result, the integrity of the surface layer is impaired due to the formation of pores, whereas the activity of the mass is almost fully preserved, unlike in the case of the method whereby a binder is introduced into the active mass. This substantially raises the utilization factor of the active iron, accounts for a high mechanical strength of the electrode and completely rules out the possibility of the active mass being washed out during the lifetime of the accumulator. As a result, it has become possible to reduce by 30% the iron content per 1 ampere-hour, reduce the electrode's weight two-fold, prolong the service life and raise the reliability of the accumulators.

Permanent Magnet

S. Takahashi and Y. Miyake; U.S. Patent 4,000,004; December 28, 1976; assigned to Agency of Industrial Science & Technology, Japan describe an anode for an alkaline storage battery which uses a permanent magnet in combination with an iron-electrode of the conventional type. As a result of the incorporation of the permanent magnet to the iron-electrode, the iron-electrode is magnetized throughout and the magnetic field formed around the iron-electrode manifests strong magnetic force. Thus, the electrode of this process is not collapsed even when it is exposed to such high current density of charging and discharging as would readily bring forth breakage of the conventional iron-electrode. Consequently, possible detachment of fragments from the electrode is repressed, so that the electrode is hardly deteriorated.

The material for the permanent magnet to be used in combination with the iron-electrode has only to satisfy the requirement that it has a high degree of coercive force. Examples of the materials which satisfy this requirement include such steel alloys as chromium steel, tungsten steel, cobalt steel and vanadium steel, aluminum-nickel-cobalt alloys and a sintered mass of extremely fine iron powder. Of the metalic materials enumerated above, aluminum-nickel-cobalt alloys and other similar substances which are susceptible to corrosion by alkaline electrolytes should have their surfaces coated with alkali-resistant films as by nickel plating.

The metallic materials of the above type having high coercive force may be combined with a material for the iron-electrode and thereafter fabricated to a desired shape and magnetized, or the metallic material which has been magnetized in advance may be joined fast with the iron-electrode. Otherwise, the metallic material may be joined fast with the iron-electrode and subsequently magnetized.

Fluid Connector for Gas Lock

According to a process described by *O.B. Lindström; U.S. Patent 4,008,099; February 15, 1977* a battery includes a plurality of bipolar electrodes which include a wall, positive electrode material disposed on one side of the wall, and negative electrode material disposed on another side of the wall. Electrolyte chambers separate the electrodes. At the top of each electrolyte chamber a gas lock is provided. The gas locks are fluidly interconnected by fluid conduits. The conduits communicate with the gas locks at a level disposed below the top of the electrolyte chamber. Pressure build-ups in one electrolyte chamber can thus be distributed to the gas locks of other electrolyte chambers. Hydrophobic plugs can be disposed within the conduits which permit the passage of gas, but restrict the passage of electrolyte.

The particularly characteristic feature of the process is that at least one fluid connection is arranged between the upper parts of the electrolyte spaces in the pile with bipolar electrodes. These connections are on such a level that during the normal operating position of the pile the connections are below the highest level in the electrolyte spaces so as to produce a gas lock, or gas space, between the highest level in the electrolyte spaces and the level of the connection. Several connections can, of course, be arranged on about the same level, as well as further connections on lower levels than the lowest level of the gas lock, which latter connections in this case primarily would serve as transporting means for electrolyte between the electrolyte spaces. Every bipolar electrode in the battery pile is thus furnished with at least one fluid connection through the electrode so that all the electrolyte spaces thereby are in fluid connection with each other.

Figure 3.11 shows in principle the disposition of the characteristic elements of the process. The bipolar electrodes **1** each contain a separating wall **2** which is at least point-wise electronically conducting but impermeable for the electrolyte. That is, the wall itself may be electronically insulative but must then contain wires which connect the positive electrode material of the associated electrode to the positive electrode materials of other electrodes (and which connect the negative electrode material of the associated electrode to the negative electrode material of the other electrodes).

FIGURE 3.11: ALKALINE BATTERY

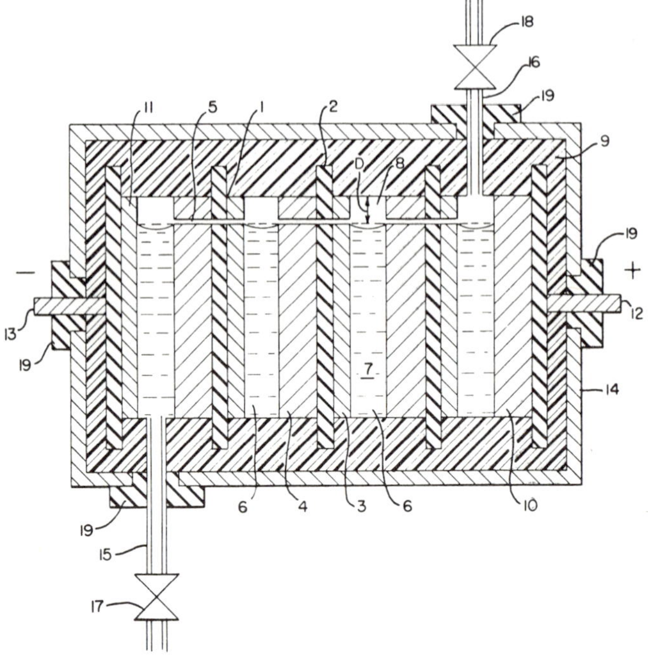

Source: U.S. Patent 4,008,099

The positive electrode material is arranged on one side **4** and the negative electrode material on the other side **3** of each separating wall. The fluid connection which is characteristic for the process is designated **5** and is in this example arranged as a conduit between the electrode and the neighboring electrolyte **6**, separated by the bipolar electrode.

The electrolyte space also contains a mass of electrolyte **7** and may also contain separating means such as distance elements, etc., to prevent short circuits. The gas lock **8** is located in the electrolyte space or chamber **6** between its top and the level of the connection **5**. In Figure 3.11 the conduit **5** is depicted as communicating with the gas lock at a level just above the electrolyte. The electrolyte space may also contain a separator which can occupy the space entirely, except for the pores of the separator itself. The separator may also be partially gas-filled in the part which is situated in the gas lock and electrolyte-filled in its lower part.

The pile of electrodes is arranged in an isolating envelope **9**. This could be a continuous body, for instance of cured resin, or could be a body which is formed by frames at the periphery of each porous bipolar electrode, which are put together as described in the Swedish Patent 217,054. The pile has monopolar electrodes with positive material **10** and negative material **11** in its two ends. The two end electrodes are joined to the pole bolts **12** and **13**. The isolating envelope can advantageously be disposed in a vessel **14**, for instance made of high pressure vessel steel.

The pole bolts are isolated from the vessel by means of isolating connections **19**. The connections for electrolyte filling **15** and venting **16**, which are equipped with shut-off valves **17** and **18**, respectively, are isolated from the vessel by the connections **19**. The isolating connection **19** is similar to the isolating connections discussed later.

Gases produced during battery charging contact the electrode materials located in the gas lock. Drying out of this electrode material, which might lead to a too violent reaction, is prevented by the proximity of the electrolyte material. These electrode materials are porous and thus have a good suction power since they contain fine pores, frequently smaller than 100 mm. Also those parts of the electrodes which are situated in the gas lock will, therefore, contribute to the discharge current and become regenerated during charging.

The distance between the highest level of the electrode and the level of the connection should, however, be smaller than 10 cm to eliminate the risk of the electrode drying out. In practice, it is frequently sufficient for a distance between 0.5 and 5.0 cm to provide for sufficient gas elimination. The distance should not be made too small since then the gas locks will be too small and the pressure surge too big. (This distance **D** has been marked with a measuring arrow in the figure).

A significant advantage of the gas locks is that they interconnect the electrolyte chambers so as to provide larger space for accommodating changes in electrolyte volume in one or more of the electrolyte chambers. Also, overpressurization by surges in gas pressure in one or more of the electrolyte chambers is prevented since gas pressure can be exhausted to others of the electrolyte chambers.

The part of the electrode area assigned to the gas lock should frequently exceed about 1% and be below 20%. A particularly useful value is between 5 and 15%. In certain cases it may, however, be justified to use larger gas locks, for instance with quite large electrodes, in order to reduce the design requirements on the outer vessel.

Hydrogen is consumed in a slow reaction with the positive electrode material, and oxygen in a faster reaction with the negative material. A proper oxy-hydrogen recombination can also take place by the catalytic effect of the electrode materials. It may sometimes be necessary to reduce the load on the electrode materials by disposing special catalysts, for instance on a noble metal basis, so as to accelerate the direct reaction between hydrogen and oxygen which is sometimes practiced with so-called sealed cells.

The connection **5** according to the process can, in its simplest case be a hole straight through the electrode as shown in Figure 3.11. A useful diameter of such a hole is between 0.3 and 2.0 mm. It is, however, possible to use holes with smaller dimensions down to one tenth of a millimeter or slightly below. Holes larger than 2 mm can be used in conjunction with large electrode dimensions (e.g., above 100 cm^2). The holes can have a different cross-section than circular in shape. It is often advantageous to provide the fluid connection internally of the electrode, that is, straight through the electrode at or near the vertical line through the midpoint of the electrode in much the same way as shown in Figure 3.11. The connections **5** may, however, also be arranged near the periphery of the electrode.

Hot Melt Adhesive Pressure Vent

According to a process described by *A. Fitchman and T.D. Wyatt; U.S. Patent 4,008,354; February 15, 1977; assigned to P.R. Mallory & Co., Inc.* an extrudible hot melt adhesive material is utilized in a low cost vent for an electrochemical cell to relieve abnormal pressure formed during improper use of the cell. The extrudible hot melt adhesive material is suitably joined or bonded to a member used to provide a cell top, and covers and fills a vent hole formed in the member. Material is extruded through the vent hole by abnormal pressure to open the vent hole and allow the abnormal pressure to be released through the opened vent hole. The venting pressure can be predetermined by varying vent hole diameter, amount of adhesive, and the composition of the filler adhesive material with respect to viscosity.

Figure 3.12a illustrates a vent **10** with a vent hole **3** at the periphery of a shallow depression **4** formed in member **5** to be used as a cell top. Other placement relationships of the vent hole in the shallow depression are possible. Figure 3.12b shows the member after an extrudible hot melt adhesive material **6** has been applied to the surface of the shallow depression of the member and to the vent hole. The penetration of extrudible hot melt adhesive material into the vent hole should be substantially complete in order to maintain good predictability and reliability in the venting pressure required to extrude the material through the vent hole to open the vent hole and relieve the venting pressure.

When an abnormal internal pressure, as shown by the arrows **8**, builds up, extrudible adhesive material in the vent hole and extrudible adhesive material in the shallow depression of the member are extruded or forced out of the vent hole **3** by the pressure at a predetermined abnormal level to open the vent hole, and the cell vents the abnormal pressure through the open vent hole.

FIGURE 3.12: PRESSURE VENT SEAL

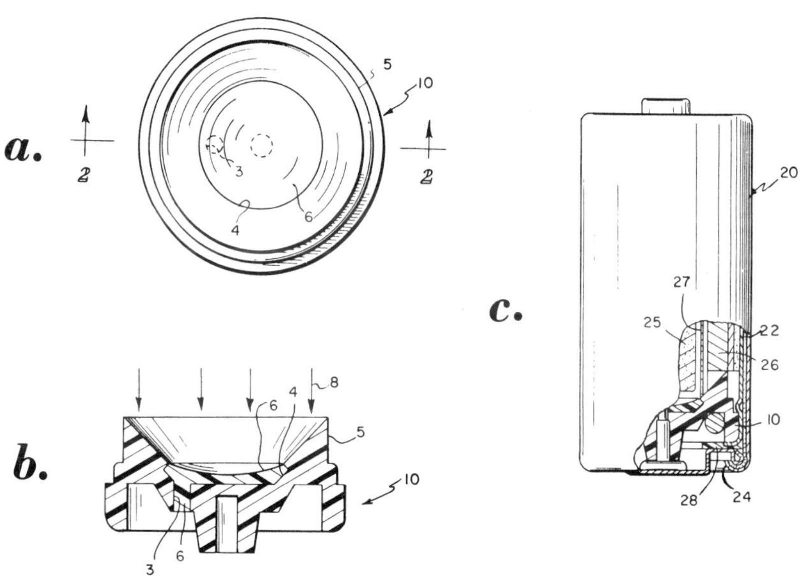

(a) Plan view of vent
(b) Section view of the vent taken along line **2–2** of Figure 3.12a
(c) Partial section view of an electrochemical cell using the vent of Figures 3.12a and 3.12b

Source: U.S. Patent 4,008,354

Figure 3.12c is a partial section view of an electrical device **20** (an electrochemical cell) having a container **22** with an open end **24** closed by the vent **10** of Figures 3.12a and 3.12b. The vent should be positioned in the open end of the container so that the shallow depression **4** is interior of the electrical device. If internal pressures within the electrical device reach an abnormal level, extrudible hot melt material **6** in the vent hole **3** and over and joined or bonded to the surface of the shallow depression of the member **5** will be extruded or forced through the vent hole by the abnormal pressure thereby allowing such pressure to be released through the vent hole without harmful rupture of the container.

The electrical means of the electrical device (an alkaline cell of AA size) includes a positive electrode **26** of a suitable metal oxide, a negative electrode **25** of a suitable substantially pure metal, an alkaline electrolyte in absorbent **27**, and an insulating film **28**. One type of alkaline electrochemical cell uses a nickel dioxide (NiO_2) positive electrode, and an iron (Fe) negative electrode with an electrolyte including potassium hydroxide (KOH) and distilled water (H_2O).

Example: A nylon 101L member intended for use as part of a cell top in alkaline size AA cells is formed with a small diameter vent hole **3** of about 0.040 inch. The vent hole extends from the shallow depression **4** of the member **5** to the opposite surface with a penetration of about 0.030 inch. The shallow depression of the member is preheated to about 160°F, though the preheating temperature can range to about 200°F, prior to the application of extrudible hot melt adhesive material **6** (a thermoplastic polyamide base material) having characteristics of a viscosity of about 280 cp at about 380°F and about 184 cp at about 420°F and a preferred application temperature between about 380° to 415°F. It is applied to the vent hole in the member and the preheated surface of the shallow depression of the member adjacent the vent hole.

The amount of material applied to the inner surface of the member is in the range of from about 0.050 to 0.75 g. The material and the surface of the shallow depression are joined or bonded together. A cell constructed with this specially constructed vent released at a pressure of 600±60 psi. The extrudible hot melt adhesive material closing the vent hole must join or cohere to the adjacent surface of the shallow depression of the member.

Joining the extrudible hot melt adhesive material to the surface of the shallow depression and to the side walls of the vent hole is important for several reasons. One of the reasons is that in order for the extrudible hot melt adhesive material to withstand pressures less than abnormal pressure without extruding through the vent hole and yet provide the feature of extruding through the vent hole when subjected to abnormal pressure, the material must be joined to the surface of the shallow depression and to the side walls of the vent. Another reason is that a joint between the material and the adjacent surface of the shallow depression helps to confine electrolyte within the cell and to prevent the ingress of harmful contaminates to the interior of the cell.

The member **5** of the vent may be of any suitable substantially rigid material such as metal, metal alloy, ceramic, thermosetting resin or thermoplastic resin which is unaffected by the nature of the elements of the electrical device. For low cost, high electrical insulation, low gas permeability, strength and durability, thermoplastic resins such as polyamides (nylon 101L of the example), polypropylene, polyethylene and modified polystyrenes, or thermosetting resins such as epoxies and most fluorocarbons are preferred.

OTHER PROCESSES

Valve Design

F. Sperandio; U.S. Patent 3,935,030; January 27, 1976; assigned to SAFT-Societe des Accumulateurs Fixes et de Traction, France describe a common exhaust valve for several cells of the same battery which will permit venting of gas independently from each cell irrespective of gas conditions in other cells of the battery. One form of a valve comprises a chamber communicating freely with a first space and communicating, moreover, with a second space. The chamber contains an elastically deformable part. This part is capable of stopping up the passage between the chamber and the second space, e.g., in a cell, when the excess of pressure prevailing, e.g., in the second space in relation to the pressure prevailing in the chamber is lower than a first predetermined value; it frees the passage by elastic deformation when the excess of pressure in the second space is higher than a second determined value, thus enabling the escaping of a fluid from the second space towards the first space. The second value is that at which the valve opens, when the pressure in the second space increases.

The first value is that at which the valve closes, when the pressure again decreases. These two values may be identical or different. In the latter case, when the pressure is comprised between them, the valve may be open or closed according to the previous variation in the pressure. A valve according to the process is characterized more particularly in that the chamber communicates with at least a third space, e.g., a second cell, the corresponding passage therefrom to the chamber being either stopped up or freed by the same elastically deformable part under like pressure value conditions as is the passage between the chamber and the first named second space. Thus, the single part in the chamber is responsive to pressure differentials between the chamber and those in the second and third spaces independently.

In relation to conventional constructions having a valve for each storage cell, the process enables a reduction of costs, since only one single deformable part is required with a consequent saving of space which is all the more appreciable as the dimensions of the storage cells are smaller. A valve according to the process may be applied to a one-piece battery comprising any number of compartments provided respectively with means for leading the gases up to the chamber of the valve which is common to all the cells. It may be used, to great advantage for four storage cells having a parallelepipedical shape with a common ridge. It is then placed at one of the ends of that ridge and the leading means are constituted simply by orifices made through the walls. In a battery having more than four storage cells, as many valves as there are sets of four storage cells may be provided.

Figures 3.13a and 3.13b show the casing made, for example, of a plastic material, of a storage cell battery consisting of a container **10** closed by a lid **11** and separated by dividing walls **5, 6, 7,** and **8** into four inner compartments **1, 2, 3,** and **4**, intended for respectively containing four storage cells. Valve means according to the process is housed in the center of the cover **11**. That valve means comprises a chamber **12** in the shape of a cup provided in the cover and communicating by openings or passages **21**, **22**, **23**, and **24** with the respective compartments **1, 2, 3,** and **4**.

The chamber contains a washer **15** made of an elastomer, supported on the bottom of the cup formed by the lid. The washer is held in position by means of a valve lid **14** rigidly fixed, for example, by gluing or welding, in a recess or bore **18** provided in the top surface of the cover. The valve lid has exhaust orifices **16** and **17** intended for keeping the chamber in constant communication with the atmosphere.

The valve lid when mounted in the recess or bore **18** exerts on the washer a pressure which is determined as a function of the maximum tolerable pressure in the compartments **1, 2, 3,** and **4**.

FIGURE 3.13: SEALED STORAGE CELL BATTERY VALVE DESIGN

a.

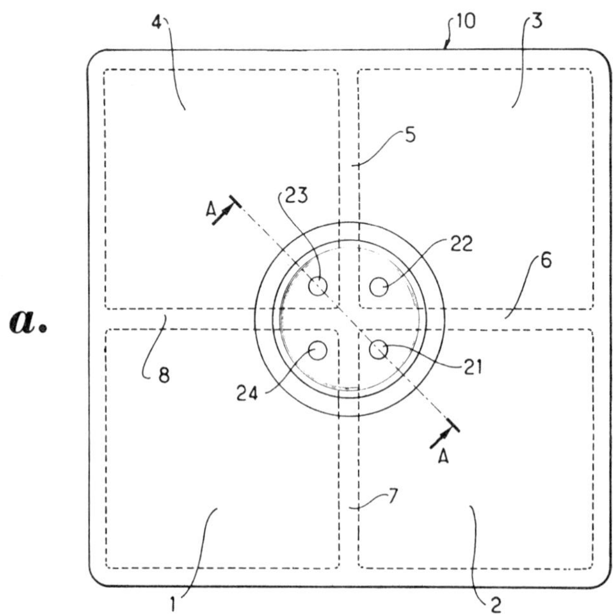

b.

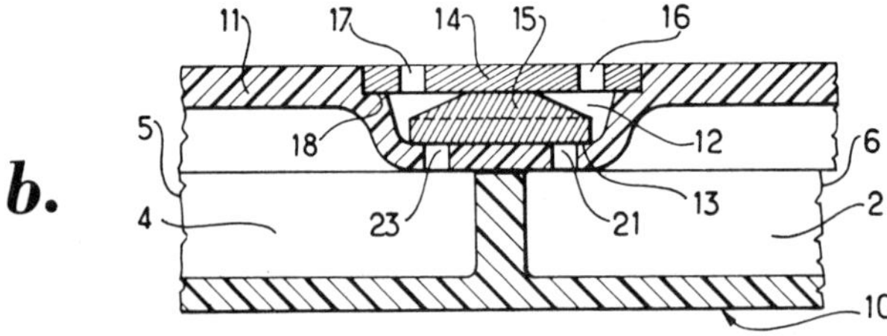

(a) Top view of the casing of a one-piece battery comprising four storage cells and from which the valve has been removed

(b) Partly sectional view through line **A–A** in Figure 3.13a, in which the valve has been shown

Source: U.S. Patent 3,935,030

By way of an example, when the internal pressure in the compartment **4** becomes greater than its maximum tolerable pressure, the washer **15** is deformed and allows the gases to escape to atmosphere through the orifices **23** and **17**, irrespective of pressure conditions in compartments **1, 2** and **3**. The same applies when the pressure prevailing in the compartment **2** becomes greater than the tolerable pressure therein and likewise for compartments **1** and **3**. The vent responsive operation of the single washer **15** is identical for each one of all the storage cells of the battery independently of pressure conditions in respective other cells.

Gas-Tight Casing

According to a process described by *C.A.J. White; U.S. Patent 3,969,140; July 13, 1976; assigned to Chloride Alcad Limited, England* an electric storage cell, of gas-tight type, has flat rectangular positive and negative plates intercalated with one another in a cylindrical (or part-cylindrical) gas-tight casing, in parallel planes parallel to the axis. The outer plates are narrower than the middle ones (in a direction perpendicular to the axis) so that their edges are stepped to roughly follow the cylindrical shape of the casing, and are of pocket type with the active perforated area extending substantially to their edges, so that portions of plates projecting beyond their neighbors provide exposed active areas for recombination of gases evolved from neighboring plates. The plates may be separated in various ways, conveniently by separators formed from a strip of separator material folded zig-zag fashion around one corresponding edge of each plate of one polarity and the opposite corresponding edge of each plate of the other polarity.

Referring to Figure 3.14a, the construction is generally as described in U.S. Patent 3,856,575 in that the plates **10** and **16** of each polarity are provided with a plate strap **12** or **18** of foil strip one passing circumferentially around and being welded to the corresponding edges of plates of one polarity and the other passing similarly around and being welded to the opposite edges of the plates of the other polarity. In this case the edges of plates that are welded to a plate strap may lie upon a substantially cylindrical surface, while some or all of the plates are narrower than would be required to bring their opposite edges to the same cylindrical surface.

The plates may be manually intercalated by pressing against the edges which are welded to the plate straps until they approach the cylindrical configuration, at which point the opposite edges will be spaced from the cylindrical surface so as to provide additional gas cavities. Each plate is of pocket type built up from one or more pockets of standard width, and the pockets of a plate are arranged side-by-side with their lengths extending parallel to the axis in a cylindrical casing. In the specific embodiment there are five positive plates **10** comprising respectively two, three, three, three and two pockets intercalated with six negative plates **16** comprising respectively one, two, three, three, two and one pockets.

As shown in 3.14b the selvages **31** of the steel strips **30** forming the pocket are folded so that when two or more pockets are assembled together they interlock to form a stable assembly and to provide electrical continuity between neighboring pockets, but the plates are not provided with the steel frame normally used on pocket type plates and which encloses the vertical edges of the plate. The whole area of the plate is perforated as shown at **32**, up to the selvages of the pocket strip and up to the tops and bottoms which are bent over together to close the pocket, so that the areas of the plates extending right up to their edges can be regarded as being active from the point of view of diffusion of gases such as oxygen.

FIGURE 1.14: GAS-TIGHT ALKALINE CELL

a.

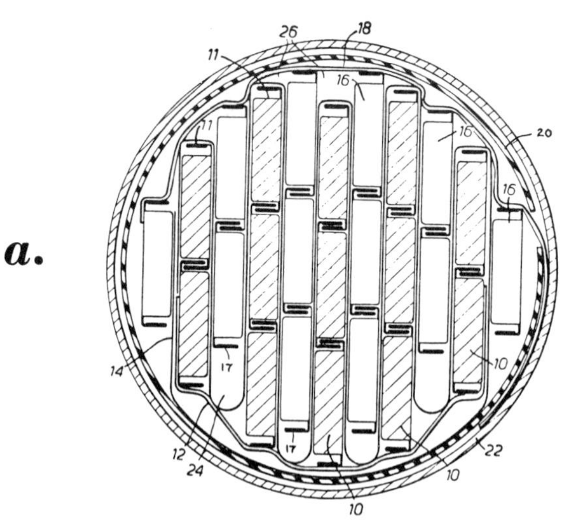

b.

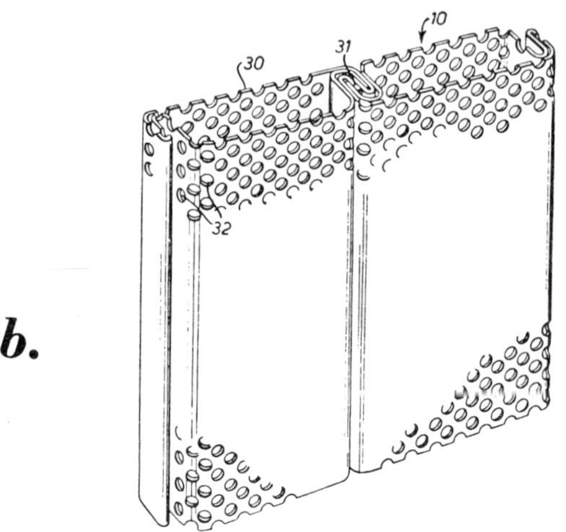

(a) Diagrammatic sectional plan view of an electric storage cell
(b) Fragmentary perspective view of two adjacent pockets of a plate showing how they are crimped together and perforated

Source: U.S. Patent 3,969,140

In assembling the element comprising the plates **10** and **16** and a strip of separator material **14** the first operation is to assemble the five positive plates **10** with one edge of each in a common plane, and to weld a connector strap **12** across the coplanar edges as described in U.S. Patent 3,856,575. As described, the plates are mounted in a jig with the spacing between neighboring middle plates less than that between neighboring outer plates to allow for the inclination of the plate strap when the plates are adjusted to approximate to a cylindrical surface. A strip of plastics fabric separating material is then wrapped around the positive plates **10** passing along the coplanar edges and being wrapped around the opposite edges and folded into the spaces between neighboring plates.

The six negative plates **16** are correspondingly welded to their connecting strap **18** and are then meshed with the positive plates **10** and the separator. The two middle negative plates, being of the same width as the three middle positive plates, are brought into register with them, while the outer plates remain retracted. The complete element is then manipulated by finger pressure to a roughly cylindrical form. It is found that in this operation the connecting strap tends slightly to withdraw the central positive plate, and the middle two negative plates terminate just short of being in register with the next two positive plates. The assembly is completed by fitting a shrink-on plastic sleeve **20** and placing the plates in a cylindrical container **22** in a conventional manner.

The result of this slightly inaccurate assembly of the plates is that parts of the face areas of the large and intermediate negative plates towards one edge are completely exposed, and, with the separator over the adjacent edges of the positive plates **10** and the plastic sleeve, will define gas spaces **26**. In addition, of course, substantially the whole of one face of each outer negative plate is exposed towards the plastic sleeve. The meshing of the negative plates, with the separator already wrapped and folded into the positive plates, causes the separator material to be drawn down firmly over the free edges **11** of the positive plates **10**, but leaves an open loop **24** over the free edges **17** of the negative plates, even if these are accurately positioned.

In practice, however, resulting from the inexact method of approximating a circular configuration, the positive and negative plates are never accurately aligned in register with each other, and this results in further freedom of the loop around the free edges of the negative plates. The loops thus formed are open at the top and bottom so as to provide unrestricted passage for the diffusion of oxygen around the negative plate edges in addition to any gas which may penetrate the separator itself.

Thus the arrangement affords a number of fully exposed parts of the negative plate surfaces providing the areas of most effective gas recombination, in addition to a number of further negative plate areas within the separator loops which constitute additional, although somewhat less effective, areas for recombination.

Plate Holder

A. Wittmann; U.S. Patent 4,000,350; December 28, 1976; assigned to Hughes Aircraft Company describes a special combination of a battery vessel, battery plates having central apertures, electrolyte, a battery plate holder, and a battery plate retainer. The battery plate holder comprises a base retainer and a nonconductive central tube protruding from the base plate. The battery plates are placed onto the battery plate holder by inserting the central tube through the aperture in the battery plates. During assembly of the stack of battery plates, a separator is placed between each plate. The battery plate retainer is mounted to the free end of the stack and holds the plates in compression onto the battery plate holder. The completed assembly is mounted in the battery vessel by coupling only the base retainer of the battery plate holder to the interior of the battery vessel. To complete the electrical assembling, the positive and negative plates are coupled respectively to separate terminals which feed through the battery vessel.

Figures 3.15a and 3.15b are respectively an elevation view and a plan view of a battery plate holder **2**. The plate holder comprises a base retainer **4**, a central tube **6** protruding from the base retainer, and a metallic weld ring **5**. The base retainer and the tube are made from a nonconductive material. The tube is coupled at one end to the base plate and the other end of the tube is partially closed and contains screw holes **8**. Furthermore, the tube has two opposing longitudinal slots **10** and **12** therein. The weld ring is coupled at one edge to the base retainer and has perforations **7** equally spaced about its circumference.

Figure 3.15c is a plan view of a battery plate **14**. The plate is a disc having a central aperture **16** therein. The battery plate further has a tab **18** protruding into the aperture. Figure 3.15d is a partial section of an elevation view of a completed battery cell **20**. The plates are mounted onto battery plate holder **2** by inserting the central tube through the apertures of the plates. The tabs of the negative and positive plates protrude respectively into slots **10** and **12** of the tube. Also, electrolyte impregnated separators, not shown, are situated between the plates. The stack **22** of the battery plates comprises a stack of alternating positive and negative plates **14**.

Furthermore, battery plate retainer **24** of substantially dish shape is coupled to one end of the tube by two nuts **27** and two bolts **26** through holes **8** in the tube and holes **28** in the battery plate retainer. The longitudinal height or the pressure on the stack is set by tightening nuts **27** and bolts **26**. Electrical leads **30** and **32** (not shown), coupled respectively to positive and negative plates **14**, are combined into single leads **34** and **36** (not shown).

Figure 3.15e is a section of an elevation view of a battery **38**. Cell **20** is inserted into cylindrical portion **40** of battery pressure vessel **42**. The other edge of weld ring **5** of battery plate holder **2** is engaged with the circumferential edge of cylindrical portion **40** of a hemispherical portion **44**. The cylindrical portion, the hemispherical portion and the weld ring are coupled together by a single circular weld **46** thereby forming a completed pressure vessel. Furthermore, since the weld ring has equally spaced perforations **7**, the base retainer is flexibly coupled to the interior of the vessel.

Since the cell is flexibly coupled to the interior of the vessel at a single end, the cell is substantially independent of radial and axial growth of the vessel, thereby minimizing mutual mechanical or structural loading and is otherwise free to move relative to the vessel. Leads **34** and **36** are coupled respectively to feedthrough terminals **48** and **50**. The feedthrough terminals are electrically insulated from the vessel. Furthermore, pressure vessel **42** is filled with a pressurized gas **52** via fill valve **54**.

In practice, the positive and negative battery plates **14** are conventional, and may be made of any suitable material. Some satisfactory materials for the positive and negative battery plates are described in Volume 122, No. 1, of the *Journal of the Electrochemical Society,* January 1975, "The Sealed Nickel-Hydrogen Secondary Cell" by Jose Giner and James D. Dunlop, pp. 4-11.

The electrolyte can be a base such as potassium hydroxide. Furthermore, the separator can be made from a material which has high wettability, high gas permeability and high resistance to degradation by the electrolyte. Such a material can be a fibrous asbestos felt. In addition, vessel **42** can be made from a high strength nickel base alloy such as Inconel 718. Also, the pressurized gas can be hydrogen.

FIGURE 3.15: BATTERY DESIGN

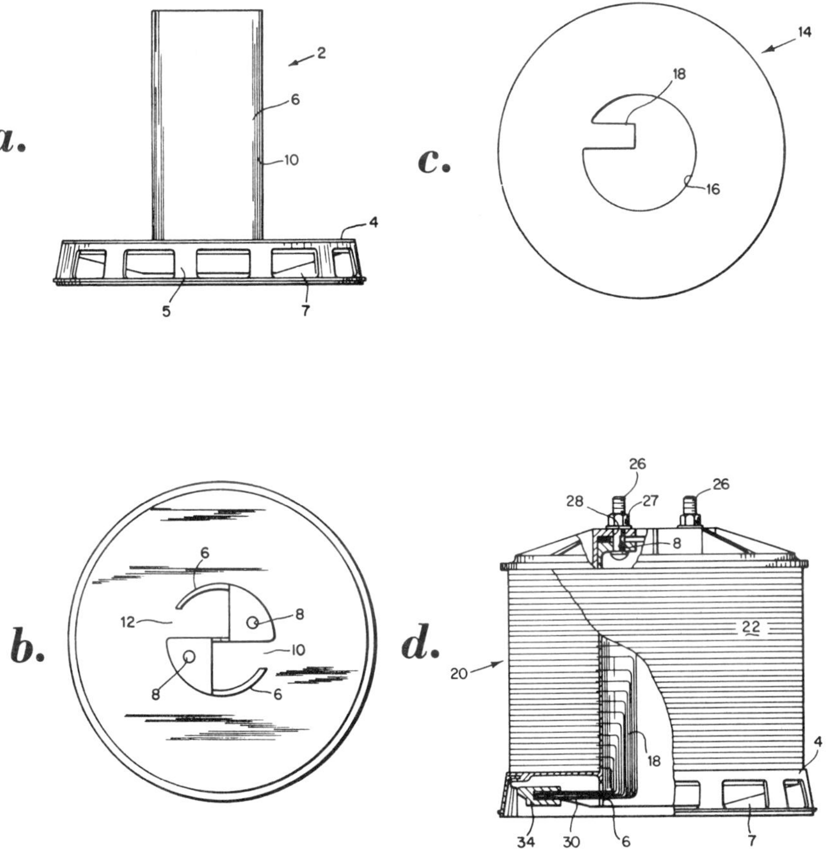

(continued)

FIGURE 3.15: (continued)

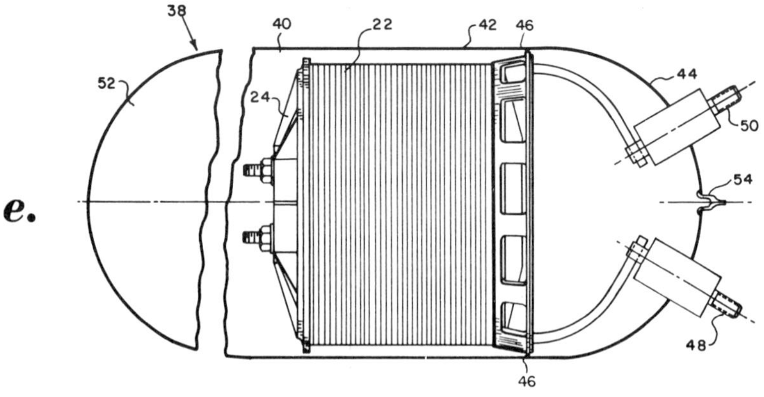

(a) Elevation view of a battery plate holder
(b) Plan view of a battery plate holder
(c) Plan view of a battery plate
(d) Partial section of an elevation view of a completed battery cell
(e) Section of an elevation view of a battery

Source: U.S. Patent 4,000,350

Metal Oxide and Hydroxide Electrolyte Additives

A process described by *K. von Benda and W. Lohnert; U.S. Patent 3,923,542; December 2, 1975; assigned to Firma Deutsche Automobilgesellschaft mbH* involves the stabilization of the capacity of hydrogen storage electrodes in galvanic cells. This is attained by providing that the electrolyte is alkaline and contains in addition to the alkali metal ions, ions of at least one oxy salt of a metal not belonging to the group of the alkali metals, and the deposition potential of the metal added is at least 50 millivolts more negative than the reversible hydrogen potential in the same electrolyte.

Preferably, the alkaline electrolyte contains aluminum (i.e., aluminate ions) as the additional ions. Furthermore, a particularly good stabilization of the capacity results if the electrolyte contains, as the additional ions, ions of the oxy salts of zinc, tin, vanadium, molybdenum, tungsten, beryllium, or gallium. In this connection, the electrolyte contains, for example, aluminum as the aluminate ions, zinc as the zincate ions, or beryllium as the beryllate ions.

Example 1: A hydrogen storage electrode having a weight of 35 g was periodically charged in half-cell measuring arrangement with a current of 5 amperes and was discharged with the same amperage. The hydrogen storage electrode was a sintered plate made by sintering together a mixture of 50% by weight of 300 mesh nickel powder and titanium hydride. The electrolyte employed was 6 M KOH with an addition of 0.3 percent by weight of aluminum hydroxide, dissolved to form aluminate ions in the electrolyte. Only minimal capacity fluctuations occurred during 250 charge-discharge cycles. In contrast thereto, for a hydrogen storage electrode of equal weight, operated with 6 M KOH as the electrolyte free of metal additive, the loss of capacity was 30% of the initial capacity under the same experimental conditions and after the same number of cycles.

Example 2: A hydrogen storage electrode of a weight of 35 g was periodically charged and discharged in a half-cell measuring arrangement under the conditions of the experiment set forth in Example 1. The electrolyte employed was 6 M KOH with an additive of 1.2% by weight ZnO, dissolved in the form of zincate ions. During 200 charge-discharge cycles, only minimal fluctuations in capacity occurred. By way of comparison, for a hydrogen storage electrode of equal weight, operated with 6 M KOH free of additive, the loss of capacity was 28% of the initial capacity, under the same experimental conditions and after the same number of cycles.

Examples 3 through 8: Additional experiments were conducted in the same manner as Example 1. In these experiments, a compound of tin, vanadium, molybdenum, tungsten, beryllium and gallium, respectively was added to the 6 M KOH electrolyte. As shown in the following tabulation, in each case the additive metal of the process served to stabilize the capacity of the hydrogen storage electrode.

Example No.	Metal Compound Added	Amount Added (%)	Cycles Tested Without Substantial Fluctuation	Loss of Capacity Without Additive (%)
3	Sodium stannate	2.0	110	22
4	Vanadium pentoxide	0.7	160	25
5	Molybdenum trioxide	2.0	160	25
6	Sodium tungstate	3.4	160	25
7	Beryllium hydroxide	0.2	258	30
8	Gallium hydroxide	0.6	210	28

Nickel-Cobalt

A process described by *H.E. Ewe and E.W. Justi; U.S. Patent 3,986,892; Oct. 19, 1976* relates to porous electrodes for alkaline accumulators which because of their structure provide both higher specific capacities and also greater current densities of charging and discharging, in line with the present need for avoiding exhaust gases and noise by changing over to accumulator cars. The process is specially suitable for negatives consisting of porous cobalt in combination with the standard positives on a nickel hydroxide basis or with air-breathing oxygen cathodes in hybrid batteries.

A suitable cobalt electrode according to the process has a pore interval of 0.15 to 1 μ. This knowledge leads to the further understanding that it is not the pore width which primarily determines the magnitude of the specific capacity which can be reached, but that these only secondarily have an adverse effect by taking up too great a volume and restricting the electrolyte flow. The process is therefore based on the fact that the parameter which is of primary importance is instead the specific internal surface area such as occurs in the transient pores of suitable width just mentioned. The thickness of this hydroxide layer, multiplied by the inner area in square meters per gram covered by it and therefore electrochemically active gives the volume of the storage mass and when this is multiplied by the Faraday constant it gives the specific capacity in Ah/cm^3 or when multiplied by the density 4.1 g/cm^3 gives the capacity in Ah/g.

Example 1: Boxes or tubes made of nickel-plated sheet steel about 0.1 mm thin, such as were used previously for holding the nickel hydroxide compound and the conductivity material in positive accumulator electrodes, are instead filled without the addition of conductivity substance with cobalt powder of 6.9 square meters per gram so that one obtains a volume porosity of between 30 and 85%. The cobalt electrode so obtained is combined as the negative in an alkaline accumulator with any nickel hydroxide positive or an air-breathing cathode in 6 N potassium hydroxide as electrolyte. With C/3 discharge the electrode gives a capacity of 0.55 Ah/g reckoned on the cobalt mass, or 0.15 Ah/g reckoned on the total weight of this negative electrode.

Example 2: 5 g of cobalt powder with an internal surface area of 2.5 square meters per gram are briefly prepressed in a hollow cylindrical matrix with an internal diameter of 40 mm with a pressure of 0.10 to 0.05 ton per square centimeter at room temperature and the resultant tablet is sintered at 860°C for about 1 hour in a stream of hydrogen. One obtains an electrode with sufficient mechanical strength and a porosity of 79% which when subjected to subsequent C/5 discharge in the alkaline electrolyte against any nickel hydroxide positive gives a specific capacity of 0.35 Ah/g. If required, it is possible inside the cobalt powder to insert an inactive nickel mesh or sieve and press it with it, to which the current lead can be welded after the electrode has been completed.

SODIUM-SULFUR SOLID ELECTROLYTE BATTERIES

GENERAL CELL DESIGN

Ceramic Spacer

L.S. Evans and J.R. Harbar; U.S. Patents 4,035,553; July 12, 1977; and 4,044,194; August 23, 1977; both assigned to the UK Secretary of State for Industry, England describe an electric cell comprising a solid electrolyte which on one side partially bounds a compartment containing liquid sodium and its other side partially bounds a compartment containing liquid sulfur impregnated in a conductive felt, and an electrode means extending along the liquid sulfur compartment in spaced relationship to the solid electrolyte.

The improvement comprises insulating spacer means disposed in the liquid sulfur compartment adjacent to the solid electrolyte, the spacer means being adapted to provide both a region of relatively high electrical resistance substantially free from electrochemical reactions between the solid electrolyte and the electrode means, and a region of the solid electrolyte through which relatively few sodium ions flow during recharge of the cell.

Referring now to Figure 4.1, the battery shown is of circular form in section and comprises an inner tubular β-alumina solid electrolyte **1** disposed lengthwise and substantially concentrically within an outer tubular β-alumina solid electrolyte **2** to define a space in which a stainless steel (for example AISI 316) tubular bipolar electrode **4** is disposed lengthwise and substantially concentrically to define subspaces **5** and **6**.

The outer solid electrolyte **2** is disposed lengthwise and substantially concentrically within a tubular metal casing **7**, mainly of low-alloy steel, to define a tubular space or gap **8** between the solid electrolyte and the casing. The inner solid electrolyte **1**, the outer solid electrolyte **2**, the bipolar electrode, and the casing are all closed at one end. A molybdenum current collecting rod **9** extends lengthwise and substantially concentrically into the space **10** inside the inner solid electrolyte. The battery is shown with its longitudinal axis disposed in a horizontal position but can operate in a vertical or intermediate position as well.

Liquid sodium is disposed in the subspace **6** and the space **8**. Graphite felt partially impregnated with liquid sulfur is disposed in the subspace **5** and the space **10**, thereby providing a suitable distributed voidage within the graphite felt to allow for expansion caused by the formation of polysulfides from the combination of sodium ions and sulfur when the battery discharges. The graphite felt also acts as a wick to constrain the liquid sulfur to "wet" the surfaces of the inner and outer solid electrolytes **1** and **2**. The lengths of the inner solid electrolyte, and the bipolar electrode **4**, the outer solid electrolyte, and the casing **7**, are such as to leave spaces **11** and **20** to act as reservoirs for liquid sodium.

FIGURE 4.1: SODIUM-SULFUR BATTERY

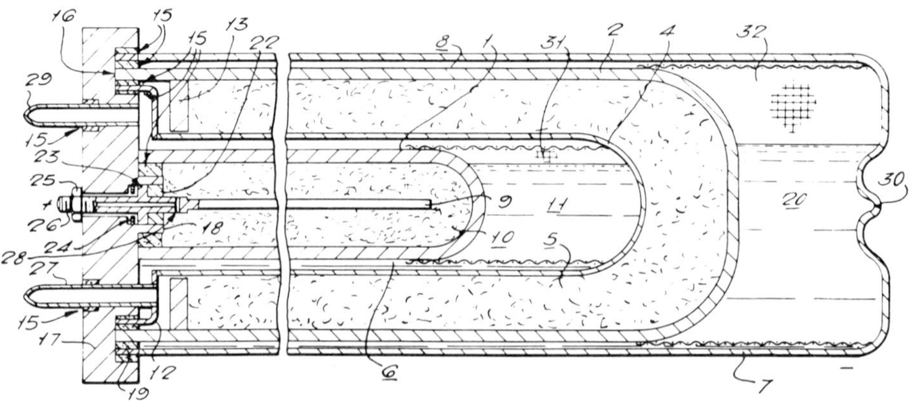

Source: U.S. Patent 4,035,553

Stainless steel wire mesh wicks **31** and **32** in the spaces **11** and **20** constrain the liquid sodium to wet the ends of the solid electrolytes **1** and **2** by capillary action. The radial gaps between the inner solid electrolyte **1** and the bipolar electrode **4**, and between the outer solid electrolyte **2** and the casing **7**, are such as to constrain the liquid sodium by capillary action to wet the surfaces of the inner solid electrolyte **1** and the outer solid electrolyte **2**, although the gaps are shown as wide gaps in Figure 4.1 for clarity.

In greater detail, the bipolar electrode **4** has a shouldered cylindrical portion **12** of Kovar nickel/iron alloy at its open end, butt-welded to the main portion of the electrode, which locates inside the outer solid electrolyte **2**. The graphite felt in the subspace **5** is capped with a loose-fitting spacer means in the form of an α-alumina annular plug-like spacer **13**. The casing **7** at its open end has a short portion of Kovar nickel/iron alloy which is butt-welded to the main low-alloy portion of the casing. The portion **19**, the solid electrolyte **2**, and the bipolar electrode are all located within an annular groove **16** in an end plate **17** of insulating material, for example, α-alumina, and are sealed thereto, and between each, with glass frit seals **15**. The inner solid electrolyte **1** locates onto a spigot **18**, which projects from the plate **17**, and is sealed thereto with a glass frit seal **15**. The graphite felt inside the space **10** in solid electrolyte **1** is capped with an α-alumina annular spacer **22** which butts against one side of a flanged portion **23** of the rod **9**. A Grafoil gasket **24** between the other side of the flanged portion and the end plate is compressed by a nut **25** on the threaded end **26** of the rod.

A Kovar nickel/iron alloy filling tube **27** projecting from the shouldered portion **12** of the bipolar electrode **4**, and a Kovar nickel/iron alloy filling tube **29**, extend through the end plate and are sealed thereto by glass frit seals **15**. A hole extends through the threaded end **26** of the rod and leads to a transverse hole **28** within the space **10**. The filling tube **29** terminates flush with the inside face of the end plate at a position adjacent to the shouldered portion **12** of the bipolar electrode **4**. A short tube **30** is recessed into the closed end of the casing. The tube **27**, and the hole in the threaded end **26** are used to feed liquid sulfur into the subspace **5** and the space **10** respectively.

The tube **29** and the short tube **30** are used to feed liquid sodium into the subspace **6** and the space **8**. When the battery has been filled with liquid sulfur and liquid sodium, the tubes **27**, **29** and **30**, and the threaded end **26** are sealed by crimping and welding. A negative terminal is welded to the side of the casing, the positive terminal being provided by the threaded end **26**.

The bipolar electrode **4** in effect separates two electric cells of the kind described and acts as a series connection between them. In operation, each of the electric cells develops a potential difference, and because of the series connection provided by the bipolar electrode, the total potential difference developed across the battery will be equal to the sum of the potential difference of each cell.

The Kovar nickel/iron portion **12** is provided to permit glass frit seals **15** to be used to join the bipolar electrode to the end plate **17** since it has a controlled expansion. The spacer means in the form of the α-alumina spacer **13** in the liquid sulfur compartment is therefore provided to protect the Kovar nickel/iron portion **12** by keeping the graphite felt away from the Kovar nickel/iron portion so that no corrosive electrochemical cathodic reactions can take place near it. The α-alumina spacer additionally protects the region of the glass frit seal by providing a region substantially free from electrochemical reactions and inhibiting the migration of sodium polysulfide ions therein, since the glass frit seal is likely to be adversely affected and short-circuit the cell if sodium ions are allowed to flow through a portion of the solid electrolyte **2** contiguous to the glass frit seal.

The glass frit seal may in such adverse circumstances become locally detached from the solid electrolyte **2**, or crack as a result of the electric potential existing across it between the charged sodium polysulfide ions and the liquid sodium compartment causing sodium ions to track through it.

It has been found that many cases of cell failures have been due to the failure of the glass seals. The distortion of the flow of sodium ions through the solid electrolyte brought about by the presence of the glass seal can lead to such a concentration of sodium ions at the surface of the solid electrolyte adjacent to the liquid sodium compartment that failure of the solid electrolyte occurs. For this reason, it is important that the spacer means **13** should also extend beyond the extremities of the glass sealing means in the liquid sodium compartment.

External Storage for Liquid Electrolyte

F.A. Ludwig and S.A. Weiner; U.S. Patent 4,038,465; July 26, 1977; assigned to Ford Motor Company describe a secondary battery or cell which is effective in promoting distribution of reactants during charge and discharge and which provides for an increased supply of reactants, thereby increasing the ampere-hour capacity of the device.

The process accomplishes the above by including as part of the secondary battery or cell (a) one or more storage zones for the liquid electrolyte cathodic reactant which zones are separate from one or more cathodic reaction zones of the device and (b) means for flowing the liquid electrolyte from at least one of the storage zones, through the one or more cathodic reaction zones and out of the cathodic reaction zones to at least one of the storage zones. More particularly, the improvement comprises the use of such storage zones and means for flowing in combination with several preferred battery or cell designs.

Figure 4.2a shows a vertical cross section schematic view of a cell illustrating the process. The cells generally indicated at **2** comprise:

(1) an anodic reaction zone which is the region internal to tubular cation-permeable barrier **4** and which contains a molten alkali metal reactant-anode **6** which is in electrical contact via lead **8** with an external circuit;

(2) a reservoir **10** containing molten alkali metal which is supplied to the anodic reaction zone internal to cation-permeable barrier **4**;

(3) a cathodic reaction zone located between cation-permeable barrier **4** and tubular cell container **12** and filled with liquid electrolyte **14**; and

(4) one or more storage zones **16**, for the liquid electrolyte adapted to supply and/or receive the electrolyte to or from the cathodic reaction zone through conduits **18, 18′** or **18″**.

Referring to the cell of Figure 4.2a, there is shown a means for flowing the liquid electrolyte **14** from storage zone **16** through the cathodic reaction zones and back to storage zone **16** comprising conduits **18′** and **18″** connecting storage zones or container **16** with the cathodic reaction zone such that liquid electrolyte **14** is circulated from storage zone **16** through one of conduits **18′** or **18″** to the cathodic reaction zone and back to storage zone **16** through the other of the conduits **18′** or **18″** by free convective flow. The term "free convective flow" as used herein refers to movement of a fluid in a gravitational field as a result of differing densities therein.

Whenever there are two different phases present in liquid electrolyte **14** during the charge/discharge cycle of the cell, i.e., sulfur and polysulfide salts of the alkali metal anodic reactant **6**, free convective flow will normally occur since the polysulfide salts are of greater density than elemental sulfur.

FIGURE 4.2: SODIUM SULFUR BATTERY WITH EXTERNAL STORAGE

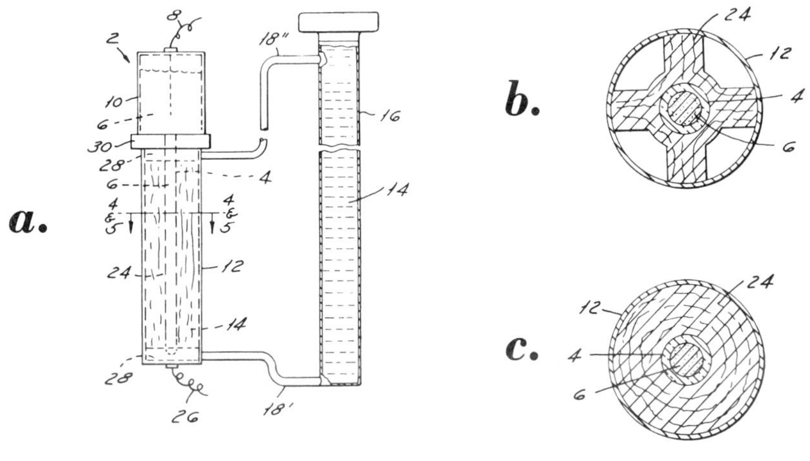

(a) Schematic view of cell.
(b)(c) Cross sections of the device of Figure 4.2a taken along line 4&5–4&5.

Source: U.S. Patent 4,038,465

Such a device may be adapted to create free corrective flow when there is none normally occurring or to enhance any normally occurring free convective flow by heating or cooling storage zone **16**, which may be heated to a temperature above that of the cathodic reaction zone during the discharge cycle of the device and cooled to a temperature below that of the cathodic reaction zone during the charge cycle of the device. It will be noted that during the two-phase portion of the charge/discharge cycle of the cell, such heating and cooling will merely enhance the normally occurring free flow thus aiding in the circulation of reactants.

During the one-phase portion of the charge/discharge cycle of the cell, such heating and cooling induces free convective flow by creating different densities. By heating during discharge, sulfur present in storage zone **16** of the fully charged cell is made less dense than sulfur in the cathodic reaction zone and rises passing through conduit **18″** to the cathodic reaction zone while denser sulfur in the cathodic reaction zone falls and passes through conduit **18′** to storage zone **16**. Of course, as greater and greater amounts of polysulfide salts are formed, the normal differences in density between the salts and the remaining elemental sulfur result in free convective flow which is enhanced by the heating of storage zone **16**.

Finally the discharge proceeds to the one-phase region in which all elemental sulfur has been discharged and only polysulfide salts remain to be discharged. The free convection of the polysulfide salts is maintained by the temperature difference between the cathodic reaction zone **14** and the storage zone **16**. During charge the opposite flow would occur if storage zone **16** is cooled. By cooling during charge, polysulfide salt in storage chamber **16** is made more dense than polysulfide salt in the cathodic reaction and falls passing through conduit **18'** to the cathodic reaction zone, while less dense polysulfide in the cathodic reaction zone rises and passes through conduit **18''** to storage zone **16**. As greater and greater amounts of elemental sulfur form, differences in density of the two phases result in free convective flow which is enhanced by cooling of the storage zone.

A preferred manner for inducing free convective flow when the liquid electrolyte is single phase in nature during either the charge or discharge cycles of the cell is by constantly cooling storage zone **16**. Cooling of course, takes less energy than heating because the storage zone can be designed to radiate, convect and conduct heat to the ambient environment, e.g., using cooling fins. It will be understood that cooling of the battery or cell cannot effectively be achieved by use of cooling means in direct proximity with the reaction zones because of the detrimental effects of uneven cooling therein.

It is an advantage of this process that uniform and effective cooling of the reaction zones can be achieved. Also, since it is contemplated that cells or batteries used at least in electric utility load levelling operations will have to be cooled anyway, the use of cooling to induce free convective flow is ideal.

Various types of cells varying in the configuration and nature of the cathodic and anodic reaction zones may be employed with the various storage and flowing means within the purview of the process.

One example of many such cells which may be employed is shown in Figure 4.2a which has been discussed in some detail above, particularly with respect to the storage zones and flowing means. The cell shown has a preferred tubular configuration with the cation-permeable barrier **4** being a tube concentrically disposed within tubular container **12**. The container, as is known to those in this art, may be conductive or nonconductive, but in any event must be a material which will withstand prolonged exposure to molten alkali metal polysulfide. The cathodic reaction zone contains a porous conductive material **24** which serves as the electrode which may be graphite felt or some other material which will withstand exposure to molten alkali metal polysulfide.

The material **24** is in electrical contact with both the cation-permeable barrier **4** and with the external circuit via lead **26**. The porous conductive material **24** within the cathodic reaction may have several configurations. Figures 4.2b and 4.2c, are horizontal cross sections of the cell of Figure 4.2a taken along line 4&5–4&5. Figure 4.2c shows one configuration wherein porous conductive material **24** fills substantially the entire cathodic reaction zone, i.e., the entire space between the container **12** and the cation-permeable barrier **4** which encloses anodic reactant **6**.

Of course, the porous conductive material may fill only a portion of the cathodic reaction zone. Figure 4.2b shows the cross section of such a cell wherein the material is disposed so as to have arms which are normal to the tubular axis of the cell. Such devices comprise a cathodic reaction zone in which porous conductive material **24** is disposed such that there are a plurality of channels and/or spaces within the zone which are free of the material and which in combination with the material are adapted to allow flow within the cathodic reaction zone of the liquid electrolyte **6**. Still another example of this type of cell configuration would comprise a tubular cell wherein the cathodic reaction zone is filled with porous conductive material except for a plurality of vertical channels, the axes of which are substantially parallel with the vertical axis of the cell.

It should be noted that cells shown in Figure 4.2a include manifold regions **28** into which conduits **18′** and **18″** enter and into which porous material **24** does not extend. The upper manifold **28** is located just below cover **30** which is sealed to the cation-permeable barrier. Generally, in cells of the design of Figure 4.2a, but without storage zones, i.e., a self-contained unit, the distance between the container **12** and cation-permeable barrier **4** would be greater than 1 cm. Although cells employing such spacing may be employed within the purview of the improvement of the process, it is not necessary. Since the liquid electrolyte is supplied from storage zone **16**, the spacing may be less than about 0.5 cm and is preferably less than about 0.2 cm.

Extended Surface Area Collectors

According to a process described by *L.S. Evans and J.R. Harbar; U.S. Patent 3,932,195; January 13, 1976; assigned to the UK Secretary of State for Industry, England* in a sodium-sulfur battery cell, an extended surface area current collector is sandwiched between portions of solid electrolyte, also of extended surface area, which provide between them a space for the liquid sulfur. The spacing between the current collector and each of the portions does not exceed the maximum distance that the sodium ions can diffuse in the liquid sulfur during a specified recharge time. A plurality of portions defining a plurality of spaces for the liquid sulfur may be provided, each space having an extended surface area current collector sandwiched between the portions defining the space.

A plurality of these battery cells may be arranged so that they are separated one from another by a bipolar electrode to provide a series connection.

Referring now to Figure 4.3a, the electric cell shown is of circular form in plan and comprises a low-alloy-steel casing **2**, having within it liquid sodium **3** as the anode, liquid sulfur impregnated in a graphite felt **4** as the cathode, and a solid electrolyte **5** of β-alumina having hollow cylindrical portions **6** and **7** disposed in contraposition so as to provide between them an annular space **8** for the liquid sulfur and graphite felt. Spaces for the liquid sodium lie both internally and externally of the space **8** being an outer annular space **9** between the casing **2** and portion **6** and an inner cylindrical space **10**. A current collector **12** of graphite is provided, and has a lower hollow cylindrical part sandwiched between the portions **6** and **7** in the space **8** for the liquid sulfur so as to divide the space **8** into two subspaces **13** and **14**.

In further detail, the radial dimension of the annular space **9** is such as to constrain, by capillary action, the liquid sodium **3** within it to "wet" the outside surface of the portion **6** as the level of the liquid sodium falls in operation of the cell. A wick structure **15** of low-alloy-steel mesh constrains the liquid sodium **3** in the cylindrical space **10** to "wet" the adjacent surface of the portion **7** in a similar manner. Alternatively, a wick structure can be introduced in the annular space **9** to constrain the liquid sodium **3** to "wet" the outside surface of the portion **6** if a large annular space **9** is needed to act as a reservoir for liquid sodium **3**. The graphite felt **4** also acts as a wick to constrain the liquid sulfur to "wet" the surface of the portion **6**.

The upper part of the current collector **12** has a perforated radially disposed section **17** to maintain the graphite felt **4** in position, and to provide for expansion into a space **18** communicating with the liquid sulfur space **8** when polysulfides are produced by the combination of the sodium and sulfur ions. The current collector may generally but not exclusively be of perforated form, or of a suitable mesh construction, to allow the equalization of any pressures generated in the subspaces **13** and **14** by the volume changes due to the formation of the polysulfides while retaining the graphite felt **4** in position.

The liquid sulfur is retained in the solid electrolyte **5** by an α-alumina disc **19** joined to the portion **6** by a glass frit seal **20** fitted between the outside edge of the disc **19** and the inside surface of portion **6**. The solid electrolyte and the liquid sodium **3** are enclosed within the casing **2** by a low-alloy-steel end cap **21** welded to the side wall at the open end of the casing. The positive terminal is provided by a flanged rod **23** which extends through a hole **24** in the α-alumina disc **19** to connect with the current collector, and is retained by a compression collar **26** using an insulating collar **27** and sealing gaskets **16**. (The collar and insulating collar are shown spaced apart for clarity.) The negative terminal **28** is attached at the closed end of the casing.

FIGURE 4.3: SODIUM-SULFUR BATTERY

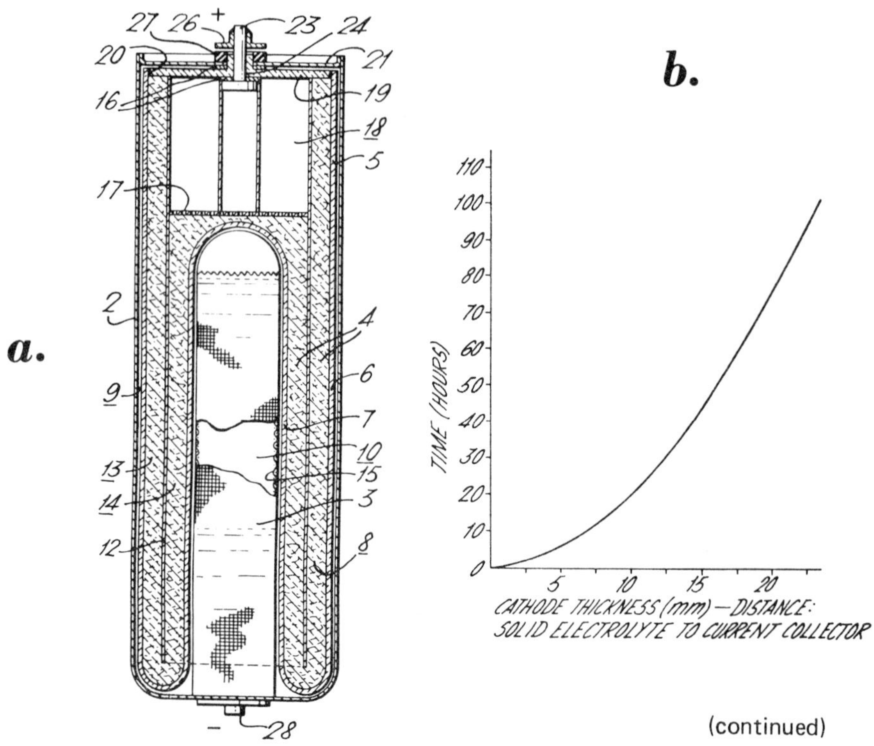

(continued)

FIGURE 4.3: (continued)

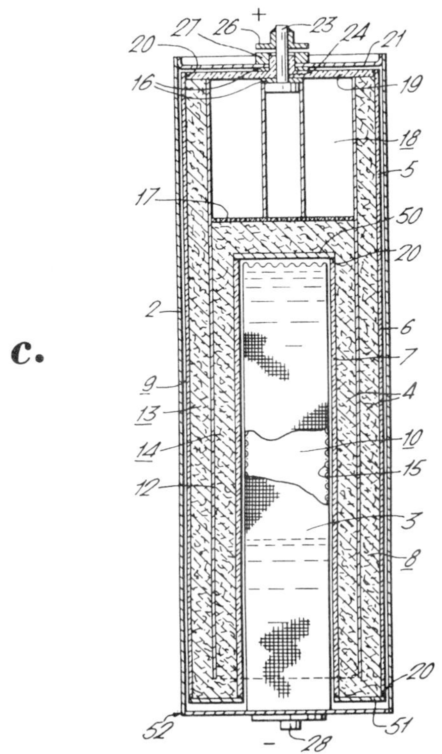

(a) Axial sectional view of a cell having a circular annular solid electrolyte configuration.

(b) Graph of the distance between the current collecting means and the solid electrolyte plotted against recharge time.

(c) Sectional view of a cell similar to that shown in Figure 4.3a but having an alternative solid electrolyte configuration.

Source: U.S. Patent 3,932,195

For maximum volume utilization of the liquid cathode and to enable fast recharge times to be achieved, the radial dimension of each of the subspaces **13** and **14** should not exceed the maximum distance that the liquid anode ions can diffuse through the liquid cathode during a specified recharge time, since the rate of the diffusion decreases with increasing distance.

Theoretically, the time taken for the liquid sodium ions to diffuse through the liquid sulfur during the charging of the electric cell is not directly proportional to cathode thickness (distance between the current collector and the solid electrolyte) but increases as a function of the square of the cathode distance as can be seen from Figure 4.3b to which reference is now made, which represents the theoretical plot of cathode thickness in millimeters against recharge time in hours based on diffusion calculations. The process, therefore, enables the cathode thickness to be kept to a minimum while providing an extended surface of solid electrolyte to maximize the current density. In practice, depending on the cell configuration other factors such as the effects of cathode convection and wicking provided by the graphite felt **4**, may for example, reduce the theoretical recharge time. Cathode thicknesses of up to 10 mm have been found satisfactory to provide convenient recharge time; for example, in one application a radial cathode thickness of 6 mm provided a convenient recharge time.

Furthermore, the extended surface area of that part of the current collector **12** sandwiched between the portions **6** and **7** in the space **8** should be such that it is at least equal to a major proportion of the area of each of the surfaces of the portions **6** and **7** that it faces, and preferably, matches the area of these surfaces for the most efficient operation of the cell.

In operation, when the electric cell discharges through an external circuit, the liquid sodium **3** in contact with the solid electrolyte **5** is ionized with the release of electrons and forms the corresponding positive sodium ions. The electrons leave the cell through the negative terminal **28** to the external circuit, while the sodium ions are conducted through the solid electrolyte to the liquid sulfur. The electrons from the external circuit are eventually conducted by the current collector **12**, and through the graphite felt **4**, to the liquid sulfur, forming negative sulfur ions which combine with the sodium ions to form polysulfides.

To recharge the electric cell, a current is provided to feed electrons through the negative terminal **28** to the liquid sodium, the other lead of the charging circuit being connected to the rod **23** to effect a reversal of the potential difference across the cell. The polysulfides dissociate and the sodium ions released flow through the solid electrolyte **5** to the spaces **9** and **10** containing liquid sodium.

The solid electrolyte shown in Figure 4.3a is of one-piece construction, but to simplify the fabrication of the solid electrolyte, it may be made initially in separate pieces and assembled together using glass frit seals **20** as shown in Figure 4.3c.

Referring to Figure 4.3c, the solid electrolyte has initially been made as hollow cylindrical pieces **6** and **7**, and joined together by glass frit seals **20** in the desired configuration using a flat circular piece **50** and a flat annular piece **51** of β-alumina to close the end of the inner cylindrical space **10** and annular space **8**, and to provide the required radial spacing between the hollow cylinders **6** and **7**. The closed end of the cell casing **2** has been provided with square corners **52** to accommodate the square corners of the solid electrolyte. If the loss of surface area of solid electrolyte can be tolerated, pieces **50** and **51** may be made from a nonconducting ceramic, for example, from α-alumina.

Elongated Holes in Separator

A process described by *T. Kogiso and H. Hayashi; U.S. Patent 3,915,741; Oct. 28, 1975; assigned to KK Toyota Chuo Kenkyusho, Japan* provides a sodium-sulfur cell having a sodium reservoir containing most of the molten sodium, with a special separator being provided with a plurality of elongated small holes extending in side-by-side relation in the separator in parallel relationship with the surfaces of the separator. A small connecting passage is provided which communicates the molten sodium from the reservoir into the small holes of the separator. The reservoir is made of metal such as stainless steel and contains the remaining molten sodium. The connecting means has a relatively long through passage with a small cross-sectional area, one of the ends of the passage opening in the reservoir and the other being connected to the small holes of the separator to transfer the molten sodium from the reservoir to the separator or vice versa. Molten sulfur and an electric current collector are immersed in molten sulfur in a separate casing.

The sodium-sulfur cell is intended to operate at about 300°C. The wall of the separator, i.e., the portion between the inner surfaces of the elongated small holes and the outer surfaces of the separator borders on a solid electrolyte. Sodium contained in the elongated small holes passes through the wall of the separator as sodium ions to the side containing molten sulfur. Sodium in an amount equal to that consumed is supplied from the reservoir through the connecting means so that a constant amount of sodium may be maintained in the elongated small holes of the separator. The sodium passing through the wall reacts with the sulfur to produce sodium polysulfide which remains in the molten sulfur and is maintained there.

The electric current which is produced in the cell is taken out through a positive lead connected to the electric current collector and through a negative lead connected to the reservoir. Sodium has a high electrical conductivity, so the current is transferred through the molten sodium itself from the elongated small holes to the reservoir.

The sodium appearing in the elongated, small holes forms, in effect, a thin sodium layer just inside the surfaces of the separator. The outer surfaces of the separator function as the effective surface of the solid electrolyte. This construction of the separator requires a smaller amount of sodium to form a given amount of surface area of the solid electrolyte compared with the conventional tube-shaped separator. This reduction of the amount of sodium necessary in the separator improves the safety and performance of the cell. The reduction of the amount of sodium in the separator reduces the amount of energy produced by the reaction of sodium with sulfur which could result from the breakage of the separator.

Since the molten sodium is contained separately in a plurality of elongated, small holes, most of the molten sodium will not effuse from the separator at the same time, even if the separator is broken due to impact. The sodium contained in the separator is usually under a reduced pressure or a substantial vacuum so that molten sulfur would follow into the separator in such an event. However, the molten sulfur which appears in the elongated, small holes, reacts with molten sodium remaining therein to produce sodium polysulfides which have high melting points, while the heat produced thereby may be immediately removed through the inner wall surface of the separator.

Consequently, the sodium polysulfides will be solidified within the elongated small holes to thereby block the holes and preclude contact of molten sodium from additional amounts of molten sulfur while preventing reverse flow of the molten sulfur into the sodium reservoir. Thus, the reservoir which contains a greater part of the molten sodium may be maintained safely. The reduction of sodium in the separator also improves the performance of the cell. If all of the sodium in the reservoir were consumed and the remaining sodium in the separator begins to be used, the output current of the cell begins to decrease sharply. The reduction of the sodium in the separator contributes directly to the decrease of the weight in the cell without lowering the functions of the cell.

The area of solid electrolytic partition walls of the separator which contacts the molten sodium can be equal to that of a solid electrolytic partition used in conventional cylindrical separators. With conventional sodium-sulfur cells, the strength of the separator depends on the distance between the wall surface which contacts molten sodium and the surface which contacts molten sulfur, i.e., the wall thickness of the separator such that the thickness of the partition wall is inherently limited from strength requirements. In contrast, the strength of the separator depends on the thickness of the separator and is partially affected by the relative positions of the elongated small holes in the cross section of the separator and by the ratio of the holes occupied to the cross-sectional area of the separator.

The distance between the inner surface of the elongated small holes and the outer wall surface of the separator which contacts the molten sulfur may be reduced to a considerable extent without lowering the strength of the separator greatly.

Multiplicity of Parallel-Connected Electrode Spaces

A process described by *F.-J. Rohr; U.S. Patent 4,038,462; July 26, 1977; assigned to BBC Brown Boveri & Company Limited, Germany* involves making the solid-electrolyte battery in the form of a multiplicity of parallel-connected, tubular anode and cathode spaces bounded by the solid electrolyte and alternatingly arranged in close proximity to one another so that each electrode space presents reaction surfaces simultaneously to at least two neighboring electrode spaces of the opposite polarity.

By splitting up the anode and cathode space into a multiplicity of parallel-connected electrode spaces, their reaction areas are considerably increased. Since here the electrode spaces are completely surrounded by solid electrolyte, no problems about additional materials arise. Each of the electrode spaces has at least two neighboring electrode spaces of opposite polarity to its own, and the specific output is therefore additionally increased. The electrode spaces are packed close together and their inner cross section is made as small as possible compared with their surface, but not so small as to interfere with the supply of reactants or removal of reaction products or with current input or output.

A cross-sectional area of the electrode spaces of preferably 10 mm^2 and a length of these spaces of about 100 mm should not be exceeded in sodium-sulfur batteries, for example. Since all electrode spaces are surrounded by solid electrolyte, no special external sealing of the battery is usually necessary.

The main advantage of the solid-electrolyte battery configuration is that the reaction area is increased, thereby reducing the current loading of this area. The reaction area is increased here preferably by a factor of 3 to 10, but larger values can also be attained. The current loading of the reaction area is thus reduced by the same factor. If the original value for the current loading is used, then the specific output of the battery is increased by the corresponding factor. This is particularly significant if high outputs, for example, over a short period of time, are to be obtained from relatively small batteries.

All or nearly all single-electrolyte solid-electrolyte batteries can be made in the manner of the process. This form is especially advantageous, however, for batteries based on alkali metal and sulfur which are intended for mobile use, e.g., in electric-drive vehicles.

Figures 4.4a and 4.4b show a solid-electrolyte battery in the form of a fuel cell. It consists of the block-shaped solid electrolyte **7** in which the anode **3** and cathode spaces **5** are circular cylinders. The electrode spaces **3,5** are uniformly distributed over the solid electrolyte and as close together as possible, so that each of the electrode spaces presents large reaction areas **26** to as many adjacent electrode spaces of the opposite polarity as possible. The electrode spaces **3,5** penetrate the solid electrolyte and at their unblocked ends open into collection spaces **9,39** provided at the ends of the solid-electrolyte block **7**. Since fuel cells usually operate with poor electrically conducting gases, their reaction surfaces are provided with electrodes **27** for collecting the current.

The electrodes here usually consist of porous silver or nickel. The collecting spaces **9,39** are hollowed out in a plate **24** of solid electrolyte which is hermetically sealed to the solid electrolyte **7** with glass or ceramic solder, for example. The open ends of the cathode spaces **5** open directly into the corresponding collecting space **39**, while the open ends of the anode spaces **3** communicate through spacer holes **25** with the corresponding collecting space **9** which is situated above and beneath the other collecting space **39**. Feed or discharge tubes **21** open into the collecting spaces **9,39** for the supply of reactants and the removal of reaction products. The feed tubes **21** serve at the same time for carrying the current out of the fuel cell.

In order to connect the feed tubes **21** electrically with the porous electrodes **27** at the reaction surfaces of the electrode spaces, porous matrices **22**, e.g., of graphite or metal, are provided in the collecting spaces and extend from the vicinity of the feed tubes **21** to the electrodes **27**. The electrodes, of course, could just as well be connected with the feed tubes by metal wires.

For the operation of the battery fuel, e.g., hydrogen, is fed into the anode spaces **3** and oxidant, e.g., air, into the cathode spaces **5** through the feed tubes. The resulting electric current is conducted away by the electrodes and the feed tubes while the reaction products are carried off at the bottom of the battery. The reactions occurring in a fuel cell are generally known, so further discussion of them is unnecessary. In this example, the solid electrolyte is zirconium dioxide.

In the example of Figures 4.4c and 4.4d, where a sodium-sulfur battery is shown, the anode spaces **3** and cathode spaces **5** are made as straight, cylindrical holes in the cylinder-shaped solid electrolyte **7**. As is particularly apparent from Figure 4.4d, these electrode spaces **3,5** alternating in polarity from one to the next, are distributed uniformly and close together in great number over the solid electrode spaces run parallel to one another and to the long axis of the cylindrical solid electrolyte **7**. The cathode spaces **5** and anode spaces **3**, here, are closed at one end **15,13** while the open ends of the electrode spaces of the same type are on the same side, i.e., the anode spaces open on one side and the cathode spaces on the opposite side. The electrode spaces **3,5** open into the equalizing spaces **11,41**. These equalizing spaces are each formed with a cup-like cover **19** sealed at its rim to the solid electrolyte. A glass or ceramic powder is used as the sealant. The outer diameters of the solid electrolyte **7** and the cover **19** are equal. The material for the cover can be aluminum, stainless steel or ceramic, e.g., α-aluminum oxiide (α-A_2O_3), while the solid electrolyte consist of β-aluminum oxide (β-Al_2O_3).

Instead of hollowing out the electrode spaces in the solid electrolyte, it is also possible to assemble the battery from separate anode and cathode spaces. Here the electrode spaces are each surrounded by solid electrolyte and formed as separate parts which fit together with large contact surfaces.

The electrical terminals **8** are mounted in the centers of the covers **19**. They each consist of a feed pipe **21** leading from outside into the equalizing spaces **11,41**. This tube can be of aluminum or stainless steel, but, for reasons of cost, aluminum is preferred just as for the cover **19**.

FIGURE 4.4: SOLID-ELECTROLYTE BATTERY

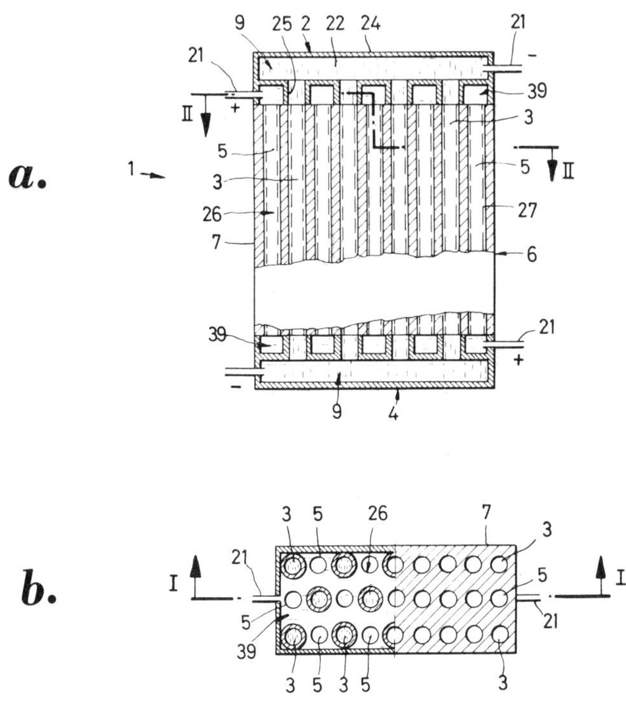

(continued)

FIGURE 4.4: (continued)

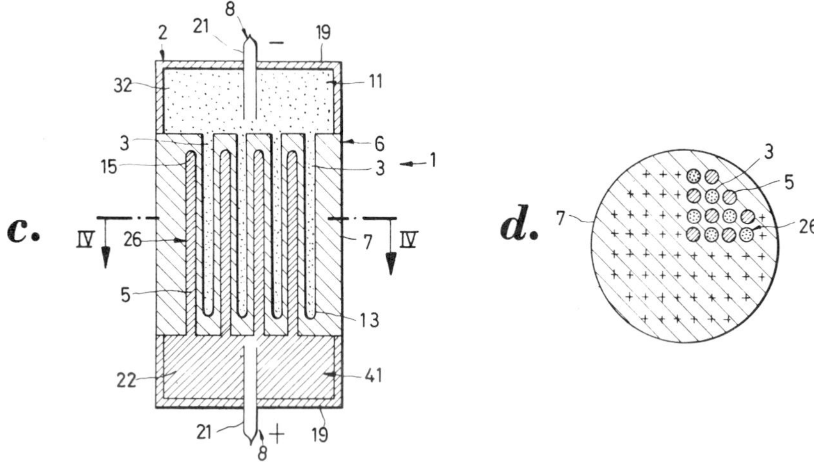

(a) Longitudinal section through a solid-electrolyte battery for the generation of electrical energy (fuel cell) along the line I-I of Figure 4.4b.

(b) Plan section through the battery of Figure 4.4a along the line II-II of Figure 4.4a.

(c) Central, vertical section through an electrochemical storage battery with alkali metal and sulfur as reactants (sodium-sulfur battery) with a cylindrical outer shape, the electrode spaces in the form of holes being closed at one end and having equalizing spaces for the reactants at their opposite, open ends.

(d) Plan section through the battery of Figure 4.4c along the line IV-IV, not all electrode spaces being shown.

Source: U.S. Patent 4,038,462

The cathode spaces **5** as well as the associated equalizing space **41** are filled with a porous matrix **22**, e.g., of graphitized felt. With this matrix, the input or output current is conducted from the electrical terminal **8** to the reaction surfaces **26**, since the reactant in the cathode space **5**, e.g., sulfur, is a poor electrical conductor. In general this is not necessary for the anode spaces **3** and the associated equalizing space **11**, since the sodium filling them is a good electrical conductor. If this were not the case, the anode space would also have to be filled with a suitable matrix **32**, such as a metal wool.

The reactants are introduced by way of the feed tubes **21**. Sodium is fed into the anode spaces **3** and the associated equalizing space **11**, while sulfur or sodium polysulfide is fed into the cathode spaces **5** and the associated equalizing space **41**. This is best done by distillation in vacuum. After the reactants have been introduced, the feed tubes **21** are closed off, e.g., by pinching off their outer ends. For normal operation of this battery, a temperature of about 300°C is required. For charging, the two electrical terminals **8** are connected with proper polarity to a dc supply. The reaction taking place in the battery is:

$$5Na_2S_3 \longrightarrow 3Na_2S_5 + 4Na$$

When the battery is charged, a load can be connected to the electrical terminals and current drawn. The following is the reaction taking place:

$$3Na_2S_5 + 4Na \longrightarrow 5Na_2S_3$$

Since sodium is used up during the discharge process, the sodium level drops in the equalizing space **11**. Care must be taken, therefore, that the corresponding electrical terminal always remains in contact with the sodium, i.e., it must extend far enough into the equalizing space, since, otherwise, the discharge process would be interrupted. During the discharge process the proportion of sulfur in the cathode spaces **5** and the associated equalizing space **41** increases. It is necessary, therefore, to match the quantities of reactants introduced to one another. The transport of the sulfur from the equalizing space to the reaction surfaces of the cathode spaces is by capillary action of the cathode spaces and the porous matrix **22**, e.g., graphitized felt, therein.

In this way, the battery is also to be operated in the position shown by Figure 4.4c, which is purely accidental. If, on the other hand, the battery is to be operable in any position, there must likewise be a porous matrix **32** in the equalizing space for the sodium as well as in the anode spaces **3** for the transport of the sodium. This matrix can consist of metal wool for example. By making the equalizing spaces **11,41** large, a large quantity of reactants can be stored and thus the operating time arbitrarily increased.

Modular Battery

A process described by *G. Desplanches, Y. Lazenned and A. Wicker; U.S. Patent 4,041,216; August 9, 1977; assigned to Compagnie Générale d'Électricité SA, France* provides a battery formed by several sodium-sulfur generating elements, such a battery being capable of providing the propulsion of an electric vehicle, while having a simple structure as well as minimum bulk and weight. The battery formed by several cells or electrochemical generating elements of the sulfur-sodium type comprises:

(a) A cathode tank containing a cathode reagent which is liquid at operating temperature and is chosen from the group formed by sulfur, phosphorus, selenium and alkaline salts of those elements.

(b) At least one solid electrolyte closed at its lower end, containing an anode reagent which is liquid at operating temperature and is constituted by an alkaline metal more particularly sodium and arranged in the cathode tank so as to be immersed in the cathode reagent, the wall of the tube being constituted more particularly by alkaline β-alumina.

(c) A ceramic insulating support for keeping the electrolyte tube in the cathode tank, the connection between that support and that tube being provided by means of a glass part.

(d) An anode tank containing a supply of the anode reagent and arranged above the cathode tank, so that the electrolyte tube opens at its upper part into that anode tank, the plate separating the open ends of the anode tank and of the cathode tank.

The battery comprises m groups of n modules each comprising p elements, characterized in that those elements are electrically interconnected in parallel in each of the modules by means of plates connected to the bottoms of the cathode tanks and anode tanks respectively, each of the plates comprising at least one substantially central bore.

Complete details of the operational structure of the cells and battery unit are provided. The batteries produced according to the process have a minimum weight and bulk for a given electrical power and/or capacity and such batteries can very easily be integrated in electric vehicles.

Partition Means in Sodium Chamber

A process described by *T. Nakabayashi and H. Kagawa; U.S. Patent 4,027,075; May 31, 1977 assigned to Agency of Industrial Science & Technology, Japan* relates to a sodium-sulfur storage battery including sulfur as a cathodic reactant, sodium as an anodic reactant and a nonporous solid electrolyte, the storage battery having a partition means in a chamber of sodium. According to this process, even if a part of the solid electrolyte is broken, direct reaction between both reactants is controlled to a small scale, thus preventing the spread of electrolyte deterioration.

Referring to Figure 4.5a, a sodium-sulfur storage battery, according to this process, is composed of various elements, i.e., a tubular cell vessel **1** which is made of stainless steel containing molten sulfur as cathodic reactant **2**; a graphite electron conductor **3** with an annular cross section; a solid electrolyte **4** made of β-aluminum oxide (β-Al_2O_3) in the shape of a tube **4'** which is provided with an upper partition means B, e.g., a stainless steel cover **21** with a small hole **21'** in it, the diameter of which is 20 to 200 μ; a sodium reservoir **5** of α-aluminum oxide (α-Al_2O_3); a cover **6** (α-Al_2O_3); a sodium injecting pipe **7** (of stainless steel) which also functions as a negative terminal post; an exhaust pipe **8**; an anticorrosive metal cover **9** which is affixed outside the reservoir **5**; a stainless steel positive terminal post **10**; and sodium as an anodic reactant **11** contained in a chamber **11'**.

The electrolyte tube **4'** is closely surrounded by the graphite conductor **3** which contains sulfur as cathodic reactant therein, and is affixed to the reservoir **5** at its upper end, in liquid-tight relationship, by glass solder or the like together with partition **21** in which partition a hole **21'** is provided to connect chamber **11'** with the interior of reservoir **5**. Tube **4'** thus assembled with partition **21** and reservoir **5** is positioned in cell vessel **1**, and after additional molten sulfur has been poured in the cavity formed between tube **4'** and cell vessel **1**, the cover **9** and positive terminal post **10** are welded to the cell vessel **1** in the same liquid-tight relationship, and then glass solder is applied at an annular junction between cover **9** and reservoir **5**.

Other junctions formed between reservoir **5**, cover **6**, sodium injecting pipe **7** and exhaust pipe **8** are also sealed with glass solder or the like. The pipe is closed after a predetermined amount of sodium is injected in the chamber **11'**

FIGURE 4.5: SODIUM-SULFUR STORAGE BATTERY

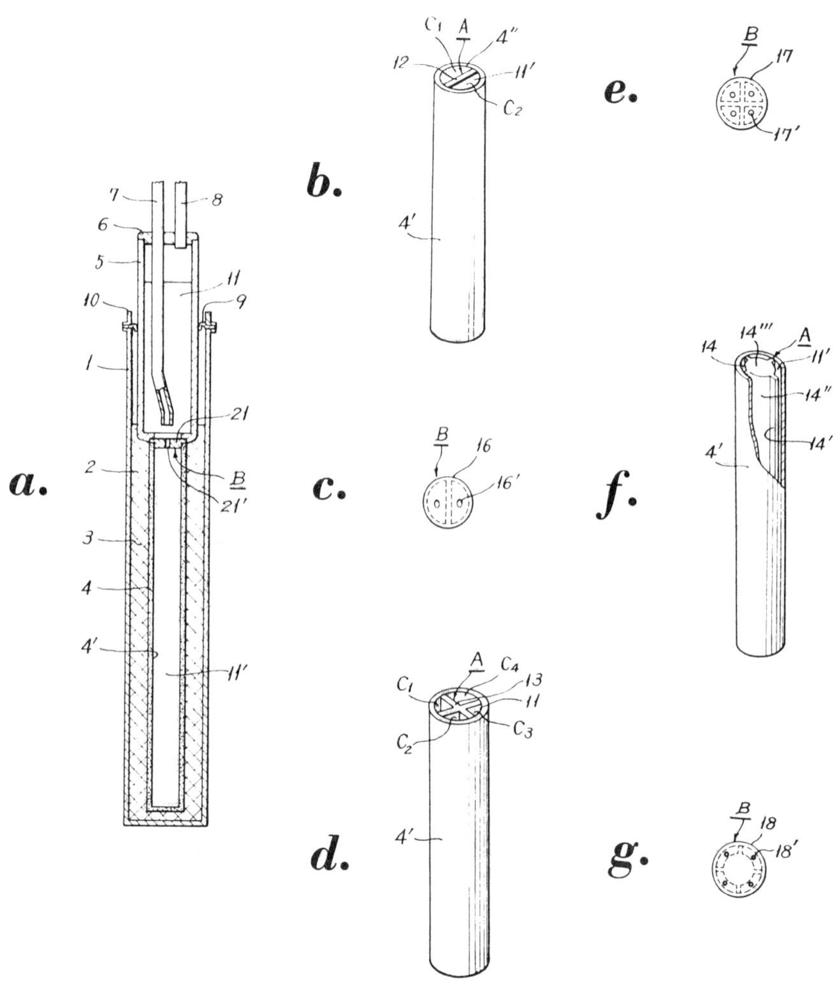

(a) Central sectional and elevational view of an elementary sodium-sulfur storage battery.

(b)(d)(f) Perspective views of modified tubes made of solid electrolyte applicable in place of a tube shown in Figure 4.5a.

(c)(e)(g) Plan views of upper partition means each of which is attached to the corresponding tube as shown in Figures 4.5b, 4.5d and 4.5f.

Source: U.S. Patent 4,027,075

and reservoir **5** so as to function only as a negative terminal post. The lower end of the pipe **7** is always dipped in the reactant **11**. The exhaust pipe **8** which opens inside upper space of the reservoir is closed after the assembling process has been completed.

Partition **21**, an example of the upper separating means **B**, divides molten sodium **11** into two parts, i.e., a lower part contained in the tube **4'** and an upper part in the reservoir, both parts communicating by means of small hole **21'** which prevents rapid fall of sodium during operation. Therefore, the hole is adapted to pass molten sodium under the influence of a slight pressure, and to prevent leak or infiltration if the pressure is eliminated.

During discharging process of the battery, sodium **11** in the chamber **11'** infiltrates and passes through electrolyte tube **4'**, and decreases in the amount as the reaction between both reactants proceeds thus forming a cavity in the upper part of the tube. On account of a vacuum developed in the upper cavity, a corresponding amount of sodium in the reservoir is supplied to the cavity through the hole **21'** to fill the chamber **11'** with sodium. On the other hand, during charging process, sodium in charge of the preceding reaction returns to the chamber and raises the inside pressure to send the excessive amount of sodium back to the reservoir.

During continued operation, when electrolyte tube **4'** has become partially broken down due mainly to the aforementioned fine cracks or pinholes, anodic reactant **11** and cathodic reactant **2** react directly at the broken part developing a large amount of heat and pressure until anodic reactant **11** (sodium) is used up for explosive reaction. The great amounts of heat and pressure thus developed fatally damage not only the battery in question but nearby normal batteries.

According to the process, anodic reactant **11** (sodium) contained in the reservoir never joins or participates in the accidental reaction because of the partition **21** as an upper partition means which prevents upper sodium from pouring in the chamber **11'** on account of high positive pressure in the latter developed by the direct reaction. Therefore, the amount of sodium which attends to the direct reaction is limited to that contained in the chamber thus restricting the accident to a smaller scale.

Electrolyte tube **4'** illustrated in Figure 4.5a may be replaced with the counterpart shown in Figure 4.5b. The tube **4** in Figure 4.5b, is provided with a longitudinal partition **12** as one example of inner partition means **A** which snugly fits on the inner surface of the tube and is sealed by glass solder to completely divide the chamber **11'** into two chambers C_1 and C_2. Affixed on upper openings **4''** of the chambers is a partition **16** (Figure 4.5c) as another example of upper partition means **B** which is provided with two small holes **16'** providing communication between chambers C_1 and C_2 and the reservoir (Figure 4.5a), in the same manner as explained in connection with hole **21'** (Figure 4.5a).

The partition **16'** is also provided with annular and lateral grooves on the lower surface as shown in dotted line on Figure 4.5c, in which the upper end portions of corresponding tube **4'** and partition **12** (Figure 4.5b) are received with glass solder between.

In an experiment, an electrolyte tube in the shape shown in Figure 4.5b, which was confirmed by an optical microscope to carry a fine crack on its side wall, was used together with the partitions **12, 16** to build up a sample battery. The battery operated in the normal manner during the first test cycle consisting of a charging and a discharging process. But in the second cycle of the test including an additional thermal cycle, fatal short circuiting occurred when the temperature around the battery reached about 290°C.

After the damaged battery was cooled, it was overhauled for a thorough checkup. As a result, it was found that the electrolyte tube **4'** was broken at the upper part where the original crack was located and direct reaction between cathodic reactant **2** and anodic reactant **11** was detected due to the fact that nothing was found in the sodium chamber $\mathbf{C_1}$ facing the crack except for a small amount of black reaction products remaining therein. Besides, a hollow part was formed on the vertical surface of the partition **12** opposing the crack, and the vertical surface was burnt and scorched.

On the contrary, sodium in the adjacent longitudinal chamber $\mathbf{C_2}$ and upper reservoir **5** did not participate in the direct reaction, thus remaining as it was. Thirty batteries with electrolyte tubes as shown in Figure 4.5b, were put to the test which includes repeated charging-discharging cycles and additional cycles of heating up to about 350°C, then cooling down to normal (room) temperature at the predetermined rate of 100°C per hour. The result is shown in the following table.

Batch	Numbers	Broken part electrolyte	on tube	Partition's condition	Sodium in broken chamber
Broken in		upper	9	normal	a little remained in two tubes
0 – 5	14	middle	4	partially abnormal	nothing remained
cycles		lower	1	partially abnormal	nothing remained
Broken in		upper	4	normal	nothing remained
6 – 15	6	middle	1	normal	nothing remained
cycles		lower	1	partially abnormal	nothing remained
Normal after 16 cycles	10				

As will be noted from the table, only a divided portion of sodium participated in direct reaction with the cathodic reactant even if a part of the electrolyte tube breaks under thermal stresses, in that internal partition means A, such as longitudinal partition **12** (Figure 4.5b) and/or upper partition means B such as partition **16** (Figure 4.5c) are employed. Therefore, accidental damages are confined to a local limited scale and its spreading is effectively prevented.

Example 1: Referring to Figure 4.5d, there is shown a longitudinal partition **13** with a cross section as another example of internal partition means A, the partition being adapted to devide the electrolyte tube **4'** into four laterally adjacent chambers C_1, C_2, C_3 and C_4. These chambers are closed by a partition **17** (Figure 4.5e) provided with four holes **17'** each communicating with the corresponding chamber. According to this construction, since each chamber has one-half volume compared with the counterpart shown in Figure 4.5b, the direct reaction decreases about one-half in scale. A test revealed that there was only a slight change in color on the inner surface of the broken chamber.

Example 2: In Figure 4.5f, electrolyte tube **4'** is divided into four longitudinal chambers by an internal partition means A such as a tubular partition **14** made of stainless steel and comprising a hollow tubular body **14''**, four longitudinal flanges **14'** on the outer surface of the body **14''** and two end plates **14'''** all in one body. The partition is affixed to the tube's inside surface with glass solder or the like. A stainless steel partition **18** (Figure 4.5g) with four small holes **18'** for sodium to pass is affixed and soldered at the upper end together with the partition. Thus the sodium content in each chamber decreases in volume because sodium is maintained in the shallow longitudinal space with an arcuate cross section.

A battery comprising the tube (Figure 4.5f) was overhauled after the same battery test. The surface of the partition was found almost normal even at the part which opposed a broken part of electrolyte tube **4'**; there was no damage in the adjacent chambers where sodium was left in the normal condition. Many small holes may be provided on the tubular body **14'** to keep supplementary sodium in it. Then end plates **14'''** are not necessary.

Capillary Flow

According to a process described by *G. Robinson and I.W. Jones; U.S. Patent 3,922,176; November 25, 1975; assigned to The Electricity Council, England* in an electrochemical cell having a liquid metal electrode and a solid electrolyte, capillary means are provided against the face of the electrolyte arranged to draw the alkali metal upwardly to lie over the surface of the electrolyte despite changes in the level of the alkali metal.

The capillary means may be constituted by a narrow spacing between the electrolyte and a housing or between the electrolyte and another element, e.g., a metal conductor for effecting electrical connection to the alkali metal or by porous or fibrous or other material forming a wick.

The cell may have an electrode chamber sealed against one face of the electrolyte and containing the capillary means. The chamber may initially be filled or substantially filled with the alkali metal. However, if the level of the alkali metal in the chamber should fall due to passage of ions through the electrolyte, the capillary means against this face of the electrolyte will maintain liquid alkali metal in contact with the whole region covered by the capillary means despite any fall in the level of the alkali metal in the chamber.

It will be noted that this construction may obviate any need to have a reservoir containing the alkali metal. For example, it has commonly been the practice, in sodium sulfur cells of tubular form, to have a sodium reservoir to maintain the liquid level and thereby avoid the effects of a falling level. The presence of the capillary means overcomes this problem. A reservoir need only be provided if the further alkali metal is required for utilization in the electrochemical reaction.

Alternatively, however, the alkali metal may be arranged in a reservoir at one end of the capillary means. In one example, the capillary means are constituted by a narrow annular region between the cylindrical electrode and an outer housing and an alkali metal reservoir is at one end of the electrode. A further advantage of this arrangement arises because the electrolyte surface is kept wet by capillary action which is not dependent on gravity and hence the cell now no longer need be operated in a particular position with the reservoir above the electrolyte, but will operate in any position. Hence, there is no restriction on the orientation of the cell, e.g., cylindrical cells may be operational with their axes horizontal.

The elimination or reduction in size of the reservoir or electrode chamber for the liquid alkali metal moreover leads to a further advantage in that it now becomes conveniently possible to provide a reservoir for the cathodic reactant above the level of the electrolyte thereby enabling an increase in the space available for storage of the cathodic reactant to be obtained. Such an increase in storage space for the cathodic reactant gives an increase in the capacity of the cell.

Heater Unit

A process described by *H. Kagawa; U.S. Patent 4,024,319; May 17, 1977; assigned to Agency of Industrial Science & Technology, Japan* relates to a sodium-sulfur storage battery comprising a sodium reservoir made of metal incorporating a heater between double walls, a solid electrolyte, an anodic reactant and a cathode reactant. The battery housing which contains the abovementioned components is sealed at the upper part thereof. According to this process, a sodium-sulfur storage battery having superior performance, longer service life, and a low price is obtainable.

In Figure 4.6a, **1** is a double-walled sodium reservoir made of metal, double walls of which comprise an inner wall **1'** made of metal such as Ni-Fe alloy having sufficient resistance to chemical and electrochemical attack by molten sodium, and an outer wall **1''** made of stainless steel or the like. The inner and outer walls are electrically insulated from each other. **2** is a nonporous solid electrolyte made of ceramic such as β-alumina, which permits only sodium ions to pass. **3** is an electric conductor of carbon felt which is impregnated with a cathodic reactant **4** such as sulfur or sodium polysulfide, and arranged at the outside of the solid electrolyte **2**.

5 is sodium for the anodic reactant contained in the solid electrolyte, and in the metallic reservoir **1** for sodium metal **6** is a cathodic collector made of stainless steel or molybdenum steel in strip form, plural pieces of which are fixed to, and around, the electric conductor **3** made of carbon felt.

7 is a battery housing made of stainless steel also serving as a cathode. At an outward flange of an upper part thereof, an upper flanged portion of the outer wall **1''** of the sodium reservoir **1** made of metal and an upper bent portion of the cathodic collector **6** are welded to each other. **8** is an insulator disposed in the double walls of the sodium reservoir, a terminal of a heater **10** being led out from a part thereof. **9** is a jointing part of the sodium reservoir and the solid electrolyte **2**, an enlarged view of the jointing part being illustrated in Figure 4.6b.

FIGURE 4.6: SODIUM-SULFUR BATTERY

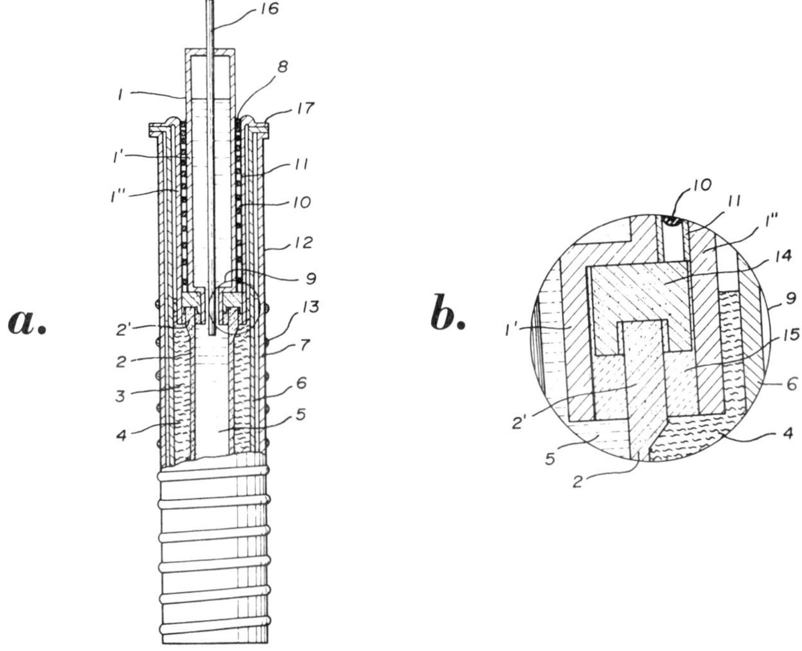

(a) Vertical sectional view of a sodium-sulfur storage battery.
(b) Enlarged view of an essential part of Figure 4.6a.

Source: U.S. Patent 4,024,319

10 is a heater made of nickel-chromium alloy wire, which is incorporated between the inner wall **1′** and the outer wall **1″** of the sodium reservoir **1**, the heater being insulated from both walls by a heat-resisting insulator **11**. The heat-resisting insulator may be either a double-walled mica cylinder incorporating the ribbon-type heater **10**, or a glass tube incorporating a sheathed heater. In the former case, the ribbon-type heater is spirally wound around an inner mica core and is covered by another outer mica shell. In the latter case the sheathed heater inserted into the glass tube is spirally wound around the inner wall **1′** of the sodium reservoir **1**. The heater is connected to an electric source which is not shown. **12** is a glass tube around the battery housing **7**, and **13** is an outer heater spirally wound around the exterior surface of glass tube **12**.

14 is an insulating ring which may be formed of α-alumina, and **15** is a solder glass. They are arranged at the jointing part **9** of the sodium reservoir and the solid electrolyte **2**. At the bottom end of the double-walled sodium reservoir **1**, the inner wall **1′** bends to the center and diminishes in diameter, then extends downward. Thus, a gap between the double walls forms a space sufficient to accommodate a thick wall part **2′** mounted on top of the solid electrolyte **2**. **16** is an anodic terminal, and **17** is a cathodic terminal made of stainless steel.

A feature of the process is the heater **10** incorporated between the double walls of the metallic reservoir for sodium. Another feature relates to the heater, windings of which are relatively close together on the upper part and relatively farther apart on the lower part, with the winding pitch thereof changed gradually and smoothly between the extremes.

An additional feature of this process relates to the solid electrolyte **2**, the upper part of which is formed into a thick wall as shown by the thick wall part **2'** in the figures. According to the process, prominent advantages are obtainable as follows:

(1) Since the heater **10** is provided within the battery, the thermal energy generated from the heater is effectively utilized to melt the cathodic and anodic reactants **4** and **5** without being released outward; thus, temperatures are able to be raised to a working temperature of the battery in a relatively very short time.

(2) Since the anodic and cathodic reactants are heated simultaneously, these reactants are molten, and volume changes due to expansion of the reactants are absorbed into and cushioned by an upper space; thus stresses tending to accumulate on the solid electrolyte, having a relatively small mechanical strength, are removed.

(3) Since the anodic reactant **5** having superior thermal conductivity to the cathodic reactant **4** is molten faster than the cathodic reactant, and since the cathodic reactant **4**, which exists around the solid electrolyte **2** made of β-alumina at the same time is still solidifying or maintained in the solid state, but is heated and molten from inside, troubles caused by collapse of the solid electrolyte **2** as recognized hitherto is almost completely obviated. Further, since the upper part of the solid electrolyte is formed into the thick wall part **2'**, a collapse is still further avoided.

Tubular Construction

A process described by *A.R. Tilley; U.S. Patent 4,024,321; May 17, 1977; assigned to Chloride Silent Power Limited, England* relates to alkali metal-sulfur cells, e.g., sodium-sulfur cells, and more particularly to such cells in which the anode compartment, i.e., the alkali-metal-containing compartment, is defined by the space between an outer tubular member and an inner tubular member constituting the solid electrolyte of the cell, in which the cathode compartment, i.e., the sulfur-polysulfide-containing compartment is defined by the interior of inner tubular member and in which a current-collecting pole extends into the cathode compartment.

Referring to Figure 4.7a, the cell is shown in its operating orientation and has a mild steel outer case 1 with an internal coating 2 of vitreous enamel and an inner coaxial electrolyte tube **3**, both of which are closed at their upper ends. The space **4** between the casing **1** and tube **3** constitutes the anode compartment and contains sodium. The casing **1** extends above the tube to form a sodium reservoir **5** which is separated from the compartment **4** by a sodium-restrictor **6**. The case **1** forms the negative current collecting pole of the cell and has a terminal **7** connected to it.

FIGURE 4.7: TUBULAR ALKALI METAL-SULFUR CELL

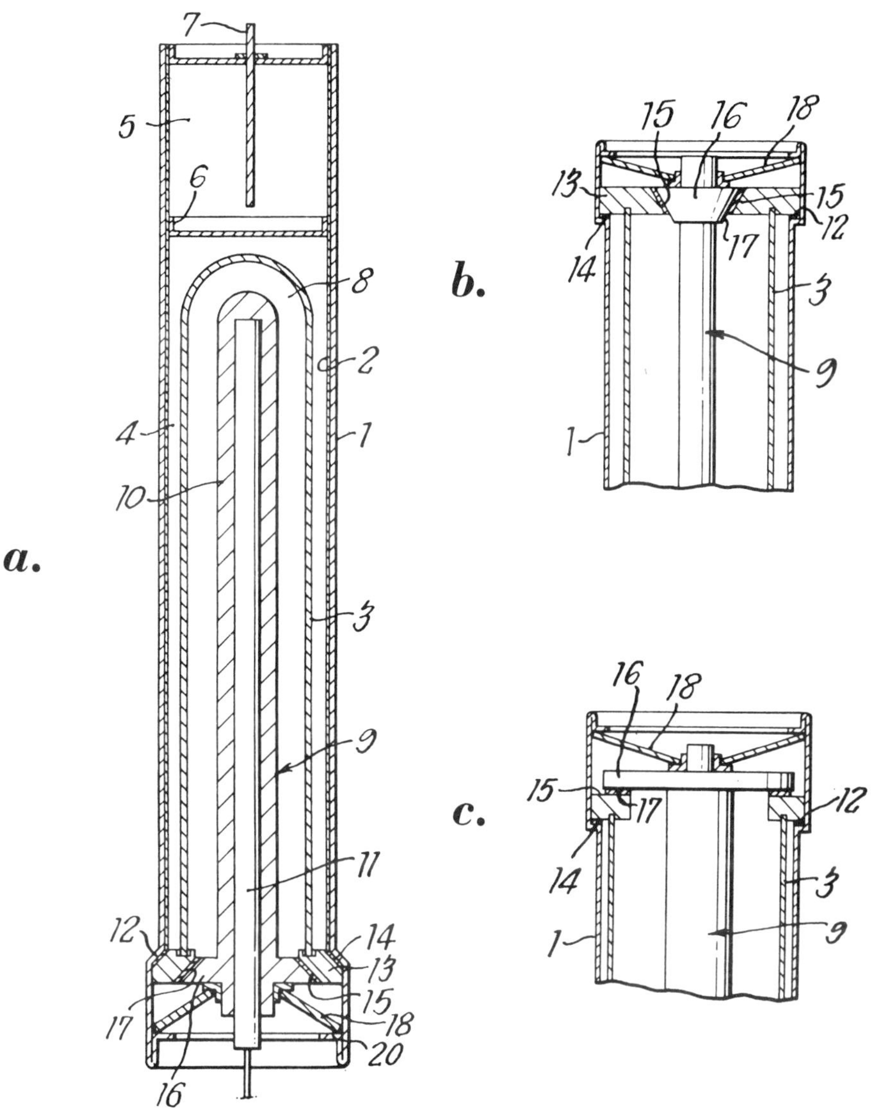

(a)-(c) Sectional elevations.

Source: U.S. Patent 4,024,321

The interior **8** of the electrolyte tube **3** constitutes the cathode compartment of the cell and contains sulfur/polysulfides in proportions dependent upon the state of charge of the cell. Extending axially within the tube **3** is the positive collecting pole **9**. This comprises an outer tubular graphite member **10** and an inner rod **11** of high conductivity metal which is in intimate contact with the internal surface of the member **10**.

The lower open ends of the casing **1** and tube **3** are closed off and sealed by a compression seal arrangement. For this purpose, the outer casing **1**, the tube **3** and current collecting pole **9** are provided with shoulder formations which abut one another. Thus the casing **1** has an internal sloping shoulder **12** which is formed by shaping of the casing **1**.

The tube **3** has an alpha-alumina flange **13** whose outer and inner peripheries have sloping shoulders **14** and **15**, the shoulder **14** abutting the shoulder **12** of the casing **1**. The graphite member **10** of the current collecting pole **9** is formed with a flange **16** whose outer periphery is formed as a sloping shoulder **17** which abuts the shoulder **15** of the alpha-alumina flange **13**. An aluminum gasket is positioned between the shoulders **12** and **14** as sealing means, and a Grafoil gasket is positioned between the shoulders **15** and **17** as sealing means.

The shoulders are urged towards one another by annular disc spring **18** which acts in the axial direction of the tube **3**. The inner periphery engages the graphite member **10** of the current collecting pole **9** through insulating ring **19** and the outer periphery of the spring **18** engages beneath a shoulder **20** formed by turning the lower edge of the case **1** in upon itself. Alternatively a flange could be welded to the case **1** to provide the shoulder **20**.

It will be appreciated that the sloping shoulder formations produce a wedging effect. The construction shown in Figure 4.7b is substantially the same as that of Figure 4.7a except that the abutting shoulder formations **12** and **13** on the casing **1** and the alpha-alumina flange extend at right angles to the axes of the casing **1** and tube **2**. The construction of Figure 4.7c is further modified compared with that of Figure 4.7a in that all the abutting shoulder formations **12**, **14**, **15** and **17** extend at right angles to the axes of the casing **1**, tube **2** and current collecting pole **9**.

In a further modification, the Figure 4.7c construction could have two disc springs **18**, disposed back-to-back, so that they abut each other at their inner peripheries. Then the outer periphery of the lower spring **18** would engage the flange **16** adjacent its outer edge and would thus apply an axial compression force substantially in line with the sealing gaskets.

To summarize, an alkali metal-sulfur cell of tubular form, has an inner tubular member constituting the solid electrolyte, an outer tubular member and a current collecting pole extending into the interior of the inner tubular member which forms the cathode compartment of the cell. The outer tubular member, the inner tubular member and the current collecting pole have shoulder formations which abut one another through sealing means and are urged towards one another by axially acting spring means.

Alumina Tubular Unit

According to a process described by *B.A. Partridge, T.R. Jenkins and M. McGuire; U.S. Patent 3,982,959; September 28, 1976; assigned to U.K. Secretary of State for Industry, England* in a sodium-sulfur battery cell of tubular form, the sulfur compartment is provided inside a tubular solid electrolyte. The sodium compartment is provided by the annular space between the solid electrolyte and a metal casing. A carbon current collector extends into the sulfur compartment. A stainless steel mesh disposed in the sodium compartment adjacent to the solid electrolyte acts as a wick for sodium.

Referring to Figure **4.8**, a sodium-sulfur battery cell **1** comprises a metal casing **2**, a solid electrolyte **3** of beta-alumina ceramic and of tubular form disposed lengthwise within the casing **2** so as to define an inner compartment **4** for sulfur and an outer, annular, compartment **5** for sodium which bounds the inner compartment **4**, and a circular-section carbon rod **6** disposed in the inner compartment so as to serve as means for collecting current generated by electrochemical reaction between the sulfur and the sodium.

In further detail, the casing **2** comprises a blind-ended tube of stainless steel. The electrolyte **3** is of the same form and a stainless steel mesh **7** of tubular form is disposed adjacent the outer surface of the electrolyte **3** to serve as a "wick" in drawing up sodium by capillary action to "wet" those parts of the membrane **3** above the level of sodium as the level falls. Sulfur-impregnated carbon felt is disposed between the carbon rod **6** and inner surface of the electrolyte **3** to assist the carbon rod in collecting current.

The upper end of the carbon rod **6** is a close sliding fit in an end cap **8** of alpha-alumina. A metal flange **14** is welded to the inside of the casing **2**, and is bonded between the end cap **8** and an alpha-alumina backing ring **9** using a ceramic/metal seal. The end cap also supports the electrolyte **3**, the two components being attached to each other by a glass frit seal. A metal seal **15** is also bonded to the end cap **8** between an alpha-alumina backing ring **9** using a ceramic/metal seal.

Sulfur is supplied to the inner compartment **4** by way of a filling hole **10** formed in the center of carbon rod **6** connecting with a transverse hole **17**. After insertion of the sulfur the hole **10** is sealed by a carbon plug **11**, and a chromium-steel plug **12** is inserted in and welded to a screwed connector **13** of the same material fitted to a recessed section at the top of the carbon rod **6** and welded to the metal seal **15**. The outer compartment **5** is filled with sodium by way of a tube **16** subsequently sealed by a weld-joint.

When the cell **1** discharges, sodium ions are conducted through the tubular electrolyte **3** (which is conductive to sodium ions only) to combine with sulfur ions, resulting in an electrochemical reaction whereby a current is generated which is picked up by the carbon felt and carbon rod **6** in the inner compartment **4**. The casing serves as a negative electrode and the plug as a positive electrode.

The arrangement has the advantage that the corrosive sulfur is kept out of contact with structural metallic components of the battery cell thus avoiding metallic corrosion. Instead, the sulfur-contacting materials of the cell are alpha-alumina, beta-alumina and carbon, all of which have good resistance to corrosion. The parts of the cell in contact with sodium are not subjected to such a corrosive

FIGURE 4.8: TUBULAR SODIUM-SULFUR BATTERY CELL

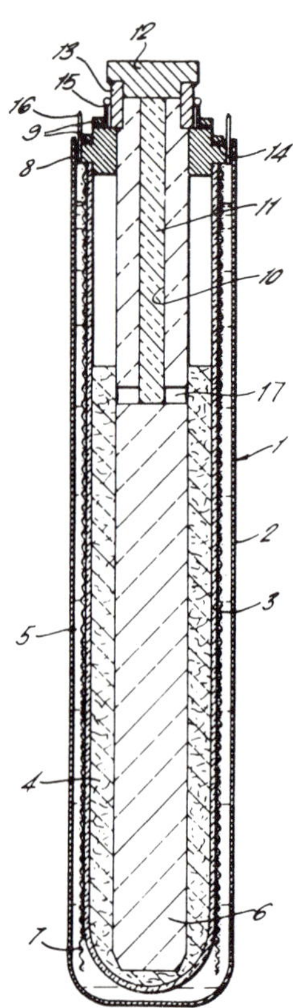

Source: U.S. Patent 3,982,959

environment as that in the inner sulfur compartment **4**, and therefore metals may be used without serious risk of corrosion. Although carbon has a higher electrical resistance than metals previously used as current collecting means, this is acceptable for many applications of sodium-sulfur battery cells.

Molybdenum Tube Barrier

L.S. Evans, T.L. Markin, R.M. Dell and A.G. Montgomery; U.S. Patent 4,029,858; June 14, 1977; assigned to U.K. Secretary of State for Industry, England describe an electric cell having a solid electrolyte and employing sodium as the liquid anode and sulfur as the liquid cathode. The liquid anode is contained in two compartments separated by a corrosion-resistant barrier in the form of a molybdenum tube. Liquid anode is arranged to flow between the compartments through a flow sufficient to meet the normal charge and discharge requirements of the cell but limits higher rates of flow.

One of the compartments has a very limited capacity for liquid sodium and is disposed about one side of the solid electrolyte to define a wicking space to constrain liquid sodium to flow over the solid electrolyte. The other compartment provides a reservoir for containing the bulk of the liquid sodium. In the event of damage to the solid electrolyte, only a very limited amount of liquid sodium is readily available to mix and react with the liquid sulfur.

According to a related process described by *L.S. Evans, R.J. Bones and J.R. Harbar; U.S. Patent 4,044,191; August 23, 1977; assigned to U.K. Secretary of State for Industry, England* in an electric cell having a solid electrolyte which partially bounds on one side a compartment for a liquid anode and on the opposite side a compartment for a liquid cathode, the compartment for liquid anode is made of very limited capacity so as to contain only a very limited proportion of the liquid anode required by the cell and is shaped so that it generally follows the contour of the solid electrolyte to present a shallow space for liquid anode normal to the surface of the solid electrolyte.

A reservoir is provided in which the bulk of the anode liquid is arranged to be stored, a duct is connected between the liquid anode compartment and the reservoir for a restricted feed of liquid anode. The duct is adapted to meet that amount of liquid anode required during normal discharge of the cell but to limit higher rates of feed of liquid anode. A barrier means separates the liquid anode compartment and the reservoir adapted to prevent liquid sulfur which might leak into the liquid anode compartment in the event of damage to the solid electrolyte from reaching the reservoir. The barrier means may be a structural member resistant to the corrosive attack of hot sodium polysulfides, or may be a space between a cell having an external reservoir.

ELECTRODE CONSTRUCTION

Easily Wettable Electrode Materials

F.A. Ludwig; U.S. Patents 3,993,503; November 23, 1976; 3,966,492; June 29, 1976; and 3,994,745; November 30, 1976; all assigned to Ford Motor Company describes a secondary cell or battery of the type comprising at least one molten alkali metal anode, at least one cathode, a liquid electrolyte electrochemically reversibly reactive with the alkali metal and in contact with the cathode, and a cation-permeable barrier to mass liquid transfer interposed between and in contact with the anode and the liquid electrolyte.

The main feature of this process comprises employing as the electrode in these devices a composite electrode which comprises:

(a) at least one first portion of porous conductive utilizing material which is disposed adjacent to the cation-permeable barrier and is in electrical contact therewith, is contiguous at least in part, with the barrier and is formed of a material which, during operation of the device, is more readily wettable by molten polysulfide salts than it is by molten sulfur; and
(b) at least one second portion of porous conductive material which is disposed adjacent to the first portion or portions and is contiguous therewith, is in electrical contact with the cation-permeable barrier and is formed from a material which is more readily wettable by molten sulfur than by molten polysulfide.

During operation of the battery or cell of this improved composite electrode the one or more first portions serve primarily as the electrode during charge and serve primarily to wick molten polysulfide away from the second portion or portions during discharge. The one or more second portions, on the other hand serve primarily as the electrode during discharge and serve primarily to wick sulfur away from the one or more first portions during charge. The result is a battery or cell which exhibits increased energy efficiency on both charge and discharge as well as an increased power and energy density as compared to prior art devices of this type.

The measure of wettability of substrate by a liquid material is the contact angle formed between the liquid and the substrate. If the liquid wets the substrate completely the contact angle will be 0°. If the liquid beads up completely on the the substrate, the contact angle will be 180°. Thus, the lower the contact angle between the liquid and the substrate the greater the wettability of the substrate by the liquid.

For example, in helium at 318°C the contact angle formed by molten Na_2S_4 on graphite is approximately 100° while the contact angle formed by molten sulfur on graphite is approximately 25°. Thus, graphite is preferentially wet by sulfur as opposed to polysulfide salts. As such, a graphite material such as a graphite felt is ideal for use as the second or discharging portion of the electrode and unsuitable unless modified to make it preferentially wettable by polysulfide, as the first or charging portion(s) of the electrode.

It has been found that by employing conductive materials which are preferentially wettable by polysulfide salts and by sulfur as the first or charging and second or discharging portion(s) of the electrode respectively it is possible to substantially reduce or eliminate electrode polarization while either charging or discharging. As a result, the batteries or cells made in accordance with the process demonstrate good electrical efficiency on both charge and discharge, and as such, are ideal for a number of uses such as for electric utility load levelling.

Among the numerous materials which are preferentially wettable by polysulfide salts, and which are therefore preferred materials for use as the first or charging portion(s) of the electrode discussed above are: (1) Metals, which as used herein shall include alloys as well as such metals or alloys having an oxidized surface(s). A preferred metal is stainless steel. It has been found, for example, that no electrode polarization occurs at a stainless steel AISI No. 446 electrode at 330°C while charging in the two phase region. The contact angle in helium at 318°C formed by Na_2S_4 on AISI No. 446 stainless steel is 0° to 5°, while the contact angle formed by sulfur on stainless steel is approximately 25°.

(2) Materials having a surface consisting of and including materials formed completely of a composition of a polar or ionic character or with unfilled d-orbitals. Such compositions include oxides or sulfides of metals selected from the group consisting of (a) metals of Group I, II and III of the Periodic Table of Elements, (b) Transition Series Metals, and (c) tin, lead, antimony and bismuth. Preferentially, the metal salts or oxides are highly insoluble in the sulfur and polysulfide phases. Preferred materials are: aluminum oxide (Al_2O_3); molybdenum disulfide (MoS_2); chromium trioxide (Cr_2O_3); lanthanum chromite ($LaCrO_3$); calcium doped lanthanum chromite ($La_{1-x}Ca_xCrO_3$); antimony pentoxide doped tin oxide (Sb_2O_5-SnO_2); lithium doped nickel oxide ($Li_xNi_{1-x}O$); titanium doped iron oxide ($Ti_xFe_{2-x}O_3$); and tantalum doped titanium oxide (Ta_2O_5-TiO_2).

(3) Surface oxidized graphite—Graphite oxide can be prepared by the standard methods (e.g., graphite in a 1:2 v/v mixture of concentrated nitric and sulfuric acids with solid potassium chlorate added or graphite in sulfuric acid containing sodium nitrate and potassium permanganate). Treatment of the graphite must be brief so that only the surface is oxidized. When this material is used as an electrode in an alkali metal/sulfur cell and heated in the presence of the cathodic reactant to operating temperatures, the graphitic oxide surface converts to graphite sulfide which is preferentially wet by the polysulfide.

(4) Electrically conducting intercalated graphite—Graphite bromide is formed by exposure of graphite felt electrodes to either liquid bromine or bromine vapor. Considerable bromine is retained in the graphite at the operating temperature of the cell. The graphite bromide surface is more wettable by polysulfides than the untreated graphite. Many materials can be reversibly intercalated in graphite. These materials all tend to make graphite more wettable by polysulfides. The intercalated graphite is prepared by heating the materials with the graphite.

(5) Graphite which bears a continuous or discontinuous coating of one or more of the materials of (1), (2), (3) or (4).

(6) A combination or mixture of the material of (1), (2), (3), (4) or (5).

The various materials which will exhibit the required preferential wettability by sulfur and which, therefore, are suitable as the second or discharging portion(s) of the electrode will be apparent to those skilled in the art. However, some preferred materials include graphite felt or foam, porous graphite, vitreous carbon foam, pyrolytic graphite felt or foam, or materials which have been covered or coated with the above carbon materials.

Arsenic and Other Metal Salt Additives

F.A. Ludwig; U.S. Patent 4,002,807; January 11, 1977; assigned to Ford Motor Company describes an improved secondary battery or cell of increased capacity. The improved cells or batteries serve to overcome difficulties caused by formation of elemental sulfur on the porous electrode surface near the solid ceramic electrolyte. The improvement is based on the principle that the addition of a third substance to a system of two partially miscible liquid components increases the mutual solubility of the two components if the added substance dissolves in both liquids.

The improvement comprises increasing the charge/discharge capacity of the battery or cell by dissolving in the cathodic reactant an additive from the group consisting of: (1) elements which will react with molten sulfur to form polysulfide salts which are soluble in and form a liquid solution with both molten sulfur and molten polysulfide salts of the alkali metal anodic reactant, e.g., sodium; (2) polysulfide salts of these elements, which salts are soluble in and form a liquid solution with both molten sulfur and the molten polysulfide salts of the alkali metal anodic reactant; and (3) mixtures of the elements and polysulfide salts.

Since the additive is soluble in both molten sulfur and molten alkali metal polysulfide salts, these two reactant components show increased mutual solubility. As a result, elemental sulfur does not form at the porous electrode surface near the solid electrolyte during charging and the battery or cell can be recharged much more efficiently than prior art devices.

Of the preferred additives discussed above the most preferred are those selected from the group consisting of arsenic, polysulfide salts of arsenic and mixtures thereof. If the elemental form of arsenic is added to the molten sulfur cathodic reactant, it will react with sulfur to form polysulfide salts. Alternatively, polysulfide salts of arsenic such as arsenic trisulfide (As_2S_3) and arsenic pentasulfide (As_2S_5) may be added to the melt or mixtures of the salts and/or the element may be added. The liquidus temperatures for arsenic-sulfur compositions are below 320°C for arsenic contents up to 50 atom percent based on the total atoms of arsenic and sulfur in the composition. Thus, the arsenic and/or arsenic polysulfide additive may be employed in any amount up to about 50 atom percent arsenic.

When arsenic or arsenic polysulfide salts are employed as additives, it is also desirable to dissolve in the cathodic reactant, in addition to the additive, between about 0.5 and 10, preferably about 3, atom percent of thallium based on the total atoms of thallium and sulfur in the reactant. The addition of this element in the amount specified serves to reduce the viscosity of the cathodic reactant, thereby promoting flow of reactant and mixing of sulfur and alkali metal polysulfide.

By employing the technique of this process it is possible to eliminate or substantially reduce the formation of elemental sulfur on the porous electrode of the cell or battery during the charging cycle.

In related work *N.K. Gupta and F.A. Ludwig; U.S. Patent 4,002,806; January 11, 1977; assigned to Ford Motor Company* describe a process for increasing the charge/discharge capacity of sodium-sulfur batteries or cells by including in the cathodic reactant of the cell between about 0.001 and 0.1 weight percent based on the total weight of the cathodic reactant of an additive selected from certain metals, metal salts and metal oxides. The additive is preferably added as particles, which may range in size up to approximately 1,000 microns. Many useful powders have an average particle diameter of 1 to 5 microns.

The additives useful in the improvement of the process may be selected from:

(1) metals selected from the group consisting of (a) metals from Groups I, II and III of the Periodic Table of Elements, (b) Transition Series Metals, and (c) antimony, lead, tin and bismuth;
(2) alloys comprising the metals of (1);
(3) salts of the metals of (1);
(4) oxides of the metals of (1);
(5) phosphides, arsenides, antimonides, carbides and nitrides of the metals of (1); and
(6) mixtures of (1) through (5).

The metal salts useful in the process are preferably selected from the groups consisting of halides, nitrates, nitrites, thiocyanates, sulfates, sulfides (or polysulfides), hydroxides and mixtures thereof. The salts are most preferably sulfides or polysulfides.

Corrosion-Resistant Metal Plating

J.G. Gibson and J.L. Sudworth; U.S. Patent 4,024,320; May 17, 1977; assigned to Chloride Silent Power Ltd., England describe a current collecting pole which is associated with an alkali metal sulfur cell.

According to the process, the current collecting pole associated with the cathode compartment comprises a first layer of an electronically conducting material, which is resistant to the corrosive action of sulfur and alkali metal polysulfides (e.g., carbon or graphite) and which defines a continuous surface in contact with the sulfur and alkali metal polysulfides in the cathode compartment and a second layer of a higher electronically conducting material, for example low resistance metals such as copper, silver, gold nickel or brass in electrical contact with the first layer over the surface of the first layer remote from the sulfur and sodium polysulfides.

Thus, by means of this arrangement the low resistance metal presents a virtually equipotential surface along the full length of the pole and the only ohmic contribution of the corrosion-resistant material is that arising from the thickness of the first layer.

When used in a tube cell in which the interior of the electrolyte tube constitutes the cathode compartment, the current collecting pole may be in the form of a tube of the corrosion resistant material projecting down into the electrolyte tube and closed at its end within the electrolyte tube, a layer of low resistance metal being provided on the inside of the tube. Alternatively the inside of the tubular current collecting pole may be filled with a metal wool of a low resistance metal.

When used in a tube cell in which the space between the inner and outer tubes constitutes the cathode compartment, the outer tube is made of the corrosion resistant material and is provided on its outer surface with a layer of the low resistance material. The outer tube could then, if necessary, be sheathed in a metal tube for mechanical protection and strength.

When used in a single flat plate cell or the terminal cell of a layer-type battery the outer surface of a plate-form current collecting pole of the corrosion resistant material is covered with a layer of the low resistance metal. In the case of the intermediate cells of a layer type battery in which the intermediate plate-form poles of the corrosion resistant material are serving as bi-poles, the layer of low resistance metal would be on the surface of the pole in the anode (i.e., sodium) compartment of the adjacent cell. The low resistance metal used would be selected so that it formed a barrier layer between the sodium and the corrosion resistant material, particularly when this is of carbon or graphite; copper and aluminum would be suitable metals. The layer of low resistance metal may be provided by standard plating methods.

Impermeable Conductive Carbon

A process described by *I.W. Jones, G. Robinson and T.L. Bird; U.S. Patent 3,982,957; September 28, 1976; assigned to The Electricity Council, England* relates to sodium-sulfur cells and is concerned more particularly with the construction of the cathode current collector.

There is provided a current collector in contact with the cathode reactant and formed of an impermeable carbon or graphite tube containing a deformable metallic conductor extending over and in contact with the internal surface of the tube. The deformable conductor may be steel wool and, in this case, a conductive solid metal core is provided within the carbon or graphite tube, the deformable conductor forming a conductive interface between the inner surface of the graphite tube and the core. Most conveniently however the deformable conductor is a metal which is liquid at the operating temperature of the cell.

It will be seen that, by this construction, the only material in contact with the cathodic reactant is graphite. Carbon is not significantly attacked by sulfur or sodium polysulfides and this arrangement therefore greatly reduces or eliminates the corrosion problems. The carbon or graphite tube is impermeable and thus the sulfur cannot penetrate it; the tube may conveniently be made impermeable by pyrolytic impregnation.

The cathode current collector described above is of a composite construction. The electrical conductivity of carbon is poor and a simple carbon rod could not be used as the cathode current collector as its high resistivity would impair the cell performance. By providing a conductive member in the graphite tube in the form of a liquid metal or a metal rod with a suitable interface between the rod and the tube, the internal surface of the carbon or graphite tube is connected directly to a low resistance electrical path for the cathode current of the cell. The current path through the graphite tube is therefore merely through the thickness of the tube and not along the length of the tube. The problem of the resistivity of carbon is thus overcome giving thereby a form of cathode collector having high conductivity and good corrosion resistance while permitting easy manufacture.

It would not generally be possible, in the conditions of a sodium sulfur cell, to make use of a cathode current collector comprising a solid metal rod with a coating of carbon. During the heating and cooling operations involved in manufacturing, filling and operating a cell, the materials of such a composite may undergo different rates of thermal expansion and the resulting stresses will tend to cause any coating to separate from the substrate metal. The provision of the liquid metal or the other deformable interface overcomes these problems.

Liquid metal forms a particularly convenient interface between a solid core and the surface of the carbon or graphite tube ensuring a conductive path to the core from the whole surface of the tube which is in contact with the liquid metal. This liquid metal may be any suitable metal which is liquid at the operating temperature of the cell and which does not react with the graphite of the solid core. The operating temperature is typically about 350°C and would be within the range of 280° to 400°C.

There is a wide range of metals which are liquid at these temperatures, for example mercury, gallium, sodium, lithium, indium, potassium, tin and cadmium and alloys and amalgams between these and/or other metals. The core may be made of any convenient metal of good electrical conductivity. With an aluminum core, it is preferred to use a soft solder (a tin lead alloy) for the liquid metal. If a solid metal core is omitted, the preferred liquid metal is tin.

The external electric connection to the current collector may be made by means of a clamp connector around the carbon or graphite tube or by means of a connection to the core or liquid metal within the tube, for example through an end plug for sealing the carbon or graphite tube.

Graphite Intercalation Compounds

A process described by *M.B. Armand; U.S. Patent 4,041,220; August 9, 1977; assigned to Agence Nationale de Valorisation de la Recherche (ANVAR), France* relates to mixed compounds derived from graphite compounds of formula

$$C_n(M'_yX_z)$$

wherein M' represents a transistion metal ion such as Ti, V, Cr, Mn, Fe, Mo; X represents a nonmetallic electronegative atom such as O, S, F, Cl or Br; y and x have values corresponding to indices defining the relative proportions of metal and nonmetallic atoms respectively in the formula of the products M'_yX_2 incorporated in the composition of the graphite derivatives; n has a value equal to or higher than a minimum value which is a function of the initial binding energy between graphite and the compound M'_yX_2 under consideration, this value not exceeding however that corresponding to the formation of continuous monomolecular graphite layers interposed between the layers of the compound M'_yX_z in the corresponding compound $C_n(M'_yX_z)$. These compounds $C_n(M'_yX_z)$ will be referred to as "intercalation compounds" of graphite and products M'_yX_z.

To this category of graphite derivatives belong in particular products of the type described by Croft (*Austr. J. Chem.,* 1956, 9, 201). The main members of this group consist, e.g., of intercalation derivatives of graphite and of the compounds $FeCl_3$, $CoCl_3$, CrO_3, CrO_2Cl_2, the chemical similarity of which is due to the organometallic type bonds between the orbitals π of the graphite and the partially empty orbitals d of the transition metal. Compounds described by Croft are characterized by a conductivity which is solely electronic. The following examples illustrate the process.

Example 1: An intercalation compound of graphite and chromic oxide (C_8CrO_3) is prepared as follows. 0.96 gram of pure graphite (Ultracarbon) and 0.998 gram of chromium trioxide were weighed and mixed in a glove-box swept with dry nitrogen obtained from evaporation of liquid nitrogen. The fineness of the particles of the graphite powder enabled the mixture to be prepared in a mortar. The powder obtained was transferred to a graphite crucible and heated within 2 hours up to 200°±5°C. The purpose of this rather slow temperature rise was to prevent graphite from being oxidized by chromium oxide, the temperature being maintained for 48 hours. The cooling of the graphite/chromic oxide intercalation compound did not require any special care.

Example 2: Na_xCrO_3 was prepared by reacting the compound C_8CrO_3 with a solution of sodium in a solution of naphthalene in 1,2-dimethoxyethane. The 1,2-dimethoxyethane used had been held previously in contact with sodium flakes and the naphthalene with calcium chips for the purpose of rendering them perfectly anhydrous.

Several preparations were made. The amount of sodium used in each of the preparations was between 0.3 and 0.5 gram. In all instances, the amount of naphthalene used was in excess of 3 grams with respect to the stoichiometric proportion required for preparing the naphthalene-sodium compound. These reactions were effected with 75 ml of solution in a three-necked vessel (250 ml), the inner volume of which was swept by a dry nitrogen stream.

The amount of the intercalation compound C_8CrO_3 added to the sodium (within 1 hour) and reacted therewith was dependent on the amount of sodium dissolved in the solution and also on the stoichiometric proportions with respect to the value of x in the compound $Na_xC_8CrO_3$ sought to be obtained. The mixture was continuously stirred. The completion of the reaction was indicated by the disappearance of the color of the initial naphthalene-sodium complex. The mixed graphite compound finally obtained was filtered on sintered glass of fine porosity and washed with pure anhydrous 1,2-dimethoxyethane. The final product was dried under vacuum and stored in a dry atmosphere.

Such procedure yielded compounds of formula $Na_xC_8CrO_3$ in which the values of x were respectively 0; 0.5; 1; 1.5; 2; 2.5; and 3.

The theoretical capacities of these storage batteries are considerable. Taking as an example the compound $Na_xC_8CrO_3$ in which x may vary from 0 to 3, calculation shows that the maximum capacity obtainable is 1.1 kWh/kg, when it is assumed that the average available emf is of 3.5 volts. This theoretical capacity is obviously considerable when compared to that of a conventional lead battery which is of about 0.05 kWh/kg. The sodium-sulfur batteries have a theoretical capacity of about 0.8 kWh/kg. Since such batteries operate only at a temperature above 300°C, such theoretical values need be divided by a factor of about 2, to take into account the required thermal insulation.

Metal Wool Wick

J.L. Sudworth and A.R. Tilley; U.S. Patent 3,939,007; February 17, 1976; assigned to British Railways Board, England describe a sodium-sulfur cell comprising inner and outer tubes, the inner tube constituting the solid electrolyte for the cell. The interior of the inner tube defines the anode compartment of the cell and the space between the inner and outer tubes defines the cathode compartment of the cell. The inner tube contains metal wool and/or a plate spaced from the inner surface of the inner tube to cause molten sodium to be distributed over the inner surface of the inner tube by capillary action.

The inner and outer tubular members are closed at one end by a compression seal assembly. The seal assembly comprises a flange on the inner tubular member engaging the end of the outer tubular member to close the end of the outer tubular member, a closure member engaging the flange to close the inner tubular member, and a clamping member fitting over the closure member and being held in position by the outer tubular member.

The cell shown in Figure 4.9 comprises coaxial inner and outer tubes **1** and **2**. The inner tube **1** constitutes the solid electrolyte of the cell and is made of β-alumina. The interior of the tube **1** constitutes the anode compartment of the cell and is filled with a metal wool **3** which is inert to the electrochemical reaction of the cell.

The outer tube **2** is made of stainless steel. The annular space **4** between the inner and outer tubes constitutes the cathode compartment of the cell. The inner tube is formed with a closed end **5** and the outer tube has an end plate **6** which is welded into position after formation of the compression seal assembly **7** at the other end of the tubes **1** and **2** and loading up of the cathode compartment with positive active material.

To form the compression seal assembly **7**, the inner tube is provided with an alpha-alumina flange **8** which engages an internal flange **9** on the outer tube

FIGURE 4.9: TUBULAR SODIUM-SULFUR CELL

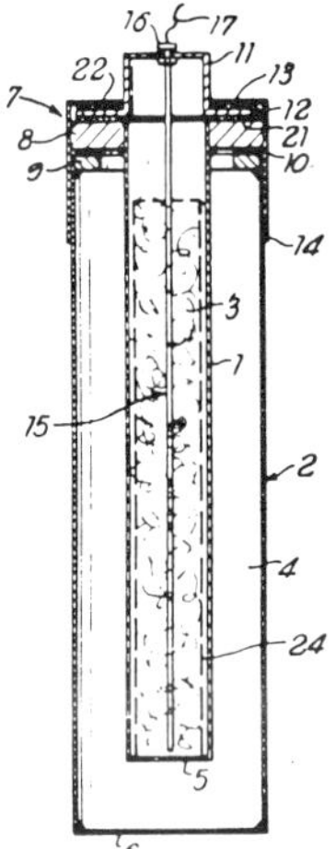

Source: U.S. Patent 3,939,007

through an aluminum or "Grafoil" sealing ring **10** to close the end of the outer tube. The end of the inner tube is closed by closure member **11** which has an out-turned rim **12** which engages the flange **8** through aluminum sealing gasket **21**. To improve the sealing, the rim **12** may have a rib extending around it which engages the gasket **21**. The assembly is held in compression by a clamping member **13** which fits over and engages the rim **12** of the closure member **11** through a mica or asbestos insulating washer **22** and is spot welded at **14** to the outer tube.

The compression seal is made before the positive and negative active materials are loaded into the cell. The positive active material is loaded through the open end of the outer tube before fitting end plate **6**. To load the negative active material, i.e., the sodium, an aluminum tube **15** passes through the closure member **11** and extends part way down the inner tube. Liquid sodium sufficient for the electrochemical reaction of the cell is passed into the tube **1** through the aluminum tube **15**. The outer end of the aluminum tube **15** is closed by an aluminum plug **16**. The tube **15** acts as one electrode of the cell and has aluminum lead **17** extending from it. The other electrode of the cell is the outer tube.

In one mode of operation of the cell the metal wool maintains electrical continuity between the solid electrolyte and the electrode tube **15**. Here equally important however it wicks the liquid sodium to the inner surface of the tube **1**, i.e., to the surface of the electrolyte. Hence a sodium reservoir for maintaining the tube **1** filled so that it contacts a substantial part of the inner surface of the tube **1** is not necessary.

In a second mode of operation the cell is used upside down and a reservoir provided by the inner space of closure **11** is used which feeds sodium into the metal wool packing **3**. Thus at any one time there is only just sufficient sodium in the tube necessary for the reaction.

In an alternative construction of a cell, the metal wool **3** is omitted and the cell is operated in a horizontal position in contrast to the vertical position shown in the drawing. The liquid sodium creeps by capillary action around the inner surface of the tube **1**. Hence a sodium reservoir is again obviated.

In an alternative construction shown in Figure 4.9 a stainless steel plate is sprung into the tube **1** so that it extends around the tube as shown at **24** in the drawing. The plate **24** has dimples (not shown) which space it from the inner surface of the tube **1** against which it is urged by its inherent resilience. The sizes of the dimples set the spacing between the tube **1** and plate **24** so that a capillary action causes the sodium to flow over the inside surface of the tube **21**.

Metal Felt Carrier

W. Baukal, R. Knodler and W.H. Kuhn; U.S. Patent 4,038,464; July 26, 1977; describe a rechargeable galvanic battery which contains liquid sodium, as the negative electrochemically active material and liquid sulfur, as the positive electrochemically active material, and a ceramic solid electrolyte, which is capable of conducting sodium ions.

The sodium is completely absorbed in a fine-pored metal felt or mat. The sulfur is completely absorbed in a graphite felt or mat. The metal felt has open pores which face toward the solid electrolyte. The pore structure of the metal felt is undisturbed even in the transition area between the metal felt and the solid electrolyte. The metal felt fills the entire anode space of the battery. At least at the operating temperature of the battery, an intimate contact over a large surface area exists between the metal felt and the solid electrolyte. The maximum distance between the metal felt and the solid electrolyte is in the order of the magnitude of the pore width of the metal felt.

Porous Electrode Design

A process described by *F.A. Ludwig, R.W. Minck and S.A. Weiner; U.S. Patent 3,980,496; September 14, 1976; assigned to Ford Motor Company* relates to electrical conversion devices of improved design wherein a porous conductive electrode disposed within the cathodic reaction zone of the device is modified so as to promote and enhance the rate of feeding and extracting of reactants and reaction products to and from reactive sites within the cathodic reaction zone.

The process involves electrical conversion devices which contain: (1) an anodic reaction zone containing a molten alkali metal reactant-anode in electrical contact with an external circuit; (2) a cathodic reaction zone containing a cathodic reactant comprising a liquid electrolyte which is electrochemically reactable with the anodic reactant and an electrode of porous conductive material which is at least partially immersed in the cathodic reactant; and (3) a solid electrolyte which is a cation-permeable barrier to mass liquid transfer and which acts as a reaction zone separator between the anodic and cathodic reaction zones, the porous conductive material being in electrical contact with both the cation-permeable barrier and the external circuit.

The improvement comprises designing the cathodic reaction zone such that there are a plurality of channels and/or spaces within the zone which are free of the porous conductive material or electrode and which are thus adapted to allow for free flow of the cathodic reactant and cathodic reaction products during operation of the device.

Most generally, the porous conductive material or electrode will be in physical contact with both the solid electrolyte or reaction zone separator and the container wall surrounding the cathodic reaction zone. More particularly, the porous conductive material or electrode has a large surface area and substantially fills the entire cathodic reaction zone with the exception of such channels and/or spaces through which the cathodic reaction products and cathodic reactant may freely flow.

Tubular or cylindrical electrical conversion devices made in accordance with this process thus comprise a cathodic reaction zone which completely surrounds the solid electrolyte or reaction zone separator. The porous conductive material or cathodic electrode substantially fills the entire space between the inner solid electrolyte tube and the outer container tube except for the channels and/or spaces which promote convective flow of the cathodic reactant and reaction zone products.

This flow results from free convection within the channels and/or spaces, and from wicking of cathodic reactants or cathodic reaction products within the conductive porous material. Flow in the channels is mainly the result of free convection forces, but if the cells are oriented so as to decrease free convection, flow in the channels still occurs in response to wicking forces. Complete details of the construction of the cathode electrode are provided.

Screening Graphite Electrode

A process described by *W. Fischer, W. Haar, H. Kleinschmager and G. Weddigen; U.S. Patent 4,029,857; June 14, 1977; assigned to Brown, Boveri & Cie. AG, Germany* relates to an electrochemical storage cell or battery based on sodium and sulfur.

In the process, corrosion in the cell wall is eliminated or minimized during current in an electrochemical storage cell or battery based on sodium and sulfur having a chamber containing sulfur material as a reactant chamber containing sodium as a reactant separated from each other by a dividing wall capable of conducting sodium ions.

Each electrode chamber has an electrode with an associated current connection. An electrode is disposed in the chamber containing sulfur material as a reactant between the dividing wall and the cell wall which defines the outside of the chamber to electrically screen the cell wall to minimize corrosion of the cell wall.

In this manner, the electric field lines or the ion transport, do not run from the inner dividing wall, which is formed as solid sodium conducting electrolyte, to the cell wall, but end at the screening electrode before reaching the cell wall. The space between the cell wall and the screening electrode which is filled with sulfur or sodium sulfide is therefore without current.

To make certain that there are no stray fields between the dividing wall and the screening electrode which fields eventually increase the corrosion of the metallic cell wall, it is recommended to carry the screening electrode insulated through the base of the cell wall or outer housing. However, in some cases it may also be sufficient to do without insulation, the advantage being that no special electrical feed-through is required and the current connection can be effected on the housing itself.

The screening electrode may be made of any suitable conductor material which is relatively inert to the corrosive effect of sulfur or polysulfide in which it is immersed in the chamber. Metals or metal alloys may be used but the electrode is advantageously made of graphite, which exhibits sufficient corrosion resistance with respect to sulfur or polysulfide.

An advantageous example of an electrochemical storage battery consists of an arrangement of several disc-shaped cells in sequence with each cell consisting of two disc-shaped electrode chambers. In this construction a separating wall made, for example, of beta-aluminum oxide and a cell wall made, for example, of steel, alternate with each other.

A single storage cell need not be disc-shaped but can also be cup-shaped. In the case of the latter, the cell wall forms the cup-shaped outer housing in which is disposed the cup-shaped sodium conducting dividing wall. In the preferred configuration the dividing wall is a tube of beta-aluminum oxide which is closed at one end. Positioned between the outer housing and the dividing wall is the screening electrode which is also cup-shaped. Cells of this kind, can be electrically connected externally, thus forming a battery.

In general, it is recommended to construct the screening electrode in such a manner that some mixing of the sulfur or sodium polysulfide-electrolyte is made possible. In the case of a screening electrode made of graphite, for example, holes will be provided; in the case of a screening electrode made of metal, a metallic net may be used.

BETA-ALUMINA PROCESSING

Electrophoretic Process

R.W. Powers; U.S. Patent 3,900,381; August 19, 1975; assigned to General Electric Company describes a method of forming beta-alumina articles which includes providing a suspension of beta-alumina particles the majority of which have a diameter in the range of 1 to 2 microns in an organic liquid vehicle, electrophoretically depositing the particles from the same vehicle onto a mandrel, drying the deposited material, removing the deposit from the mandrel, and sintering the dried material.

Reduction in size of the starting powder and electrical charging of the particles were carried out simultaneously by milling under the vehicle, i.e., the organic liquid in which the beta-alumina particles are suspended during deposition. Various kinds of milling were examined in the course of this work. Very satisfactory results were obtained with ball milling.

Milling has been carried out in 32-ounce wide-mouth polyethylene bottles. Their use as mill jars reduces contamination since any abraded material is burned out during sintering. These containers were half-filled with grinding media, either 1,100 grams of alumina media or 1,800 grams of zirconia. When suspensions were milled the volume of the vehicle was 200 ml. The amount of starting powder in the charge ranged from 35 to 200 grams. Milling was done on a commercial jar mill.

The following is a preferred milling process. A friable powder such as Alcoa XB-2 beta-alumina is used. Milling is carried out at a high powder concentration, e.g., 200 grams per 200 ml n-amyl alcohol vehicle. The suspension is milled in a clean polyethylene jar. Zirconia is used in preference to alpha-alumina media. If slight contamination by zirconia is intolerable, a 91% alumina media is employed.

In Figure 4.10 there is shown a partial sectional view of the electrophoretic deposition apparatus for carrying out the process. The equipment is shown generally at **10** which comprises a stainless steel vessel **11** which functions as a counter-electrode filled with a suspension designated generally at **12** which includes the milled beta-alumina particles, and the n-amyl alcohol vehicle. This suspension is transferred from the polyethylene jar after the grinding or milling has been accomplished.

A stainless steel mandrel **13** has a deposition portion **14**, tape **15** surrounding a nondeposition portion, a portion **16** fitted within an electrically insulating cover **17** for vessel **11**, and a portion **18** extending outwardly from cover **17**. A high voltage source **19** is connected by a negative (-) lead **20** to a clamp **21** attached to vessel **11** thereby making vessel **11** the negative counter-electrode. Portion **14** of mandrel **13** is made the positive electrode (+) by connection of a positive lead from source **19** to portion **18** of mandrel **13**. A current meter, a current recorder and a charge indicator (not shown) are connected in series with power source **19**.

The high voltage source is a standard commercial device which has been used to supply voltages in the range 200 to 1,000 V between mandrel **13** and counter-electrode **11**. Such voltages correspond to fields between 100 and 10,000 V/cm. It is preferred practice to use the highest fields permitted by the power supply. This procedure usually reduces deposition times to about a minute and makes stirring of the suspension unnecessary during deposition.

Deposition in electrophoretic forming is carried out as is conventional in the art by applying a dc voltage from power source **19** between portion **14** of mandrel **13** and counter-electrode **11**. Tape portion **15** can be of a different length or eliminated depending on the beta-alumina closed end tube desired. Suitable tapes include 3M Company No. 471 plastic tape and Teflon polymer. Difficulties were encountered with paper masking tape as well as silicone rubber. Open end tubes are produced by applying the same tape on the lower end of a nontapered mandrel.

Other beta-alumina articles have been produced in disc and pocket-shaped configurations by using a correspondingly shaped mandrel under similar conditions. In addition to stainless steel, the mandrels have been made of nickel and Invar alloy. Other suitable materials for deposition vessels include polyethylene and glass which would require a counter-electrode therein. Plexiglass plastic containers were unsuitable. After a deposition time of approximately 1 minute, a beta-alumina article is formed on portion **14** of mandrel **13**.

The properties of sintered beta-alumina depend on details associated with each of the three major fabrication steps: the preparation of starting powders, the forming of greenware, and finally the sintering procedure. When sintering is carried out at temperatures between 1725° and 1825°C for times ranging from a few minutes to a few hours in an oxygen oxidizing atmosphere, such as air

FIGURE 4.10: PREPARATION OF BETA-ALUMINA ARTICLES

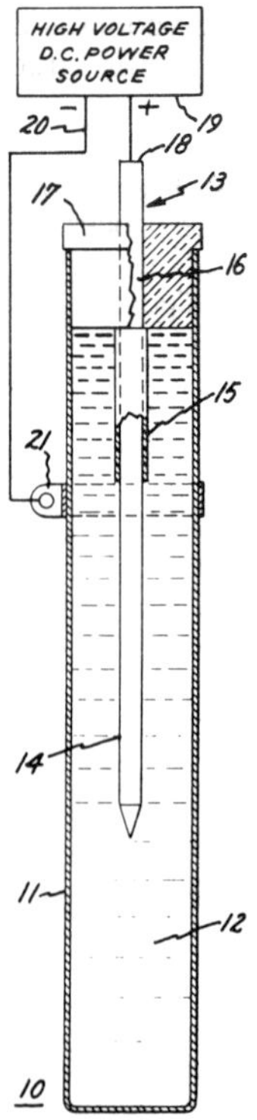

Source: U.S. Patent 3,900,381

or oxygen, densities ranging from 2.98 to 3.26 g/cc are obtained on single-phase material. These values correspond to residual closed porosity from 10 to less than 1%. Examples of beta-alumina articles made in accordance with the process are as follows.

Examples 1 through 18: In Examples 1 through 18, 18 closed end tubes of beta-alumina were prepared. Three separate suspensions were prepared from each of which six closed end tubes were made. Each suspension contained 150 grams of Alcoa XB-2 powder in 300 ml of n-amyl alcohol. These materials were milled for 18 hours using 1,800 grams zirconia cylinders in a polyethylene jar. Subsequently, each suspension was transferred to a separate stainless steel vessel as described above and as shown in Figure 4.10.

A mandrel was employed as shown in Figure 4.10 with an exposed surface portion having a length of 4.5 inches. A dc electric field of 1,980 volts per centimeter was applied initially from the power source across the mandrel as the positive electrode and across the vessel as the negative electrode. The tubes, which are numbered 1 to 18, correspond to Examples 1 through 18.

In the table below, the tubes, deposition time, suspension, and amount of deposited beta-alumina are set forth. Six tubes were made from each suspension.

Tube No.	Deposition-Time-Seconds	Suspension No. & Mass in Grams	
1	40	1	12.49
2	46	1	13.16
3	53	1	13.38
4	64	1	14.06
5	82	1	14.78
6	115	1	14.92
7	40	2	12.39
8	46	2	12.95
9	53	2	13.11
10	64	2	13.51
11	82	2	13.17
12	115	2	14.67
13	40	3	12.08
14	46	3	12.66
15	53	3	13.14
16	64	3	13.67
17	82	3	14.59
18	115	3	14.96

After the above depositions each closed end tube was air dried for 24 hours on its respective mandrel. Each closed end tube was then removed from its mandrel and sintered as above described at a temperature of 1800°C for a period of 5 minutes. The resulting closed end tubes were beta-alumina articles made in accordance with the process.

In related work *S.P. Mitoff, R.W. Powers and R.N. King; U.S. Patent 3,896,019; July 22, 1975; assigned to General Electric Company* describe a method of forming beta-alumina articles which includes sintering beta-alumina greenware between 1650° and 1825°C in a preheated oxygen oxidizing atmosphere at a controlled passage rate from one-half to four inches per minute. Such articles are useful as containers for sodium and as solid electrolytes in sodium-sulfur and sodium-halogen batteries.

Calcining and Sintering Process

H.C. McGowan, J.A. Male and M.G. McLaren; U.S. Patent 3,895,963; July 22, 1975; assigned to Exxon Research & Engineering Co. describe a process for the preparation of low electrical resistivity beta-alumina-type ceramics. The process comprises forming a finely divided, homogeneous mixture of alpha-alumina, sodium oxide and dopant, preferably Li_2O or MgO, where the alumina/sodium oxide mol ratio, in the mixture, ranges between about 5 and 11 and calcining the mixture at a temperature below about 1150°C and preferably ranging between about 85° and 1100°C.

The calcined mixture is then milled and formed into a green body of high density containing a major amount of alpha-alumina and, thereafter, fired, in a sintering operation, at a temperature ranging between about 1400° to 1700°C, thereby yielding the desired beta-alumina-type ceramic containing a major amount of beta-alumina.

The ceramic is characterized by having an electrical resistivity, measured at 25°C, of less than about 250 and preferably less than about 100 ohm-centimeters, a substantially uniform thickness ranging between about 1 and 250 mils, and a density greater than about 85% of theoretical density and preferably between about 90 and 100% of theoretical. Thus, for example, the density may range between about 2.75 and 3.25 g/cm.

Tube Production

I.W. Jones; U.S. Patent 3,950,463; April 13, 1976; assigned to The Electricity Council, England describes a method of producing a tube of beta-alumina ceramic which comprises the steps of forming a tube of compressed powder of the required composition and then moving this tube through a furnace so that a short length of the tube is raised to the sintering temperature, the tube being continuously moved so that the heated zone is gradually moved along the length of the tube.

The compressed powder may, in accordance with known techniques for the production of beta-alumina ceramic, be homogeneous and composed of the final beta-alumina ceramic in finely divided crystalline form or it may be a mixture of the oxides in the required proportions or may be a mixture of alpha-alumina and compounds of sodium (e.g., sodium aluminate) and of magnesium (e.g., magnesium oxide) and/or lithium (e.g., lithium carbonate) which together will form on heating the required ceramic composition. In the latter case the sintering includes the chemical reaction to form the required homogeneous chemical and crystal structure.

The furnace has a temperature profile, i.e., relationship between temperature and position along the length of the furnace, in which the temperature increases from one end to a maximum temperature between 1600° and 1900°C and then decreases towards the other end. The length (l) of the sintering zone, which is the region within 10°C of the maximum temperature in the furnace, and the velocity (v) at which the tube is traversed are such that the time l/v during which any point on the tube is in the sintering zone is between 12 seconds and 2 minutes. Furthermore, the temperature profile and the rate of traverse of the tube are such that a point in the tube is heated up to the sintering temperature and cooled from the sintering temperature at a rate between 200° and 2400°C per minute.

With this method, a very short sintering and heating time is employed while still maintaining the required close control of temperature. The heating time depends on the length of the furnace and on the rate at which the tube is passed through the furnace. It is possible therefore to use a very short furnace, shorter than the length of the tubes to be produced. More particularly however very long lengths of tube can readily be made with a small furnace.

Independently of the length of the sintering zone in the furnace, the time in that zone can be controlled by controlling the rate of movement of the tube through the furnace. The zone sintering process can be readily controlled so that the final grain size is very close to that of the original powder particles. Grain size in the preferred dense ceramic can be kept as low as 1 to 2 μm because of the rapid heating and cooling and short exposure to sintering temperature that is possible with the method of this process.

Alternatively, using different conditions, recrystallization phenomena can be controlled to produce grain sizes intermediate between 1 μm and 20 to 50 μm. (20 to 50 μm is the grain size of completely recrystallized zone sintered material.)

The rate processes which cause sintering increase exponentially with temperature. In the method of this process, the maximum temperature in the furnace is made between 1600° and 1900°C; whatever temperature is chosen, the region in the furnace within 10°C of the maximum temperature is referred to as the sintering zone since it is the region at or near the maximum temperature which is of importance for sintering and which will largely determine the grain growth. The duration of sintering is therefore defined as the length (l) of the zone where the temperature is within 10°C of the maximum divided by the traverse speed.

In this process, this time is very short compared with the many hours used in prior techniques. The calculation l/v tends to overestimate the duration of sintering at high traverse speeds through short hot zones, because the finite thermal mass of the tube and its finite thermal conductivity cause its time-temperature profile to lag behind the static temperature profile of the furnace.

Conveniently an inductively heated tube furnace is employed. This may have an overall length shorter than the length of the ceramic tube to be sintered. The zone at the sintering temperature may, for example have a length not more than four times and typically only twice the diameter of the furnace bore through which the tube is passed.

The sintering zone may be maintained at a temperature between 1600° and 1900°C, typically at 1700°C, and each portion of the tube may typically remain in the sintering zone for a time of less than 2 minutes. By using a high speed of traverse, the time may be very much shorter than this and tubes of high density have been made with a time in the sintering zone as short as 12 seconds. With such a construction having a short sintering zone, it is possible to control the sintering zone temperature closely to a required value. It may readily be controlled to ±10°C.

By the technique described above it is possible to produce tubes of beta-alumina ceramic with a bulk density of 3.1 to 3.2 g/cc. It is readily possible, using the process, to obtain densities some 4 to 5% higher than the highest values previously known. For a solid electrolyte, open porosity must be totally eliminated from the material during the sintering process in order that the electrolyte should be impervious to metallic sodium and elemental or ionic sulfur. This requires that the density must closely approach the theoretical maximum density. With the method of the process, this very high density is combined with the achievement of fine grain size. These two factors result in a material of high strength and high physical integrity.

OTHER PROCESSES

Molten Sodium Haloaluminate Electrolyte

A process described by *J.C. Sklarchuk; U.S. Patent 3,988,163; October 26, 1976; assigned to ESB Incorporated* relates to secondary electrochemical cells utilizing a molten sodium negative reactant, a solid ionically-conductive separator, a molten sulfur and mixture of molten metal or metal-like halides positive reactant, and a molten sodium haloaluminate electrolyte.

The electrolyte utilized in the process makes possible a substantial reduction in the operating temperature of molten sulfur batteries as compared with those commonly known and used in the art. For example, the operating temperature of the normal molten sodium-molten sulfur secondary battery wherein the molten cathodic electrolyte is an ionized combination of sodium and sulfur, i.e., ions of sodium polysulfide, Na_2S_5, is generally recognized to be well above 200° and closer to 300°C, the melting temperature of sodium pentasulfide being as high as 265°C.

In marked contrast, the battery of this process utilized an electrolyte on the positive reactant side of the solid separating member which comprises molten sodium haloaluminate and permits the operation of the battery to be carried out at temperatures of about 150° to 225°C. The advantages of operating at a reduced temperature are of course apparent, e.g., reduced corrosion, increased battery life, and reduced cost of containers and seals, e.g., silicone rubber may be used as a negative gasket seal.

By the term molten sodium haloaluminate as used herein is meant materials which include sodium halides, as for example, chlorides, bromides, fluorides, or iodides or sodium, and aluminum halides, for example chlorides, bromides, fluorides or iodides of aluminum. All of these metal halides will form the corresponding sodium haloaluminate electrolytes of the process. The preferred electrolyte is sodium chloroaluminate. The following examples illustrate the process.

The following information applies to all cells. Unless otherwise indicated, all quantities are by weight. The makeup of the cells in the examples comprised a Pyrex glass positive container, a carbon steel negative container and a beta-alumina separator disc.

The negative seal was silicone rubber, and the positive seal was Teflon. The negative current collector was the steel sodium container. The positive current collector varied as to type and configuration, as described in the examples, but in all cases it was sealed to the Pyrex glass with a Monel Gyralok fitting containing Teflon ferrules. In all cases the carbon was Cabot Vulcan XC-72R. This carbon has a relatively low surface area of approximately 200 square meters per gram.

Example 1: A tungsten coil current collector was sealed to a one-fourth inch glass tube then inserted into the positive Pyrex glass container. The cell was then assembled and sealed to the beta-alumina separator. The following positive ingredients were then added in powder form, after which the Pyrex container was sealed with a Monel Swagelok:

0.5 g of sulfur
0.5 g of XC-72R conductive carbon
0.5 g of NaCl
2.5 g of $AlCl_3$
1.15 g of $SbCl_3$

The cell was heated to and held at 210°C then put on discharge. The capacity of this cell equalled 1,250 joules per gram of positive mix. The turnaround efficiency was 88%.

Cycle	Mode	Current	Polarization	Capacity
	C = charge D = discharge	(MA)	(MV)	(Ah)
1	D	50	190	0.75
	C	50	190	0.55
2	D	50	190	0.60
	C	50	190	0.62
3	D	50	190	0.62
	C	50	190	0.63
4	D	50	190	0.63
	C	50	190	0.64
5	D	50	190	0.63
	C	50	190	0.63

Example 2: The cell assembly procedure was the same as described in Example 1. The following positive ingredients were added in powder form:

0.5 g of sulfur
0.75 g of XC-72R conductive carbon
1.25 g of NaCl
4.50 g of $AlCl_3$
1.25 g of $SbCl_3$

The cell was heated to 210°C and discharged. The capacity of this cell equalled 850 joules/gram of positive mix. The turnaround efficiency was 85%.

Cycle	Mode	Current	Polarization	Capacity
	C = charge D = discharge	(MA)	(MV)	(Ah)
1	D	50	205	0.78
	C	50	205	0.70
2	D	50	205	0.71
	C	50	205	0.69
3	D	50	205	0.69
	C	50	205	0.69
4	D	50	205	0.69
	C	50	205	0.69
5	D	50	205	0.69
	C	50	205	0.69

In related work *J. Werth and J.C. Sklarchuk; U.S. Patent 3,969,138; July 13, 1976; assigned to ESB Incorporated* describe a secondary battery which utilizes a molten sodium negative reactant, a sulfur-aluminum halide positive reactant melt having a carbon powder dispersed within the melt, a molten sodium haloaluminate electrolyte, and a selectively ionically-conductive separator positioned between the negative and positive reactants.

Sodium Polyaluminate and Aluminum Oxide Electrolyte

A process described by *K.R. Kormanyos, H.L. McCollister and P.L. White; U.S. Patent 4,041,215; August 9, 1977; assigned to Owens-Illinois, Inc.* is directed to sodium polyaluminate electrolyte structures finding utility in sodium sulfur batteries.

The process for preparing the solid electrolyte composite comprises contacting a surface of a porous, anhydrous, crystalline body consisting of alpha-Al_2O_3 with a precursor of an ionic conductive crystalline sodium polyaluminate and heating the contacted surface, which of course will have thereon the precursor, at a temperature and for a time sufficient to convert the precursor to a crystalline ionic conductive membrane of sodium polyaluminate.

The term sodium polyaluminate comprehends a composition of the formula $ZMO_Y \cdot Na_2O \cdot XAL_2O_3$ wherein MO_Y is a metal oxide present in an amount sufficient to thermally stabilize the polyaluminate composition, Y is an integer dependent upon the valence of M, X is an integer between 5 and 11 and Z is a molar amount equivalent to from zero or 0.5 to 5% by weight of the total composition. One exemplary sodium polyaluminate is $Na_2O \cdot 11Al_2O_3$; another, a stabilized specie, is $0.24LiO_2 \cdot 1Na_2O \cdot 6.3Al_2O_3$.

Other suitable stabilizing metal oxides include for example calcium oxide and magnesium oxide. On a weight basis, based on the total aluminate composition, the amount of lithium oxide or magnesium oxide or calcium oxide will be from about 0 up to 5% by weight and more typically, when employed, these stabilizing oxides will be present in an amount between about 0.5% by weight to about 5% by weight.

As indicated, the precursor once applied to the surface of the porous, anhydrous, crystalline alpha-Al_2O_3 body will be heated at a temperature and for a time sufficient to convert the precursor to a crystalline ionic conductive sodium polyaluminate and to sinter the sodium polyaluminate so as to form a barrier layer, or membrane portion, on the alpha-Al_2O_3 body which, for example, is impermeable to the passage of molten sulfur. The following example illustrates the process.

Example: A precursor of a $1Na_2O \cdot 11Al_2O_3$, sodium polyaluminate, was prepared by first reacting about 2.3 grams of sodium metal with about 50 milliliters of ethylene glycol monomethyl ether (methyl Cellosolve) to form a sodium cellosolvate. To this solution there was then added about 271 grams of aluminum tri(secondary-butoxide) and about 1,500 ml of benzene. With vigorous stirring approximately 50 grams of water was then added to hydrolyze the hydrolyzable aluminum and sodium compound and this resulted in the formation of a precipitate. The precipitate was filtered and dried by heating at a temperature of 500°C for about 8 hours to produce a particulate, amorphous precursor of a crystalline ionic conductive sodium polyaluminate.

The dried hydrolyzed precursor was then heated at a temperature of 1200°C in air for about 24 hours and x-ray analysis showed the presence of beta-alumina and alpha-Al_2O_3 with the beta-alumina being around 70% and the alpha-alumina around 30%. Similar results were obtained when heating in air at about 1450°C for about 30 minutes. When heated for 20 minutes, while being encapsulated in platinum, at about 1585°C the dried hydrolyzed precursor showed the presence of around 80% of beta-Al_2O_3 and around 20% of alpha-Al_2O_3.

A crystalline, porous, tubular body consisting of alpha-Al_2O_3, having a porosity of about 50%, is then employed to form a composite solid electrolyte as contemplated herein. The dried hydrolyzed precursor is contacted, or applied, to one of the surfaces of the tube, and then the contacted surface is heated, for example in the range of about 1500° to 1650°C for about 15 minutes, to form a layer, or membrane, of a crystalline ionic conductive sodium polyaluminate. This structure is ideally suited for use in sodium-sulfur batteries.

Hermetically Sealed Casing Using Alpha-Alumina Ceramic Ring

S.P. Mitoff, R.W. Powers and M.W. Breiter; U.S. Patent 3,960,596; June 1, 1976; assigned to General Electric Company describe a battery casing and a sealed sodium-sulfur battery wherein the battery casing includes separate seals of low temperature melting glass joining a portion of an inner casing of a solid sodium ion-conductive material to the interior surface of a ceramic ring, and joining two opposed outer metallic casings to the ceramic ring. A sealed sodium-sulfur battery has the above type of casing with a sodium anode in one casing and a cathode of sulfur in conductive material in the other casing.

In Figure 4.11a there is shown generally at **10** a battery casing which has a ceramic ring **11**, an inner casing of a solid sodium ion-conductive material **12** with one open end **13**, and a glass seal **14** sealing a portion of the outer wall **15** of inner casing **12** adjacent its open end **13** within and to the ceramic ring **11**. A first outer metallic casing **16** with opposite open ends **17** and **18** has a glass seal **19** sealing one end **17** of first outer metallic casing **16** to ceramic ring **11**. First outer casing **16** surrounds inner casing **11** and is spaced therefrom. A removable closed end **20** for opposite open end **18** of first outer casing **16** has a fill opening **21** in removable closed end **20**, and a fill tube **22** affixed to removable closed end **20** and in communication with the fill opening **21**.

A second outer metallic casing **23** with opposite open ends **24** and **25** has a glass seal **26** sealing one end **24** of second outer metallic casing **23** to ceramic ring **11**. Second outer metallic casing **23** extends in an opposite direction to first outer metallic casing **16**.

A second removable closed end **27** for opposite open end **25** of second outer casing **23** has a fill opening **28** in second removable closed end **27**, and a fill tube **29** affixed to second removable closed end **27** in communication with fill opening **28**.

Thus, it has been found that one could form a battery casing by providing a ceramic ring, for example, of alpha-alumina. An inner casing of a solid sodium ion-conductive material of sodium beta-alumina with one open end has a portion of its outer wall adjacent its open end sealed within and to the ceramic ring with a glass seal. A first outer metallic casing with opposite open ends is sealed to the ceramic ring by a glass seal. In this manner, the first outer casing surrounds the inner casing and is spaced therefrom. A removable metallic closed end is provided for the opposite open end of the first outer casing. This closed end has a fill opening therein and a fill tube affixed thereto and is in communication with the fill opening.

A second outer metallic casing with opposite open ends is sealed to the ceramic ring by a glass seal. In this manner, the second outer metallic casing is spaced from and extends in an opposite direction to the first outer metallic casing. A second metallic removable closed end is provided for the opposite open end of the second outer casing. This closed end has a fill opening and a fill tube and is in communication with the fill opening.

FIGURE 4.11: HERMETICALLY SEALED BATTERY CASING

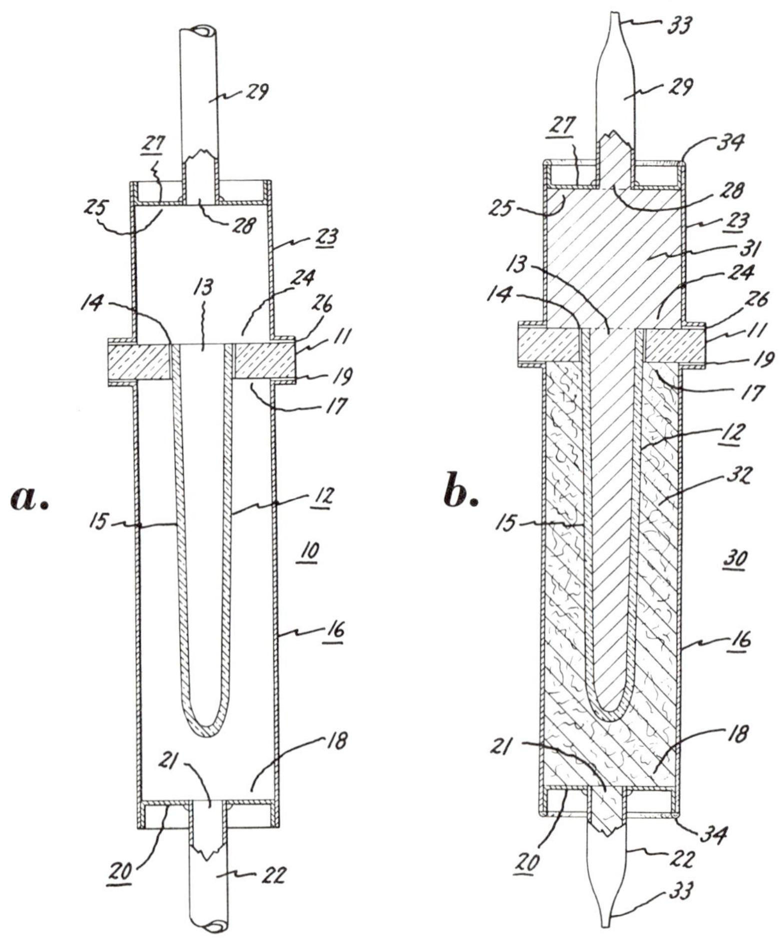

(a) Sectional view of a battery casing
(b) Sectional view of a battery

Source: U.S. Patent 3,960,596

The preferred metal for the outer casings and removable closed ends is a Rodar alloy which is an alloy of iron-nickel-cobalt. Such an alloy consists of 53 weight percent iron, 29 weight percent nickel, 17 weight percent cobalt, and 1 weight percent of impurities. Additionally, the casing, when used to contain a sulfur compound in a battery structure, can be formed of tantalum or stainless steel. The casing, when used to contain sodium in a battery structure, can be formed of an alloy of 20 weight percent nickel, 17 weight percent cobalt, 0.2 weight percent manganese, and the balance iron.

For each of the above seals, glass is provided in the form of a washer. Each washer is positioned between adjacent associated parts to be sealed. The glass washers are made of a suitable sodium and sulfur resistant glass such as Corning glass No. 7052, General Electric Company glass No. 1013, Sovirel glass No. 747 or Kimble glass No. N-51A. The associated components with the glass washer therebetween are heated to a temperature in the range of 950° to 1250°C in an argon atmosphere whereby the glass seals together the associated components. The upper portion of the outer wall of the inner casing is sealed in this manner within and to the interior surface of the ceramic ring. Each outer casing is sealed separately to the ceramic ring in the same manner. The end of each outer casing can be provided with an outwardly or inwardly extending flange for such sealing to the ceramic ring. Further, the end of each outer casing can be sealed to the perimeter of the ceramic ring.

In Figure 4.11b, there is shown a hermetically sealed sodium-sulfur battery **30** employing battery casing **10** of Figure 4.11a. An anode **31** is positioned preferably within inner casing **12**. Anode **31** is sodium metal. A cathode **32** of sulfur and sodium polysulfide in a conductive material is positioned preferably within outer casing **16** and is in contact with outer wall **15** of inner vessel **12** and with the interior surface of outer casing **16**. Fill tubes **22** and **29** are shown closed in any suitable manner such as by respective welds **33**. Closed ends **20** and **27** are affixed to outer casings **16** and **23**, respectively, as by welding **34**. The resulting structure is a hermetically sealed sodium-sulfur cell.

Example 1: A battery casing was assembled as above described and as shown in Figure 4.11a by providing an alpha-alumina ceramic ring, an inner casing of a solid sodium ion-conductive material of sodium beta-alumina with one open end, and a portion of its outer wall adjacent its open end sealed within and to the ceramic ring with No. 1013 glass. This seal was separately made in air at 1250°C. A first outer metallic casing of Rodar alloy with opposite open ends is sealed to the ceramic ring by No. 7052 glass. In the manner, the first outer casing surrounded the inner casing and was spaced therefrom.

A removable Rodar alloy closed end was provided for the opposite open end of the first outer casing. This closed end had a fill opening and a fill tube and was in communication with the fill opening. A second outer Rodar alloy casing with opposite open ends was sealed at one end to the ceramic ring by No. 7052 glass. In this manner, the second outer metallic casing is spaced from and extended in an opposite direction to the first outer metallic casing.

A second Rodar alloy removable closed end was provided for the opposite open end of the second outer casing. This closed end had a fill tube and was in communication with the fill opening. For each of the above seals, the glass was provided in the form of a washer which was held between associated parts. Each assembly was heated to a temperature of 950°C in an argon atmosphere thereby sealing together the associated parts.

Example 2: A hermetically sealed sodium-sulfur battery was assembled as above described and as shown in Figure 4.11b. The battery casing was assembled as described above in Example 1 except that the removable closed end for the sulfur compound compartment was not positioned at the end of the casing as shown in Figure 4.11a. Prior to inserting the removable closed end, a conductive material such as carbon felt was inserted within the outer casing and surrounded the inner casing.

The removable closed end was then positioned in the open end of the outer casing. The removable closed end was then welded to the end of the casing. The other removable closed end was then placed in the open end of the opposite casing and welded to the end of the casing. A molten sulfur compound of sulfur and sodium polysulfide was then added through the fill tube into contact with the carbon felt. The fill tube was then closed by crimping. Molten sodium was then added to the other fill tube after which the fill tube was closed by crimping. The resulting device was a hermetically sealed sodium-sulfur battery.

Example 3: At operating temperature, the hermetically sealed sodium-sulfur battery of Example 2 exhibited the following polarization behavior which is shown below in the table.

Current Density (mA/cm^2)	Cell Voltage (V)
1.0	2.068
3.0	2.055
10.0	2.020
30.0	1.45
100.0	1.735

In related work *M.W. Breiter; U.S. Patent 3,959,013; May 25, 1976; assigned to General Electric Company* describes a cathode cell casing portion, a cell casing, and a hermetically sealed sodium-sulfur cell. The metallic cathode cell casing portion is made of one of several specific metals with a corrosion resistant and electronically conducting layer adhering to the inner surface of the casing.

The cell casing includes the above cathode cell casing portion and an opposed anode cell casing portion joined to a ceramic ring supporting an inner casing of a solid sodium ion-conductive material. A hermetically sealed sodium-sulfur cell has the above type of casing with a sodium negative electrode in the inner casing and a positive electrode of sulfur in conductive material in the cathode casing portion surrounding the inner casing.

Liquid Metallic Layer on Solid Electrolyte

A process described by *R.R. Dubin, F.G. Will and W.L. Mowrey; U.S. Patent 3,933,523; January 20, 1976; assigned to General Electric Company* relates to solid sodium ion-conductive electrolytes and, more particularly, to such electrolytes with a liquid metallic layer on one surface.

The process provides a composite article consisting of a liquid metallic layer on one major surface of a solid sodium ion-conductive electrolyte. When the improved composite article is employed in a cell including a sodium-containing reactant, the liquid metallic layer is in contact with the sodium-containing reactant resulting in substantially improved wetting at the sodium and electrolyte interface.

It has been found that one could adhere intimately a liquid mercury-indium alloy layer on one major surface of the electrolyte disc, tube or other configuration. Intimate adherence or wetting of the alloy with electrolyte surface can be obtained by applying the alloy on the electrolyte surface and rubbing the alloy onto the surface. The alloy layer is generally from 1.0 to 100.0 microns in thickness.

Such alloys include mercury-indium, mercury-indium-thallium, mercury-sodium-indium and mercury-sodium-indium-thallium alloys. The mercury-indium alloy contains from 40 to 99 weight percent mercury and from 1 to 60 weight percent indium.

The mercury-indium-thallium alloy contains from 40 to 98.9 weight percent mercury, from 1 to 59 weight percent indium, and from 0.1 to 1.0 weight percent thallium. The mercury-sodium-indium alloy contains from 85.0 to 99.998 weight percent of sodium and mercury, and from 0.002 to 15.0 weight percent indium. The mercury-sodium-indium thallium alloy contains from 84.85 to 99.997 weight percent of sodium and mercury, from 0.002 to 15.0 weight percent indium, and from an amount of less than 0.001 to 0.15 weight percent thallium.

The above preferred alloy of 56.0 weight percent mercury-43.8 weight percent indium-0.2 weight percent thallium is available commercially as Viking LS232 liquid metal alloy. This alloy has the unique ability to "wet" virtually all materials, nonmetallic as well as metallic, to form contacts of very low electrical and thermal resistance. Further, typical materials wet by this alloy include ceramics of barium titanate, lead-zirconium titanate and ferrites and glass of boron, Pyrex, Vycor, quartz fiber glass and arsenic.

The composite article is useful in a cell using a sodium-containing reactant whereby improved cell performance results. Previous untreated beta-alumina electrolyte often exhibited unsatisfactory cell performance resulting from a large polarization occurring at the sodium or sodium mercury amalgam and electrolyte interface. Such large interfacial polarization resulted from the poor wetting characteristics of the liquid sodium or liquid sodium mercury amalgam in contact with the untreated electrolyte surface.

The improved composite article improves the wetting characteristics of the liquid sodium reactant and the beta-alumina surface in intimate contact with the mercury-indium layer. This improved wetting is employed advantageously in a cell whereby performance is improved. The beneficial effects of the alloy layer are maintained during cell operation from ambient to about 350°C temperature. The effect is more pronounced at the lower end of the temperature range.

LITHIUM BATTERIES

METAL CHALCOGENIDE CATHODES

Impregnated Porous Matrix

R.C. Saunders and L.A. Heredy; U.S. Patent 3,925,098; December 9, 1975; assigned to Electric Power Research Institute describe a positive electrode for a lithium-molten salt-transition metal chalcogenide electrical energy storage device having a greater cycle life and higher energy density than can be achieved using the known positive electrodes. In accordance with the process a rechargeable electrical energy storage device is provided which includes an anode containing lithium, a fused salt electrolyte containing lithium ions, and an improved cathode containing at least one selected transition metal chalcogenide. The improved cathode comprises a porous pliable felt matrix formed from resilient carbon fibers or filaments, the matrix being impregnated with particles of the selected transition metal chalcogenide.

The preferred positive electrode materials include copper sulfide, iron sulfide, nickel sulfide, and nickel oxide. The term "chalcogenides" refers to the Periodic Table Group VI electronegative elements in combined form, namely, the oxides, sulfides, selenides, and tellurides. Of these, because of ease of preparation, greater availability, lower molecular weight, and other specific properties, the sulfides and oxides are generally preferred. The sulfides are particularly preferred.

Because of the need for a rechargeable power-producing secondary cell having a high current density and a low internal resistance, the chalcogenides, which are solid at the temperature of operation of the molten salt cell, must be made readily available in a finely divided form presenting a high specific surface. Generally, the chalcogenides will have a median particle size of from about 20 to 150 microns and preferably a median particle size of 37 to 63 microns. Obviously, the finely divided chalcogenides must be confined within some specific volume and advantageously, they will be substantially uniformly distributed throughout such specific volume provided. Further, the products resulting from discharge of an electrical energy storage device utilizing a chalcogenide as the active cathode material will occupy approximately twice as much space as the original chalcogenide. Therefore, the volume of space provided for the chalcogenide must be sufficiently large to allow for such expansion. Also, about 20 to 70% of the free volume must be allotted for molten salt electrolyte to provide for satisfactory ionic conduction inside the electrode. Accordingly, the chalcogenide ordinarily may only occupy approximately 15 to 40% of the volume provided.

In accordance with the process, a porous felt body of resilient carbon or graphite fibers is used as the matrix. It has been found that such a body of woven or nonwoven fibers of carbon or graphite is a particularly effective matrix for retaining the finely divided particles of chalcogenide. Suitable graphite and carbon fiber materials form the subject of U.S. Patent 3,107,152 and are commercially available as felts. The felt body or matrix is further characterized as being readily compressible. The compressive strength of the felt should be such that a load of from 0.2 to 5.0 psi (120 to 3,500 kg/m^2), will cause at least a 10% deformation of the felt. The particularly preferred felts are those that are deformed at least 10% by a compressive load of from about 0.2 to 1.0 psi (140 to 700 kg/m^2). It also generally is preferred that the felt matrix have a thickness when unrestrained (free from any applied load) of from about 0.13 to 2.5 cm with a thickness of from 0.25 to 0.75 cm being particularly preferred.

The cathode structure preferably is formed by placing the porous matrix in a suitable container or housing, e.g., of metal, ceramic, or dense graphite, and substantially covering the matrix with finely divided particles of the selected chalcogenide. The container, porous matrix, and chalcogenide then are vibrated to cause the finely divided particles of chalcogenide to permeate through the interstices of the porous matrix. Periodically, the porous matrix may be removed and the excess chalcogenide shaken off. The matrix then is weighed to determine the amount of chalcogenide contained therein. Obviously, the amount of chalcogenide contained in the porous matrix should not occupy more than about 35% of the void volume of the porous matrix. Generally, it is preferred that the chalcogenide occupy from about 25 to 30% of the void volume of the matrix.

It is not critical that substantially complete uniform distribution of the chalcogenide throughout the porous matrix be obtained initially. More particularly, when the cathode structure is used, for example, in a battery, which is alternately charged and discharged, the resilient carbon fibers yield and deflect to accommodate expansion of the reacted active materials and movement of the active materials which are displaced by the volume changes of adjacent materials. In those areas where a large concentration of active material exists, such material during the discharge cycle is displaced throughout the matrix by the increased volume of the formed discharge reaction products, which deflect the resilient carbon fibers. Thus, after several cycles, the material is substantially uniformly distributed throughout the flexible felt-like matrix.

Niobium Tetraselenide

D.W. Murphy and F.A. Trumbore; U.S. Patent 4,035,555; July 12, 1977; assigned to Bell Telephone Laboratories Incorporated describe a nonaqueous secondary cell in which the positive electrode contains as the active material the compound niobium tetraselenide with nominal formula $NbSe_{4+x}$ in which x may vary from 0 to 0.5. This composition range refers to a single phase of niobium tetraselenide. The negative electrode is conventional and is lithium because of its large negative electromotive force and light weight. The electrolyte is also conventional and may contain various additives to improve recyclability. A typical electrolyte is $LiClO_4$ dissolved in propylene carbonate. Other substances may be added including tetraglyme. Batteries employing the positive electrode exhibit high energy density and are extensively rechargeable under operating conditions. In addition, when cycled at low currents these cells exhibit constant voltage over much of their discharge curve which is highly advantageous for many applications.

Example 1: A mixture of 4.767 grams of selenium shot (5N purity) and 1.403 grams of niobium powder (30 μm, 99.95% pure) was heated in a sealed vitreous silica ampoule for 3 days at 550°C. The resulting $NbSe_4$ contained fibers of $NbSe_3$, as observed in the scanning electron microscope (SEM), at levels of less than ~5% as shown by the absence of $NbSe_3$ lines in the x-ray powder pattern. A pocket (3 cm x 1 cm x 50 mils) was constructed of perforated nickel into which was stuffed 0.314 gram of the $NbSe_4$-$NbSe_3$ mixture. A cell was constructed using this pocket cathode sandwiched between two 15-mil-thick lithium anodes, all three electrodes being contained in polypropylene separator bags wetted with tetraglyme. An electrolyte of 1 M $LiClO_4$ in propylene carbonate was used.

The cell was then cycled between 2.8 and 1.0 volts at a discharge current of 5 milliamperes and a charging current between 3.6 and 5 milliamperes. The initial discharge capacity was 80.5 mAh compared with a theoretical capacity of 82 mAh assuming 4 lithium atoms per mol of $NbSe_4$. This cell was cycled 28 times with an average depth of 35 mAh.

Example 2: Niobium selenide in the range $NbSe_{4.0}$ to $NbSe_{4.5}$ was prepared by heating stoichiometric amounts of niobium and selenium in sealed ampoules from 400°C to 500°-510°C at a rate of about 25°C per day and holding at the higher temperature for 2-3 days. The resulting products were free from any significant $NbSe_3$ contamination as determined from SEM and x-ray observations. A mixture of 150 mesh $NbSe_{4.5}$, prepared in this manner, with 325 mesh graphite powder and polyethylene powder (Microthene 500) in the ratio 65.3:29.6:5.1 weight percent, respectively, was rolled in a jar mill for 16 hours. The mixture was then pressed to 6,500 psi at 130°C for 3 minutes onto an expanded nickel screen to form a composite cathode 3 cm x 3 cm x 25.5 mils thick with a weight of 1.61 grams, exclusive of the nickel screen.

A cell was constructed, again by sandwiching the cathode between two Li electrodes, all electrodes in polypropylene separator bags, and placed horizontally in a Petri dish-type container with 8 ml of 1 M $LiClO_4$ in propylene carbonate. This cell was cycled between 2.8 and 1.0 volt with a discharge current of 10 milliamperes and a charge current of 5 milliamperes. The initial discharge capacity was 109.5 mAh compared to a theoretical capacity of 252 mAh based on 4 Li atoms per mol of $NbSe_{4.5}$.

Titanium Disulfide

M.S. Whittingham; U.S. Patent 4,009,052; February 22, 1977; assigned to Exxon Research and Engineering Company describes a battery in which the anode contains as the anode-active material a metal selected from the group consisting of Groups I-a, I-b, II-a, II-b, III-a and IV-a metals (lithium is preferred); the cathode contains as the cathode-active material a chalcogenide of the formula MZ_x

where M is an element selected from the group consisting of titanium, zirconium, hafnium, niobium, tantalum and vanadium (titanium is preferred); Z is an element selected from the group consisting of sulfur, selenium and tellurium; and x is a numerical value between about 1.8 and 2.1. The electrolyte is one which does not chemically react with the anode or the cathode and which will permit the migration of ions from the anode-active material to intercalate the cathode-active material. A highly useful battery may be prepared utilizing lithium as the anode-active material, titanium disulfide as the cathode-active material and lithium perchlorate dissolved in tetrahydrofuran (70%) plus dimethoxyethane (30%) solvent as the electrolyte.

Alternatively, the battery may be constructed in the discharged state of a cathode containing as the cathode-active material an intercalated chalcogenide in which the intercalating species is derived from the same materials previously mentioned as being useful for the anode-active material and the chalcogenide would be the same as that previously described. In this alternative type of battery, the anode may merely be a substrate (e.g., a metal grid) capable of receiving the intercalating species deposited thereon.

Lithium-Aluminum Anode

According to a process described by *B.M.L. Rao; U.S. Patent 4,002,492; Jan. 11, 1977; assigned to Exxon Research & Engineering Company* a high energy density electrochemical cell comprises an anode consisting essentially of between 63 and 92% lithium, on an atomic basis, and the balance essentially aluminum, a cathode and a nonaqueous electrolyte. Advantageously, the cathode is an electrochemically active transition metal chalcogenide, such as titanium disulfide, and the nonaqueous electrolyte is an organic solvent, such as dioxolane, having at least one lithium salt, preferably lithium perchlorate, dissolved therein.

Lithium-aluminum alloys containing between about 63 and 92% lithium contain significant amounts of dilithium aluminide and anodes made of these alloys display many of the advantages of pure lithium anodes while minimizing the adverse electrochemical effects of large amounts of aluminum. These advantages are as follows: (1) decreased sacrifice of half-cell voltage; (2) increased half-cell energy density; and (3) decreased polarization voltage during charge-discharge. These improvements contribute to increased cell voltage, energy densities, and decreased cell polarization.

Powdered Graphite

L. Herédy and L.R. McCoy; U.S. Patent 3,898,096; August 5, 1975; assigned to Rockwell International Corporation describe a high-temperature lithium-molten salt power-producing secondary cell having improved cycle life on repeated charge and discharge cycles. The cell utilizes a selected transition metal chalcogenide as the electrochemically active material of the positive electrode. Preferred positive electrode materials include copper sulfide, iron sulfide, nickel sulfide, and nickel oxide.

Because of the need for a rechargeable power-producing secondary cell having a high current density and a low internal resistance, the chalcogenides, which are solid at the temperature of operation of the molten salt cell, must be made readily available in a finely divided form presenting a high specific surface. In one method of preparing such a positive electrode, the chalcogenide (other than the oxide) in finely powdered form is mixed with finely divided powdered graphite, and the powdered mixture is enclosed within a graphite container having a porous ceramic or carbon face so that the molten salt may make ready contact with the chalcogenide. Since the oxide will react with carbon, the positive electrode containing the metal oxide is generally prepared by using the oxide powder alone, or the oxide in admixture with the transition metal per se in powdered form. In certain applications it is found to be convenient to mix the fusible electrolyte in finely powdered form with the graphite and the metal chalcogenide. Upon the cell attaining operating temperature, the electrolyte is then in the molten state and makes ready contact with the positive electrode material.

In another method, a lattice of porous graphite is used, and the lattice is impregnated using a slurry of the chalcogenide in a liquid such as alcohol. The porous graphite is then baked to evaporate the liquid leaving the chalcogenide in the form of fine particles distributed throughout the interstices of the porous graphite matrix.

Example 1: A lithium-molten salt cell was prepared using CuS as the active material of the positive electrode. A mixture was prepared consisting of CuS, carbon black, and a binary salt (LiCl-KCl eutectic) at a weight ratio of 30:50:20, respectively. The depth of the cathode cavity was 0.05 inch, and a total of 0.49 gram (0.042 ampere-hour theoretical capacity calculated for Cu_2S) of the cathode mixture was hard pressed into a dense graphite holder. Calcium zirconate (28% porosity) was used to cover the mixed cathode material as a separator. A magnesia cylinder was used to contain the molten lithium pool serving as the negative electrode. The cell was operated at 400°C. After initial cycling, it was found that saturating the electrolyte with Li_2S (2 g/l) prevented the escape of active sulfide material from the cathode and retained the cathode capacity over a long period of time.

The cell was cycled for more than 250 cycles using a charge cut-off of 2 volts and a discharge cut-off at 1 volt. Cycling was discontinued following mechanical failure (cracking) of the calcium zirconate separator. After the 97th cycle, the cell was left on open circuit overnight for 16 hours; the cell voltage remained steady at 1.68 volts. A discharge current at 50 mA was then applied; the ampere-hour efficiency then decreased from 100 to 82.5%, indicating some self-discharge.

Example 2: A lithium-molten salt cell having 0.38 ampere-hour storage capacity was prepared using copper sulfide as the active material of the positive electrode. The pores of a porous graphite disk (90% porosity) were filled with fine (–400 mesh) CuS powder. This disk was then encased in a dense graphite electrode holder. The graphite electrode structure was sealed with a porous alumina separator, and the electrode was immersed in a LiCl-KCl eutectic salt saturated with Li_2S (0.2 wt %) as electrolyte. Molten lithium absorbed on a porous metal substrate was used as the negative electrode.

More than 400 charge-discharge cycles were run at 50 mA/cm^2. Discharge was at a lower voltage cut-off of 1.0 volt, with charge being at an upper voltage cut-off of 2.0 volts. The open-circuit voltage of the cell was 1.68 volts. The discharge time was 95 minutes compared with a charge time of 100 minutes, showing 95% coulombic efficiency. The average discharge voltage was 1.35 volts, with an average charging voltage of 1.90 volts. While CuS was initially utilized, it was found that after the first discharge cycle the CuS was converted to Cu_2S and then to metallic copper. On charge, it was noted that the Cu was converted to Cu_2S, and this indicated the end of the charge cycle. Therefore Cu_2S appeared to be the active electrode material of the rechargeable copper sulfide-containing positive electrode.

Example 3: A lithium-molten salt cell utilizing K_2MoS_4 as positive electrode active material was prepared. A mixture of the molten salt (KCl-LiCl eutectic) with the molybdate and carbon black was prepared. The weight ratio of K_2MoS_4:carbon black:binary salt was 50:30:20. Both the K_2MoS_4 and the carbon black were prior vacuum heated overnight to remove any moisture present. Two grams of the mixture was then pressed into a disk cavity similar to that used in Example 1. The lithium electrode was prepared in a similar manner to that of Example 1. An alumina separator impregnated with electrolyte was used to cover the mixed cathode as diffusion barrier. The cell was operated at 370° to 380°C.

The cell was run for a total of 301 cycles. The cell operation reached steady-state conditions after 200 cycles with a coulombic efficiency of 55%, and 25% of theoretical capacity being utilized calculated on MoS_3 as the active component. X-ray diffraction analysis of the discharged mixed cathode compartment at the end of the run showed that the major components were LiCl, KCl, and MoS_2. The run was made without saturation of the electrolyte with Li_2S, which ordinarily would be expected to enhance the retention of sulfur in the cell thereby giving a greater coulombic efficiency.

METALLIC SULFIDE CATHODES

Iron Powder and Lithium Sulfide

According to a process described by *H. Shimotake, L.G. Bartholme and J.D. Arntzen; U.S. Patent 4,006,034; February 1, 1977; assigned to the U.S. Energy Research and Development Administration* a secondary electrochemical cell is assembled in an uncharged state for the preparation of a lithium alloy-transition metal sulfide cell. The negative electrode includes a material such as aluminum or silicon for alloying with lithium as the cell is charged. The positive electrode is prepared by blending particulate lithium sulfide, transition metal powder and electrolytic salt in solid phase.

The mixture is simultaneously heated to a temperature in excess of the melting point of the electrolyte and pressed onto an electrically conductive substrate to form a plaque. The plaque is assembled as a positive electrode within the cell. During the first charge cycle lithium alloy is formed within the negative electrode and transition metal sulfide such as iron sulfide is produced within the positive electrode.

In more specific aspects of the process, the transition metal powder within the positive electrode mixture can comprise by weight a major portion of iron and a minor portion of copper for the formation of FeS and copper sulfides on electrochemically charging. In one other procedure a minor portion by weight of cobalt can be combined with a major portion of iron to provide FeS_2 and cobalt sulfides. The positive electrode is assembled within the electrochemical cell along with negative electrodes containing a material such as aluminum or silicon for alloying with lithium and any other alkali metal or alkaline earth metal additive. The cell with added electrolyte is electrically charged to produce a transition metal sulfide such as ferrous sulfide within the positive electrode and a lithium alloy such as lithium-aluminum within the negative electrode. The initial and subsequent charges during cell operation are at sufficiently low voltages to prevent formation of free sulfur or metal sulfides of higher potential. In the case of the FeS positive electrode, charge voltages of less than about 1.8 V are used to prevent FeS_2 formation with accompanying corrosion of iron current collector and structural components. This is of particular importance where excess Li_2S over iron powder is included in the initial positive electrode composition.

Where additives such as lithium carbide are within the initial composition and are electrochemically charged to lithium-aluminum alloy in the negative and carbon particles within the positive electrode, discharge voltages are maintained at a sufficiently high level, e.g., above 0.8 V for lithium carbide, to prevent reformation of the additive.

Example: (Cell R-7) — A solid particulate mixture containing 104 grams Li_2S, 101 grams Fe powder, 57 grams Cu powder, 112 grams LiCl-KCl eutectic were blended at ambient temperature and ground to pass 351 micrometer openings (-42 mesh). The mixture was divided into two parts and each part poured into a die containing an iron backing sheet covered with an attached iron mesh with about 1 cm openings. The die was heated to 360°C which is above the melting point of the electrolytic salt and the mixture was pressed at approximately 2.8 MPa (400 psig) for 10 minutes while maintaining that temperature. The resulting plaques that were formed were spot welded together at the backing sheets and encased within layers of zirconia and boron nitride cloth. The boron nitride cloth was prewetted with a methyl alcohol solution of LiCl-KCl electrolyte. The assembled plaques were then equipped with an electrode terminal and enclosed within an interlocking frame assembly.

The negative electrodes were prepared by placing a stainless steel screen between layers of aluminum wire and compacting at about 6.9 MPa (10,000 psig) and 200°C. The cell was assembled and operated for an extended period of time. One significant improvement shown in the performance of this and other cells prepared in accordance with the process over previous cells was in the high utilization of active material (75 to 85%) in the initial and early operating cycles. Other characteristics of the cell R-7 are shown in the table below along with data for cells R-5 and R-6, prepared in similar manner.

Cell No.	R-5	R-6	R-7
Positive Electrode			
Active Material	Li_2S-Fe	Li_2S-Fe	Li_2S-Fe
Additive Material	None	10 Wt. % Cu	20 Wt. % Cu
Electrode Area, cm^2	278	278	278
Theo. Capacity, A-hr	69.2	56.3	105
Current Collector	Iron sheet	Iron sheet	Iron sheet
Initial Thickness, cm	0.95	0.68	1.08

(continued)

Cell No.	R-5	R-6	R-7
Negative Electrode			
Material	Al	Al	Al
Electrode Area, cm^2	320	320	320
Theo. Capacity, A-hr	69.2	56.3	105
Current Collector	Al wire-SS strap	SS screen	SS screen
Initial Thickness, cm	0.4/plaque	0.4/plaque	0.26/plaque
Test Results			
Charge Cutoff Voltage (IR-free), V	1.65 V	1.65	1.65
Discharge Cutoff Voltage (IR-free), V	1.0	1.0	1.0
Discharge Current, A	4	5	5
Capacity A-hr	53.66	44	86
Utilization, %	77.5	78	71
Cell Temp., °C.	450	450	450
Charge Current, A	4	5	5
Ah Efficiency, %	99.5	93	98
Wh Efficiency, %	73.2	80	85
Cell Resistance, M	60	4	4
Cell Life			
Test Duration, hr	>3500	>3000	>2500
Number of Cycles	>250	>200	>150

In some related work *T.D. Kaun, D.R. Vissers and H. Shimotake; U.S. Patent 4,011,373; March 8, 1977; assigned to the U.S. Energy Research and Development Administration* describe an uncharged positive-electrode composition which contains particulate lithium sulfide, another alkali metal or alkaline earth metal compound other than sulfide, e.g., lithium carbide, and a transition metal powder. The composition along with a binder, such as electrolytic salt or a thermosetting resin is applied onto an electrically conductive substrate to form a plaque.

The plaque is assembled as a positive electrode within an electrochemical cell opposite to a negative electrode containing a material such as aluminum or silicon for alloying with lithium. During charging, lithium alloy is formed within the negative electrode and transition metal sulfide such as iron sulfide is produced within the positive electrode. Excess negative electrode capacity over that from the transition metal sulfide is provided due to the electrochemical reaction of the other than sulfide alkali metal or alkaline earth metal compound.

W.J. Walsh, C.C. McPheeters, N.-P. Yao and K. Koura; U.S. Patent 3,992,222; November 16, 1976; assigned to the U.S. Energy Research and Development Administration describe an active material for use within the positive electrode of a secondary electrochemical cell which includes a mixture of iron disulfide and a sulfide of a polyvalent metal. Various metal sulfides, particularly sulfides of cobalt, nickel, copper, cerium and manganese, are added in minor weight proportion in respect to iron disulfide for improving the electrode performance and reducing current collector requirements.

Iron Mesh and Lithium Sulfide

According to a process described by *N.-P. Yao and W.J. Walsh; U.S. Patent 3,947,291; March 30, 1976; assigned to the U.S. Energy Research and Development Administration* a completely discharged, secondary electrochemical cell assembly is provided with a positive and a negative electrode along with an electrolyte including a lithium salt. The positive electrode includes one or more layers of metallic mesh or screen having an intimate mixture of an alkali metal sulfide such as lithium sulfide and the electrolyte embedded into the mesh. The mesh as thus impregnated, is held within a porous containment structure that is permeable to molten electrolyte. An electrode terminal penetrates the containment to make electrical contact with the mixture and the metallic mesh.

In the negative electrode is a porous mass of aluminum metal impregnable by the electrolyte in molten state and an electrode terminal in electrical contact with the mass of aluminum. The terminals to both the positive and negative electrodes are electrically accessible from outside the cell housing for connection to an external electrical load or circuit. By passing an electrical current through the cell, a metal sulfide is produced in the positive electrode from reaction of the lithium sulfide and the metallic mesh while lithium-aluminum alloy is formed within the negative electrode.

Metal Sulfide-Carbon Porous Structures

According to a process described by *E.C. Gay and F.J. Martino; U.S. Patent 3,933,520; January 20, 1976; assigned to the U.S. Energy Research and Development Administration* particulate electrode reactants, for instance transition metal sulfides for the positive electrodes and lithium alloys for the negative electrodes, are vibratorily compacted into porous, electrically conductive structures. Structures of high porosity support sufficient reactant material to provide high cell capacity per unit weight while serving as an electrical current collector to improve the utilization of reactant materials. Pore sizes of the structure and particle sizes of the reactant material are selected to permit uniform vibratory loading of the substrate without settling of the reactant material during cycling.

Example 1: Positive Electrode — Particulate FeS_2 in sufficient amount to provide a theoretical capacity density of 0.92 Ah/cm^2 (16.1 grams) is loaded into a porous vitreous carbon substrate of about 98% porosity in the manner described above. The substrate is about 5 cm in diameter and 0.5 cm thick and is tightly fitted into a porous graphite cup of greater strength and lower porosity. This positive electrode is operated opposite a negative electrode of similar diameter and approximately 0.32 cm thick. The negative electrode is electrochemically prepared by a method in which sufficient lithium is electroplated onto a porous aluminum structure with a stainless steel current collector interwoven therein to provide a theoretical capacity density of 0.43 Ah/cm^2 based on lithium.

The cell was operated at about 400°C for over 360 hours and 35 cycles. Its most impressive performance was that of 0.147 Ah/cm^2 (corresponding to 32.2% of the Li-Al theoretical capacity density) at a discharge current density of 0.425 ampere per square centimeter. This high current density was sustained for 20.7 minutes.

Example 2: Negative Electrode — The same positive electrode was employed opposite a negative electrode having 60 atom % lithium in Li-Al alloy that was metallurgically prepared, vibratorily loaded into a substrate of about 95% porosity and 1.62 cm thick nickel foam. Sufficient lithium was provided for a theoretical capacity density of 1.55 Ah/cm^2. The cell was operated for 7 cycles and 144 hours.

Copper Sulfide

R.K. Steunenberg, A.E. Martin and Z. Tomczuk; U.S. Patent 3,941,612; March 2, 1976; assigned to the U.S. Energy Research and Development Administration describe a high-temperature, secondary electrochemical cell which includes a negative electrode containing an alkali metal such as lithium, an electrolyte of molten salt containing ions of that alkali metal and a positive electrode containing a mixture of metallic sulfides. The positive electrode composition is contained within a porous structure that permits permeation of molten electrolyte and includes a mixture of about 5 to 30% by weight Cu_2S in FeS.

In providing the electrode compositions, a uniform and homogeneous blend of FeS and Cu_2S is prepared. Fine powders, e.g., about 100 to 400 micron size, of each of these constituents can be thoroughly blended into a uniform mixture. It is of importance that no large domains of FeS are left to react with the electrolyte to form the solid J phase.

One manner of ensuring a near uniform composition of FeS and Cu_2S is to melt powders of the materials in the desired proportions at about 1200°C and form a molten solution. After thorough blending, the melt is frozen rapidly by casting in a cold metal mold of sufficient mass and heat capacity to cool the melt to about 900°C in about 3 minutes. The rapid solidification produces a solid ingot with a minimum amount of macroscopic nonuniformity. The ingot contains a solid solution of Cu_2S in FeS and one or more ternary Cu-FeS phases such as chalcopyrite, $CuFeS_2$ and bornite, Cu_5FeS_4. The cooled ingot, having a near uniform composition, is ground to a fine powder of about 100 to 400 microns particle size and blended thoroughly. Most of the individual particles will contain both a solid solution of Cu_2S in FeS and one or more ternary phases, since the crystal structure formed by this method will be of smaller scale than these particle sizes.

In order to demonstrate the process, a number of cells were set up and operated with various cathode compositions. The first of the cells employed only FeS powder as the active cathode material, another employed only Cu_2S and the remaining cells included a mixture of Cu_2S and FeS. One of the cells designated (K-1) used a commercial matte containing both Cu_2S and FeS as the positive electrode material. This type matte is routinely produced in an early separation step in the recovery of copper metal from ore. The ore is melted and iron silicate slag removed, leaving the heavier molten matte of primarily Cu_2S and FeS with a few percent of other metallic sulfides and impurities.

Each of the cells had a negative electrode area of about 5 cm^2 and a positive electrode area of about 4.75 cm^2. The negative electrodes included a substrate and current collector of porous stainless steel (known as Feltmetal) impregnated with molten lithium metal at the cell operating temperature. The interelectrode spacing in each cell was about 1 cm.

Tables 1, 2 and 3 are given below to summarize the characteristics of these electrochemical cells. Table 1 gives the physical characteristics of the overall cell, while Table 2 shows the composition and source of the positive electrode materials. Table 3 gives some of the electrical performance characteristics for each of these five cells.

TABLE 1

	Cell Designation				
	Li/FeS (1)	Li/FeS-Cu_2S (2)	Li/FeS-Cu_2S (K-1)	Li/Cu_2S (3)	Li/FeS-Cu_2S (4)
Negative electrode					
Weight Li, g (approximately)	1.3	1.2	1.0	1.2	1.3
Theoretical capacity, Ah (approximately)	5.0	4.6	3.9	4.6	5.0
Positive electrode					
Weight of active material, g	1.7	1.8	2.0	1.5	2.0
Theoretical capacity, Ah	1.04	1.0	0.90	0.5	1.12
Theoretical capacity density, Ah/cm^2	0.22	0.21	0.19	0.1	0.21
Powdered carbon, g (current collector)	none	0.5	0.2	1.0	none
Electrolyte-LiCl-KCl, g	99.7	87.1	95.7	89.8	92.6
Operating temperature, °C	387-436	389-437	412-430	380-433	420

TABLE 2

Cell Designation	Source of Electrode Material	Composition of Electrode Material Cu_2S (wt %)	x*	Equivalent Weight
Li/FeS (1)	FeS powder	0	0	43.95
Li/FeS-Cu_2S (2)	FeS and Cu_2S powders melted to form an ingot which was ground to form a powder	20.0**	0.121	48.27
Li/FeS-Cu_2S (K-1)	Powder matte	31.8***	0.216†	51.64††
Li/Cu_2S (3)	Cu_2S powder	100	1.0	79.57
Li/FeS-Cu_2S (4)	FeS and Cu_2S powders, physically mixed	16.8**	0.100	47.52

*The value of x in the general formula: $Fe_{1-x}Cu_{2x}S$.
**Balance from 100 is FeS.
***Weight percent FeS = 63.9, balance of 4.3% = impurities.
†The impurities were ignored in calculating x for this matte.
††The equivalent weight for the FeS-Cu_2S component of the matte.

TABLE 3

	Cell Designation				
	Li/FeS (1)	Li/FeS-Cu_2S (2)	Li/FeS-Cu_2S (K-1)	Li/Cu_2S (3)	Li/FeS-Cu_2S (4)
Current density, mA/cm^2					
Discharge	30-40	30-40	21-38	~30	25-38
Charge	30-40	30-40	21-38	~30	25-38
Cutoff voltage, V					
Discharge	1.06	0.95	1.02	1.0	1.05
Charge	2.0	1.927	1.975	1.92	1.96
Number of completed cycles	9	17	9	23	8
Average capacity, % of theoretical*					
Discharge	77.3	79.5	**	46	89.2
Charge	79	83.2	**	47	95.0
Cell termination voltage	2.3	1.93	1.70	—	1.96
Phases observed in fully charged positive electrode	J, FeS	FeS, J	Li_2FeS_2	—	FeS, Li_2FeS_2

*Based on the reactions: 2 Li + FeS → Li_2S + Fe and Cu_2S + 2 Li → Li_2S + 2 Cu.
**Data beyond cycle two were poor because of Li electrode dewetting and are not included.

The performance tests conducted with the Le/FeS (1) and other similar Li/FeS cells indicate that their characteristics vary substantially with temperature. This may be due to the formation of the J (tentatively $K_2Fe_7S_8$) phase in the fully charged electrode. Not only is the J phase a solid material at the cell operating temperature, but in its formation by removing KCl from the electrolyte, it can cause the crystallization of an approximately equivalent amount of LiCl, thus reducing the amount of molten electrolyte associated with the positive electrode. At the lower operating temperatures (e.g., 380° to 450°C), pockets or crusts of solidified LiCl could be formed within the positive electrode to dampen its performance.

One particular characteristic associated with the J phase and observed with the Li/FeS cells is that at sufficiently high charge voltages it can be converted to FeS. At 380°C this appeared to occur at about 2.2 volts (IR included), while at 490°C a charge voltage of only about 1.7 volts (IR included) appeared to produce FeS. Consequently, in plots of charge voltage versus time the lower temperature cells show two plateaus, one for the conversion of the cell reaction product to J phase and the second plateau representing the conversion of J phase to FeS. Both of these conversions appear to occur at about the same voltage for cell temperatures of 490°C and above. The higher charge voltages, however, are undesirable within a cell containing iron as current collector material, because of its possible oxidation to $FeCl_2$. In addition, the higher operating temperatures are to be avoided in order to extend the life of electrical insulators and electrode feedthrough components.

An examination of Table 3, showing the results of the operation of the five cells, reveals substantial improvement for the cells including small weight percentages of Cu_2S additive to the FeS positive electrode material. In cell (1), having only the FeS as the positive electrode composition, the fully charged positive electrode included both FeS and J phase, but it was necessary to employ elevated charge voltage at the cell termination (2.3 volts) to obtain conversion to FeS. In cell (2), having 20 weight percent Cu_2S, some J phase was observed, but in a substantially reduced amount to that found in previous FeS electrodes charged at the same voltages and temperatures. In cells (K-1) and (4), having approximately 32 and 17 weight percent Cu_2S, no J phase was detected even at the reduced termination voltages.

It is also seen from Table 3 that improvement in the utilization of the positive electrode material, that is, the average capacity as a percent of theoretical, results from the addition of minor percentages of Cu_2S to the FeS positive electrode material. This is a completely unexpected result, as the average capacity of cell (3) employing Cu_2S alone as a positive electrode reactant exhibits rather poor utilization of the positive electrode material. It is also to be noted that these increases in utilization occur at low current densities, e.g., 20 to 40 mA/cm^2. More substantial improvements are to be described below in conjunction with tests conducted at higher current densities.

Lithium Sulfide as Cathode Reactant Product

E.C. Gay and F.J. Martino; U.S. Patent 3,907,589; September 23, 1975; assigned to the U.S. Energy Research and Development Administration describe a method for preparing a high-temperature, electrochemical cell including a transition metal sulfide as a cathode reactant, alkali metal as an anode reactant and an electrolyte including ions of the alkali metal. The method comprises mixing the transition metal sulfide and the sulfide product of the cell reaction into a flocculent homogeneous mass with electrolyte and a particulate current collector. The mass is enclosed within a cathode structure and assembled with the electrolyte and anode as an electrochemical cell. The cell is heated to a temperature sufficient to melt the electrolyte. An electric current is applied from cathode to anode wherein the sulfide reaction product is electrolyzed in the cathode to form additional cathode reactant other than the initial, transition metal sulfide. After thus charging the cell, it is then ready for use as a source of electrical voltage.

Thus, it has been found that, by including critical concentrations of the alkali metal sulfide that is the reaction product of the cell, e.g., lithium sulfide, within the cathode mixture, beneficial results are produced. Sufficient lithium sulfide is included within the cathode material to provide a total molar, that is atomic, concentration of sulfur in stoichiometric excess of the total atomic concentration of iron or other transition metal. In the case of iron pyrite, FeS_2, the atomic concentration of sulfur is more than twice the atomic concentration of iron within the cathode material. The following examples are presented to illustrate the cathode composition and the method of the process. The table below summarizes the physical characteristics of the cells presented in Examples 1 through 3.

	Example Number		
	1	2	3
Cell number*	S-48	S-52	S-50
Lithium electrode			
area, cm^2	22.8	167.2	12.9
Li or Li alloy, g	10.2	19.9**	2.5**
Electrolyte	LiCl-KCl	LiF-LiCl-KCl	LiF-LiCl-KCl
Sulfur electrode			
area, cm^2	16.5	50.8	5.1
Active material	FeS_2	FeS_2	FeS_2 + Li_2S
Weight, g	11.6	40.0	5.0
Composition, wt %			
FeS_2	60.0	60.0	33.5
Li_2S	2.2	2.2	26.6
Electrolyte	29.3	29.3	32.5
Carbon black	7.0	7.0	6.6
Fe	1.5	1.5	0.8
Equivalents, S			
FeS_2	11.6	40.0	2.8
Li_2S	0.56	1.9	2.9
Fe (as FeS_2)	0.60	2.1	0.14
Theoretical capacity			
density, Ah/cm^2	0.67	0.44	0.60

*The interelectrode distance for these cells was 0.6 cm, the temperature range was 380° to 400°C; a boron nitride sheath was used for all of the cells.

**These anodes contained about 25 wt % copper and 75 wt % lithium impregnated within a porous metallic substrate.

Example 1: (Cell S-48) – A little more than 11 grams of cathode constituents are blended together to form a flocculent mass of cathode material at the cell operating temperature of about 400°C. The cathode composition comprises about 60 wt % FeS_2, 2.2 wt % Li_2S, 29.3 wt % electrolyte (LiCl-KCl), 7 wt % carbon black and 1.5 wt % of particulate iron filings. The amount of iron is more or less stoichiometrically equivalent to the amount of sulfur within Li_2S, such that on charging this cell, all of the sulfur can be combined to form Fe_2S. The iron pyrite, lithium sulfide and iron filings are all commercial-grade products and, in some instances, contained inert impurities. For example, up to 15 wt % silicon oxide was noted in the iron pyrite samples. Particle sizes of about 0.015 μm to 0.2 mm have been found to be sufficient for use in the cathode composition. The carbon black is a finely divided commercial product of about 0.015 μm particle size. This carbon black material is of low porosity, that is, individual carbon black particles have low surface areas in respect to volume of the particles.

The cell constituents are thoroughly blended together within a vitreous carbon crucible at about 400°C. The resulting mixture appeared as a flocculent but homogeneous mass with intimate wetting of both the iron pyrite and carbon black particles with the electrolyte. The mixture is in contrast to previous cathode compositions containing no lithium sulfide in which separation of the iron pyrite and carbon black appeared.

On assembling the cathode composition within the cathode chamber of an electrolytic cell, capacity densities greater than 0.3 Ah/cm^2 were measured at 0.1 A/cm^2 to a 1.0 V cutoff. This corresponds to a specific capacity of about 0.5 Ah/cm^3 of sulfur electrode volume. However, in spite of these promising results, a substantial volume increase was noted in the cathode materials.

Example 2: (Cell S-52) – A somewhat larger electrochemical cell with a cathode composition substantially the same as that shown in Example 1 was prepared. As in Example 1, the cell size and characteristics are set forth in the above table. This cell, cell (S-52), employed a cathode structure made of a molybdenum mesh basket enclosed with boron nitride cloth. A coiled molybdenum mesh current collector was placed within the basket and in electrical contact with the cathode composition. The electrode assembly was covered with boron nitride cloth and compressed between two molybdenum nuts. The cathode as thus prepared was assembled within a sealed electrochemical cell containing an anode as described above, except that approximately 25 wt % copper was blended into solution with the lithium metal.

The initial operation of the cell was interrupted by a short circuit between the electrodes caused by expansion of the cathode composition. After the electrical short was cleared up, the cell was operated for 165 hours and 21 cycles.

Example 3: (Cell S-50) – Cell (S-50) was prepared in much the same manner as cell (S-48) and its characteristics are also shown in the table above. This cell includes a substantial concentration of lithium sulfide within the initial cathode composition, thus providing a considerable stoichiometric excess of sulfur over that required to form iron disulfide. This cell demonstrated a much higher power density than the previous cells having only small quantities of Li_2S in the initial cathode composition. Power densities even for a short time are of importance in evaluating a cell or battery for use as a power supply within an electric automobile or other vehicle. Short-time power densities as much as 1.4 watts per cm^2 with power densities of 0.6 watt per cm^2 sustained for up to 50 seconds were obtained from cell (S-50). This is in contrast to short-time power densities of only 0.5 watt per cm^2 obtained in cell (S-52) and cell (S-48). Additional detailed performance data for cell (S-50) are given in the table below.

Performance Data for Cell S-50*

Cycle No.	Discharge Current Density (A/cm^2)	Charge Current Density (A/cm^2)	Discharge Capacity Density (A/cm^2)	Percent of Theoretical Capacity Density (%)	Average Discharge Voltage (V)	Watt-Hour Efficiency (%)
3	0.11	0.05	0.25	41.7	1.5	57.8
21	0.11	0.05	0.32	53.3	1.4	65.5
28	0.31	0.05	0.14	22.6	1.2	15.3
31	0.06	0.10	0.21	35.2	1.5	73.3
34	0.11	0.20	0.18	29.2	1.3	64.7
39	0.20	0.20	0.16	26.9	1.1	44.3
71	0.10	0.05	0.36	60.3	1.5	60.0

*The charge and discharge cutoff voltages were 2.37 and 1.0 V, respectively.

It can be seen from the above table that at least some of the sulfur introduced into the cathode composition as LiS_2 is converted into a cathode reactant. The theoretical capacity density employed in computing the fifth column of the above table is based on reaction of all of the sulfur in both the FeS_2 and in the Li_2S. In Example 3 the amount of sulfur required to combine with all of the iron as FeS_2 is about one-half the total sulfur in the cathode composition. (See the first table for equivalents of sulfur.) Phrased differently, the atomic concentration of sulfur in this cathode is about four times the atomic concentration of iron.

Thus, a reaction of all the FeS_2 to form Li_2S in any one cycle, which is most improbable, would provide a percentage of theoretical capacity density of only 50%. In view of the above percentages near and in excess of 50%, it is only reasonable to conclude that a cathode reactant other than FeS_2 is present within the charged cell of this example.

Another aspect of cell (S-50) performance shown in the above table is the reproducible capacity density with cycling. Cycles 3, 21 and 71, all having approximately the same charge and discharge currents, show a slight increase in capacity with cycling. Intervening cycles shown in the table were either charged or discharged at increased currents which accounts for their decreased capacity densities.

Porous Matrix of Thermosetting, Carbonaceous Material

T.D. Kaun; U.S. Patent 4,011,374; March 8, 1977; assigned to the U.S. Energy Research and Development Administration describes an electrode for use in a secondary electrochemical cell. The electrode includes a porous matrix of thermosetting, carbonaceous material with solid particles of the electrode active material fixedly embedded within the matrix. The matrix can also contain embedded electrically conductive materials for electrode current collection.

The process also comprehends a paste composition that can be molded into desired electrode shapes. The paste composition includes a thermosetting, carbonaceous material, particulate active material and a solid volatile, all blended into a generally uniform paste mixture. After forming the desired electrode shape and structure with the paste composition, the paste is heated to a sufficient temperature to transform the volatile to vapor phase, by sublimation or decomposition, while curing the thermosetting carbonaceous material to form a rigid but porous matrix.

The improved electrode structure is suitable for both positive and negative electrodes within secondary cells. It is particularly well suited for electrodes that include solid, particulate active material and are to be operated in an orientation, e.g., vertical, tending to produce slumping or drifting during cycling. Examples of contemplated active materials in the positive electrodes include the chalcogenides, i.e., the oxides, sulfides and selenides of metals such as iron, cobalt, nickel, copper, lead, zinc, antimony, and manganese. In the negative electrode, examples of active materials include solid alloys of lithium, calcium or possibly sodium with such relatively inert elements as aluminum, boron, beryllium or magnesium. Also metals that remain solid at the cell operating temperature, e.g., lead in a lead-sulfuric acid battery, can be employed.

Example 1: (Cell TK-3) – About 20 to 25 grams of a paste composition including 5% by volume phenolformaldehyde resin, 45% FeS_2 particles, about 60 to 230 micrometers particle size, and 50% particulate ammonium carbonate, about 40 micrometers particle size, was prepared by blending the constituents together into a uniform mixture. A thin layer of a few millimeters thickness of this paste was spread over an expanded molybdenum mesh screen within a graphite cup. A layer of carbon cloth was then embedded into the exposed face of the paste mixture. The paste was cured in air by slowly heating to a temperature of about 60°C over a period of about 2 hours and then heating to 120°C which was maintained for about 16 hours.

This procedure produced a rigid porous substrate structure including about 50% porosity with much of the active material, FeS_2, exposed to the interstitial volume. The substrate was assembled in an experimental cell opposite a Li-Al electrode of excess capacity and operated at about 450°C with LiCl-KCl eutectic salt as electrolyte. The cell operated for over 350 hours and 30 cycles using about 86% of theoretical capacity at 40 mA/cm^2 current density and 78% at 60 and 80 mA/cm^2 current densities. The positive electrode exhibited no apparent deterioration during the test.

Example 2: As a proposed alternative to the positive electrode described in Example 1, copper acetylacetone is substituted for ammonium carbonate as the volatile material in preparing the paste mixture. After partially curing the paste, the electrode is disposed in an inert gas atmosphere and heated to about 1000°C for about 7 minutes in order to carburize the thermosetting resin. Further temperature increase to about 2800°C for about 8 hours graphitizes the structure to form an electrically conductive carbon matrix. During the early portions of the heating procedure, the volatile is driven off to ultimately form a porous, graphite matrix with embedded FeS_2 particles exposed to interconnecting interstitial volumes.

Example 3: The paste composition of Example 1 is altered by substituting an epoxy resin of epichlorohydrin, bisphenol A and diethylenetriamine for the phenolformaldehyde and by substituting ammonium bicarbonate for the volatile. In addition to the other base constituents, approximately 1.5 grams of graphite powder, of less than 40 micrometers particle size, is included into the paste mixture to impart added current collection.

Example 4: (Cell KK-1) — About 250 grams of a paste including by volume about 5% phenolformaldehyde, about 45% FeS particles and about 50% ammonium carbonate was prepared. The paste was spread in two 5-mm thick layers on two perforated iron sheets and cured at about 50°C for 18 hours in air and at about 110°C under vacuum for 6 hours to insure removal of all volatiles. These two portions of the positive electrode were assembled with a sheet of carbon cloth between the two halves contacting the steel sheets and with zirconia cloth and stainless steel cloth assembled around the periphery as retainers. This positive electrode was assembled along with conventional negative electrodes within a cell having the characteristics shown in the table below.

The cell was operated at 450° to 525°C for over 1,300 hours and 62 cycles at more than 75% energy efficiency and 60% active material utilization. Current densities between 25 to 150 mA/cm^2 were obtained. Inspection of the positive electrode after operation indicated that the FeS had remained fairly uniformly distributed within the porous, carbonaceous matrix.

General*	
Matched electrode capacity	105 Ah
Theoretical energy density	120 Wh/kg
Positive Electrode	
Dimensions	7.6 x 12.7 cm
Half thickness	0.5 cm
Electrode surface area	190 cm^2
FeS loading density (50% void vol)	1.15 Ah/cm^3
Current collector	0.04 mm thick steel
Negative Electrode – LiAl in Iron Foam	
Electrode thickness (each electrode)	0.8 cm
Electrode area (each electrode)	95 cm^2
Separator	BN cloth with yttria on negative side
Electrolyte	LiCl-KCl eutectic

*Two negative and one intermediate positive electrode in vertical layers

Lithium-Aluminum Alloy Anode

Z. Tomczuk, T.W. Olszanski and J.E. Battles; U.S. Patent 4,011,372; March 8, 1977; assigned to the U.S. Energy Research and Development Administration describe a negative electrode that includes a lithium alloy as active material which is prepared by briefly submerging a porous, electrically conductive substrate within a melt of the alloy. Prior to solidification, excess melt can be removed by vibrating or otherwise manipulating the filled substrate to expose interstitial surfaces. Electrodes of such as solid lithium-aluminum filled within a substrate of metal foam are provided.

Example 1: (Cell Li-Al/FeS No. 1) — Two foamed nickel discs of about 96% porosity with about 800 micrometer pore size were immersed for about 15 seconds within a tantalum crucible containing 40 atom % lithium and 60 atom % aluminum as molten alloy at about 700°C. The substrate was removed and gently shaken manually within a helium gas atmosphere of about 20° to 30°C. Because of surface tension effects the liquid alloy more or less completely filled the foamed metal structure and only an estimated 5 to 10% of the porosity was regained prior to solidification of the alloy. The two discs filled with solid lithium-aluminum alloy were assembled as a two-layered negative electrode within a cell. Other characteristics of the cell are shown in the table below.

Physical Characteristics of Cell Li-Al/FeS (No. 1)

Positive Electrode	
Area, cm^2	4.8
Active material	FeS
Weight of active material, grams	0.81
Theoretical capacity, Ah	0.5
Theoretical capacity density, Ah/cm^2	0.1
Current collector	none
Negative Electrode	
Area, cm^2	3.1
Weight of Li-Al, grams	~2.4
Current collector	Ni foam
Theoretical capacity, Ah	~1.3
Other Characteristics	
Weight of electrolyte (LiCl-KCl), grams	127
Interelectrode distance, cm	~1
Operating temperature, °C	440

The cell was operated for 21 discharge-charge cycles and about 300 hours at current densities varying from 20 to 30 mA/cm^2 (based on the negative electrode area) before it was voluntarily terminated. Cutoff voltages of about 1.0 on discharge and 1.8 on charge were employed. After about the seventh or eighth cycle, cell capacities of about 0.5 Ah were obtained enough through the remainder of the test with a charge to discharge current efficiency of in excess of 99%. Somewhat lower capacities during the first five or six cycles are attributed to the later development of additional porosity within the negative electrode.

Example 2: A negative electrode was prepared in essentially the same manner as in Example 1 except that the nickel foam was immersed in the lithium-aluminum melt for only about 5 seconds and was vigorously shaken after removal until a porosity of about 70% was obtained.

Example 3: As a proposed alternative to the negative electrode in Example 1, the nickel foam is immersed for about 1 second in a melt of about 6 atom % lithium and 94 atom % aluminum at about 680°C. It is then withdrawn and manipulated to remove excess melt before solidification occurs. The resulting electrode structure is assembled as a negative electrode within an electrochemical cell having a positive electrode containing Li_2S and iron. This cell is charged to a sufficient cutoff of about 2.2 volts to provide FeS_2 in the positive electrode and about 50 atom % lithium and 50 atom % aluminum as solid alloy in the negative electrode.

The process permits uniform and controlled loading of lithium-aluminum alloy into a porous current collector structure. Porosity and loading of the substrate can be conveniently controlled by manipulating or shaking the substrate after immersing it within molten lithium-aluminum alloy.

Pelletized Electrolyte and Electrode

According to a process described by *B.A. Askew and R. Holland; U.S. Patent 4,013,818; March 22, 1977; assigned to National Research Development Corp., England* a high temperature secondary battery comprises a stack of cells each cell having a pellet of immobilized electrolyte sandwiched between a negative electrode pellet composed at least partially of lithium and a positive electrode pellet composed at least partially of a metal sulfide, and an intercell sheet adapted to prevent direct chemical action between the electrode pellets of adjacent cells while maintaining electrical contact between them. The electrolyte pellet is preferably prepared from a mixture of alkali metal halides at least one of which is a lithium salt and is immobilized by inclusion of a finely divided high melting powder which is inert to the reactants. The preferred immobilizing powder is lithium fluoride.

The negative electrode pellet is preferably prepared from a mixture of a lithium alloy (e.g., lithium-aluminum) and the electrolyte material. Alternatively metallic lithium impregnated into a porous nickel matrix may be used.

The positive electrode pellet is prepared at least partially from a metal sulfide powder. The metal sulfide powder is iron sulfide, in which case the positive electrode is prepared from a mixture of iron sulfide and electrolyte material. The intercell sheet may conveniently be a sheet of stainless steel, though other metals may be used.

The stack of cells is encased in a close-fitting tube of chemically inert electrically insulating material, such as hot-pressed lithium fluoride. One or more such stacks may be enclosed in a hermetically sealed container with appropriately insulated and sealed terminals.

Batteries according to the process operate typically in the temperature range 375° to 450°C. The heat required to raise the battery to this temperature is supplied initially from an external source such as an external electrical supply to heating coils. Thermal insulation around the battery is used to minimize heat loss and further heat to maintain the temperature may be supplied from the external source, from the battery output or by heat generated within the battery during charge and discharge.

Partitioned Electrode Structure

According to a process described by *D.R. Vissers, H. Shimotake, E.C. Gay and F.J. Martino; U.S. Patent 4,029,860; June 14, 1977; assigned to the U.S. Energy Research and Development Administration* electrodes for secondary electrochemical cells are provided with compartments for containing particles of the electrode reactant. The compartments are defined by partitions that are generally impenetrable to the particles of reactant and, in some instances, to the liquid electrolyte used in the cell. During cycling of the cell, reactant material initially loaded into a particular compartment is prevented from migrating and concentrating within the lower portion of the electrode or those portions of the electrode that exhibit reduced electrical resistance.

In Figure 5.1a, an electrochemical cell is shown with vertically arranged electrodes of a type illustrating one form of the process. The cell is contained within an outer housing **11** containing a centrally located positive electrode **13** between two negative electrodes **15a** and **15b**. The housing is filled, except for space for expansion, with liquid electrolyte **17** that permeates a porous electrically insulative fabric **19** intermediate the positive and negative electrodes. In high temperature cells, molten salt eutectics such as LiCl-KCl and others described in U.S. Patent 3,716,409 can be selected as electrolyte.

FIGURE 5.1: COMPARTMENTED ELECTRODE STRUCTURE

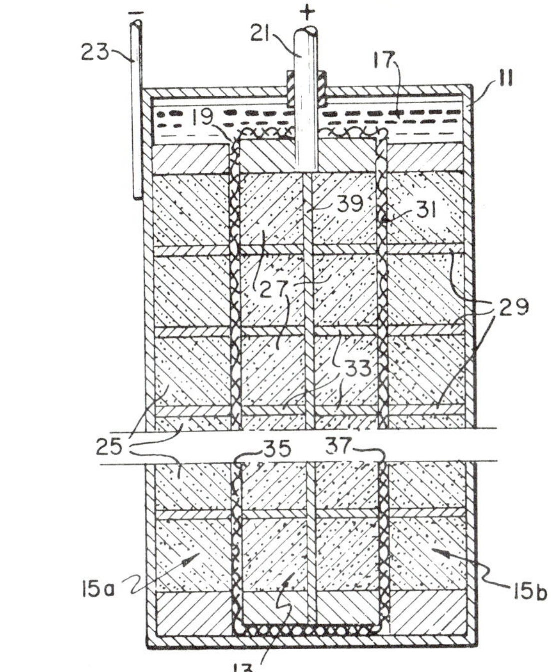

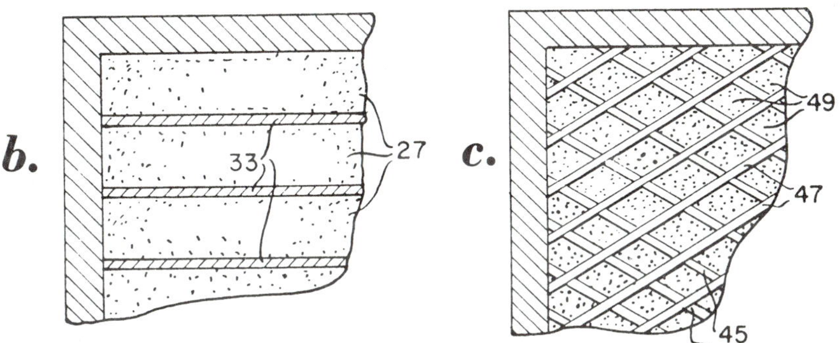

(a) Elevation view in cross section of an electrochemical cell.

(b) Fragmentary side view of an electrode of the type illustrated in Figure 5.1a.

(c) Fragmentary side view of one other electrode configuration.

Source: U.S. Patent 4,029,860

Conductor **21** penetrates the cell housing **11** and separator fabric **19** into contact with the positive electrode, while the negative electrodes **15a** and **15b** each make electrical contact to the cell housing **11** which is shown provided with conductor **23**.

The positive electrode **13** and the negative electrodes **15a** and **15b** are shown each with a plurality of compartments **25** in the negative electrode and **27** in the positive electrode. Compartments **25** within the negative electrode are shown separated and divided from one another by horizontal partitions **29** that extend across the electrode from their inwardly facing, major surfaces **31**. Similarly, compartments **27** in positive electrode **13** are illustrated as divided by partitions **33** that run horizontally from each of the positive electrode major surfaces **35** and **37** to the electrode center. In some embodiments partitions **33** can extend completely across the electrode between surfaces **35** and **37** but in the Figure 5.1a illustration electrode **13** includes a vertically disposed central partition **39**. Partition **39** divides the positive electrode into two vertical portions each having a plurality of compartments **27** defined between the partitions **33**.

It will be understood that the partitions and compartments in both the positive and negative electrodes can be of various designs and configurations. Figure 5.1b illustrates a fragmentary side view of the electrode illustrated in Figure 5.1a. The partitions serve as shelves to separate the electrode compartments from the front to the back. It will be clear that vertical partitions, not shown in Figure 5.1b, can also be included at generally normal dispositions to partitions **33** and **39** to further subdivide compartments **27**.

Figure 5.1c presents one other example of an electrode configuration. Partitions **45** and **47** course in intersecting alignments so as to intermesh into a honeycomb or network that define compartments **49**. Compartments **49** can be of various shapes including diamond, square, honeycomb, etc. Merely by way of example, compartment dimensions of about 0.5 to 2.5 cm may be used. Each of these partitions **45** and **47**, as in the electrode configuration illustrated in Figure 5.1a, extends from a major electrode surface to the opposing major electrode surface or to an intersecting central partition such as partition **39** in Figure 5.1a.

It will also be seen that in a particular electrochemical cell either one or more of the electrodes can include one of the contemplated configurations of partitions and compartments. For instance, in some electrochemical cells, it may only be desirable to compartment only the negative or only the positive electrodes, while in others both positive and negative electrodes can be partitioned. Where both electrodes are partitioned, the compartments in opposing electrodes may be in alignment as illustrated in Figure 5.1a or in overlapping and other noncorresponding relationships including compartments of different size.

Partitioned electrodes are most useful where particulate electrode reactants are employed to prevent drifting and redeposition of the reactants. The electrode compartments can contain solid reactant particles of such as, in the negative electrode, lithium-aluminum, lithium-silicon, or lithium alloyed to form binary and ternary compositions with other elements from Groups II-A, III-A or IV-A of the Periodic Table. Particulate metals such as lead, nickel, cadmium, etc. might also be employed as reactants.

In positive electrodes, the compartments can contain metal chalcogenides, that is oxides, sulfides and selenides. Particulate iron, nickel, copper and cobalt sulfides and their mixtures have been found well suited as positive electrode reactants opposite lithium alloy negative electrodes.

Example 1: (Cell S-82) – An electrochemical cell of the general configuration shown in Figures 5.1a and 5.1b including a central, partitioned positive electrode between two negative electrodes was assembled and tested. The positive electrode included particulate FeS_2-15 wt % CoS_2 composition vibratorily loaded into layers of vitreous carbon. The layers were separated by horizontal sheets of carbon cloth coated with a resin of polymerized furfuryl alcohol. The negative electrodes were made up of foamaceous nickel metal loaded with particles of about equal atom proportions of lithium-aluminum alloy. No partitions were used in the negative electrode. The cell was operated for over 115 cycles and 2,800 hours at cutoff voltages of about 1 on discharge and 2.2 on charge at a capacity density of about 0.6 to 0.7 Ah/cm^2. Additional data regarding the cell are given in the table below.

Cell S-82

Effective area, cm^2	193.6
Theoretical capacity, Ah	
Positive	269.6
Negative	192.0
Current density, A/cm^2	
Charge	0.075
Discharge	0.052
Cutoff voltage, IR-included, V	
Charge	2.27
Discharge	0.90
Typical capacity, Ah	130.0*
Typical capacity density, Ah/cm^2	0.67
Percentage utilization	
Li-Al	67
FeS_2-CoS_2	48
Ah, efficiency, %	95
Temperature, °C	435-463

*Last 1,000 hours

Example 2: (Cell LP-1) – An electrochemical cell was prepared with the negative electrode arranged in horizontal compartments. Lithium-aluminum alloy strips were electrochemically prepared by reacting lithium metal with compacts of aluminum fiber. The compacts were assembled into a frame having horizontal stainless steel partitions to form the separate compartments of the electrode.

Each compartment with lithium-aluminum alloy strips included an about 0.6 x 6 cm surface facing the positive electrode and about 0.3 cm depth towards the partitions and frame. The positive electrode contained particulate FeS and Cu_2S loaded into a porous stainless steel current collector which was assembled with LiCl-KCl salt and the negative electrode to form the electrochemical cell. The cell was cycled through about 13 charges and discharges at 450° to 500°C, at 0.5 to 2.0 A and about 6 to 8 Ah.

Compact Cell

R.C. Saunders; U.S. Patents 4,012,562; March 15, 1977 and 3,997,363; Dec. 14, 1976; both assigned to Electric Power Research Institute, Inc. describes a compact modular electrical energy storage device comprising one or more cell modules in which certain components of the device are maintained in a substantially fixed position in the device by utilizing a resilient porous body of compressed carbon fibers to resiliently urge the components into physical contact with one another. The process further provides means for maintaining the cell components in such fixed position during cell operation at elevated temperatures for extended periods of time.

A preferred compact cell module construction particularly suitable for use in a high-temperature, high-energy-density lithium battery includes two positive electrode assemblies positioned in opposing relationship by a U-shaped spacer member maintained in contact with a separator member of each positive electrode assembly by a resilient porous body of compressed graphite fibers, a unitary double-faced negative electrode assembly being positioned between the two positive electrode assemblies.

Referring to Figures 5.2a and 5.2b there is shown a preferred modular single cell form of the electrical energy storage device of the process. The cell is particularly adaptable for use in a battery containing a plurality of similar such cells, each of which is electrically connected in series to provide a higher desired battery potential, in parallel to provide a desired ampere-hour capacity, and combinations thereof. The cell **10** comprises a housing **12** which contains an electrolyte **13**, a negative electrode assembly **14**, and a positive electrode assembly **16**. Each of the electrode assemblies is substantially immersed in the electrolyte.

The positive electrode assembly **16** includes a material-holding member **17** which defines a cavity for receiving a body of particulate active cathode material **18**. In this particular embodiment the material-holding member **17** is a body of dense graphite. The active material **18** is retained in place by a porous separator plate **20**. Separator plate **20** is suitably and preferably formed from an electronically nonconductive material such as, for example, a ceramic.

Advantageously, the selected ceramic should have a high degree of purity, i.e., be substantially free of any impurities. The selected separator material may have a porosity of from 10 to 90%, that is to say the material may have an apparent density of from 90 to 10% of that of the base material. A particularly preferred porosity range is from about 20 to 40%.

The pore size should be sufficiently small to prevent the escape of the particulate active material **18**, while still permitting the free passage therethrough of electrolyte ions. The separator member should therefore have a median pore size within the range of from 5 to 500 microns, and preferably from 20 to 100 microns for use with the preferred molten salt lithium battery. The two positive electrode assemblies **16** each are provided with electrical conductor leads **22** which are in electrical communication with one another.

FIGURE 5.2: COMPACT MODULAR CELL

a.

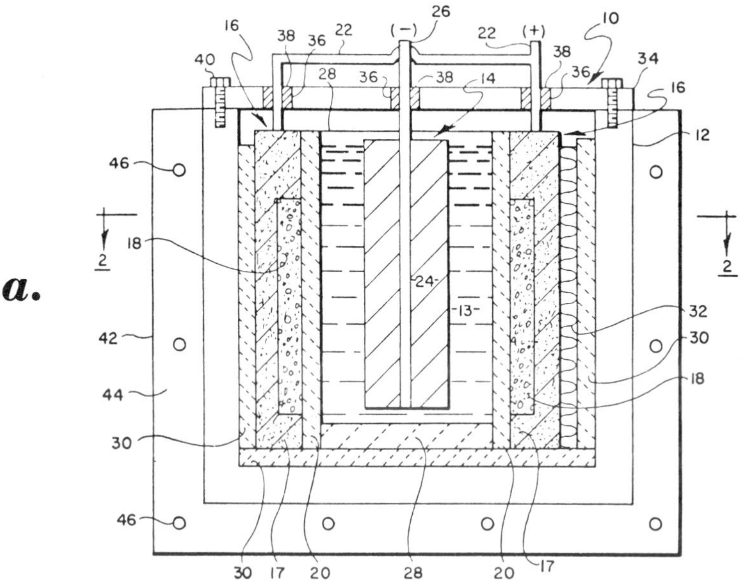

b.

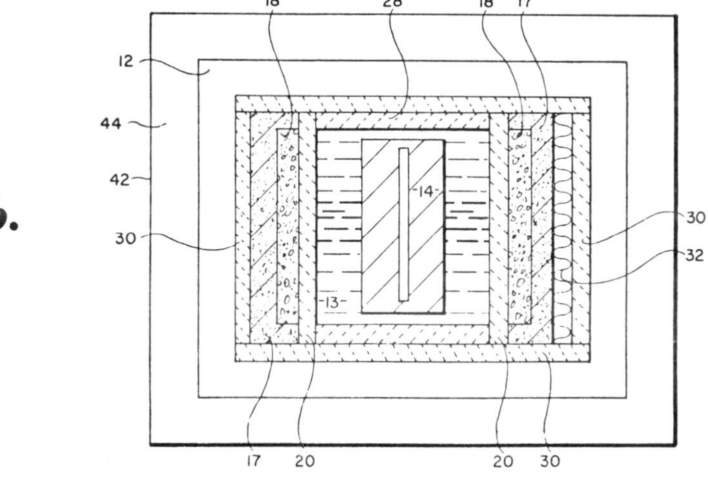

(a) Sectional view in elevation of a compact modular cell.

(b) Sectional top plan view of the modular cell taken along the lines **2–2** of Figure 5.2a.

Source: U.S. Patent 4,012,562

The negative electrode assembly **14** comprises a body **24** of a lithium material and is provided with an electrical conductor lead **26**. The body **24** of lithium material may be either a lithium alloy or a porous substrate impregnated with lithium, or lithium alloyed with another material to enhance the wetting of the substrate.

Intermediate the two positive electrode assemblies **16** and abutting porous separators **20** is a spacer member **28**. The spacer member depicted in the drawing is shown as a unitary U-shaped member. Obviously, however, the spacer member could comprise a plurality of adjoining discrete parts, for example, two vertical side portions and a bottom portion. The material for the spacer member preferably also is a ceramic.

Also supplied in the housing **12** is a resilient carbon body **32**, preferably of woven partially compressed graphite fibers, for resiliently urging the spacer member and separator members into contact with one another so that the active material **18** is retained in the material-holding member. The body of woven graphite fibers has an apparent density, when the body is unrestrained of from 3 to 50% and preferably from 3 to 15%. The term "partially compressed" means that the body in its unrestrained state occupies a volume in excess of that which is provided when present in the electrical energy storage device.

The woven graphite fibers preferably have a diameter within a range of from about 10 to 150 microns, and particularly from 20 to 75 microns. Advantageously, the body of porous woven graphite fibers is compressed to from 25 to 50% and preferably from 30 to 40% of its unrestrained volume when it is placed in the electrical energy storage device. It will be appreciated that the exact amount of compression will vary depending upon, among other things, (a) the amount of force that it is desired to exert against the components, (b) upon the apparent density of the selected material, and (c) the number of such bodies of woven or nonwoven carbon and graphite fibers used.

In the preferred form depicted in Figures 5.2a and 5.2b, the active components of electrical energy device **10** (active lithium body, active cathode material, and electrolyte) are electronically isolated from housing **12** by a plurality of insulator members **30**.

Advantageously, the active components of the cell also are isolated from the atmosphere, for example, by a cover member **34**, which encloses and is in sealing engagement with housing **12**. Cover member **34** is retained by fastening means such as a plurality of threaded fasteners **40**. The cover member also includes a plurality of apertures **36**, each of which is provided with seals **38** through which extend electrical conductor leads **22** and **26**. In Figure 5.2a, the two conductor leads **22** of the positive electrode assembly are shown connected together external to the cover. It generally is preferred, however, that the interconnection of the electrode assemblies be internal to reduce the number of openings required in the cover member, particularly when a number of cells are interconnected to form a battery having a desired voltage and ampere-hour capacity.

In a preferred example of the device, the electrolyte is a salt which is molten at the operating temperature of the battery. For such examples, some means must be provided to heat the device to its operating temperature. Heat may be provided by an external source such as by placing the device in a furnace or other enclosure.

OTHER ELECTRODES AND PROCESSES

Lithium-Silicon Alloy

A process described by *S.-C. Lai; U.S. Patent 3,969,139; July 13, 1976; assigned to Rockwell International Corporation* relates to a lithium electrode and an electrical energy storage device using such an electrode. The lithium electrode comprises an alloy of lithium and silicon in intimate contact with a supporting current-collecting matrix, thereby generally providing a unitary or composite electrode structure. The term "alloy" as used herein is defined as an intimate mixture of the two metals in which the metals may form mixed crystals, solid solutions, or chemical compounds. It is an essential feature of the process that the alloy contain from about 80 to 28 weight percent lithium and from about 20 to 72 weight percent silicon. A preferred alloy is one containing from about 60 to 40 weight percent lithium, with the balance consisting essentially of silicon.

The lithium-alloy electrode structure further includes a supporting current-collecting matrix in intimate contact with the alloy. Suitable materials for the supporting current-collecting matrix are those materials resistant to attack by lithium or lithium-silicon mixtures. Examples of such materials include iron, steel, stainless steel, nickel, titanium, tantalum, and zirconium. The purpose of providing a matrix in intimate contact with the alloy is to provide for substantially uniform current density throughout the alloy and also to provide structural support for the alloy.

It has been determined that the lithium-silicon alloy utilized in the process lacks structural integrity when used in an electrical energy storage device as the sole component of the negative electrode, particularly in a molten salt electrolyte at its high operating temperature. To function for any significant length of time without disintegration, therefore, it is essential that the lithium alloy be provided with a supporting matrix. The matrix may be in the form of an electronically conductive porous substrate having an apparent density of from about 10 to 30% of that of the base material. Advantageously, the substrate will have a median pore size within the range of from about 20 to 500 microns and preferably from about 50 to 200 microns. A particularly preferred form of such a substrate is formed from woven or nonwoven wires pressed together to a desired apparent density and then sintered.

Such pressed and sintered wire structures are known and commercially available as Feltmetals. The porous substrate then is impregnated with the alloy, for example, by immersion in a molten bath of the alloy followed by removal and cooling. Alternatively, the alloy may be cast about a matrix formed from a wire screen or expanded metal.

In a preferred case, this electrode is formed by surrounding the matrix with the alloy in a molten state, for example, by immersing a porous substrate in a molten body of the alloy. The alloy may be formed by mixing particulate lithium and silicon and heating such a mixture to a sufficiently elevated temperature to form a melt. In accordance with a preferred method however, the lithium first is heated, in an inert atmosphere, to a temperature above the melting point of lithium, and thereafter the silicon is added in an amount to provide the desired weight percent for the alloy. In such latter method, the exothermic reaction between the lithium and silicon will provide substantially all of the heat required to form a melt of the alloy.

Lithium-Aluminum Alloy Composition

J.L. Settle, K.M. Myles and J.E. Battles; U.S. Patent 3,957,532; May 18, 1976; assigned to the U.S. Energy Research and Development Administration describe a method of producing an electrode material for use within a high-temperature and high specific energy electrochemical cell. The electrode comprises a uniform alloy composition of lithium and aluminum having at least 50 atom percent lithium. The alloy is prepared by melting lithium metal at a temperature substantially below the liquidus temperature of the intended alloy composition. Solid aluminum is dissolved in the melt while increasing the temperature at a sufficiently high rate to maintain a liquid phase until the intended aluminum concentration is reached.

After sufficient aluminum is included in the melt, the temperature is elevated above the liquidus temperature at the intended alloy composition with agitation to insure dissolution of the remaining aluminum. The melt is rapidly solidified by splat casting onto a surface to minimize development of nonhomogeneous regions. The solid is heated to a temperature just below the solidus temperature at which liquid formation begins and maintained at that temperature for a sufficient period to complete chemical homogenization. The homogenized solid is comminuted to particles. The particles can be pressed to form a compact of the desired alloy composition having essentially uniform lithium-aluminum distribution.

Example 1: A weighed charge of lithium metal was melted within a tantalum crucible and heated to about 300°C. An aluminum rod of sufficient weight to form a 50-50 atom percent lithium-aluminum alloy was placed in the melt and allowed to dissolve and react. During dissolution the temperature was gradually increased at a sufficient rate to maintain a liquid phase in contact with the solid. Expressed differently, a liquid phase was maintained which closely followed the liquidus temperature, within about 10°C, as the aluminum concentration was increased within the melt from 0 to 50 atom percent.

After most of the aluminum had dissolved and the temperature was raised to the liquidus temperature corresponding to the desired alloy composition, that is 718°C, the melt was superheated about 30°C more and mechanically agitated for a period of about 5 minutes to insure a homogeneous molten mixture. A plaque of about 0.5 cm thickness was next formed by splat casting the melt onto a smooth slab of stainless steel at about 20°C. After cooling, the plaque was broken into fragments of roughly a centimeter in size and placed into a fresh tantalum crucible for reheating.

The fragments were heated to about 700°C and held at that temperature for 1 hour to complete homogenization. After cooling to room temperature, the fragments were mechanically ground to particles having an average size of between 100 and 500 micrometers. An amount of LiCl/KCl eutectic equal to about 20 weight percent of the total was blended with the particles and the mixture pressed within a mold at 4,000 psi and 350° to 450°C for about 2 minutes in order to form a compact.

Two compacts prepared as in Example 1 were assembled on opposite sides of a FeS_2 electrode structure within an electrolytic cell. Each negative electrode was disk-shaped with a surface area of about 20 cm^2 and made electrical contact with the cell housing to which the negative terminal of the cell was attached. The cell was operated for over 1,300 hours and 125 charge and discharge cycles. During this period only a minor portion of the lithium-aluminum alloy was lost from the electrode. The utilization of lithium within the initial cycles was about 67% at a charge cutoff voltage of 2.3 and a current density of 0.024 ampere per square centimeter. After 4 weeks of continuous operation, the utilization had decreased to 44% of lithium within the electrodes.

Example 2: Several other electrodes were prepared as in Example 1 with 50 atom percent lithium and 50 atom percent aluminum, except the compacts were prepared of particles having a distribution of between 44 and 74 microns size. An electrochemical cell was assembled with a stack of four disk-shaped negative electrodes of this type in alternating sequence with five parallel positive electrodes containing FeS_2. Each negative electrode had a surface area of about 20 cm^2. The cell was operated for over 2,000 hours and 200 cycles, during which time it was required to replace one of the four anodes which disintegrated into the electrolyte. Otherwise the cell performed well for most of this period with a percent utilization of lithium generally above 50% until the last cycles. The cell was operated at 0.0025 to 0.0088 A/cm^2 with a cutoff of 0.5 to 0.7 volt on discharge and at 0.0013 to 0.025 A/cm^2 with a cutoff voltage of 1.6 to 2.3 volts on charge.

Lithium-Aluminum Sandwich Construction

According to a process described by *J.C. Schaefer; U.S. Patent 3,981,743; September 21, 1976; assigned to ESB Incorporated* a lithium-aluminum negative electrode for an electrochemical cell is prepared by forming a sandwich comprised of lithium and aluminum such that lithium is disposed between the aluminum layers of the sandwich. The sandwich is heat soaked at a temperature below the melting point of lithium while pressure is applied to the sandwich thereby causing the lithium and aluminum to chemically react to form a lithium-aluminum alloy.

Example 1: In a glove box having an argon atmosphere, a sandwich of aluminum-lithium-aluminum was prepared such that the total percentage of lithium in the sandwich was equivalent to about 12 to 18 weight percent, based on total composition. The sandwich was placed between the heatable platens of a small Burton Press and heat and pressure were applied to the sandwich. The sandwich was heat soaked at a temperature below, close to, but not equal to the melting point of lithium (179°C). Sufficient pressure was applied to assure contact between the abutting faces of the sandwich. It may be pointed out here that the amount of pressure applied to the sandwich is not critical other than it should be enough to assure contact.

In this example, the pressure applied was approximately 30 psi. The time of heat soaking was sufficient for the lithium to alloy with the aluminum. Of course, the time of heat soaking will vary with the thickness of the layers of the sandwich. In this particular example, the aluminum sheets were about 32 mils thick and the lithium about 40 mils thick and it required about 175°C for approximately 8 hours.

Example 2: In a glove box having an argon atmosphere, an aluminum-lithium-aluminum sandwich was fabricated. The lithium in the sandwich was equivalent to about 17.4 weight percent, based on total composition and the sandwich was approximately 4" x 4". The sandwich was placed in a room temperature furnace and the temperature was gradually raised to approximately 450°C which took about 20 minutes. Thereafter, the sandwich was cooled and examined. Examination showed that there was no uniformity in alloy formation, the sandwich was badly misshapen and unusable as a cell electrode. It is believed that these results were effected because there was no containment of the sandwich other than the sandwich package itself and because the 450°C temperature was above the melting point of the lithium contained in the sandwich.

Example 3: A complete cell was fabricated in an argon atmosphere and was sealed completely, leak checked and was removed to a normal room atmosphere. The anodes were fabricated as in Example 2 with a lithium content of about 8% by weight. The cell was placed in an oven at room temperature with a set point of about 170°C. The temperature at the cell's surface was monitored as the oven warmed. It took about 105 minutes for the 170°C set point of the oven to be reached. A reaction was apparent at about 153°C and the temperature rose to 176°C although heat was turned off at 153°C. The cell was cooled and then placed in a furnace set-up to cycle cells through charge and discharge cycles. The cell was cycled for 13 cycles, found to perform adequately and to reach a rating of 10 ampere hours.

Electrodes or anodes fabricated by the method afford the lowest possible contamination by oxides, nitrides, carbonates, etc. since the lithium sheets can be the purest state of lithium, and the aluminum sheets can be virgin, electrical grade, aluminum of 99.5 minimum percent aluminum. Initial low contamination is not the only advantage for this process method. Cost of manufacture of lithium-aluminum electrodes is drastically reduced since the extra costs of grinding, casting etc., normally attendant with metallurgical processes for forming lithium-aluminum electrodes are eliminated. The simplicity of sandwich making, and unattended heat soaking of the sandwich further reduces costs by avoiding the, only partially successful, anode cleaning procedures required of prior art lithium-aluminum anodes prepared by the metallurgical processes of melting and casting the alloy constituents.

Natural Graphite-Lithium Fluoride

According to a process described by *D.N. Bennion, R.K. Hebbar and S.L. Deshpande; U.S. Patent 4,009,323; February 22, 1977; assigned to the U.S. Secretary of the Navy* a positive electrode of a storage battery is formed by applying an electrical charge through a nonaqueous lithium perchlorate solution to an electrode formed of powdered Madagascar-type graphite and lithium fluoride.

As shown, the Figure 5.3a cell includes a lithium negative electrode, a positive electrode formed of graphite and lithium fluoride, an electrolyte which is lithium perchlorate dissolved in dimethyl sulfite and a semipermeable membrane separating the positive and negative electrodes. As will be recognized, the electrode spacing in this cell is quite limited so that the amount of the electrolyte per unit volume of the electrode also is limited. The battery system itself can be expressed in the manner shown on the following page.

$Li^+ + e^- \rightleftharpoons Li$	$LiClO_4$, dimethyl sulfite	cation exchange membrane	$LiClO_4$, dimethyl sulfite	$xnC + xLiF \rightleftharpoons (C_nF)_x + xLi^+ + xe^-$	Pt

The lithium which forms the negative electrode, can be a commercial grade lithium foil. The lithium perchlorate-dimethyl sulfite electrolyte solution is between 1 and 3 molar, and it is carefully purified to reduce water content below 100 ppm. The final water removal is performed by mixing the solution with lithium chips. The membranes generally used for the test runs have included various types of commercially available, radiation crosslinked, radiation-grafted polypropylene or polyethylene cation exchange films.

As to the positive electrode, best results have been obtained with finely divided (1 micrometer characteristic dimension) and purified Madagascar graphite. This material is a naturally occurring graphite which geologically has been subjected to very high temperature and pressure to the extent that it is a crystalline material having few imperfections. Other tested materials have produced rather disappointing results to the extent that it appears that the use of a Madagascar-type graphite, whether naturally occurring or otherwise, provides a significant contribution. The purifying of the graphite consists mainly of washing it with various strong acid solutions to dissolve silicate and other metal oxide impurities. Another location of natural graphite is in Northern New York.

FIGURE 5.3: GRAPHITE-LITHIUM FLUORIDE ELECTRODE

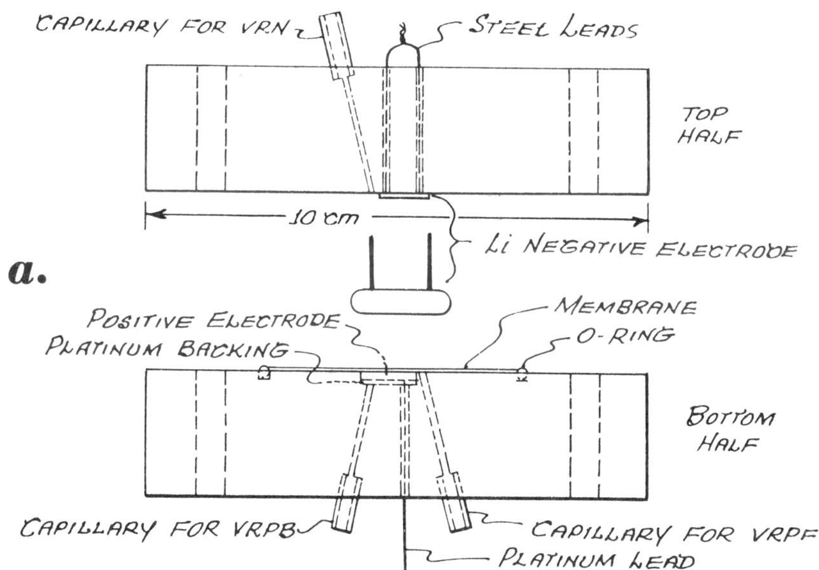

(continued)

FIGURE 5.3: (continued)

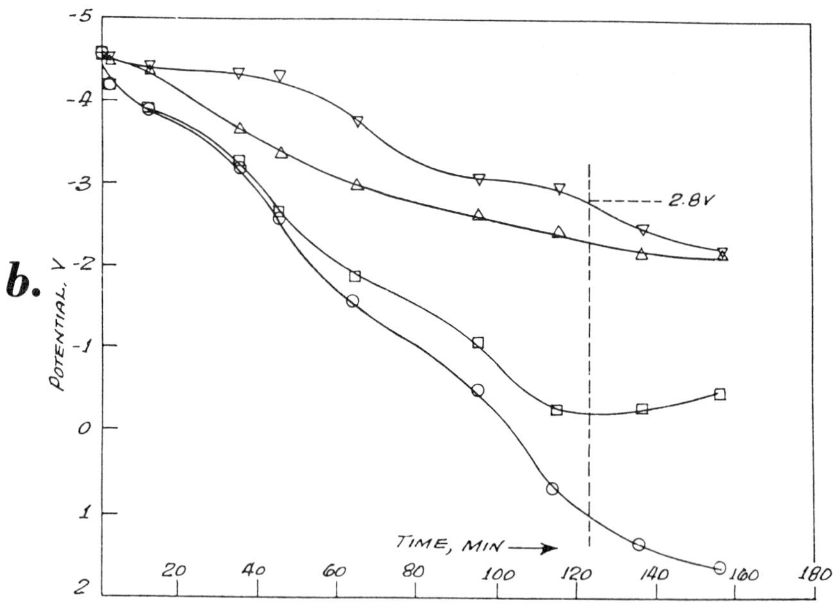

Discharging curve for Li-CF_x cell. Q = 92.17 x 10^{-6} equiv. ○ VC, □ VRN, △ VRPF, ▽ VRPB.

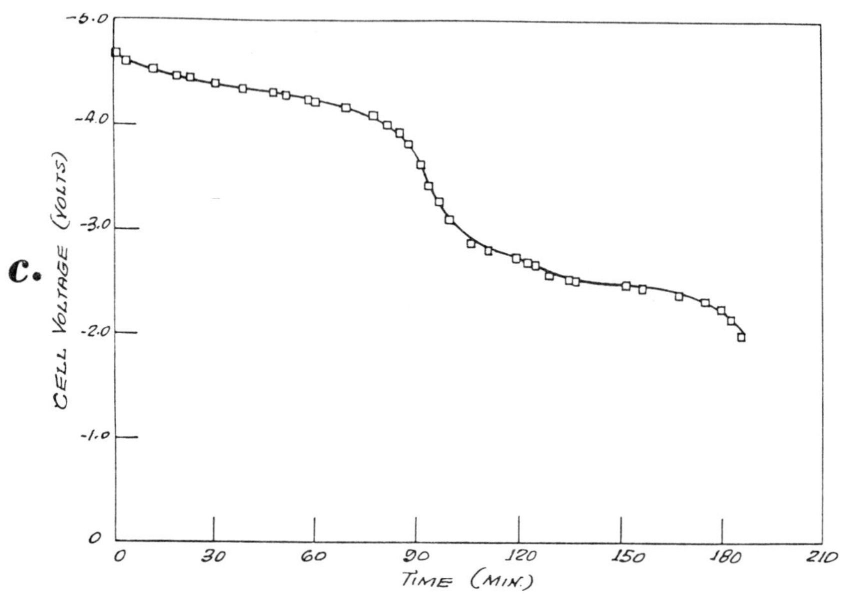

Discharge curve, Li/$LiClO_4$, DMSU/C, LiF, 1.2 M $LiClO_4$ in DMSU.

(a) Schematic view of a test cell.
(b) (c) Discharge curves for the test cell of Figure 5.3a.

Source: U.S. Patent 4,009,323

The lithium fluoride is a commercially available, purified material which has been powdered (1 micrometer characteristic dimension) and vacuum dried at high temperature (800°C). The graphite and lithium fluoride may be mixed together in approximately equal molar ratios although the exact proportion, instead of being critical, appears to be a matter of assuring a sufficiency of graphite to provide good matrix conductivity. As indicated, graphite cloth or platinum may be used as a current collector. In some runs, such as that illustrated by the discharge curve of Figure 5.3c, the positive electrode is bound into a porous, solid plate and electrical contact made directly to the plate above the cell.

To make a plate, the natural graphite-lithium fluoride mixture is mixed with an organic material to make a paste. The paste is formed into an electrode and fired in an inert atmosphere at about 800°C. In other runs the graphite and lithium fluoride mixture is contained in a cup or a recess and electrical contact is assured by mechanical pressure. The overall cell reactions appear to be predominately as follows:

Negative electrode: $Li^+ + e^- \rightleftharpoons Li$

Positive electrode: $xLiF + nxC \rightleftharpoons xLi^+ + (C_nF)_x + xe^-$

The charge (upper arrows) of the positive electrode takes place at between 4.0 and 4.5 volts relative to a reversible lithium electrode in the same solution. Discharge (lower arrows) takes place at between 2.8 and 3.5 volts relative to a reversible lithium electrode in the same solution, although, as shown in Figure 5.3b a part of the discharge curve takes place above 3.8 volts.

Before describing the test results, it should be helpful to understand in greater detail the various electrical leads which provide reference points for establishing the interior voltage of the Figure 5.3a cell. Generally, the leads were employed to locate the losses within the cell. Thus, as will be noted, the Figure 5.3a cell provides four potential differences which can be measured and these potential differences are provided by the so-called capillary leads shown by the legend of Figure 5.3a.

Referring to Figure 5.3a VRPB which identifies a voltage reference of the potential of a lithium wire at the rear of the positive electrode minus the potential of the positive electrode. VRPF is the potential of a lithium wire at the front of the positive electrode on the positive electrode side of the membrane separator minus the potential of the positive electrode. VRN is the potential of the lithium wire in the negative electrode compartment minus the potential of the positive electrode. VC is the lithium negative electrode minus the potential of the positive electrode. In principle, VRPB is the total available driving force of the cell. VRPB minus VRPF represents the potential loss through the positive electrode. VRPF minus VRN is the potential drop across the membrane. VRN minus VC is the loss due to a film and other irreversibilities at the surface of the lithium electrode. VC is the net or effective cell potential at the operating current density.

The membranes themselves constitute a continuing subject for investigation, although the research that has been conducted has shown that radiation treated porous polypropylene membranes are most appropriate. These membranes have been developed by and obtained from the RAI Research Corporation and are identified as membranes RAI 1101-28 No. 1 and RAI 1101-29 No. 2. Other membranes, although satisfactory for eliminating dendritic shorts, nevertheless, may demonstrate a large voltage drop across the membrane in the order of 5 to 10 volts. The so-called experimental membranes demonstrate significantly lower resistance losses.

Figures 5.3b and 5.3c provide test results obtained using the Figure 5.3a cell. Figures 5.3b and 5.3c represent discharge curve data for the cell. The following table provides data on a number of runs to demonstrate coulombic efficiency as well as other operating characteristics.

Summary of Cycles for the Lithium-Graphite Cell*

Run No.	Equivalents in ($\times 10^6$)	Equivalents out ($\times 10^6$)	VRP Cut Off Volt	Nominal c.d. mA/cm^2	Coul./g Out
1	341.1	286.4	−2.75	0.5	885
		428.1	−2.67	0.5	1330
2	677.0	477.0	−2.80	1.5	1480
		679.5	−2.60	1.5	2190
3	267.1	175.8	−2.62	0.3	545
4	560.0	16.1	−2.76**	1.5	50
5	133.2	64.8	−2.75	1.5	202
6	827.4	163.6	−2.71	1.0	508
7	188.2	307.2	−2.68	1.5	955
8	705.3	709.4	−2.80	1.5	2200
9	932.2	1005.7	−2.85	1.5	3120
10	1015.8	1032.3	−3.0	1.5	3200
11	1348.0	1230.1	−2.96	1.5	3820
12	1790.8	1328.0	−2.99	2.0	4130
13	1291.2	1023.8	−3.55***	1.0	3180

*The positive electrode consists of 0.0312 grams Asbury graphite powder and 0.0286 grams LiF.

**In the absence of any membrane.

***Discharge was terminated when discharge current fell below 0.50 mA.

As to the charging cycle of the cell, which is not illustrated the losses on charging at 1 milliamp (VC-VRPB) have been found to be 1 volt or less. There appear to be no irreversibilities at the lithium electrode (VC-VRN). Losses across the membrane (VRN-VRPF) begin small but get large quickly and slowly decrease. This behavior is likely due to concentration variations which arise during passage of current. The reason for the decrease is not clear and it may be due to a minor artifact arising because of shifts with time and current density distribution relative to the reference electrodes. Overall, the charging curves have appeared quite good and with improved engineering design a good battery could result based on this charging evidence alone if the electrode reactions are as assumed above. Thus, based on these assumptions, as well as an ability to control side reactions, a viable secondary battery system can be provided having a relatively low cost and high energy density.

Figures 5.3b and 5.3c show discharge curves for the battery system, Figure 5.3b being derived from experiments using the specially placed reference electrodes which already have been identified and Figure 5.3c demonstrating an overall cell performance. In Figure 5.3c, the input was 0.2 x 10^{-3} equivalent per cm^2. Discharge was 0.12 x 10^{-3} equivalent/cm^2 to a cutoff cell voltage (VC) of 2.0 V. Current density was held constant at 2 mA/cm^2 during charging and 1 mA/cm^2 during discharging. Other tests conditions and results are noted in the legend of Figure 5.3c.

Figure 5.3b provides significant evidence that a reversible positive electrode is achieved. In this run, VRPF remains high. The losses in the positive electrode, VNPB-VRPF, appear tolerable. The large losses across the membrane, as well as the losses at the negative electrode, are important and improvement should be achieved in these areas.

In general, Figure 5.3b demonstrates the existence of three distinct plateaus. Looking at VRPB, the upper plateau is from -4.5 down to -4.1 V in magnitude. The middle plateau is from -3.2 down to -2.8 V. Below -2.5 V another plateau appears to be forming. The plateau for VRPB below -2.6 V in magnitude is believed due to solvent decomposition as the reduction reaction. This plateau is known to go on for many hours and generally it is avoided except when over discharging by mistake as happened to occur at the end of the Figure 5.3b run.

Since it has been possible to identify at least one solvent reaction plateau, it appears reasonable that the plateaus at higher, more positive potential relative to lithium are not due to solvent reduction. Figure 5.3c shows only VC, but here two distinct plateaus are noted, one about -4.0 V and another between -2.3 and -2.8 V in magnitude. The lower plateau can be viewed as two plateaus close together. Thus, although problems with membranes, as well as films forming on the lithium and the making of good electrical contacts persist, the positive electrode consistently yields two discharge plateaus one just above and the other just below 3.0 V relative to a lithium reference electrode. When measuring just VC, the plateaus are shifted down in magnitude due to cell losses. Charging of the positive electrode consistently takes place between 4.5 and 5.0 V relative to a lithium reference electrode.

The interpretation of the Figure 5.3b and 5.3c observations, as well as a large number of other observations derived from test runs, is that the finely divided (1 micrometer size), purified, Madagascar graphite electrochemically forms intercalation compounds with perchlorate and fluoride ions. The perchlorate only intercalates to the extent that its composition C_nClO_4 is one in which the n is in the order of 90. This reaction only occurs above 4.0 V relative to a reversible lithium electrode. Below 4.0 V relative to lithium the fluoride does not appear to react with the carbon. It may be that the perchlorate is needed to activate the graphite so that the fluoride will react in which case there is a synergistic effect of the perchlorate. For large charge storage, the net reaction appears to be almost entirely the intercalation of the fluoride:

$$xnC + xLiF \rightarrow (C_nF)_x + xLi^+ + xe^-$$

In the above reaction, at full charge, n can be as low as 1.95 and it may be possible to take n down to 1. However, C_4F is a known stoichiometric compound of relatively high electronic conductivity. CF is a known stoichiometric compound of very low conductivity. Thus, as n goes from 4 to 1, it might be expected that internal losses within the positive electrode will get progressively larger. At n equal to 4, charge storage capabilities are many times larger than any existing commercial, storage battery. It might be noted that the subscript x as used in expressing $(C_nF)_x$ simply emphasizes that the compound is part of a polymer-type matrix.

In conclusion, experiments using specially-placed reference electrodes (Figure 5.3b) and overall cell performance (Figure 5.3c) have demonstrated that a natural graphite-lithium fluoride positive electrode can reversibly store up to 10^{-3} equivalent per square centimeter as shown in the foregoing table. It seems that the acceptable hypothesis for satisfactorily explaining the observations is that the active species formed in the positive electrode is a nonstoichiometric intercalation compound of graphite and fluorine.

Vanadium Pentoxide

J.S. Gore and C.R. Walk; U.S. Patent 3,929,504; December 30, 1975; assigned to Honeywell Inc. describe a power source capable of operating in a discharge mode and in a charge mode which includes a lithium anode, a vanadium pentoxide cathode and an electrolyte including lithium perchlorate dissolved in propylene carbonate. Also included is a means for enclosing one of the electrodes with a permeable material to permit passage of electrolyte ions and yet prevent passage of larger lithium metal formed during operation in the charge mode.

Specifically, it has been found that a lithium battery of the type described may be operated as a secondary battery if there is a means enclosing at least one electrode, and preferably the lithium anode by a porous member surrounding the lithium anode to prevent lithium metal from depositing other than on the anode during operation of the cell in the charge mode. It has been found that a significant number of cycles of charge and discharge may be obtained using the anode, cathode and electrolyte described above. Particularly preferred is one molar solution of lithium perchlorate dissolved in propylene carbonate.

The porous member which is employed to enclose one of the electrodes to permit passage of electrolyte ions and prevent migration of lithium metal can, in a preferred embodiment, comprise a flexible porous material which is inert to the anode, cathode and electrolyte. The pore size of this material should range from approximately 0.3 to 30 microns. Particularly preferred is a natural rubber material having an average pore size of about 2 microns.

Macroporous Electrode Structure

J.C. Hall; U.S. Patent 4,003,753; January 18, 1977; assigned to Rockwell International Corporation describes a rechargeable electrical energy storage device which utilizes an improved electrode structure in contact with a molten salt electrolyte. Broadly, the electrode structure comprises a unitary multicell structure including a plurality of wall members having edges and axially extending surfaces which form a plurality of cells having at least one open end, the cells having a cross-sectional area of at least about 0.04 cm^2. The edges of the wall members in the open end of the cells are aligned in a common plane to form a planar face. The axially extending surfaces of the wall members are substantially perpendicular to the planar face. The structure further includes a body of electrochemically active material disposed in the cells, the material being solid at the operating temperature of the device. The body of electrochemically active material is retained in place by an electrolyte-permeable member which is affixed to the wall members and covers the open end of the cells. Generally the electrode structure will have a planar face having a surface area of from about 25 to 300 square centimeters.

The multicell structure is essentially a macroporous or open-faced cellular structure. The individual cells may take various forms, however, such as squares, diamond shapes, rectangular, circular, octagonal, or indeed just about any geometric shape. Further, the individual cells may or may not share a common wall. The particularly preferred form is one in which the individual cells are hexagonal in shape, sharing a common wall to form a honeycomb structure. This preferred shape optimizes the void volume for retention of active material while at the same time providing a high strength to weight ratio. In some instances, however, other less complex forms such as square-shaped cells may be preferred for economic reasons.

It is an essential feature that the open cell cross-sectional area be at least about 0.04 cm^2. Further, the void volume of each cell must extend substantially perpendicular to the planar face of the multicell structure. Generally, the wall members of the structure will be formed to provide cells having a cross-sectional area of from about 0.04 to 2 cm^2 with a range of from about 0.04 to 0.2 cm^2 being preferred. An advantage of this structure over the prior art porous structures is the ease with which it can be uniformly loaded with active materials.

Referring to Figure 5.4a, there is depicted a unitary multicell electrode structure **10**, constructed in accordance with the process. The structure comprises a plurality of stainless steel wall members **12**, having axially extending surfaces and edges **14**, which form the plurality of cells which are open on each end. The thickness of the wall member **d**, and the depth or length of the cell **L**, are selected such as to provide the desired strength and rigidity. For a multicell electrode structure having an overall surface area of from about 50 to 1,000 cm^2, a wall thickness **d** of from about 0.002 to 0.02 cm and a length **L** of from 0.5 to 2 cm, has been found to be satisfactory for cells having a cross-sectional area of from about 0.04 to 0.2 cm^2.

It is a feature of this process that the cross-sectional area of the cells is substantially uniform through their length **L** and that the axis of each cell is substantially perpendicular to the planar face formed by the edges **14**. The edges **14** of the wall members form a substantially planar face which is covered by one or more conductive or nonconductive electrolyte-permeable members. Preferably, in the embodiment shown, a body of active material **16** is retained in place by two superimposed stainless steel screens **18** and **20**. One of the screens, either **18** or **20**, is selected to have a sufficient number and size of openings to retain the active material within the multicell electrode structure. The other screen is selected to have a wire size sufficient to provide the desired additional structural strength.

The screens are affixed to one another and to the wall members by welding at a plurality of points, by diffusion bonding or brazing the screens to substantially the entire surface of the exposed edges **14**. As shown, the electrode structure also includes an optional housing **22**, which circumscribes the periphery of the wall members and is fixedly attached thereto preferably in the same manner as the screen members. In a particularly preferred case, the screen member also is attached to the housing. The electrode structure also includes a bracket means **24** (also affixed to the electrode structure) for supporting the electrode structure and further providing a current-collecting and conducting means for the passage of electric current therethrough. Generally, only one electrolyte-permeable member is used rather than the two depicted in Figure 5.4a. When only one is used, it should have the wire size, number of openings per square centimeter, and size of openings hereinbefore described.

FIGURE 5.4: ELECTRODE STRUCTURE

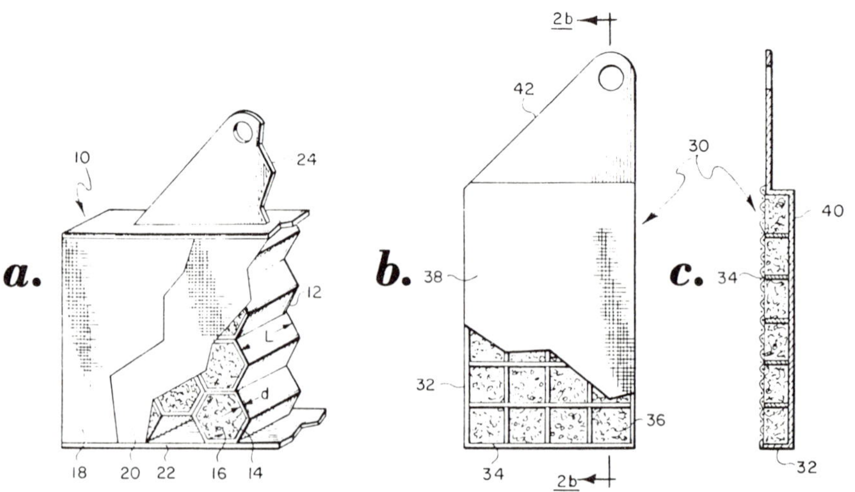

(a) Partially broken away perspective view of a multicell electrode structure.

(b) Partially broken away front elevation view of another example of the multicell electrode structure.

(c) Sectional view of Figure 5.4b taken along the line **2b–2b.**

Source: U.S. Patent 4,003,753

Referring to Figures 5.4b and 5.4c, there is depicted another example of the multicell electrode structure **30** of the process. As depicted, the cells have a single open faced end, are rectangular in shape, and are formed by a plurality of wall members **32**, having edges **34**, which edges form a substantially parallel plane. The cells share a common wall with an adjacent cell and contain a body of active material **36**. The planar face defined by wall members **32** is covered by a stainless steel wire screen **38** having a mesh size of 26 x 500, the opposite side being closed by an impervious conductive cover member **40**. A bracket means **42** is provided for support and current collection.

Example: The following tests demonstrate the superiority of an electrode formed in accordance with the process over those of the prior art. A piece of metal felt was obtained and cut to a size of 5 x 5 x 0.64 cm. The metal felt was formed from 430 stainless steel, had a median pore size of about 0.2 mm and an apparent density of about 20%. This electrode was impregnated with a lithium-silicon alloy containing 70 weight percent of lithium. The lithium was stripped electrochemically to provide an electrode containing a lithium-silicon alloy containing 55 weight percent of lithium and having a theoretical capacity of 1.2 ampere-hours per cubic centimeter of electrode void volume. This electrode was placed in an open cell containing a molten potassium chloride-lithium chloride electrolyte and alternately discharged and charged at a current density of 40 mA/cm^2 against a metal sulfide-containing cathode structure. The charge-discharge curve showed little symmetry (indicating poor reversibility), and the characteristic voltage plateaus for lithium-silicon were poorly defined in the charge mode. A recovered capacity of about 60% of the theoretical value was all that was attained.

Another similar electrode substantially the same overall size was formed except in accordance with this process. Specifically, the structure comprised a honeycomb configuration of cells similar to that depicted in Figure 5.4a. This electrode also was filled with a lithium-silicon alloy containing 70 weight percent of lithium and then stripped electrochemically to provide an electrode containing 55 weight percent lithium to provide a theoretical capacity substantially the same as the aforementioned prior art electrode. This electrode then was tested in substantially the same manner as the prior electrode. The charge-discharge curve obtained with this electrode is much more symmetrical, indicating that the electrode is highly reversible.

A capacity corresponding to 80% of the theoretical value was reached when the electrode was cycled at the same current density (40 mA/cm^2) as the prior art anode. This electrode was cycled more than 300 times over a period of time in excess of 5,000 hours with no significant loss in coulombic efficiency, utilization or structural integrity being observed.

A similar test was performed using an anode substantially as depicted in Figure 5.4a, which had a thickness **L** of only about 0.26 cm. When this electrode was cycled, at a current density of 35 mA/cm^2, 91% utilization of the active material was attained, thus providing an electrode with a usable specific capacity of 1.06 Ah/cm^2.

When the foregoing comparison test is repeated using a transition metal chalcogenide, a higher loading of active material and hence a higher usable specific capacity (Ah/cm^3) is obtainable with the electrode of the process.

Polyalkylene Glycol Ethers as Wetting Agents

J. Broadhead, T.M. Putvinski and F.A. Trumbore; U.S. Patent 3,928,067; Dec. 23, 1975; assigned to Bell Telephone Laboratories, Incorporated describe a lithium nonaqueous secondary cell in which certain dopants are added in relatively small amounts to improve recycling characteristics. These dopants are certain polyalkylene glycol ethers, certain tetraalkylammonium halides, and certain lithium salts. The polyalkylene glycol ethers have the following general formula $R(OR')_nOR''$, where R and R" are methyl or ethyl groups, and R' is a $—CH_2—CH_2—$ group (ethylene group) and n varies from 2 to 6 with tetraglyme (n = 4) preferred because of easy availability and extensive liquid range about room temperature. A concentration range of 1 to 5 weight percent is preferred because outside this concentration range recycling is less extensive. The halides in tetraalkyl ammonium halide are limited to chloride, bromide and iodide, and each alkyl group has up to 6 carbons. Tetrabutylammonium chloride and tetrabutylammonium iodide are preferred because of low cost, easy availability and the extensive recycling obtained with these substances. The concentration range should preferably be between 0.01 and 0.6 molar. The lithium salt should have a concentration between 0.01 and 2.0 molar and be soluble and ionizable in the electrolyte solvent.

The compounds LiI and $LiClO_4$ are preferred because of high solubility. A convienient way of adding the polyalkylene glycol ether and using it as a wetting agent for the separators is to wet the separators with the ether prior to assembly of the battery.

OTHER BATTERY SYSTEMS

LEAD-OXYGEN DC POWER SUPPLY SYSTEM

W.J. Britz, W.A. Boshers and J.J. Kaufmann; U.S. Patent 3,964,928; June 22, 1976 describe an electrical power supply system for supplying power to a dc electrical apparatus, such as lightweight vehicles and emergency standby power sources. The power supply system has a charging device for converting commercial ac power to dc power for charging the dc power source; and a controller connected between the charging device and the dc power source for regulating the charging current to protect the dc power source.

The controller selectively connects the dc power source to the electrical apparatus for supplying power. The dc power source includes a battery box containing an atmosphere of oxygen and containing at least one lead-oxygen battery cell comprising a charging grid member providing a separate positive electrode terminal for charging the cell, a conventional negative electrode providing a negative terminal, and a positive discharge electrode providing a positive terminal through which the battery is discharged.

A pump evacuates oxygen released from the battery cell during the charging operation and compresses the released oxygen in a storage tank for use during the discharge operation. As the released oxygen is delivered to the storage tank a condenser is provided to cool the oxygen and cause the moisture therein to condense. The condensed moisture is accumulated and stored in an accumulator tank for later return to the electrolyte in the battery cell.

Thus, a closed-loop system is provided in which water and oxygen are reused and none is lost or contaminated. In addition, a recombination electrode is provided inside the battery cell for recombining hydrogen given off during certain operations with oxygen to form water further restoring the original water supply.

The lead-oxygen power system is expected to have a long life cycle exceeding that of the lead-acid system. The life cycle is expected to be enhanced by the addition of a specially designed positive charging grid electrode used only during the charging operation correcting sulfation which normally results on the conventional positive electrode causing active material to be lost due to gassing at the electrode and gradual deterioration.

Calculations show that the lead-oxygen battery system could deliver up to 35 watt-hours per pound, whereas the conventional lead-acid battery normally found in an automobile delivers about 10 to 15 watt-hours per pound. This significant weight reduction is accomplished by removing the solid lead peroxide plate positive electrode and substituting a metal current collector positive grid electrode used only for discharge, a catalyst, and a Teflon membrane. Oxygen is introduced into the cell through this membrane.

There is little or no regular maintenance required in the lead-oxygen power system. Oxygen is reclaimed from the positive charging grid during the charging operation and compressed in a storage tank. Water evaporated from the electrolyte is condensed out of the reclaimed oxygen and is returned to the battery cell automatically. The system is charged through an ac/dc power converter plugged into a commercial ac power line outlet, or may be charged by another dc power source.

SILVER OXIDE-HYDROGEN

A process described by *D.C. Briggs and R.J. Haas; U.S. Patents 4,004,067; January 18, 1977 and 4,004,068; January 18, 1977; both assigned to Aeronutronic Ford Corporation* relates to a rechargeable electrochemical cell for a battery and, more particularly, to such a rechargeable electrochemical cell for a battery in which a reduction of the capacity of the battery is eliminated because of the loss of electrolyte from operative portions of the battery.

In the process, a rechargeable electrochemical cell for a battery includes a closed container confining therewithin at least a cathode, a gas generating and consuming anode, a separator and electrolyte storage material between the cathode and the anode, an electrolyte in the separator and electrolyte storage material, and an open volume for gas storage. The cell improvement comprises lining the inside surface of the closed container defining the open volume for gas storage with a layer of a capillary mat material.

The lining of the capillary mat material has at least a portion thereof in liquid transfer contact with the separator and electrolyte storage material located between the cathode and the anode. Any electrolyte of the cell vaporized into the gas storage volume will condense onto the capillary mat lining material. The liquid transfer contact of the capillary mat lining material and the separator and electrolyte storage material provides a liquid flow path for returning vaporized and subsequently condensed electrolyte to a storage location between the electrodes.

Referring to Figure 6.1 there is shown a single, rechargeable electrochemical cell, generally identified by the numeral **10**. This cell may be used by itself, or grouped with other cells to form a battery. In particular, the cell is a silver-hydrogen type discussed generally in U.S. Patent 3,565,691. In accordance with a preferred construction for the electrochemical cell **10**, a container **12** is provided. This container is formed from a high strength material such as type 718

A Inconel so that it resists high pressures of hydrogen when that gas is generated internally of the cell as is discussed later. The container is cylindrical in cross section at its lower portion and is provided with a domed upper portion **14** which encloses a gas storage volume **16**.

FIGURE 6.1: SILVER-HYDROGEN CELL

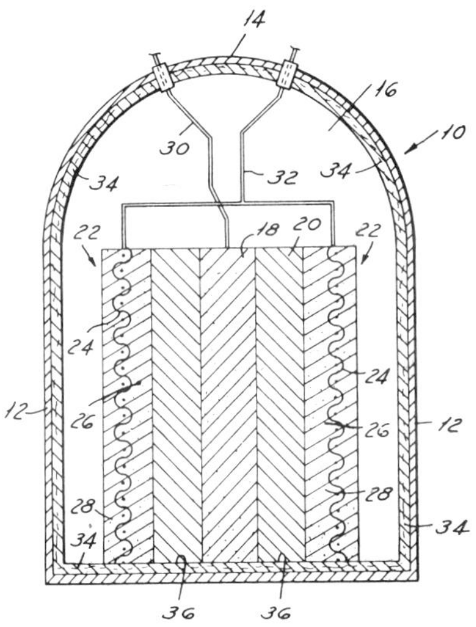

Source. U.S. Patent 4,004,067

A silver cathode **18** is provided which has a circular cross section. When the cell of the battery is charged, the silver cathode acts as an anode and is oxidized to silver oxide. Upon discharge, the silver cathode acts as a cathode and the silver oxide is reduced to silver. This silver cathode may be formed by pressing silver powder, e.g., Handy and Harman SIL Powder 130 in a mold to 2500 psi and sintering at 625°F for five minutes. In this manner the silver cathode is porous and it has a very high surface area per unit volume.

Encircling the silver cathode **18** is a separator and electrolyte storage material **20**. In the preferred case, the separator and electrolyte storage material has a ring-shaped cross section. This material may be made of a nonwoven nylon such as the material sold as Pellon-2505. This material may also be made of a fuel cell grade asbestos. The material serves two basic functions. The first function is that the material stores the aqueous electrolyte, in this case a 30% solution of potassium hydroxide, in open pores of the material so that the electrolyte is available during the charge and discharge of the cell to carry out its chemical function.

The second purpose of this material is to serve as a medium for conducting the electrolyte from one area of the cell to another along with the ions to be transferred from one electrode to another. Since this material is continuous in nature, the concentration of electrolyte in one zone will be equal to the concentration of electrolyte in another zone thereof. If any disruption of the concentration takes place in a particular zone, electrolyte will be transferred to the zone of reduced electrolyte by the capillary nature of the pore structure of the material. Thus the concentration of electrolyte is balanced throughout the cell by use of this separator and electrolyte storage material.

On the outside of the separator and electrolyte storage material **20** is found a gas generating and consuming anode, generally identified by the numeral **22**. This anode has a ring-shape cross section. This anode may be manufactured by coating a forty mesh nickel screen **24** on one side with a liquid semipermeable catalytic mixture **26** containing platinum black which is the catalyst. The other side is coated with porous Teflon **28**, the pores of which are made sufficiently large to render the layer easily permeable to gaseous hydrogen but small enough to be impermeable to the electrolyte contained in the separator and electrolyte storage material **20**. Since Teflon is not wet by water, the coating is waterproof and hydrophobic.

The active platinum black of the catalytic mixture **26** is held by a Teflon binder and is also made sufficiently porous to be permeable to hydrogen gas but semi-impermeable to the electrolyte. Thus by virtue of the anode construction, during discharges of the cell, gaseous hydrogen will contact the electrolyte at the catalytic interface that, in turn, promotes ionization of the hydrogen so that it enters into the cell reaction.

An electrical lead **30** is connected to the silver cathode **18** while an electrical lead **32** is connected to the gas generating and consuming anode **22**. These electrical leads are used during charging and discharging of the cell of the battery for the purpose of conducting electrons between the electrodes. During battery charge, the silver cathode is oxidized to silver oxide while hydrogen gas is produced at the gas generating and consuming anode **22**. The hydrogen gas fills the gas storage volume **16**, thereby building up a gas pressure in this volume.

Upon full charge of the battery, the maximum gas pressure will be found in the gas storage volume. The oxygen for oxidizing the silver electrode comes from the water of the aqueous solution. Thus the concentration of the water in the aqueous solution is reduced during the charging of the battery, the amount of reduction being determined by the charge built up in the cell. Upon discharge of the cell of the battery, hydrogen is consumed at the gas generating and consuming anode **22** and the silver oxide of the silver cathode **18** is reduced to silver. In space applications, the cell **10** may be charged through such a device as a solar cell array illuminated by the sun.

The inside surface of the closed container **12**, particularly the domed upper portion **14** thereof defining the gas storage volume **16**, is lined with a layer of a capillary mat material **34**. This material is preferably formed from a nonwoven nylon such as the Pellon 2505 previously mentioned. In the construction of an actual cell, this lining would have a thickness in the range of from 0.002 to 0.005 inch whereas the separator and the electrolyte storage material **20** would have a thickness in the range from 0.01 to 0.03 inch in thickness. The capillary mat material **34** and the separator and electrolyte storage material **20** may be one and the same material or they may be different materials. If they are the same material, the concentration per unit volume of electrolyte throughout the material generally will be the same. If the materials are different, the concentration of electrolyte in each material will be substantially uniform throughout the material but may be different than what is found in the other material depending upon the exact physical makeup of each material. The principal function of both these materials is electrolyte storage, with the separator and electrolyte storage material **20** being of greater mass and, therefore, storing the bulk of the electrolyte of the cell.

The purpose of lining the entire gas storage volume of the cell with the capillary mat material **34**, is that although the gas generating and consuming anode **22** is generally impervious to the passage of electrolyte therethrough, some of the material does pass therethrough and becomes a vapor in the gas storage volume **16**. This is particularly true in space applications where the forces of gravity are not present to retard the passage of the vapors through the gas generating and consuming anode **22**. Without the capillary mat material **34** as a lining, the vapors are free to condense against the outer surface of the container **12** and are thereafter lost for future use in the battery.

The loss of the electrolyte reduces the potential charge which may be stored in the battery, and as sufficient electrolyte becomes condensed in the storage volume of the container, other problems such as overdischarge generation of hydrogen and oxygen gas at the anode and the cathode respectively leading to a potential explosive mixture of gas in the presence of the anode catalyst can arise. The capillary mat material **34** is provided as a lining throughout the interior surface of the container **12**.

The lining material, in any event, has at least a portion thereof which is continuous from the domed upper portion **14** of the container **12** down the side walls of the container and across the bottom wall thereof to a position of liquid transfer contact **36** with the separator and electrolyte storage material **20**. By establishing such a liquid transfer contact, any material which is vaporized into the gas storage volume **16** and subsequently condensed on the lining material, will have a capillary path for being conducted back to the separator and electrolyte storage material.

The reason that at least an equivalent volume of the electrolyte is conducted back to the separator and electrolyte storage material **20** is that the condensation of electrolyte in a small zone of the lining material **34** will set up a concentration gradient which produces flow through the capillary mat because the entire mat desires to be at a uniform concentration. The concentration gradient therefore results in a mass transfer of electrolyte and eventually an equilibrium state is once again achieved in the battery cell in which proper amounts of electrolyte are contained in both the separator and the electrolyte storage material **20** and the capillary mat lining material **34**.

The reason for making the capillary mat material substantially thinner than the storage material is that this insures that under equilibrium conditions the substantial portion of the electrolyte is contained in the separator and the electrolyte storage material so that it is available to carry out its intended function during charging and discharging of the cell.

IRON-SILVER OXIDE

O.B. Lindstrom; U.S. Patent 4,011,365; March 8, 1977; assigned to AB Olle Lindstrom, Sweden describes a cell which uses silver oxides as positive electrode material and iron as negative electrode material in an alkaline electrolyte with the silver oxides being reduced to mainly metallic silver during the discharge of the cell. This cell is sealed so as to bring the gases developed in the cell into contact with the negative as well as the positive material thus eliminating the gases. This sealed alkaline cell gives outstanding performance with respect to energy and power density and cycling life. The following examples illustrate the process.

Example 1: An iron electrode is made by compacting a mixture of carbonyl iron powder and rock salt powder at 1.8 tons/cm^2. The bulk density of the iron powder is 0.8 g/cm^3 and the particle size about 5 μm. The salt has been ground to a particle size below 30 μm. One part by weight of salt is mixed with two parts by weight of iron powder. The mold could be circular with a diameter of 10 cm and the powder charge adjusted so as to give a disc with a thickness of 0.24 cm in the compacted green condition. The disc is sintered in hydrogen at 650°C during 1 hour. The salt is then leached out in water with KOH added to pH 12. After drying, electrodes of required size are punched out of the disc.

Example 2: An iron electrode is made by loose sintering of the iron powder of Example 1. The mold is filled with the iron powder to a thickness of 1.0 cm. Sintering is then taking place at 750°C in a hydrogen atmosphere during 45 minutes. The disc is then compacted to a thickness of 0.24 cm. Electrodes of the required size are punched out of the disc.

Example 3: An iron electrode is made as in Example 2 but with a powder of pigment grade iron oxide. Fe_2O_3 is substituted for the iron powder. The iron oxide powder is made by roasting $FeSO_4$ in air followed by grinding.

Example 4: An iron electrode is made by compacting the iron powder of Example 1, to which has been added 4% by weight of polyethylene powder, Microthene MN 722, to a porosity of 65%. The disc is sintered in hydrogen at 120°C during 30 minutes. Electrodes of required size are punched out of the disc.

Example 5: An iron electrode is made as in Example 1 but with 6% by weight of the iron powder substituted for the same weight of finely ground cadmium oxide.

Example 6: A silver oxide electrode is made by compacting a mixture of the silver oxide Ag_2O with a bulk density of 1.2 g/cm^3 and rock salt powder ground to a particle size below 30 μm. Three parts by weight of the silver oxide are thoroughly mixed with one part by weight of the salt powder with the mixture being finally passed through a 270 mesh screen. The compaction pressure is 1.5 tons per cm^2 and sintering is taking place at 400°C in air during 25 minutes. The salt is leached out with water during 24 hours whereafter the electrode is anodically oxidized in 1 N KOH at mA/cm^2 during 48 hours.

Example 7: The silver oxide powder of Example 6 is compacted directly into the electrode can to a porosity of 65%.

Example 8: The silver oxide powder of Example 6 is mixed with 3% by weight of polyethylene binder, Microthene MN 722, and compacted directly into the can to a porosity of 68%. The can is heated in air to 120°C during 10 minutes.

Example 9: A silver oxide electrode is made as described in Example 6 but with equal parts of silver oxide and salt powder. After leaching the disc is compacted to a porosity of 62% prior to the electrochemical formation procedure. A suitable electrolyte containing zincate is made by dissolving 2% by weight of zinc oxide, ZnO, in hot 30% solution, and is cooled and filtered prior to use.

A battery cell, with a diameter of 1.13 cm and a height of 0.52 cm thus delivers 125 mAh at a discharge current of 2 mA within the voltage range 1.4 to 0.9 V. The cell can take overcharge and deep discharge without being damaged and with no leakage of gases or electrolyte to the surroundings. The life is in excess of 500 deep cycles compared to only a few hundreds of cycles in the corresponding vented design. The corresponding commercially available nickel-cadmium cells in button cells give only 20 to 30 mAh, which illustrates the excellent properties of this power source.

IRON-AIR

A process described by *O.B. Lindstrom; U.S. Patent 4,032,693; June 28, 1977; assigned to AB Olle Lindstrom, Sweden* relates to a procedure for stabilizing an iron-air battery comprising one of several cells with a positive electrode containing at least one of the metals cobalt, nickel and silver for the reduction of the oxygen of the air during discharge and the development of oxygen during charge, an alkaline electrolyte and a negative electrode containing electrochemically active iron, whereby a sulfur-containing compound is added to the electrolyte.

Thus, it has been found that the addition to the electrolyte of a sulfur-containing compound (which forms free sulfide ions in the electrolyte at least temporarily) in a concentration of at least about 10 ppm and up to about 1,000 ppm, counted as weight of sulfur per weight active iron material, stabilizes and improves the performance of the iron-air cells to a considerable degree.

SODIUM-IODINE BATTERY

According to a process described by *M.A. Cadman, M. Voinov and H. Tannenberger; U.S. Patent 3,935,025; January 27, 1976; assigned to Etat Francais, France* an electrical storage battery, providing high energy density, comprises an anode compartment containing at least one reducer metal and a cathode compartment containing an electrically conductive solution of at least one salt of the metal in a nonaqueous solution.

The solution is in contact with at least one electron acceptor material and the two compartments are separated by a wall which is sealed to the fluids and is constituted by a solid material that allows the migration, at ambient temperature, of the reducer metal of the anode in the form of ions. The electron acceptor substance is soluble in the solution contained in the cathode compartment in the reduced form thereof, as well as in its original form.

The electron acceptor material used in the storage battery can be any substance which, at ambient temperature, has a high solubility in a nonaqueous solvent and which is able to accept at least one electron per molecule, according to the general formula

$$A + ne^- \longrightarrow A^{ne^-}$$

in which n is a whole number and A represents the original form of the substance in question, and whose reduced form A^{ne^-} is likewise highly soluble at ambient temperature in the same solvent. In particular, a halogen can be used as soluble electron acceptor material with iodine being the preferred material. An organic compound can also be used, e.g., a quinone, an oxidized benzidine and the like, or even a mineral complex, e.g., a ferricyanide.

According to the charge state of the storage battery, the electron acceptor material may be entirely in its original state or entirely in a reduced state, or even in a mixture of the two. The reducer metal is preferably sodium although it is also possible to use other metals, for example alkali metals such as lithium, potassium and rubidium, or even alkaline earth metals such as calcium, strontium and barium. It is also possible to use other reducer metals such as magnesium, lanthanum and the like.

In the preferred example, the anode compartment contains an electrically conductive solution of at least one salt of the metal of the anode in at least one organic solvent. The salt, as for example, in the case of the solution contained in the cathode compartment, can be a perchlorate, a halide, a tetraphenyl borate, a hexafluorophosphate, or a sulfocyanide of the reducer metal, e.g., $NaClO_4$; NaI; $NaPF_6$; $NaB(C_6H_5)_4$; NaCl; and NaSCN, among others. The solvent in the anode compartment can be a compound such as propylene carbonate, tetrahydrofuran, ethylene carbonate, acetonitrile, butyrolactone, dimethyl formamide and N,N-dimethyl acetamide.

The nonaqueous solvent used to form the solution contained in the cathode compartment can be an organic solvent such as propylene carbonate, tetrahydrofuran, N,N-dimethyl acetamide, ethylene carbonate, acetonitrile, butyrolactone, dimethyl formamide and the like. Preferably, the above organic solvents are used in their highest possible state of purity. The salt of the reducer metal can, for example, be a perchlorate, a halide, a tetraphenyl borate, a hexafluorophosphate, or a sulfocyanide. In the case of sodium, use may be made of one of the following compounds: $NaClO_4$; NaI; $NaPF_6$; $NaB(C_6H_5)_4$; NaCl; NaSCN.

Referring to Figure 6.2a, a storage battery is shown which comprises a cylindrical glass receptacle or container **10** and a tubular or cylindrical separator **12**, fabricated of β-sodium alumina of the approximate composition $Na_2O \cdot 11\ Al_2O_3$ and having fluid-tight walls and a thickness of 0.2 mm, placed inside receptacle **10**. Tubular separator **12** is hermetically sealed with respect to the bottom of receptacle **10** by means of a glass weld indicated at **14**.

Weld **14** can be made in any other suitable way or can be replaced by a fluid-tight joint or by bonding using an insoluble plastic material, i.e., a plastic material which is insoluble in organic solvents with high dipole moment. An annular space **16** formed between the outer wall of tubular separator **12** and the inside wall of receptacle **10** comprises the anode compartment of the battery.

FIGURE 6.2: HIGH ENERGY DENSITY BATTERY

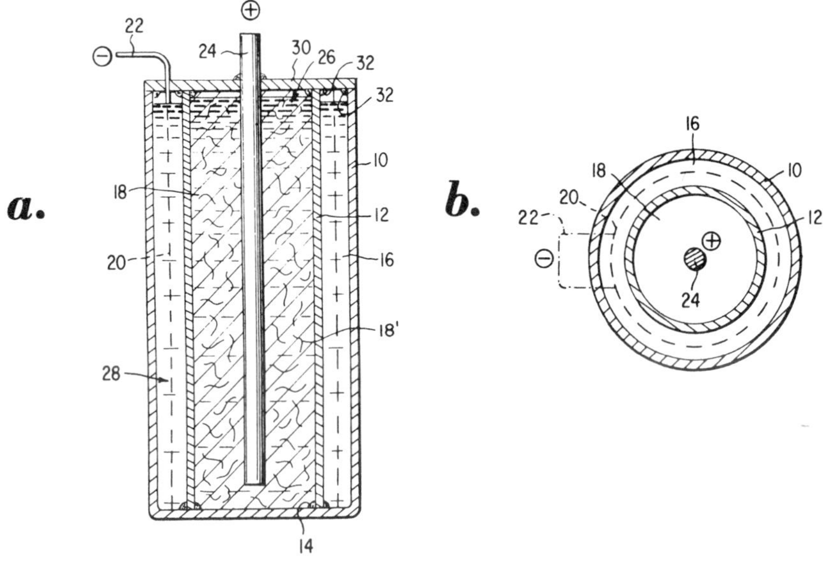

(a) Vertical sectional view of a storage battery
(b) Top sectional view of the battery of Figure 6.2a

Source: U.S. Patent 3,935,025

Similarly, a cylindrical space **18**, formed inside tubular separator **12** comprises the cathode compartment of the battery. A fine cylindrical grid **20**, made of copper wire having a diameter of 0.1 mm, interlaced as a grid with a square mesh, 0.4 mm to the side, is placed in anode compartment **16**, 1 mm from the wall of tube **2**. Grid **20** constitutes the anode current collector of the battery and is connected to negative terminal **22** of the battery. A graphite rod **24**, 3 mm in diameter in the specific example under consideration, extends into the center of cathode compartment **18** and constitutes a current collector whose free end constitutes the positive terminal of the storage battery.

An electrically conductive substance, which is indicated at **26** and is constituted by a homogeneous mixture of carbon felt **18'** and powdered sodium iodide impregnated with acetonitrile, substantially completely fills the space between rod **24** and the inner wall of tubular separator **12**. Anode compartment **16** is similarly filled with a saturated solution, indicated at **28**, comprising sodium perchlorate ($NaClO_4$) in propylene carbonate.

A glass cover **30** is hermetically sealed onto receptacle **10** and separator **12** by bonded joints of epoxy resin indicated at **32**, the inlet passages for terminals **22** and **24** being also sealed in the same way. The sealing of cover **30** is effected with the contents of compartments **16** and **18** kept under nitrogen to avoid contamination of the contents by air.

The storage battery as just described is in the electrically uncharged state thereof and can be stored this way without change for very long periods. Hence, the first operation to be effected so as to energize the battery is electric charging. To this end, terminal **22** is connected to the negative terminal of a suitable direct current voltage, e.g., 3.5 volts, and the positive terminal **24** is connected to the positive terminal of the same source of current.

With the battery so connected the physicochemical phenomena which occur in the battery can be described, with the aid of formulas, as follows: In anode compartment **16**, the sodium ions are reduced to the metallic state according to the following formula:

$$Na^+ + e^- \longrightarrow Na$$

and the metal thus formed is deposited on grid **20**. A number of sodium ions equal to that of the ions thus discharged pass through the wall of tubular separator **12** from compartment **18** and replace the ions discharged in compartment **16**. In compartment **18**, the iodide ions discharge according to the formula:

$$I^- \longrightarrow \tfrac{1}{2}\, I_2 + e^-$$

and the iodine thus formed passes into solution in the acetonitrile, presumably as complex ions I_3^- formed by the reaction

$$I_2 + I^- \longrightarrow I_3^-$$

For a storage battery in which receptacle or container **10** has a length of about 5 cm and an internal diameter of 1.3 cm, separator **12** having an outer diameter of 1 cm, compartment **18** can be furnished with a mixture constituted by 5 g sodium iodide and 0.3 g carbon felt to which is added about 1 cc acetonitrile when the mixture is placed within compartment **18**.

The storage battery presents the special advantage, among others, that the battery can be manufactured in an electrically uncharged state which thus facilitates the storage until the battery is put to use. The storage battery can be used in a range of temperatures from below 0°C, e.g., at temperatures between -30° and -20°C, to the boiling temperature of the most volatile solvent used, e.g., at temperatures between 100° and 200°C.

TITANIUM-RUTHENIUM COATINGS FOR BROMINE CELLS

According to a process described by *A. Nidola, V. de Nora and P.M. Spaziante; U.S. Patent 4,037,032; July 19, 1977; assigned to Diamond Shamrock Technologies SA, Switzerland* electric storage batteries with titanium or other film-forming metal anodes and cathodes are provided. The anodes are coated with mixed oxide coatings of a film-forming metal oxide containing a platinum group metal oxide catalyst, which coatings may contain oxides of other metals to alter the breakdown voltage and increase the oxygen overpotential of the cathodes while preserving a low bromine and chlorine overpotential.

The cathodes may be uncoated titanium or other film-forming metals or film-forming metals with a silver coating thereon. Bipolar film-forming metal cathodes and anodes are provided carrying a mixed oxide coating on the cathodic face and either no coating or a silver coating on the anodic face. The process is applicable to all types of electric storage batteries, either primary or secondary, and multiple batteries may be connected in series or in parallel to provide the desired amperage and voltage characteristics.

As noted, the oxide coatings on the cathodes may contain a mixture of titanium dioxide and a platinum group metal oxide such as ruthenium dioxide and may contain oxides of other metals as doping agents or to impart special properties to the cathode coatings for increasing the conductivity and breakdown voltage of the oxide film on the titanium or for other purposes. A particularly effective coating comprises 45 to 65% of titanium dioxide, 30 to 35% of ruthenium dioxide, 1 to 20% of stannous oxide and 1 to 5% of cobalt dioxide, the percentages being based on the weight of the metals in the oxides. In some instances the coating may contain a platinum group metal oxide alone.

Coating No. 1: Titanium trichloride in HCl solution is dissolved in methanol; the $TiCl_3$ is converted to the pertitanate by the addition of H_2O_2. This conversion is indicated by a change in color from $TiCl_3$ (purple) to Ti_2O_5 (orange). An excess of H_2O is used to insure complete conversion to the pertitanate. Sufficient $RuCl_3 \cdot 3H_2O$ is dissolved in methanol to give the desired final ratio of TiO_2 to RuO_2. The solution of pertitanic acid and ruthenium trichloride is mixed and the resulting solution is applied to a cleaned titanium battery cathode surface, which had been cleaned by boiling in a 20% solution of hydrochloric acid at a reflux temperature of 109°C for 20 minutes, by brushing or spraying.

The coating is applied as a series of coats with baking at about 350°C for five minutes between each coat. After a coating of the desired thickness or weight per unit of area has been applied, the deposit is given a final heat treatment at about 450°C for 15 minutes to 1 hour. The molar ratio of TiO_2 to RuO_2 may be varied from 1:1 TiO_2:RuO_2 to 10:1 TiO_2:RuO_2. The molar values given above corresponded to 22.3:47 weight percent Ti:Ru and 51:10.8 weight percent Ti:Ru.

Coating No. 2: A coating consisting of the ingredients shown in the table below was applied to a cleaned titanium base which had been cleaned by boiling at a reflux temperature of 109°C in a 20% solution of hydrochloric acid for 20 minutes. The coating was prepared by first blending the ruthenium, cobalt and tin salts in the required amount. $TiCl_3$ solution (15% as $TiCl_3$ in commercial solution) was then slowly added under stirring. After the salts were completely dissolved, a few drops of hydrogen peroxide (H_2O_2, 30%) were added, sufficient to make the solution turn from the blue of the commercial $TiCl_3$ solution to the brown-reddish color of a peroxyhydrate or pertitanate compound.

	Metal, mg/cm^2	Weight % Metal
Ruthenium as $RuCl_3 \cdot 3H_2O$	1.60	45% Ru
Cobalt as $CoCl_2 \cdot 6H_2O$	0.036	1% Co
Tin as $SnCl_4 \cdot 5H_2O$	0.142	4% Sn
Titanium as 15% $TiCl_3$ solution (commercial)	1.78	50% Ti

At the end a few drops of isopropyl alcohol were added to the solution after cooling. The coating, thus prepared, was applied to the working side of a battery cathode exposed to the electrolyte by brushing or spraying in 10 to 14 subsequent layers. After applying each layer, the cathode was heated in an oven under forced air circulation at a temperature between 300° and 400°C for 5 to 10 minutes, followed by fast natural cooling in air between each of the first 10 to 14 layers and after the last coat was applied the cathode was heated at 450°C for 1 hour under forced air circulation and then cooled. Any compatible electrochemical and electrolyte system, either acid or alkaline, may be used in the batteries.

METAL HALOGEN HYDRATE SYSTEM

H.K. Bjorkman, Jr.; U.S. Patent 3,993,502; November 23, 1976; assigned to Energy Development Associates describes a secondary electrical energy storage system incorporating or associated with a halogen hydrate forming and storing system which produces solid halogen hydrate in a form suitable for replenishing the hydrate reservoirs of such storage systems. The halogen hydrate forming and storing system is fully operable upon charge and discharge of the battery system by the use of a single centrifugal pump, which results in simplicity, low cost and reduced potential maintenance of the system.

Referring to Figure 6.3a, a typical flow arrangement of a rechargeable electrical energy storage system is illustrated in accordance with the preferred practice of the process. As shown, the system comprises an electrode area or stack, indicated as **S**, which is comprised of one or more, usually a plurality, of individual cells, each containing a normally positive electrode and a normally negative electrode. The stack is connected by means of an outlet pipe **10** and a return pipe **12** to a halogen hydrate storage area or receptacle, indicated at **H**, through which the electrolyte is continuously recirculated such as by means of a pump **P**.

FIGURE 6.3: METAL HALOGEN HYDRATE BATTERY SYSTEM

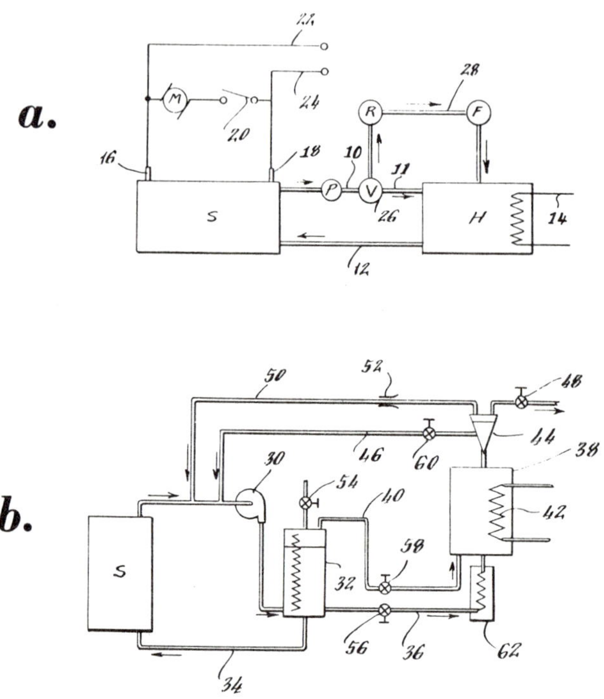

(continued)

FIGURE 6.3: (continued)

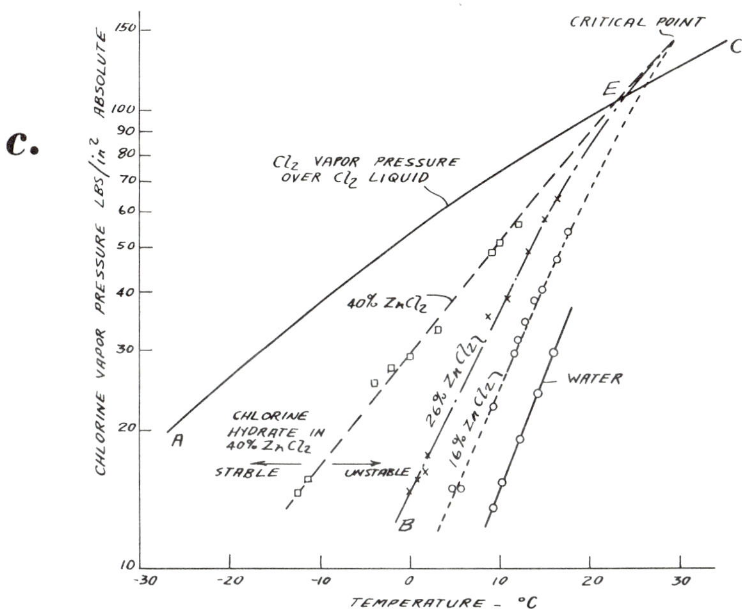

(a) Schematic flow sheet illustrating the important components of an electrical energy storage system.

(b) Schematic flow diagram in greater detail of the electrical energy storage system

(c) Phase diagram for a halogen hydrate system in which the halogen comprises chlorine, the metal comprises zinc and the aqueous solution contains zinc chloride.

Source: U.S. Patent 3,993,502

The passage of the electrolyte directly through the hydrate storage reservoir **H**, by means of valve and line **11** during a normal discharge cycle of the storage device effects a progressive decomposition of the halogen hydrate therein, whereby the liberated halogen gas is dissolved and/or entrained in the electrolyte and is conveyed by means of the return line **12** to the stack **S** for replenishment of the supply of elemental halogen at the normally positive electrodes therein.

Since the decomposition of the halogen hydrate is an endothermic reaction, the storage reservoir **H** suitably may be provided with a heat exchanger, indicated at **14**, for maintaining the halogen hydrate and the electrolyte therein at a temperature at which optimum performance of the storage battery is achieved during a discharge cycle. During the recharging cycle, a reversal of the chemical reaction is effected whereby the oxidized metal is reduced at the normally negative electrode and deposits or plates out while the halide ion becomes oxidized at the normally positive electrode, returning to the elemental state in the form of dissolved or minute gas bubbles in the electrolyte. The halogen gas thus formed is continuously removed from the cell by the electrolyte through the outlet **10** and the electrolyte is diverted, such as by a selector valve **26**, through a branch circuit **28** having a cooling or refrigeration device **R** therein and a hydrate former device **F** for effecting a regeneration of the halogen hydrate.

The halogen hydrate thus regenerated is returned by the circulating electrolyte to the storage reservoir **H** in which it is separated and retained in readiness for the next discharge cycle of the storage device; electrolyte separated from the halogen hydrate within **H** is returned to the cell, **S**, by means of line **12**.

In accordance with this process, the hydrate former **F** and the cooling or refrigeration apparatus **R** are incorporated in the system in accordance with the arrangement schematically illustrated in Figure 6.3a. A recharging of the electrical energy storage system can conveniently be achieved by connecting the wires **22, 24** to a commercial source of a rectified current, such as purchased from a local utility, effecting an in situ recharging of the system from time to time as may be necessary. Systems of this arrangement are suitable for use as the principal or auxiliary power system for the propulsion of vehicles or other mobile apparatus.

Alternatively, the hydrate former **F** and the refrigeration or cooling unit **R** can be disconnected from the system during normal discharge thereof and are located at a central processing or service station. In accordance with this latter arrangement, an appropriate amount of electrolyte is withdrawn from the electrical storage systems which are in a substantially discharged condition and the electrolyte is reprocessed through the local service station processing facility to effect a regeneration of the halogen hydrate and a recovery of the metal.

The halogen hydrate and metal can be directly inserted into the discharged storage battery, effecting a refueling thereof and a restoration of the battery to a fully charged condition. The refrigeration unit **R** and the hydrate former provide for increased efficiency due to the larger size of such units, which are adaptable for servicing a plurality of storage batteries, while at the same time providing for a proportional reduction in the weight of such storage battery systems in view of the elimination of the refrigeration and hydrate forming components.

In either event, a hydrate former constructed in accordance with the process is illustrated in Figure 6.3b. The particular temperature, pressure and operating characteristics of the hydrate former shown will vary depending upon the composition of the electrolyte or aqueous solution employed, the type of halogen utilized and with the concentration and type of metal ions present.

A phase diagram is shown in Figure 6.3c which depicts the temperature-pressure relationship for the halogen chlorine, the preferred halogen, when the gas phase is in equilibrium over the solid hydrate in aqueous solutions of zinc chloride of various concentrations. As noted in Figure 6.3c, the area to the left of any line of zinc chloride concentration encompasses those conditions for which chlorine hydrate is present and may be readily formed in an electrolyte of that concentration. The area to the right of any line encompasses those conditions for which gaseous chlorine is present in the aqueous media. Above and to the left of line **AC**, liquid chlorine is present. In the area bounded by points **A, B** and

E, chlorine hydrate is present as a solid in a 25% concentration of zinc chloride in water. It will be appreciated that the phase diagram, as shown, encompasses three phases, namely: solid, liquid and gas. The critical temperature of a halogen hydrate, such as chlorine hydrate as shown in Figure 6.3c, is defined as that temperature above which a halogen hydrate cannot exist.

Referring in detail to Figure 6.3b, a working example of the system is shown. The working system has two fluid flow loops, operating in parallel and driven by the same pump, both on charge and discharge of the system. This results in simplicity of construction, low cost and reduced potential maintenance since as few as possible moving parts are utilized in the working embodiment of the system.

On charge, the electrolyte is pumped by a centrifugal pump **30** from the stack **S** with the halogen gas dissolved or entrained in the electrolyte. The gas-liquid mixture passes through pump **30** into a filter separator unit **32**, the purpose of which is to remove particulate matter and to break the froth which comes from the stack so that the gas and liquid will separate. One stream of electrolyte is passed back to the stack through a conduit **34**; a second stream of electrolyte is passed from filter-separator unit **32** through a conduit **36** to a hydrate former-store unit **38** where a portion of the water in the electrolyte is combined with the halogen gas taken from filter-separator unit **32** through a conduit **40**, to form halogen hydrate.

A typical hydrate former-store unit **38** is shown in U.S. Patent 3,814,630. As shown in this patent, hydrate former-store unit **38** is provided with a heat exchanger coil **42**, effecting an appropriate extraction or addition of heat to the halogen hydrate/electrolyte to form solid halogen hydrate during the charging of the system.

Excess electrolyte passes through the filter in the top of the hydrate former-store unit **38** into a small secondary separator **44**. This liquid electrolyte returns from the bottom of separator **44** through a conduit **46** to the low pressure side of pump **30** where it is recycled. Excess halogen gas in the hydrate former-store unit **38** which is not converted to halogen hydrate, as well as any inert gases which are present are passed through the filter in the top of unit **38** into separator **44**, from which a small portion may be vented through a pressure valve **48**, since this gas is relatively rich in inerts, having been reduced in halogen content in unit **38**.

The remaining gas is passed back to the low pressure side of pump **30** by a gas return conduit **50** for recycling. Conduit **50** is equipped with a restrictive orifice **52** which provides restriction to liquid flow but allows gas to flow therethrough freely. When separator **44** is filled with liquid, orifice **52** tends to block the liquid flow resulting in a combined flow to pump **30** only slightly greater than that occurring when the flow is returned by the liquid line **46** alone. If gas should fill separator **44**, the gas passes back to the low pressure side of pump **30** through liquid return conduit **46** thereby causing more liquid from the stack loop to flow into the store unit **38**.

On discharge, the flows in the various conduits of the system are in the same direction, albeit different quantities are involved. Electrolyte from filter-separator unit **32** is passed through conduit **36** into the hydrate former-store unit

38 where it decomposes the stored halogen hydrate releasing halogen gas. Since the decomposition of the halogen hydrate is an endothermic reaction, coolant flow through heat exchanger **42** is stopped during discharge, unless discharge is stopped and it becomes necessary to cool to stabilize the hydrate.

The halogen gas and reconstituted electrolyte return through separator **44** and conduits **52** and **46**, respectively, to the low pressure side of pump **30** where it is circulated through the parallel stack loop and hydrate former-store loop. Secondary gas separator **44** operates as before, except during discharge, the vent is not taken through valve **48**, but through pressure valve **54** at the top of the filter-separator unit **32**, because the concentration of inerts is higher at this point. Further, during the discharge cycle, gas inlet line **40** from the filter-separator unit **32** to the hydrate former-store unit **38** will normally be closed so that gas does not pass through the unit **38** during discharge.

Valves **56**, **58** and **60** can be disposed in the various flow lines to control the rate of flow of the fluids in the system by shutting the lines down completely or to vary flow as the pressure drop across the unit **38** changes. An auxiliary heat exchanger **62** can also be provided in the liquid inlet line **36** to the hydrate former-store unit **38** to precool the electrolyte on charge depending on the hydrate requirements. On discharge, coolant flow can also be stopped through the coils of heat exchanger **62**.

A process described by *P.C. Symons; U.S. Patent 3,940,283; February 24, 1976; assigned to Energy Development Associates* is concerned with a method of conveniently storing halogen by employing halogen hydrates for use in electrical energy storage devices, such as primary and secondary batteries. Halogen hydrates provide a convenient means of storing halogens to be used in the discharge of primary and secondary batteries since they allow the concentrations of halogen and electrolyte to be controlled easily.

The halogen hydrate produced during the charging of a secondary battery is a convenient means of storing the halogen until it is used during the discharge of the secondary battery. Another aspect of the process is the production of an electrolyte from the halogen hydrate for use during the discharging of a secondary battery.

HYDROGEN-HALOGEN BATTERY

C. England; U.S. Patent 3,894,887; July 15, 1975; assigned to California Institute of Technology describes a secondary battery utilizing hydrogen and halogen as primary reactants which comprises inert anode and cathode initially contacting an aqueous solution of an acid and an alkali metal bromide. The hydrogen generated during charging of the cell is stored as gas, while the bromine becomes dissolved predominantly in the lower layers of the acid electrolyte. Preferred components comprise phosphoric acid and lithium bromide.

Referring to Figure 6.4, the secondary cell comprises the cell container **12** which is preferably designed to retain fluid at superatmospheric pressure. As illustrated, container **12** comprises the vessel **14** of chemically inert metal such as stainless steel with an insulating lining **15** of polymerized tetrafluoroethylene or the like. The cover **16** is typically of insulative material secured to vessel **14** by the bolts **17** with the joint sealed by the gasket **18**. If a metal cover plate is preferred, insulation may be provided at the bolts **17** and gasket **18**. The body of electrolyte **20** is contained in the lower portion of vessel **14**, preferably leaving an appreciable free volume **24** within container **12** and above the free surface **21** of the electrolyte for accommodating hydrogen gas.

FIGURE 6.4: HYDROGEN-BROMINE BATTERY

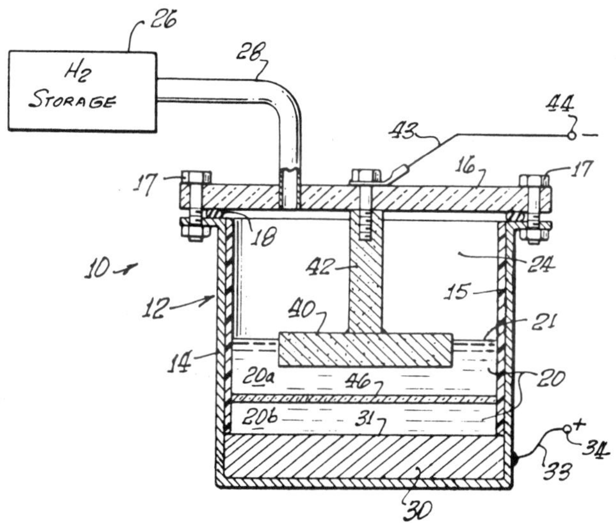

Source: U.S. Patent 3,894,887

That gas storage means may be supplemented by auxiliary storage means indicated schematically at **26**, which is typically in free communication with volume **24**, as via the conduit **28**. The volume **24**, or the combined volume of **24** and **26**, is typically designed to receive all the hydrogen that will be evolved upon charging of the cell, at only a moderate predetermined internal pressure. The numeral **10** designates the entire housing means, comprising container **12** and also auxiliary hydrogen storage means **26** and conduit **28**, if used.

The cathode **30** is formed of electrically conductive and chemically inert material, such as carbon, for example, and is mounted in the extreme lower portion of vessel **14**, typically presenting a horizontal upper surface **31** to the electrolyte. Cathode **30** is typically in electrical contact with vessel **14**, and is connected via the vessel and the conductor **33** to the positive cathode terminal **34**.

The anode **40** is mounted by the conductive post **42** from cover **16**, with electrical connection in sealed relation through that cover to the conductor **43** and the negative anode terminal **44**. Anode **40** comprises a generally horizontal structure extending both above and below the level of the electrolyte surface **21** and thus presenting a large lower surface to the electrolyte and a corresponding upper surface to the gas in volume **24**. The anode, which may be of any suitable type, is preferably formed of a porous, electrically conductive, and chemically inert material, such as carbon felt, for example. Thus, it is permeable for diffusion of the gas and the liquid electrolyte, and provides a large surface area at which they can react. Oxidation and reduction of hydrogen at the anode surface is facilitated by electrolytic deposition upon it of platinum or other suitable catalyst.

The body of electrolyte **20** is preferably divided horizontally into upper and lower portions **20a** and **20b** by the permeable separator **46**, which permits substantially free diffusion of liquid and solutes but prevents mixing by convection. The separator is typically formed of glass fiber or similar nonconductive and inert material.

The cell is initially filled to the desired level with an aqueous solution of acid and alkali metal bromide. Phosphoric and sulfuric acids are particularly suitable for that purpose. All three of the alkali metals, lithium, sodium and potassium, are satisfactory for formulation of the bromide. When the weight of the battery must be held to a minimum phosphoric acid and lithium bromide are particularly advantageous.

The concentration of acid is typically selected to provide approximately maximum electrical conductivity in the solution, and the acid solution is then typically substantially saturated with the alkali metal bromide. The free volume **24** above the electrolyte surface is preferably initially filled with hydrogen, or may be evacuated of all gases other than water vapor. The same treatment applies to the auxiliary storage volume **26** and conduit **28**, if used. That general assembly procedure results in a complete cell in fully discharged condition.

Specific electrolyte compositions in accordance with the process are given below as Examples 1, 2 and 3, but are intended only as illustration. The relative concentrations of the components are given in parts by weight.

	Parts by Weight
Example 1	
Water	70
Sulfuric acid	30
Lithium bromide	46.3
Example 2	
Water	80
Acetic acid	20
Potassium bromide	37.6
Example 3	
Water	50
Phosphoric acid	50
Lithium bromide	125.4

To charge the cell, current is caused to flow through it from cathode terminal **34** to anode terminal **44**, that is, from cathode **30** to anode **40**. During that charging operation electrons supplied from the anode combine with hydrogen ions in the electrolyte to produce gaseous hydrogen, which emerges into the free volume **24**. At the same time negative bromine ions from the electrolyte give up electrons to the cathode, forming molecules of Br_2, which are readily soluble in the aqueous solution. The resulting bromine-rich solution is appreciably heavier than the remainder of the electrolyte, and therefore tends to remain close to the electrode surface **31** at which the bromine was produced, resulting in a sharply stratified condition of the electrolyte. Separator **46** preserves that stratification against mixing by accidental convection.

As the charging process progresses, ions of bromine removed from lower chamber **20b** and hydrogen ions removed from upper chamber **20a** by electrolysis are partially replaced by downward migration of bromine ions and upward migration of hydrogen ions through separator **46**. Hence bromine ions and hydrogen ions are progressively removed from both the upper and the lower chambers. As the cell approaches fully charged condition, the electrolyte in the upper chamber becomes predominantly an aqueous solution of the salt of the selected alkali metal and the selected acid, while the electrolyte in the lower chamber approaches a solution of the same salt containing a progressively larger concentration of dissolved bromine molecules.

The charging process is preferably terminated while there is still a sufficient concentration of bromine ions to prevent significant oxidation of other components of the electrolyte at the anode, and while there is still a sufficient concentration of hydrogen ions to maintain good electrical conductivity in the electrolyte.

The latter condition is promoted by initially supplying acid and bromide to the electrolyte in such proportions that the atomic ratio of hydrogen to bromine exceeds unity by at least about 3%. Values of that ratio from about 1.03 to 1.15 are generally satisfactory, with values near the lower end of that range preferred when a high value of energy density is desired. The illustrative compositions given above correspond to an excess of hydrogen ions of approximately 5%.

During discharge of the cell, the same reactions occur as during charging, but proceeding in the opposite directions. Thus, the hydrogen gas and bromine atoms diffuse to the respective active electrode surfaces and become electrolyzed to ionic form, with current flow through the external circuit from cathode terminal **34** to anode terminal **44**.

The overall cell reaction, assuming for definiteness that the cell initially contained an aqueous solution of phosphoric acid and lithium bromide, typically as given above in Example 3, can be expressed as

$$2H_3PO_4 + 6LiBr \underset{\text{discharge}}{\overset{\text{charge}}{\rightleftharpoons}} 3H_2 + 3Br_2 + 2Li_3PO_4$$

Omitting the mass of the aqueous solvent, the theoretical energy density of the system for that particular selection of ingredients is about 103 watt-hours per pound.

CALCIUM ALLOYS IN METALLIC SULFUR CELLS

According to a process described by *M.F. Roche, S.J. Preto and A.E. Martin; U.S. Patent 3,980,495; September 14, 1976; assigned to the U.S. Energy Research and Development Administration* calcium alloys such as calcium-aluminum and calcium-silicon, are employed as an active material within a rechargeable negative electrode of an electrochemical cell. Such cells can use a molten salt electrolyte including calcium ions and a positive electrode having sulfur, sulfides, or oxides as an active material. The calcium alloy is selected to prevent formation of molten calcium alloys resulting from reaction with the selected molten electrolytic salt at the cell operating temperatures. The following examples illustrate the process.

Example 1: A secondary electrochemical cell in the discharged state was constructed in conventional fashion. The negative electrode included small diameter aluminum wire and stainless steel screen wrapped in zirconia and boron nitride cloth. The positive electrode included calcium sulfide and iron powders in an alumina cup with a foamed-iron current collector wrapped in suitable retainer screens and cloths.

The cell electrolyte was 34 weight percent sodium chloride and 66 weight percent calcium chloride, a eutectic mixture having a melting point of 506°C. The loading of the two electrodes was calculated to provide on full charge 6 Ah of $CaAl_2$ in the negative electrode and an excess of FeS in the positive electrode. Since calcium forms compounds with aluminum and sodium does not, the cell could be operated at a lower charge potential than that required to form molten calcium-sodium alloy thus preventing degradation of the electrolyte.

The cell was operated at about 550°C for ten cycles at a current of approximately 0.4 to 0.9 A charge and 0.2 to 0.6 A discharge. Cutoff voltages of about 1.7 to 2.0 volts were employed on charge and 0.6 to 1.2 volts on discharge. Open circuit voltage plateaus of 1.65 V and 1.35 V corresponding to $CaAl_4$ were found.

Example 2: Cell SP-2 – $CaAl_2$ was prepared by fusion of the elements at about 1100°C in a zirconia crucible. Metallographic and x-ray examination confirmed that the product was mostly single-phase $CaAl_2$. The material was comminuted to approximately 100 micrometer particle size and vibratorily distributed within a porous substrate of foamed stainless steel for use as a negative electrode. The positive electrode included FeS, distributed in a current-collector structure. The cell was operated through 14 discharge and charge cycles of about 0.8 volt cutoff discharge and 1.8 to 1.9 volt charge cutoff.

The electrolyte was molten $CaCl_2$-NaCl at an operating temperature of about 550°C. Currents were typically 0.6 A on discharge and 0.3 to 0.6 A on charge. The cells Ah efficiency was near 100%. The capacity and voltage indicated that the negative electrode was cycling between $CaAl_4$ and Al. The cell was terminated after the fourteenth charge cycle and metallographic examination of the negative electrode revealed a major phase of $CaAl_4$ and minor dispersals of $CaAl_2$ within the $CaAl_4$. No aluminum was detected.

ALUMINUM-SULFUR

A process described by *G. Mamantov, R. Marassi and J.Q. Chambers; U.S. Patent 3,966,491; June 29, 1976; assigned to the U.S. Secretary of the Army* relates in general to molten salt electrochemical systems and in particular to such systems using molten chloroaluminate solvents that melt below 200°C, an aluminum metal anode and positive oxidation states of either iodine or sulfur as the active cathode material. The low melting chloroaluminate electrolyte melts below 200°C and is comprised of mixtures of 51 to 20 mol percent or 31.3 to 9.9 weight percent NaCl and 49 to 80 mol percent or 68.7 to 90.1 weight percent $AlCl_3$. Other alkali chlorides such as lithium or potassium chloride may be used partly or completely in place of sodium chloride.

The electrolyte may be prepared in a number of ways one of which is as follows: $AlCl_3$ is first sublimed through a medium porosity Pyrex frit in a sealed cell. The correct amount is then weighed in an argon filled dry box and mixed with the corresponding quantity of NaCl which has been purified by double crystallization from water and drying for 24 hours at 400°C under vacuum. The mixture is then melted in a container, sealed under vacuum in the presence of aluminum metal and kept at 200° to 250°C until the melt is water-like. The following examples illustrate the process.

Example 1: This example illustrates the use of the product of the electrochemical oxidation of sulfur as a cathode in a battery. A simple H-type cell is used. In the cathode compartment, a tungsten spiral having an area of about 1 cm^2 serves as the current collector in a solution of molten chloroaluminate containing 0.5 gram of sulfur. In lieu of the tungsten spiral, a carbon crucible or rod can also be used. The anode is a spiral of aluminum in a solution of molten chloroaluminate.

An additional compartment containing pure melt is present between the other two. The same melt composition of 49.8 mol percent or 69.36 weight percent $AlCl_3$ to 50.2 mol percent or 30.64 weight percent NaCl is used in all three compartments. The experimental temperature is 175°C. Two coarse Pyrex frits serve as separators. The central compartment is added to prevent diffusion and to avoid the problems associated with the dendrite formation during the charging cycle. The cell is charged by anodizing the tungsten spiral at constant current.

Sulfur cations are produced both by direct oxidation at the electrode and chemically through the reaction of chlorine and sulfur. Upon charging, the solution in the cathode compartment turns green and then brown. The potential, as measured with respect to an external aluminum reference, rises from +1.1 volts to 2.2 volts. A discharge curve as measured at a discharge current density of 6 mA/cm^2 and obtained after charging at 30 milliamperes for two hours indicates that the cell voltage remains at an average value of 2.0 volts for about 40 minutes.

When the current is cut off, the initial open circuit voltage of about 2.1 volts is rapidly recovered. It is further noted that an increase in the charging time causes an increased capacity of the cathode. That is, discharge curves with an average cell voltage of 2.0 volts are obtained with a current density of about 20 milliamperes per square centimeter for longer than one hour.

Example 2: In this example, a cell is used with two compartments separated by a medium porosity Pyrex frit. In the cathode compartment, a large tungsten spiral or carbon conductor is dipped in a concentrated solution of sodium iodide in the chloroaluminate melt, $AlCl_3$-NaCl, 49.8:50.2 mol percent. In the other compartment, a larger aluminum electrode is dipped in the chloroaluminate melt saturated with sodium chloride. The experimental temperature is 175°C. Upon anodization of the tungsten spiral, I_2 and I^+ are formed, and the potential of the electrode is increased to progressively more positive potentials of about 1.9 volts.

REDOX FLOW CELL

L.H. Thaller; U.S. Patent 3,996,064; December 7, 1976; assigned to the U.S. National Aeronautics and Space Administration describes a bulk energy storage system including an electrically rechargeable reduction-oxidation (redox) cell divided into two compartments by a membrane, each compartment containing an electrode. An anode fluid is directed through the first compartment at the same time that a cathode fluid is directed through the second compartment, thereby causing the electrode in the first compartment to have a negative potential while the electrode in the second compartment has a positive potential. The electrodes are inert with respect to the anode and cathode fluids used and the membrane is substantially impermeable to all except select ions of both the anode and cathode fluid, whether the cell is fully charged or in a state of discharge.

Referring to Figure 6.5a, there is shown an electrical system including an electric or voltage-producing cell **10** divided by an ion selective membrane **11** into a first compartment **12** and a second compartment **13**.

FIGURE 6.5: RECHARGEABLE REDOX FLOW CELL

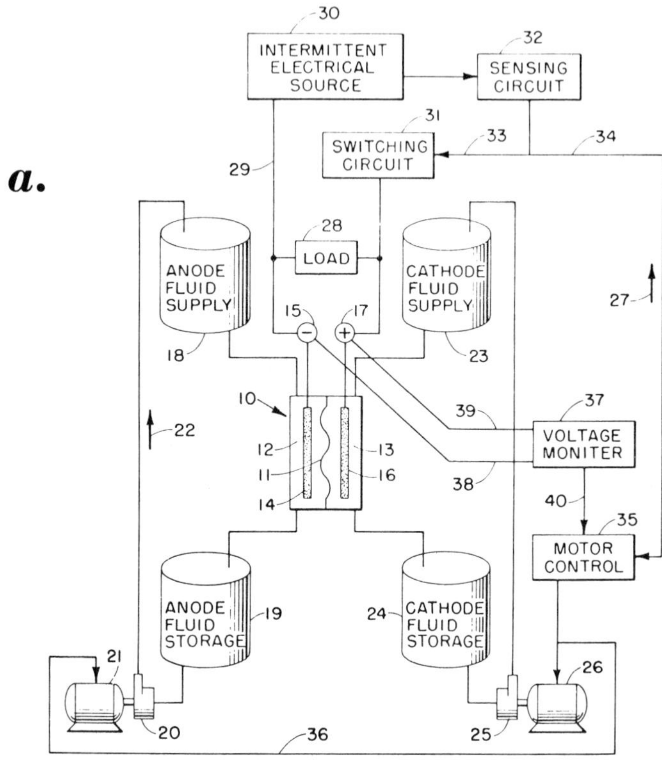

(continued)

FIGURE 6.5: (continued)

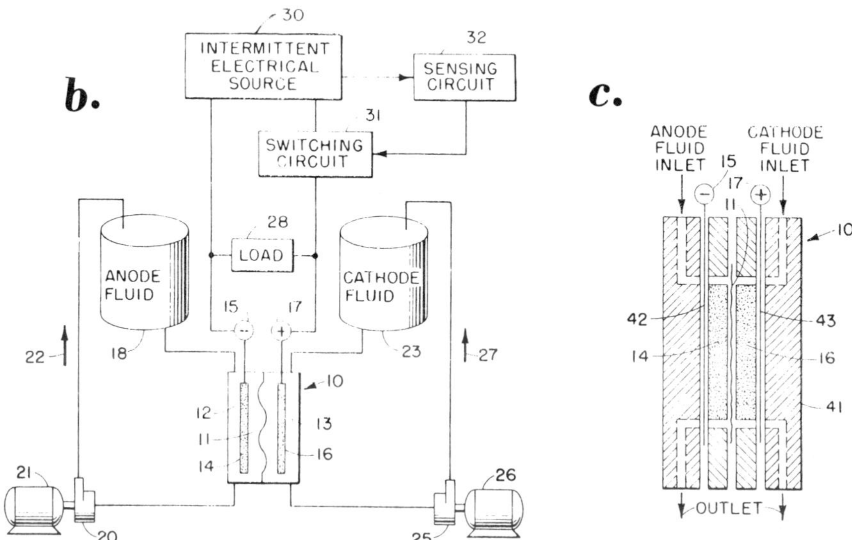

(a) Schematic drawing of electrical system

(b) Similar to Figure 6.5a but some of the components have been eliminated

(c) Schematic showing more details of the electric cell of Figures 6.5a and 6.5b

Source: U.S. Patent 3,996,064

An inert electrode **14** is disposed in the first compartment **12** and produces a negative potential at a terminal **15**. In a like manner, an inert electrode **16** is disposed in the second compartment **13** and produces a positive voltage at a terminal **17**. The difference of electric potential or voltage between the terminals **15** and **17** results from flowing an anode fluid containing a redox couple primarily in its reduced state through compartment **12**. Similarly, a cathode fluid containing redox couple primarily in its oxidized state flows through compartment **11**.

The anode fluid for compartment **12** flows from an anode fluid supply tank **18** through compartment **12** and into an anode fluid storage tank **19**. After all of the anode fluid has passed from supply tank **18** into the anode storage tank **19**, a pump **20** driven by an electric motor **21** pumps the anode fluid from storage tank **19** back to the anode fluid supply tank. Motor **21** is energized from the terminals **15** and **17** of cell **10**. Just as the anode fluid, cathode fluid flows from a cathode fluid supply tank **23** through the compartment **13** and into cathode fluid storage tank **24**; a pump **25** driven by an electric motor **26** pumps the cathode fluid back to supply tank **23** as indicated by arrow **27**. Like motor **21**, motor **26** is energized from terminals **15** and **17** of cell **10**.

The purpose of the electrical system shown in Figure 6.5a is to supply electrical power as needed to a load **28**. This load is connected between the terminals **15** and **17** of cell **10**. One side of load **28** is connected through a lead **29** to an intermittent or noncontinuous electrical source **30** as, for example, a wind-driven generator. The other side of load **28** is connected through a switching circuit to the source **30**.

To the end that the intermittent electrical source **30** may be disconnected from load **28**, and consequently from the terminals **15** and **17**, when there is little or no electrical power being generated by source **30**, there is provided a sensing circuit **32** having an input connected to source **30** and an output connected via a lead **33** to the input of switching circuit **31** and via a lead **34** to a first input of a motor control **35**. An output of the motor control **35** is connected via leads **36** to motors **21** and **26**. When the output voltage of source **30** drops below a predetermined value, sensing circuit **32** causes switching circuit **31** to open, thereby disconnecting load **28** and the terminals **15** and **17** from source **30**.

To prevent cell **10** from becoming completely discharged, there is provided a voltage monitor having an input connected by leads **38** and **39** to terminals **15** and **17**, respectively, and an output connected via a lead **40** to a second input of motor control **35**. Thus, when the cell voltage, as manifested at terminals **15** and **17**, drops below a predetermined value, voltage monitor **37** signals motor control **35** to inactivate motors **21** and **26**.

Sensing circuit **32** signals motor control **35** via lead **34** when source **30** has sufficient electrical voltage to charge cell **10**. This signal overrides the signal from voltage monitor **37** so that the pumps **21** and **26** will circulate the anode and cathode fluids when charging voltage is available.

The electrical system shown in Figure 6.5b is similar to that shown in Figure 6.5a and corresponding parts are identified by corresponding numerals. In the system shown in Figure 6.5b, the storage tanks **19** and **24** have been eliminated, together with the voltage monitor and the motor control. With this arrangement, motors **21** and **26** run continuously. However, the sensing circuit **32**, if desired, may function as in Figure 6.5a to disconnect the intermittent electrical source **30** from load **28** and terminals **15** and **17** when its voltage drops below a predetermined magnitude.

While Figures 6.5a and 6.5b show the fluid supply tanks **18** and **23** positioned higher than cell **10**, this is not necessarily true as to the actual physical location. There is some advantage to locating the supply tanks **18, 23** and the storage tanks **19, 24** above and below cell **10**, respectively. With such an arrangement, fluid will flow by gravity through cell **10** even though pumps **21** and **26** are inoperative. However, Figure 6.5b does not rely on gravity as the pumps continuously circulate the anode and cathode fluids.

Referring to Figure 6.5c, the voltage cell **10** is shown in greater detail than in Figures 6.5a and 6.5b. Parts corresponding to those of cell **10** in Figures 6.5a and 6.5b are identified by like numerals. As shown, cell **10** comprises a container **41** which may be of plastic, hard rubber or the like divided into first and second compartments by an ion selective membrane **11**. Electrodes **14** and **16** disposed in the first and second compartments, respectively, may be graphite felt pads which are inert to the anode and cathode fluids. Voltage and current are provided for terminals **15** and **17** by graphite foils **42** and **43** which contact the electrodes **14** and **16**, respectively.

The inert electrodes may also be graphite foils, cloths or felts, or platinized metal screens of the type used in fuel cells. In general, the electrodes must be inert to electrochemical or chemical reaction with the anode and cathode fluids while promoting the redox reaction on their surfaces. They must also be porous electronic conductors.

The system of Figure 6.5b is preferred slightly to that of Figure 6.5a because it is simpler and more efficient. If both systems were to cycle between 5 and 95% depth of discharge, the open circuit voltage would vary between 1.334 and 1.032, assuming the cells **10** are identical and the couples are Fe^{+3}/Fe^{+2} and Cr^{+2}/Cr^{+3}. With the arrangement of Figure 6.5b and continuously flowing fluids, the cell voltage under load would gradually decrease from the upper to the lower limit (neglecting IR and polarization losses).

With the system of Figure 6.5a the voltage would be nearly constant and closer to the lower limit as the fluids flow from the supply tanks **18** and **23** to the storage tanks **19** and **24**, respectively. Factors which must be considered regarding the electrodes **14** and **16**, membrane **11** and the anode and cathode fluids which contain reduced and oxidized species dissolved in a solvent will now be discussed. The preferred solvent is water.

Electrode Considerations: The selection of electrode materials for an electrically rechargeable redox flow cell involves several simple criteria:

(1) Both the oxidized and reduced species must be soluble.
(2) The oxidation and reduction potentials of the different species must be such that the solvent will remain inert during charging as well as on stand.
(3) Oxygen containing or demanding ions (e.g., MnO_4^-, Cr^{+3}) should be avoided to simplify hydrogen ion management problems.
(4) Complexed ions [e.g., $Fe(CN)_6^{-4}$] should be avoided to minimize overall system weight.

The first criterion simply encompasses the classical type of redox electrode such as the ferrous/ferric couple in a chloride solution,

(1) $$Fe^{+2} \longrightarrow Fe^{+3} + e^-$$

Here both the chlorides are highly soluble and the inert electrode does not become fouled or complicated by the presence of insoluble species. Charge neutrality is maintained by the diffusion into or out of the half-cell compartment of some other type of ion (i.e., H^+ or Cl^-).

The second criterion deals with the requirement of maintaining the solvent as a neutral species during charge, stand, and discharge. If water is the solvent, ions above hydrogen in the list of oxidation/reduction potentials should be avoided (Cr^{+2}, V^{+2}, Ti^{+2}), since hydrogen could be displaced. On the other end of the scale, ions such as Co^{+3}, Ce^{+4}, and Mn^{+4} could possibly displace oxygen from solution. Care must also be exercised during the charging process where the overpotentials applied could lead to the simultaneous discharge of solvent fragments along with the desired ions.

Criterion three comes into consideration when the overall charge balance and ion management schemes are arranged. Consider the permanganate ion reduction

$$MnO_4^- + 8H^+ + 5e^- \longrightarrow 4H_2O + Mn^{+2}$$

Eight hydrogen ions are required for this half-cell reaction. In a complete system, large amounts of acid would be required and pH changes would be large. This may be contrasted with the ferrous/ferric half-cell reaction (1) which requires no hydrogen ions. The fourth criterion is an extension of the third. The ferri/ferro cyanide couple will serve as an illustration.

$$Fe(CN)_6^{-4} \longrightarrow Fe(CN)_6^{-3} + e^-$$

$$Fe^{+2} \longrightarrow Fe^{+3} + e^-$$

Although the above pair of half-cell reactions meet the first three criteria, the equivalent weight of the cyanide complex is 272 as opposed to 56 for the noncomplexed iron. Thus, from an energy density standpoint the use of complexed ions is not desirable.

Membrane Considerations: The requirements and options in regard to the membrane that is used to separate the anode and cathode fluids can be illustrated by considering a simple redox cell system with the reactions written in the discharge direction.

$$\text{Anode compartment; } A^{+1} \longrightarrow A^{+2} + e^-$$

$$\text{Cathode compartment; } C^{+2} + e^- \longrightarrow C^{+1}$$

The membrane must provide an impermeable barrier to both the A and C ions in both states of charge. Further, it must provide the means by which charge neutralization is maintained. During discharge either positive ions (H^+, for example) must move from the anode compartment through the membrane to the cathode compartment or negative ions (Cl^-, for example) must move from the cathode compartment to the anode compartment.

There is an inherent disadvantage from an energy standpoint in moving positive ions from the anode compartment during discharge as opposed to the other option. The iron/chromium system will be used to illustrate this. An ideal hydrogen ion transport membrane and an ideal chloride ion transport membrane will be used in the illustration.

Case 1. - H^+ membrane

Before discharge			After discharge		
Cr^{+2} - 1.0 N		Fe^{+3} - 1.0 N	Cr^{+3} - 1.0 N		Fe^{+2} - 1.0 N
H^+ - 1.0 N		Cl^- - 3.0 N	Cl^- - 3.0 N		H^+ - 1.0 N
Cl^- - 3.0 N					Cl^- - 3.0 N

Case 2. - Cl^- membrane

Before discharge			After discharge		
Cr^{+2} - 1.0 N		Fe^{+3} - 1.0 N	Cr^{+3} - 1.0 N		Fe^{+2} - 1.0 N
Cl^- - 2.0 N		Cl^- - 1.0 N	Cl^- - 3.0 N		Cl^- - 3.0 N

In Case 1 as compared to Case 2, 1 mol of HCl is required per Faraday over and above any acid that may be required for pH adjustment needed for solution stabilization. In addition, an anion membrane would most likely have a better selectivity for preventing cross diffusion of the redox ions. Cross diffusion of such ions would represent a permanent loss of system capacity.

Redox couples which preclude the electrolysis of water during discharge of the cell must be used. The anode and cathode fluids are respective solutions between 1 and 4 molar. The anode fluid preferably contains water as a solvent having dissolved therein a chloride salt selected from the group consisting of titanium chloride, chromium chloride, tin chloride, vanadium chloride, manganese chloride and cerium chloride whereby cations in a reduced state are produced.

The cathode fluid preferably contains water as a solvent having dissolved therein a chloride salt selected from the group consisting of iron chloride, chromium chloride, vanadium chloride, manganese chloride, and cerium chloride whereby cations in an oxidized state are produced.

HYDROGEL CONTAINING AN ELECTRON TRANSFER AGENT

C.F. Cole, Jr.; U.S. Patent 3,936,318; February 3, 1976; assigned to Continental Oil Co. describes an electrical energy storage battery having a case, and at least two porous electrolyte containers positioned within the case. A solid hydrogel is positioned within the case and between the electrolyte containers, the hydrogel containing an effective amount of an electron transferring agent to render the hydrogel conductive. At least one anode is positioned within at least one of the electrolyte containers, and at least one cathode is positioned within at least one other of the electrolyte containers.

A second solid hydrogel is positioned within the container for the cathode, the gel being saturated with a compound selected from the group consisting of zinc chloride, zinc oxide, cadmium chloride, and cadmium oxide to render same conductive. A third solid hydrogen is positioned within the container for the anode, the gel being saturated with a compound selected from the group consisting of chromium chloride, chromium oxide, and alkali metal-containing chromate salts to render same conductive. The electron transferring agent employed in the solid hydrogel is selected from alkali metal salts, alkali metal hydroxides, alkaline earth metal salts, and alkaline earth metal hydroxides.

Referring to Figure 6.6a and Figure 6.6b, an example of the electrical energy storage battery **11** is shown. Battery **11** comprises a case member **12** and a cover member **13**. Case member **12** has positioned therein two porous electrolyte containers **14** and **16**. Porous containers **14** and **16** are positioned inside case **12** which also contains a solid hydrogel **17** which separates containers **14** and **16**.

An anode means **18** and a cathode means **19** are positioned in electrolyte containers **14** and **16** respectively. Anode means **18** and cathode means **19** are connected to electrical connectors **21** and **22** which are in electrical contact with an external load not shown. Container **16** containing cathode **19** contains a second solid hydrogel **23**. Porous container **14** containing anode **18** contains a third hydrogel **24**.

FIGURE 6.6: STORAGE BATTERY

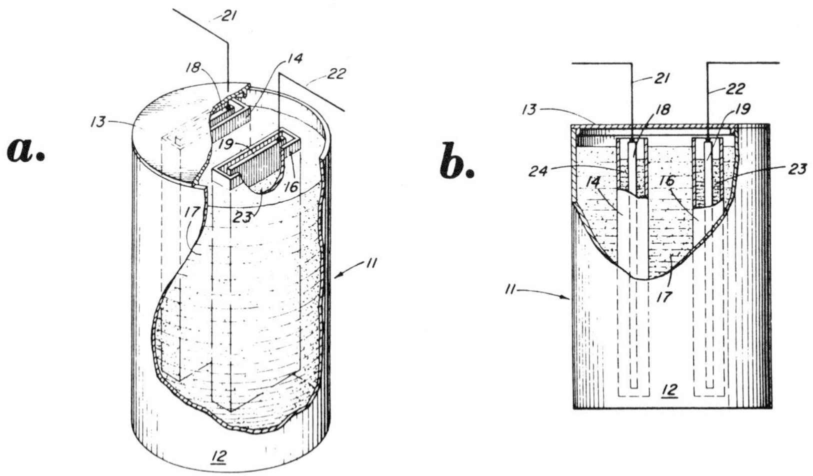

(a) Partially broken perspective view of the electrical energy storage battery
(b) Partially broken plane elevational view of the electrical energy storage battery of Figure 6.6a

Source: U.S. Patent 3,936,318

Hydrogel **17** positioned inside case **12** and between electrolyte containers **14** and **16** contains an effective amount of an electron transferring agent to render hydrogel **17** electrically conductive so that electrons can pass between anode **18** and cathode **19**. While any suitable electron transferring agent can be employed to render hydrogel **17** conductive, it is desired that such transfer agent be selected from the group consisting of alkali metal salt, alkali metal hydroxide, alkaline earth metal salt, and alkaline earth metal hydroxide.

Examples of such transfer agents are sodium chloride, potassium chloride, lithium chloride, sodium hydroxide, potassium hydroxide, lithium hydroxide, calcium chloride, barium chloride, calcium hydroxide and barium hydroxide. While the amount of transfer agent employed in the hydrogel can vary widely, it has been found that desirable results can be obtained when the amount of transferring agent employed with hydrogel **17** varies from about 20 to 70 weight percent of the transfer agent.

The term hydrogel as used herein is understood to mean those compounds which when admixed with warm water dissolve therein and upon cooling, form a solid gel. Such compounds are illustrated by the following group: agar-agar, water glass, hydroxymethylcellulose, hydroxyethylcellulose, modified cellulose, flour, starch and the like. While any of the above can be employed, especially desirable results from a physical property standpoint are obtained when the hydrogel is selected from the group consisting of agar-agar, water glass, hydroxymethylcellulose, and modified cellulose.

In addition to the electron transfer agent employed in hydrogel **17** positioned between containers **14** and **16**, it is often desirable to incorporate into hydrogel **17** an effective amount of graphite powder. By employing such graphite powder the life of the cell is improved. Such is believed to be primarily due to the lower internal resistance present in the cell when the graphite powder is incorporated along with the electron transfer agent into the solid hydrogel. While the amount of graphite can vary widely, desirable results are obtained when from about 2 to 8 weight percent of graphite is incorporated into hydrogel **17**.

As previously stated, at least two porous electrolyte containers **14** and **16** are positioned inside case **12** of battery **11**. Porous electrolyte container **16** containing cathode means **19**, is adapted to have positioned therein a solid hydrogel **23**, the gel being saturated with a compound selected from the group consisting of zinc chloride, zinc oxide, cadmium chloride, and cadmium oxide, such compounds being important to render the gel conductive so that electrons can flow therethrough into contact with conductive hydrogel **17** positioned between electrolyte containers **14** and **16**.

Electrolyte container **14** containing anode **18**, is likewise provided with solid hydrogel **24** positioned within the container, the gel being saturated with a compound selected from the group consisting of chromium chloride, chromium oxide, and alkali metal containing chromate salts, such being required to render the hydrogel conductive to electron transfer. When desirable, an effective amount of graphite can be incorporated into the hydrogen positioned in electrolyte container **14** containing anode **18**. While the amount of such graphite can vary widely, desirable results are obtained wherein the amount of graphite incorporated into the hydrogel **24** varies from about 2 to 8 weight percent.

Porous electrolyte containers **14** and **16**, which can be employed in equal numbers, i.e., 2, 4, 6, 8 and the like, or in odd numbers can be produced of any suitable material which will serve as a flow barrier between the various hydrogel compositions. However, such barriers must be porous so that charged particles such as ions, electrons and the like can readily flow between anode **18** and cathode **19** of the battery **11**. Examples of suitable materials which can be employed for the porous electrolyte container means are microporous polymeric materials such as polyethylene, polypropylene, polyvinyl chloride, and the like and ceramic materials such as microporous porcelain containers.

Example: A battery was constructed similar to Figure 6.6a and Figure 6.6b. Porous porcelain containers, having dimensions of 9 cm x 7 cm x 2 cm were employed as the electrolyte container means. Carbon plates were utilized as both the anode and cathode. The hydrogel employed between the porcelain containers was an agar gel saturated with sodium chloride. The hydrogel employed in the porcelain container containing the anode was an agar gel saturated with chromium chloride. The hydrogel employed in the porcelain container containing the cathode was an agar gel saturated with zinc chloride.

The battery was attached to an external power source and electrically charged at 0.6 ampere for four hours. After the four hour charge the power source was removed and measurements made on the battery. These measurements are as follows: open circuit, 2.457 volts; short circuit, 2.0 amperes; battery and short circuit resistance, 1.228 ohms. Measurements were then made to determine the battery open circuit voltage over a period of about four hours. The data are as follows.

Voltage, volts	Time, minutes
2.139	0
2.129	23
2.125	34
2.108	113
2.104	144
2.098	178
2.094	198
2.085	243
2.083	248

Load Voltage, volts	Current, milliamperes	Time, minutes
1.635	72	0
1.335	60	10
1.268	57	60
1.262	57	87
1.258	56.5	105
1.252	56	133
1.250	56	145
1.208	54.5	340

All of the above data were collected at 72°F.

LANTERN BATTERY

R.O. Hammel; U.S. Patent 3,972,739; August 3, 1976; assigned to The Gates Rubber Company describes a rechargeable battery having an enclosure, a self-contained battery or plurality of self-contained electrochemical cells disposed within the enclosure and having first and second terminals, output terminals mounted exteriorly on the enclosure, a recharge jack mounted exteriorly on the enclosure and having leads connected to the first and second terminals of the battery.

The improvement comprises the provision of a low resistance, high power current-limiting means connected between one of the output terminals and the corresponding polarity first or second terminal of the battery, the current-limiting means bearing, or having in close proximity, heat indicating means for indicating whether a threshold level of current corresponding to a short-circuit condition across the output terminals had been carried by the current-limiting means.

Referring to Figure 6.7, the rechargeable battery (including its enclosure) is generally designated **10** and includes a self-contained rechargeable battery **12** which alternatively may be composed of a plurality of self-contained electrochemical cells connected in desired relation, an enclosure **18, 33** in which the battery is disposed, a pair of output terminals **22, 24** mounted on the enclosure, and a recharge jack **20** shown unassembled. From the recharge jack **20** emanates a pair of leads **26, 28** which respectively connect to opposite polarity first and second terminals **14, 16** of the battery. Lamp **30** is shown connected across the output terminals of the battery although it will be appreciated that various other types of loads could be connected to the battery.

FIGURE 6.7: RECHARGEABLE BATTERY WITH CURRENT LIMITING MEANS

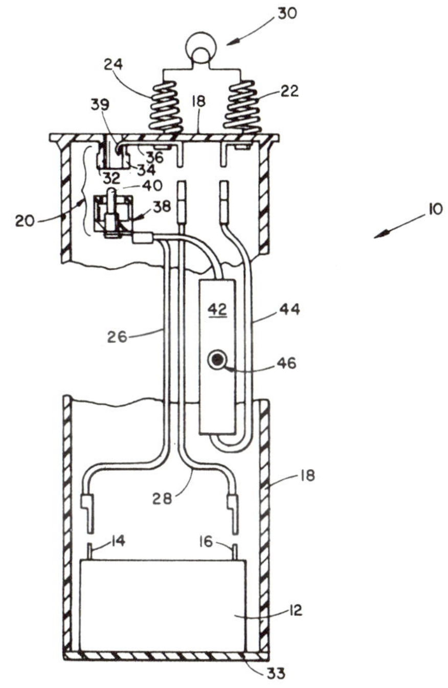

Source: U.S. Patent 3,972,739

The battery enclosure or case **18**, **33** is made of a suitable electrically insulating material, such as plastic and includes a molded single piece top and side portion **18**, including male jack portion **32**, and a bottom portion **33** welded or otherwise affixed to the top container portion **18**. The enclosure may suitably be of rectangular or cylindrical shape, for instance.

Recharge jack **20** is comprised of male cylindrical portion **32**, which is provided with slot **34** to accommodate spring contact **36** and female portion **38** adapted to snap-fit over the male portion with the aid of interlocking shoulders, or otherwise mutually attached. The female jack portion **38** is also provided with a central contact post **40** which is normally disposed in spaced relationship with respect to resilient end contact portion **39** of the spring contact **36**.

As is customary, when a recharge plug of the coaxial type is engaged in the jack **20**, the inner portion of the plug will make contact with contact post **40**, while the outer portion of opposite polarity will make contact with spring end **39** and facilitate recharging of battery **12**. Contact **39** is connected to one terminal of the battery **16** through conductive leg **28** and is also in contact through a rivet connection to output terminal **24**. Similarly, contact post **40** is connected to terminal **14** of the battery through lead **26**.

In accordance with the process, there is provided a low resistance, high power current limiting means **42** which is connected between one of the output terminals, in this case terminal **22**, and the corresponding polarity terminal of the battery, in this case terminal **14** via leads **44** and **26**. The particular resistance and power rating of the device will be chosen according to the circumstances including the particular voltage drop desired across the element **42**, the temperature withstand capability of the battery **12** which will be in close proximity to the heat generating element **42**, the maximum steady state current output desired, and size restriction, for instance. The following example illustrates the process.

Example: Lamp **30** is a 5-volt device and battery **12** has a nominal open circuit voltage of 6 volts, being composed of three series connected 2-volt sealed lead-acid cells. It is thus desired to drop approximately 1 volt across resistive element **42**. Since the standard lantern bulb draws about 500 mA to achieve a 1-volt drop, a 2-ohm resistor **42** is employed. Thus, the current through the device will be limited to about 3 amperes so that selection of a 20 watt 2 ohm ± 10% bathtub type resistor will be sufficient to dissipate the heat generated upon short circuit of terminals **22** and **24**. Under these short circuit conditions, the surface temperature of the resistor will reach about 210°C, which is compatible with the system.

Under the short circuit conditions described in the above example, it is desirable to determine, after the fact, whether the terminals of the battery had in fact been short circuited. This is accomplished by the placement of indicator means **46** on the resistive element which will reveal whether a certain threshold level of current corresponding to a short-circuit across the output terminals had been carried by the element **42**.

The indicating means **46** can include a spot of temperature sensitive paint or crayon, for instance, which undergoes a transformation upon heating above the threshold level, thereby indicating that the temperature level (corresponding to the particular amperage) had been exceeded. The paint or crayon may, for instance, undergo a color change or transform from an initially dry opaque mark to a translucent liquid smear. As another example, the indicator **46** could be an adhesive-backed temperature monitor consisting of one or more heat-sensitive indicators sealed under a transparent circular window. The encircled indicators may turn black irreversibly at the threshold or rated temperature.

Thus, element **42** with heat indicating means **46** not only limits the current output of the battery to a desired level, well below the potential maximum discharge current, but also protects the environment and/or battery from damage. Simultaneously, indicator **46** provides an irreversible monitoring means or indicator of the current generated during some previous discharge of the battery.

COMPANY INDEX

INVENTOR INDEX

U.S. PATENT NUMBER INDEX

NOTICE

Nothing contained in this Review shall be construed to constitute a permission or recommendation to practice any invention covered by any patent without a license from the patent owners. Further, neither the author nor the publisher assumes any liability with respect to the use of, or for damages resulting from the use of, any information, apparatus, method or process described in this Review.

MAGNETOHYDRODYNAMIC ENERGY FOR ELECTRIC POWER GENERATION 1978

Edited by R. F. Grundy

Energy Technology Review No. 20

The production of electricity by magnetohydrodynamic (MHD) processes is classed as direct energy conversion.

In an MHD generator the expanding working fluid (hot ionized gas or plasma) interacts with a magnetic field to produce electricity. The amount of current produced is set by the velocity and temperature of the ionized electrically conductive gas.

Motional electromagnetic induction (without mechanical contrivances) gives the MHD generator the unique capability among heat engines of delivering its output directly in electrical form.

The electrical power level at which MHD generators become more economical than mechanical turbines is generally considered to be over one megawatt. Success depends on the intended application and the availability of high-temperature heat sources required by MHD technology.

At this time the addition of an MHD stage to coal-fired power plants seems attractive, since the MHD stage may be able to operate directly on the high temperature products of coal combustion. Other sources of high temperatures are, of course, fusion and fission type atomic reactors.

This book is limited to the potential of MHD in the production of the much needed 60 Hz alternating current at less cost and less fuel consumption than by conventional power-generating systems. As such it presents an accurate status report based largely on federally-funded studies. A partial and condensed table of contents follows here.

ISBN 0-8155-0689-9

230 pages

OFFSHORE AND UNDERGROUND POWER PLANTS 1977

Edited by Robert Noyes

Energy Technology Review No. 19
Ocean Technology Review No. 6

Uncontroversial plant sites for generating electric power by any means are becoming increasingly scarce. Both nuclear and fossil fuel power plants require nearness to large quantities of cooling water and to load centers, but not on land prone to earthquakes. Consequently, utility companies are looking out to sea or beneath the surface of the earth for suitable power plant locations.

This is a most exhaustive and detailed treatise: the first six chapters being devoted to nuclear and fossil-fuel fired power plants, while the remaining six chapters deal with power generation of a more esoteric variety.

The relative costs of fixed and floating offshore plants depend on the depth of water and on the degree of earthquake resistance required. For shallow water, the fixed breakwater offshore floating station is of the most interest.

Most of the information presented in this book is based on federally-funded studies. Each chapter has its own bibliographic list, while the references at the end of the book give the full titles of the government reports on which this book is based, together with a source of their purchase. A partial and condensed list of contents follows here.

ISBN 0-8155-0680-5 **309 pages**

FUEL CELLS FOR PUBLIC UTILITY AND INDUSTRIAL POWER 1977

Edited by Robert Noyes

Energy Technology Review No. 18

Fuel cells are generators of electricity containing no moving parts except small extraneous pumps for the movement of fuel and oxidant into the cell and the products of oxidation out of the cell.

Public utilities and industrial consumers require high voltage, three-phase alternating current. In this application fuel cells must compete with turbine-driven generators which provide such current. While the output of a fuel cell is low voltage DC power, cells may be connected in various series and parallel arrangements to give whatever voltage is desired, but mechanical rotary converters or delicate electronic inverters must then provide conversion to AC (60 cycles for each phase in the USA).

The advantages of a fuel cell system over turbine-driven generators lie in greater efficiency at full load which even increases as the load diminishes, so that inefficient peaking generators are not needed. There are considerable pollution control advantages to be gained as well. Because the various suitable fuels react electrochemically rather than by burning in air, no nitrogen oxides are formed. For the same reason, emissions of unburned or partly burned gaseous and particulate products are practically nil.

Fuel cell installations using inverters have a long life with relatively little maintenance. It is now entirely feasible to have small, completely unattended fuel cell power plants using waste fuels on location (such as hydrogen from chlorine production) to produce convenient power.

This book, based on information derived from U.S. government-contracted studies, contains considerable practical down-to-earth technical information relating to fuel cells for power plants. A partial and condensed table of contents follows.

ISBN 0-8155-0676-7 **325 pages**

HOW TO SAVE ENERGY AND CUT COSTS IN EXISTING INDUSTRIAL AND COMMERCIAL BUILDINGS 1976

An Energy Conservation Manual

by Fred S. Dubin, Harold L. Mindell and Selwyn Bloome

Energy Technology Review No. 10

This manual offers guidelines for an organized approach toward conserving energy through more efficient utilization and the concomitant reduction of losses and waste.

The current tight supply of fuels and energy is unprecedented in the U.S.A. and other countries, and this situation is expected to continue for many years. Never before has there been as pressing a need for the efficient use of fuels and energy in all forms.

Most of the energy savings will result from planned systematic identification of, and action on, conservation opportunities.

Part I of this manual is directed primarily to owners, occupants, and operators of buildings. It identifies a wide range of opportunities and options to save energy and operating costs through proper operation and maintenance. It also includes minor modifications to the building and mechanical and electrical systems which can be carried out promptly with little, if any, investment costs.

Part II is intended for engineers, architects, and skilled building operators who are responsible for analyzing, devising, and implementing comprehensive energy conservation programs. Such programs involve additional and more complex measures than those in **Part I.** The investment is usually recovered through demonstrably lower operating expenses and much greater energy savings.

A partial and much condensed table of contents follows here:

PART I

PART II

Much of the technology required to achieve energy savings is already available. Current research is providing refinements and evaluating new techniques that can help to curb the waste inherent in yesteryear's designs. The principal need is to get the available technology, described here, into widespread use.

ISBN 0-8155-0638-4 **725 pages**

ENHANCED OIL RECOVERY

Secondary and Tertiary Methods 1978

Edited by M. M. Schumacher

Chemical Technology Review No. 103
Energy Technology Review No. 22

This is a book about greater oil recovery by succeeding methods; the possible techniques and their present or future capabilities for extracting more petroleum from oil fields after primary production.

These methods are also applicable to fields which were abandoned in previous decades when there was a plentiful supply of energy. Sometimes less than one-third of the crude oil in place was recovered, when only the original reservoir pressure was allowed to drive the oil into the producing well.

The concept of enhanced oil recovery applies to a whole collection of methods which are becoming increasingly profitable as energy becomes more expensive and the price of oil goes up. Each method appears to have its own claimed unique capability to extract the most oil from a particular reservoir.

After a decline in pressure has caused the oil recovery to become uneconomic, oil production can be increased by immiscible gas injection and waterflooding. These traditional "secondary" methods are now being supplemented by "tertiary" recovery methods including miscible fluid displacement, microemulsion (micellar) flooding, cyclic steam injection, controlled *in situ* combustion (fireflood) and other related techniques which are either thermal or miscible methods, thereby reducing the surface tension between oil and driving fluid.

Most of the data in this book are based on federally funded studies—a partial and condensed list of contents follows here:

ISBN 0-8155-0692-9 **207 pages**